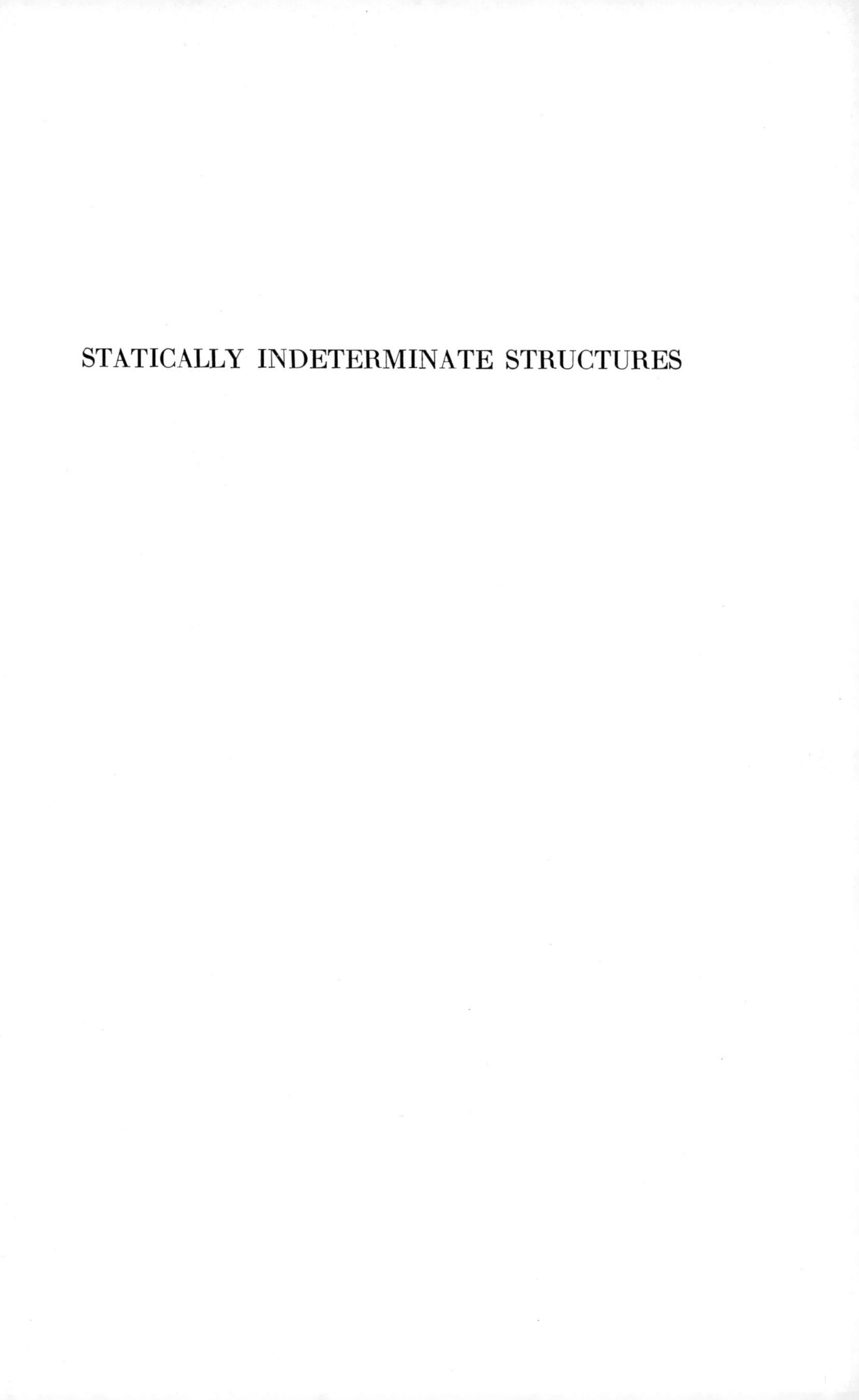

STATICALLY INDETERMINATE STRUCTURES

STATICALLY INDETERMINATE STRUCTURES

Chu-Kia Wang, Ph.D.

School of Architecture, University of Illinois
Urbana, Illinois

New York Toronto London

McGRAW-HILL BOOK COMPANY, INC.

1953

STATICALLY INDETERMINATE STRUCTURES

Library of Congress Catalog Card Number: 52-9253

V

THE MAPLE PRESS COMPANY, YORK, PA.

PREFACE

The text material and illustrations have been prepared to serve primarily as a workbook for assimilating the basic principles and methods in the analysis of statically indeterminate structures. The analysis of statically indeterminate structures consists essentially in imposing the conditions of the geometry of the deformed structure upon those of statics. The principles are generally simple, but proficiency can be achieved only through practice on a variety of problems.

Experience in teaching has shown that students still need frequent reviews in the principles and techniques of statics besides those pertaining to indeterminacy. The examples in the text may substitute for some, although not all, of the blackboard writing on the part of the teacher so that he may apportion more of his time in class to discussions on such items as the relation of analysis to design, the history or bibliography, the comparison of the relative merits of different methods, and the citation of actual structures to which the topic of the day can be applied.

An inspection of the Contents will show that deflections of statically determinate structures are treated in Chaps. II and III before the applications of finding deflections to the analysis of statically indeterminate structures are taken up in Chaps. IV and V. Some teachers may prefer to assign some articles in Chaps. II and IV together and then those in Chaps. III and V. The same is true with the slope-deflection versus the moment-distribution methods. The choice is between studying slope deflection first and then moment distribution or taking up both methods in relation to any one problem at the same time. The former order has been adopted in the writing of Chaps. VII and VIII. The author does not recommend the teaching of moment distribution only without slope deflection.

When it is not possible to cover the entire text for undergraduate seniors, some or all of the following articles may be postponed: Arts. 12, 19, 20, 26, 28, 32, 43, 44, 53, 54, 62, 63, 64, 68, 70, 71. These articles can then be included in the curriculum for the first-year graduate students.

The author is indebted to Professor Warren Raeder, head of the

Department of Civil Engineering, University of Colorado, for his kindness in editing the manuscript and making valuable suggestions.

Great appreciation should be given Mrs. Wang, not only for her typing of the manuscript in the later stage, but also for her patience in bearing with the situation when the author was more than fully occupied during the preparation of this text.

C. K. Wang

BOULDER, COLO.
OCTOBER, 1952

CONTENTS

CHAPTER VIII. The Moment-distribution Method . . . 216

CHAPTER IX. The Method of Column Analogy . . . 289

CHAPTER I

GENERAL INTRODUCTION

1. Statically Determinate Structures vs. Statically Indeterminate Structures. Most structures fall into one of the following three classifications: beams, frames, or trusses. A beam is a structural member subjected to transverse loads only, and it is completely analyzed when the shear and moment diagrams are found. A frame, or *rigid frame,* is a structure composed of members which are connected by rigid joints (welded joints, for instance). A frame is completely analyzed when the variations in direct stress, shear, and moment along the lengths of all members are found. A truss is a structure in which all members are usually considered to be connected by hinges, thus eliminating moment in the members. A truss is completely analyzed when the direct stresses in all members are determined.

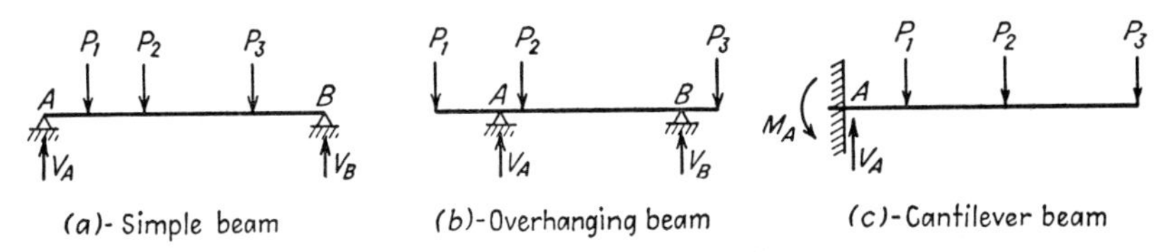

FIG. 1. Statically determinate beams.

Shear and moment diagrams of beams can be drawn when the external reactions are known. In the study of the equilibrium of a coplanar parallel-force system, it has been proved that not more than two unknown forces can be found by the principles of statics. In the case of beams these two unknown forces are usually the reactions. Thus the two reactions to simple, overhanging, or cantilever beams (Fig. 1) can be determined by the equations of statics, or these three types of beams are *statically determinate.* If, however, a beam rests on more than two supports or in addition one or both end supports are fixed, there are more than two external reactions to be determined. Statics offers only two conditions of equilibrium for a coplanar parallel-force system, and thus only two reactions can thereby be found; any additional reactions are excessive, or *redundant.* These reactions cannot be determined by the equations of statics alone, and beams with such reactions are called *statically indeterminate* beams. The *degree of indeterminacy* is given by the number of extra, or redundant, reactions. Thus the beam in Fig. 2*a* is statically indeterminate to the second degree because there are four unknown reactions and statics furnishes only two conditions or two equations of equilibrium; the beam in Fig. 2*b* is statically indeterminate to the fourth degree; the beam in Fig. 2*c* is statically indeterminate to the sixth degree.

A frame is statically determinate if there are only three external reactions, because statics offers only three conditions of equilibrium for a

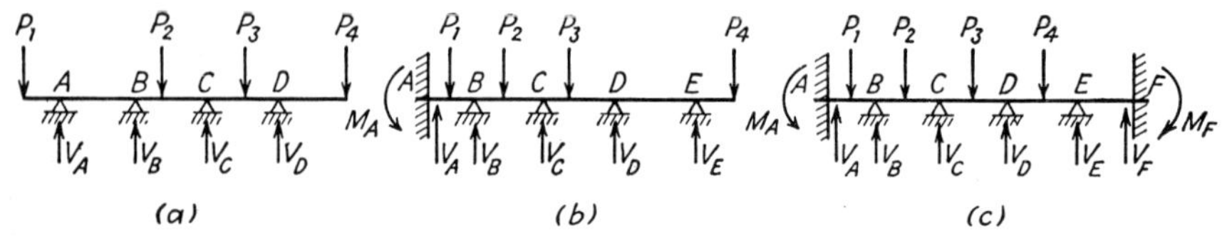

FIG. 2. Statically indeterminate beams.

general coplanar-force system. Thus the two rigid frames shown in Fig. 3 are statically determinate. If, however, a rigid frame has more than three external reactions, it is statically indeterminate, the degree of indeterminacy being equal to the number of redundant reactions. Thus the frame of Fig. 4*a* is statically indeterminate to the first degree; Fig. 4*b*, to the third degree; Fig. 4*c*, to the fifth degree; Fig. 4*d*, to the sixth degree.

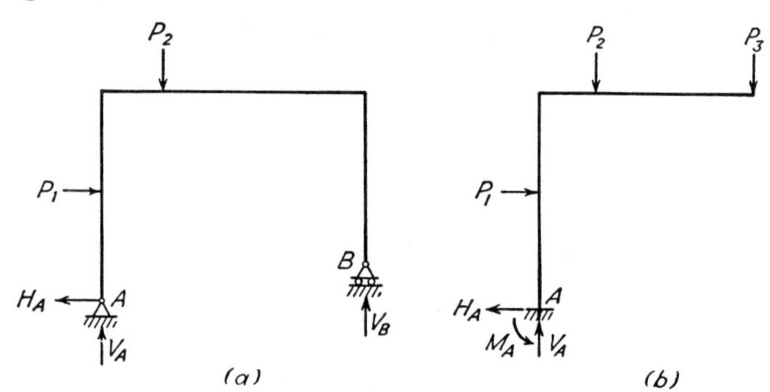

FIG. 3. Statically determinate frames.

A truss is statically determinate if it has not more than three external reactions (two in the case of parallel-force system) and not more than $(2j - 3)$ members, where j is the number of joints. While the first requirement for statical determinacy is obvious, the second requirement may need some explanation. A truss is just internally stable if it consists of a series of triangles as shown in Fig. 5. The first triangle is made up of three joints and three members; each successive triangle requires two additional members but only one additional joint. Thus, if m is the number of members in the truss and j is the number of joints, $(m - 3) = 2(j - 3)$, or

$$m = 2j - 3 \tag{1}$$

Thus the trusses shown in Fig. 6 are statically determinate. The truss shown in Fig. 7a is statically indeterminate to the second degree because

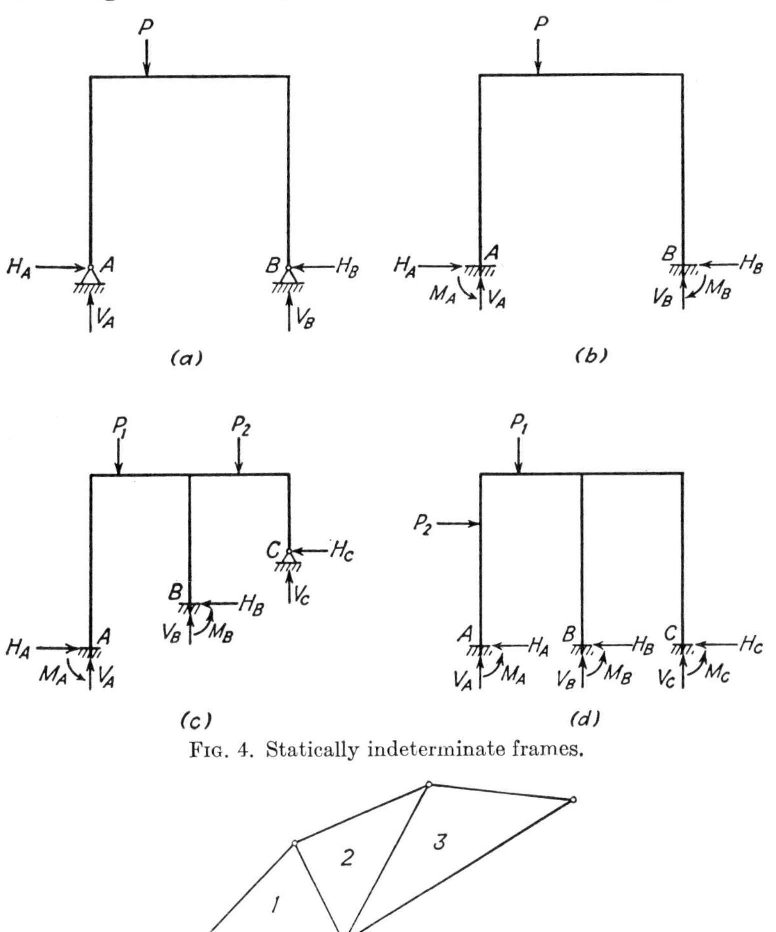

FIG. 4. Statically indeterminate frames.

FIG. 5.

it has four unknown reactions and only two equations of equilibrium are available; Fig. 7b, to the third degree because there are three redundant members $(m = 2j)$ plus three unknown reactions, whereas only three equations of equilibrium are available; Fig. 7c, to the fourth degree.

2. Conditions of Geometry. It can be seen from the preceding discussion that in analyzing statically indeterminate structures it is necessary to have as many extra conditions, in addition to those of statics, as there are redundant reactions. Or the number of "nonstatic" conditions must be the same as the degree of indeterminacy. The extra conditions are generally furnished by the *geometry* of the deformed structure. For instance, the beam shown in Fig. 2*a*, although ordinarily thought of as a continuous beam, can be considered as a simple beam, AD, on which the

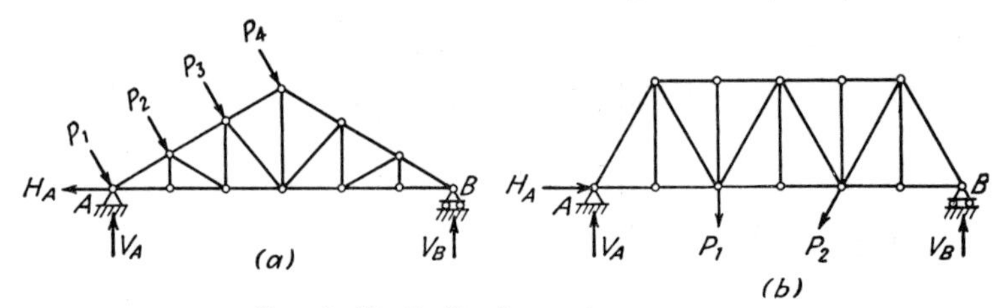

Fig. 6. Statically determinate trusses.

forces P_1, P_2, P_3, P_4, V_B, and V_C are acting. The conditions of geometry to be satisfied by the deformed simple beam AD are that the deflections at B and at C must both be zero. These two conditions of geometry, together with the two conditions of statics, furnish the necessary conditions for solving V_A, V_B, V_C, and V_D. The beam shown in Fig. 2*b* can be considered either as a cantilever beam on which the forces P_1, P_2, P_3,

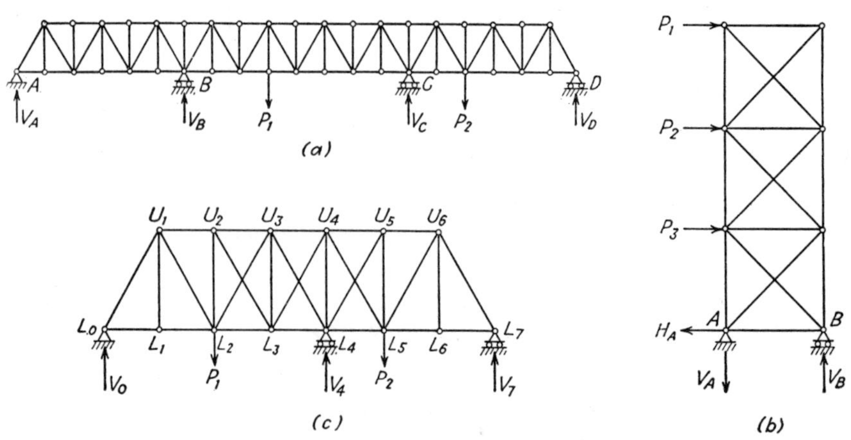

Fig. 7. Statically indeterminate trusses.

P_4, V_B, V_C, V_D, and V_E are acting or as a simple beam BE on which the forces P_1, P_2, P_3, P_4, V_A, V_C, and V_D and the moment M_A are acting. In the former case, the conditions of geometry are that the deflections at B, C, D, and E must be zero; in the latter case, the conditions are that the deflections at A, C, and D must be zero and the tangent at A must remain horizontal. The rigid frame shown in Fig. 4*b* can be considered either as a frame fixed at A and free at B on which the forces P, H_B, and V_B and the moment M_B are acting or as a frame hinged at A and supported on rollers at B on which the forces P and H_B and the moments M_A and M_B are acting. Under the first consideration the conditions of geometry are that the horizontal and vertical deflections at B must be zero and that the tangent at B must remain vertical. Under the second consideration the conditions of geometry are that the horizontal deflection at B must be zero and the tangents at A and B must stay vertical. The truss shown in Fig. 7c can be considered as a simple truss L_0L_7 with the support at L_4 removed and with three diagonals U_2L_3, U_3L_4, and U_4L_5 cut, on which the forces P_1, P_2, V_4, X_a-X_a, X_b-X_b, and X_c-X_c (Fig. 8) are acting. The conditions of geometry are that the deflection at L_4 must be zero and that the distance between deflected points U_2 and L_3, U_3 and L_4, and U_4 and

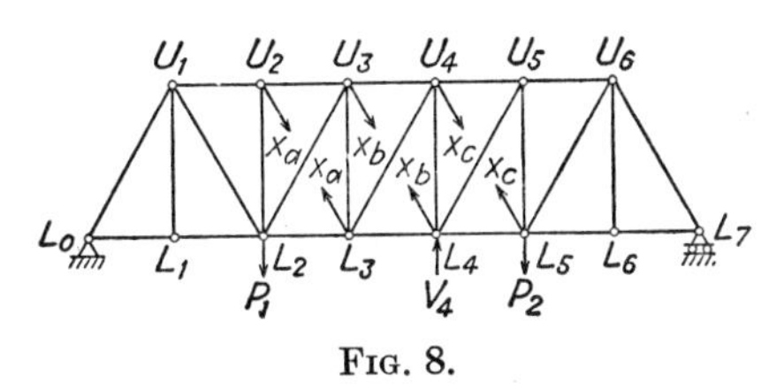

FIG. 8.

L_5 must be equal to the deformed lengths of bars U_2L_3, U_3L_4 and U_4L_5 in which the stresses are X_a, X_b, and X_c, respectively. The specific manner in which the conditions of geometry are so used is explained and illustrated in Chaps. IV and V.

3. The Method of Consistent Deformation. The most basic and most general method of analyzing statically indeterminate structures is the method of consistent deformation. The procedure consists in first setting up a basic determinate structure from the given indeterminate structure by removing the redundants and considering these redundants as loads acting on the basic determinate structure. There will always be as many conditions of geometry as there are redundants. A system of N simultaneous equations, where N is the degree of indeterminacy, can be established under these conditions of geometry with the redundants as unknowns. When the equations are solved and the redundants found, they can be put back on the given indeterminate structure and the remaining reactions solved by the equations of statics. It is to be noted that there may be several ways of choosing the basic determinate structure, as explained in Art. 2.

Before the above method can be illustrated (see Chaps. IV and V), it is necessary, therefore, to take up the various methods of finding the deflections (or rotations of tangents) of statically determinate beams, frames, and trusses. These various methods are treated and illustrated in Chaps. II and III.

It should be pointed out that the deflections of beams, frames, and trusses depend on the sizes of the constituent members in the structure. Therefore, before analyzing a statically indeterminate structure, the size of its members must first be assumed, although in most cases only the relative sizes of members are necessary. The procedure in designing a statically indeterminate structure is that of successive approximations; in other words, the structure is first assumed, then analyzed, then the design modified, then reanalyzed, etc., until the last assumed structure needs no further modification. Fortunately the structure as first assumed will in general need little or no further modification after the first analysis.

The reader may not be able to grasp fully the entire significance of this article at this time, but if he reads it again after working through Chaps. II to V, he will find the whole article very obvious and illuminating.

4. Other Methods of Analysis. The method of consistent deformation is the most general method of analyzing statically indeterminate structures. It can be used to analyze beams, rigid frames, or trusses. It is the only method of analyzing statically indeterminate trusses (see Chap. V) or "composite" structures, in which some members are primarily subjected to direct stresses and others to bending stresses (see Chap. XII). In analyzing statically indeterminate beams or frames, there are other methods, which are considerably shorter than the method of consistent deformation, namely, the three-moment equation, the slope-deflection method, the moment-distribution method, and the method of column analogy. These four methods are developed and illustrated in Chaps. VI to IX.

Two special topics of practical importance, the fixed arch and secondary stresses in trusses, are treated and illustrated in Chaps. X and XI. The methods of analyzing statically indeterminate structures developed in other chapters can be applied to these two subjects.

CHAPTER II

DEFLECTION OF STATICALLY DETERMINATE BEAMS AND FRAMES

5. Common Methods. The *double-integration* method of finding deflections of statically determinate beams is usually treated in the first course in strength of materials and will not be further discussed here. Perhaps the two methods most commonly used and applicable to both beams and frames are the *unit-load* method and the *moment-area* method. In the former only one component of the deflection at a point can be found in one application of the unit load. In the moment-area method the total deflection at any point can be found once the properties of the moment area are established. Both methods may also be used to find the rotation of the tangent at any point in a statically determinate beam or frame. The two methods, unit load and moment area, will be developed in the subsequent articles.

6. The Unit-load Method—Derivation of Basic Formula. The problem is to find the vertical deflection Δ at point C of a simple beam AB (Fig. 9*b*) subjected to loads P_1, P_2, and P_3. This set of loads causes internal stresses in the beam, for instance, a total compression of S in any fiber such as MN with cross-sectional area dA. This fiber, MN, is shortened by an amount dL. Also, of course, the loads produce deflections all along the beam, such as Δ_1 at P_1, Δ_2 at P_2, Δ_3 at P_3. The total external work done on the beam, if the loads are gradually applied, is $\frac{1}{2}P_1\Delta_1 + \frac{1}{2}P_2\Delta_2 + \frac{1}{2}P_3\Delta_3$. The total internal energy stored in the beam is equal to $\frac{1}{2}\Sigma S\,dL$. By the law of conservation of work or energy, the total external work done on the beam is equal to the internal energy stored in the beam, or

$$\tfrac{1}{2}P_1\Delta_1 + \tfrac{1}{2}P_2\Delta_2 + \tfrac{1}{2}P_3\Delta_3 = \tfrac{1}{2}\Sigma S\,dL \qquad (2)$$

Now if on the same simple beam AB a unit load of 1 lb is first gradually applied at C (Fig. 9*a*), it will cause deflections of δ at point C, δ_1 at point 1, δ_2 at point 2, and δ_3 at point 3. The loads P_1, P_2, P_3, when applied separately on the beam AB, cause deflections of Δ at point C, Δ_1 at point 1, Δ_2 at point 2, and Δ_3 at point 3 (Fig. 9*b*). If the loads P_1, P_2, P_3 are added gradually to the beam of Fig. 9*a*, on which the unit load at C is already applied, the deflections will be $\delta + \Delta$ at point C, $\delta_1 + \Delta_1$ at point

1, $\delta_2 + \Delta_2$ at point 2, and $\delta_3 + \Delta_3$ at point 3 (Fig. 9*c*), by the principle of superposition.

When the unit load at C is first applied, the relation between external work and internal energy is

$$\tfrac{1}{2}(1)(\delta) = \tfrac{1}{2}\Sigma u\, dl \tag{3}$$

where u is the total compression in pounds on any fiber MN with area dA caused by the unit load and dl is the total shortening of this fiber.

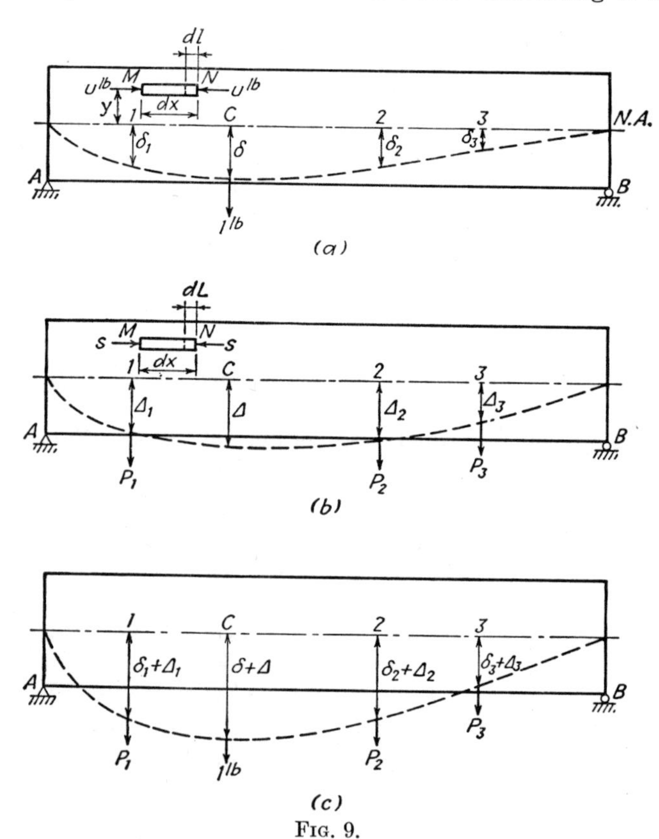

Fig. 9.

When the loads P_1, P_2, P_3 are gradually added, the additional external work done on the beam is $\tfrac{1}{2}P_1\Delta_1 + \tfrac{1}{2}P_2\Delta_2 + \tfrac{1}{2}P_3\Delta_3 + (1)\ (\Delta)$ because the *constant* 1-lb load already on the beam goes through the additional deflection Δ and the *gradually* applied loads P_1, P_2, P_3 go through the additional deflections Δ_1, Δ_2, Δ_3. The additional internal energy stored in the beam is $\tfrac{1}{2}\Sigma S\, dL + \Sigma u\, dL$. The total external work done on the beam thus far is therefore $\tfrac{1}{2}(1)(\delta) + \tfrac{1}{2}P_1\Delta_1 + \tfrac{1}{2}P_2\Delta_2 + \tfrac{1}{2}P_3\Delta_3 + (1)(\Delta)$, while the total internal energy is $\frac{1}{2}\Sigma u\,dl + \frac{1}{2}\Sigma S\,dL + \Sigma u\,dL$. Again by the law of conservation of energy,

$$\tfrac{1}{2}(1)(\delta) + \tfrac{1}{2}P_1\Delta_1 + \tfrac{1}{2}P_2\Delta_2 + \tfrac{1}{2}P_3\Delta_3 + (1)(\Delta) = \tfrac{1}{2}\Sigma u\,dl + \tfrac{1}{2}\Sigma S\,dL + \Sigma u\,dL \qquad (4)$$

Subtracting the sum of (2) and (3) from (4),

$$(1)(\Delta) = \Sigma u\,dL \qquad (5)$$

This equation (5), leading to the working formula (9) for beams, is the basic formula in the unit-load method. It can be applied to finding the deflection or rotation at any point in a beam, frame, or truss where dL may be caused by applied loadings, temperature changes, fabrication errors, or settlements of supports.

7. The Unit-load Method—Application to Beam Deflections. It is required to find the deflection Δ at point C of a simple beam AB due to

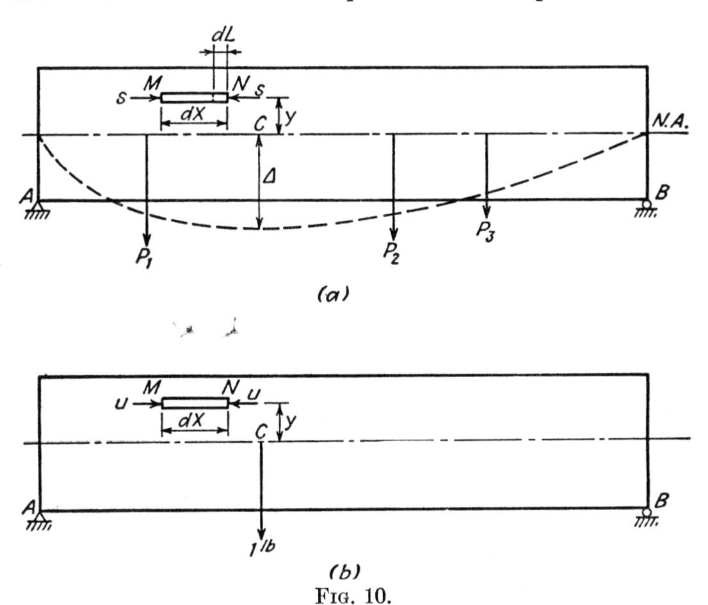

FIG. 10.

the applied loading of P_1, P_2, and P_3 (Fig. 10*a*). A unit load of 1 lb is applied at point C to the unloaded beam AB (Fig. 10*b*). By Eq. (5), $(1)(\Delta) = \Sigma u\,dL$ or $\Delta = \Sigma u\,dL$. Now let the bending moment at MN in Fig. 10*a* be M and the bending moment at MN in Fig. 10*b* be m. Let the original length of MN be dx. Then

$$u = \frac{my}{I}\,dA \qquad (6)$$

$$dL = \frac{S}{dA}\frac{1}{E}\,dx \tag{7}$$

Substituting $S = (My/I)\,dA$ in Eq. (7),

$$dL = \frac{My}{EI}\,dx \tag{8}$$

Substituting Eqs. (6) and (8) in Eq. (5),

$$\Delta = \sum u\,dL = \sum \left(\frac{my}{I}\,dA\right)\left(\frac{My}{EI}\,dx\right) = \int_0^L \int_0^A \frac{Mmy^2\,dA\,dx}{EI^2}$$

$$= \int_0^L \frac{Mm\,dx}{EI^2} \int_0^A y^2\,dA = \int_0^L \frac{Mm\,dx}{EI} \tag{9}$$

Equation (9) is the working formula by which the deflection at any point of a statically determinate beam due to the applied loading can be found.

Example 1. Find the deflection Δ_B by the unit-load method (Fig. 11).

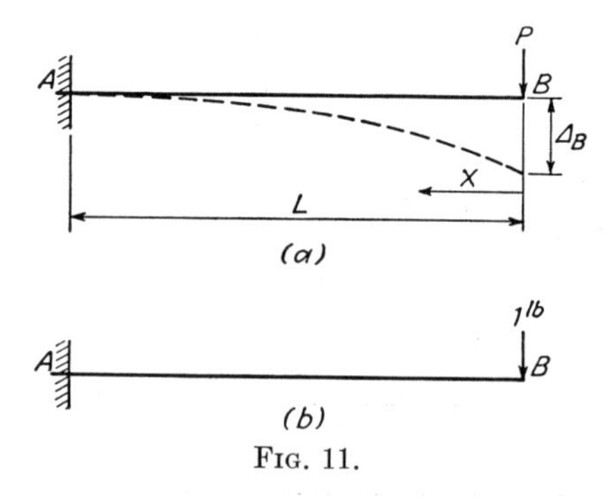

FIG. 11.

Solution

TABLE 1-1

Portion of beam	AB
Origin	B
Limits	$x = 0$ to $x = L$
M	$-Px$
m	$-(1)(x)$

$$\Delta_B = \int_0^L \frac{Mm\,dx}{EI} = \int_0^L \frac{(-Px)(-x)\,dx}{EI} = \int_0^L \frac{Px^2\,dx}{EI} = \left[\frac{Px^3}{3EI}\right]_0^L$$

$$= \frac{PL^3}{3EI} \text{ downward}$$

Note that the deflection is downward because the unit load has been applied in that direction.

Example 2. Find the deflection Δ_B by the unit-load method (Fig. 12).

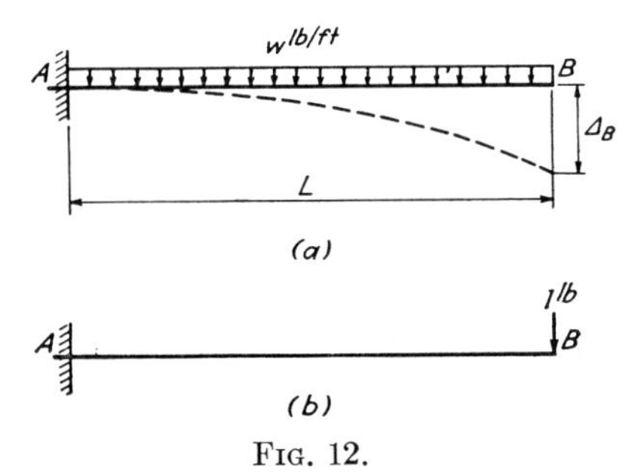

FIG. 12.

Solution

TABLE 2-1

Portion of beam	*AB*
Origin.	B
Limits.	$x = 0$ to $x = L$
M.	$-wx^2/2$
m.	$-(1)(x)$

$$\Delta_B = \int_0^L \frac{Mm\,dx}{EI} = \int_0^L \frac{(-wx^2/2)(-x)\,dx}{EI} = \int_0^L \frac{wx^3\,dx}{2EI} = \frac{wL^4}{8EI} \text{ downward}$$

Example 3. Find the deflection Δ_C by the unit-load method (Fig. 13).

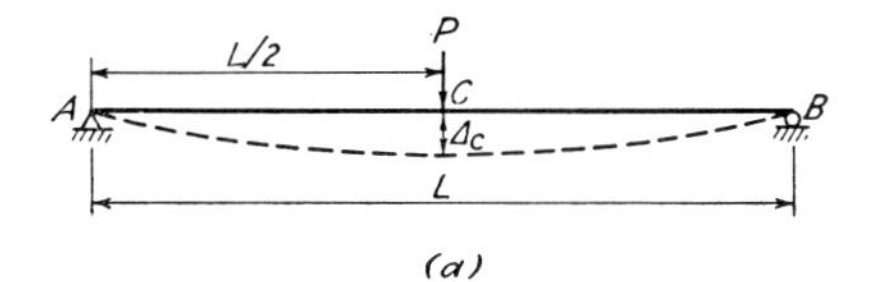

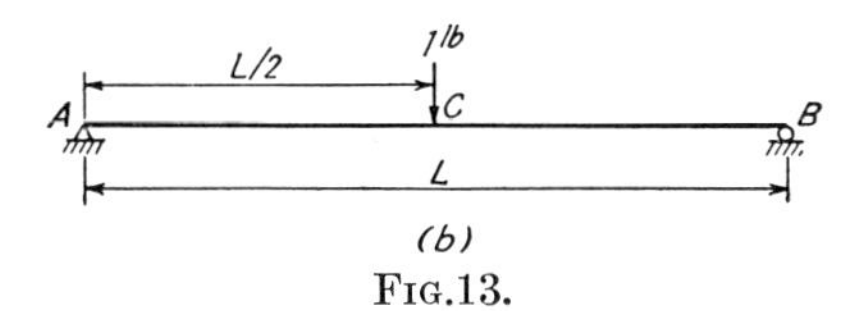

FIG.13.

Solution

TABLE 3-1

Portion of beam	*AC*	*CB*
Origin.	A	B
Limits.	$x = 0$ to $x = L/2$	$x = 0$ to $x = L/2$
M.	$Px/2$	$Px/2$
m.	$\frac{1}{2}x$	$\frac{1}{2}x$

$$\Delta_C = 2\int_0^{L/2} \frac{(\frac{1}{2}Px)(\frac{1}{2}x)\,dx}{EI} = \int_0^{L/2} \frac{Px^2\,dx}{2EI} = \left[\frac{Px^3}{6EI}\right]_0^{L/2} = \frac{PL^3}{48EI} \text{ downward}$$

Example 4. Find the deflection Δ_C by the unit-load method (Fig. 14).

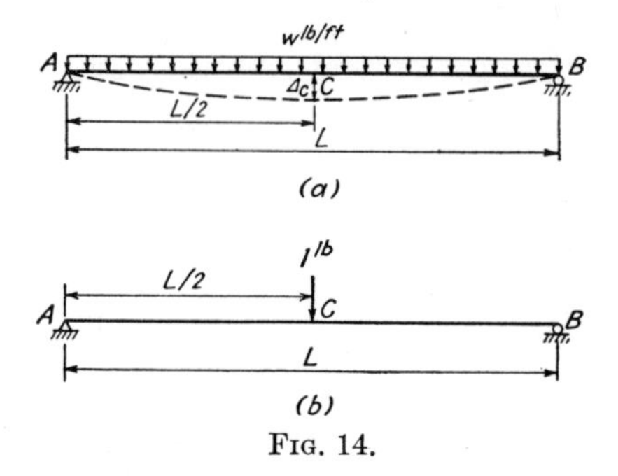

FIG. 14.

Solution

TABLE 4-1

Portion of beam	AC	CB
Origin	A	B
Limits	$x = 0$ to $x = L/2$	$x = 0$ to $x = L/2$
M	$\frac{1}{2}wLx - \frac{1}{2}wx^2$	$\frac{1}{2}wLx - \frac{1}{2}wx^2$
m	$\frac{1}{2}x$	$\frac{1}{2}x$

$$\Delta_C = 2\int_0^{L/2} \frac{(\frac{1}{2}wLx - \frac{1}{2}wx^2)(\frac{1}{2}x)\,dx}{EI} = \int_0^{L/2} \left(\frac{wLx^2}{2} - \frac{wx^3}{2}\right)\frac{dx}{EI}$$

$$= \frac{1}{EI}\left[\frac{wLx^3}{6} - \frac{wx^4}{8}\right]_0^{L/2} = \frac{wL^4}{EI}\left(\tfrac{1}{48} - \tfrac{1}{128}\right) = \frac{5wL^4}{384EI} \text{ downward}$$

Example 5. Find the deflection Δ_D in inches by the unit-load method (Fig. 15).

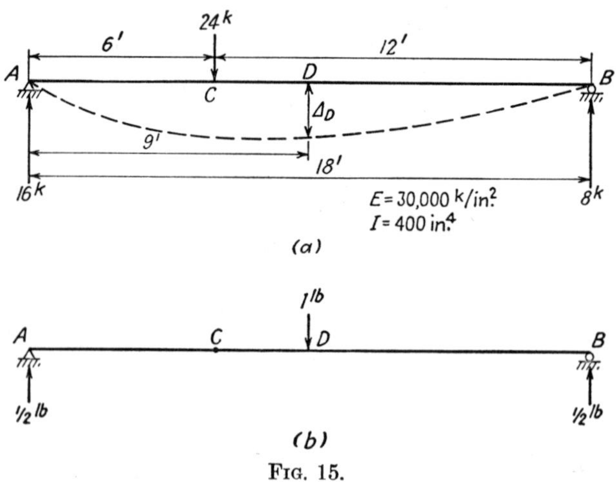

FIG. 15.

Solution

TABLE 5-1

Portion of beam	AC	CD	DB
Origin	A	A	B
Limits	$x = 0$ to $x = 6$	$x = 6$ to $x = 9$	$x = 0$ to $x = 9$
M	$16x$	$16x - 24(x - 6)$	$8x$
m	$\frac{1}{2}x$	$\frac{1}{2}x$	$\frac{1}{2}x$

$$EI\Delta_D = \int_0^6 (16x)(\tfrac{1}{2}x)\,dx + \int_6^9 [16x - 24(x-6)](\tfrac{1}{2}x)\,dx + \int_0^9 (8x)(\tfrac{1}{2}x)\,dx$$
$$= 576 + 936 + 972$$
$$= 2{,}484 \text{ kip-ft}^3$$
$$\Delta_D = \frac{2{,}484 \times 1{,}728}{30{,}000 \times 400} = 0.358 \text{ in. downward}$$

Example 6. Find the deflection Δ_D by the unit-load method (Fig. 16).

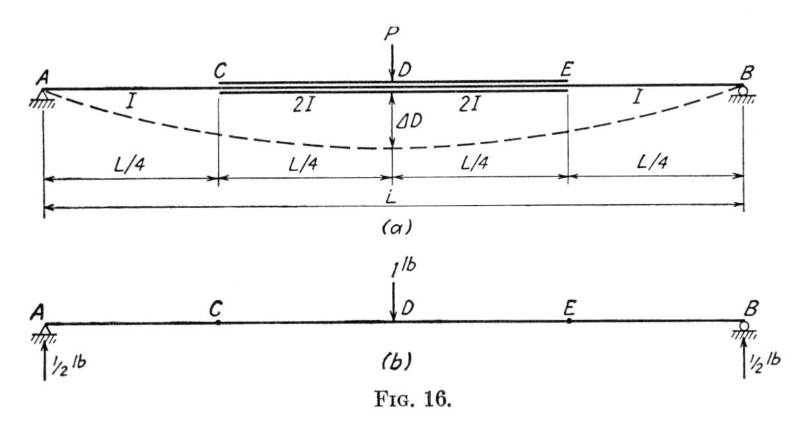

FIG. 16.

Solution

TABLE 6-1

Portion of beam	AC	CD	DE	EB
Origin	A	A	B	B
Limits	$x = 0$ to $x = L/4$	$x = L/4$ to $x = \frac{1}{2}L$	$x = L/4$ to $x = \frac{1}{2}L$	$x = 0$ to $x = L/4$
M	$\frac{1}{2}Px$	$\frac{1}{2}Px$	$\frac{1}{2}Px$	$\frac{1}{2}Px$
m	$\frac{1}{2}x$	$\frac{1}{2}x$	$\frac{1}{2}x$	$\frac{1}{2}x$
I	I	$2I$	$2I$	I

$$
\begin{aligned}
EI\Delta_D &= 2\int_0^{L/4} (\tfrac{1}{2}Px)(\tfrac{1}{2}x)\,dx + (\tfrac{1}{2})(2)\int_{L/4}^{L/2} (\tfrac{1}{2}Px)(\tfrac{1}{2}x)\,dx \\
&= \int_0^{L/4} \tfrac{1}{2}Px^2\,dx + \int_{L/4}^{L/2} \tfrac{1}{4}Px^2\,dx \\
&= PL^3\left(\tfrac{1}{384} + \tfrac{1}{96} - \tfrac{1}{768}\right) = \frac{3PL^3}{256}
\end{aligned}
$$

$$
\Delta_D = \frac{3PL^3}{256EI} \text{ downward}
$$

EXERCISES

1 to 3. Find the deflection Δ_B by the unit-load method.

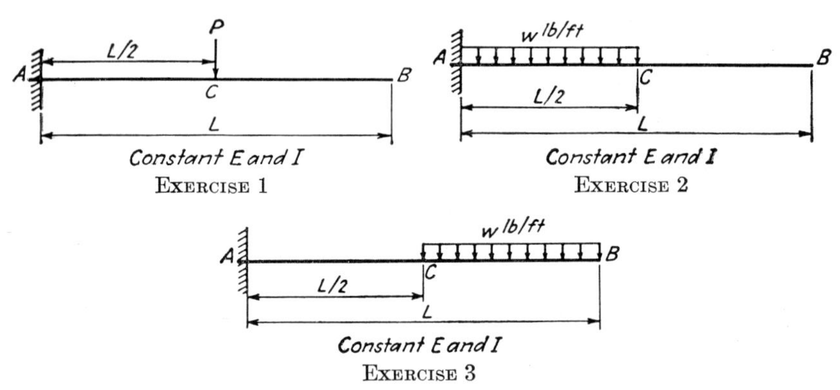

EXERCISE 1 EXERCISE 2 EXERCISE 3

4 and 5. Find the deflection Δ_D by the unit-load method.

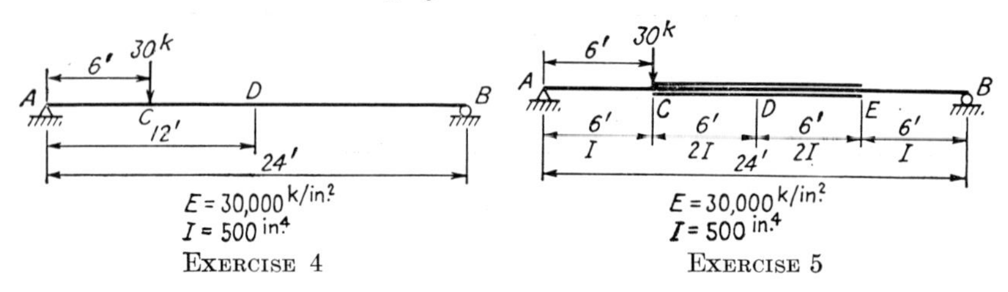

EXERCISE 4 EXERCISE 5

6 and 7. Find the deflection Δ_C by the unit-load method.

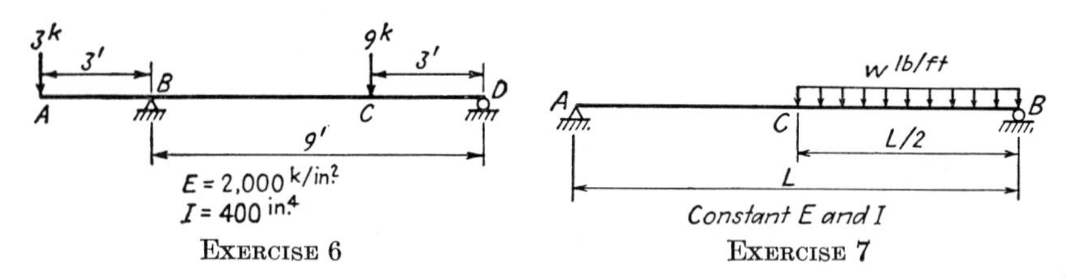

EXERCISE 6 EXERCISE 7

8. The Unit-load Method—Application to Rotation of Beams. The angle of rotation, θ, denotes the angle in radians between the tangent to the elastic curve at any point of the beam and the original tangent at the same point before loading. It will be shown that

$$\theta = \int_0^L \frac{Mm\,dx}{EI} \tag{10}$$

where M has the same meaning as in Eq. (9) and m is the bending moment at any section owing to the action of a unit moment at the point of the

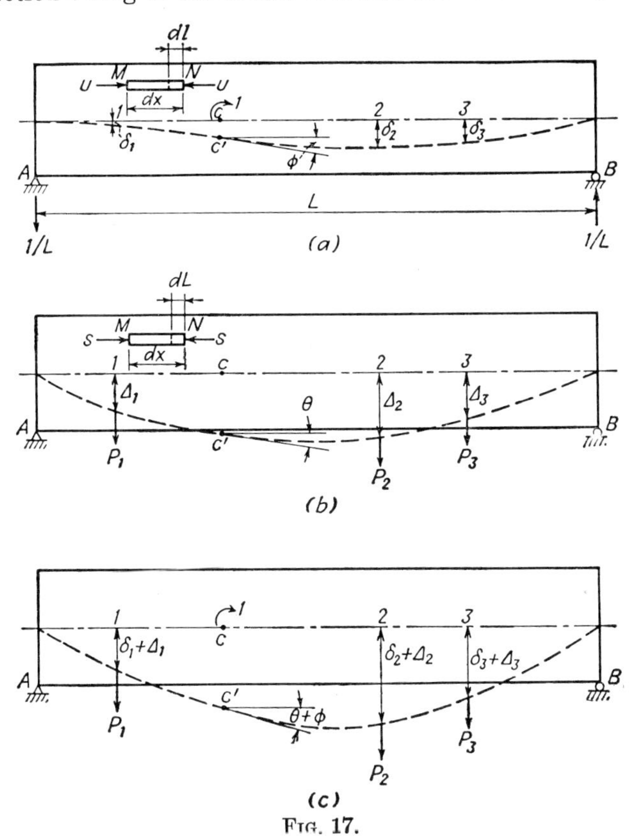

FIG. 17.

unloaded beam where the rotation is being sought. The method of deriving Eq. (10) is similar to that of deriving Eq. (9). Let Fig. 17 replace Fig. 9. Then the work done by the unit moment on the beam is $\frac{1}{2}(1)(\phi)$, which is equal to $\frac{1}{2}\Sigma u\,dl$. When the loads P_1, P_2, P_3 are added gradually, additional external work done on the beam is $\frac{1}{2}P_1\Delta_1 + \frac{1}{2}P_2\Delta_2 + \frac{1}{2}P_3\Delta_3 + (1)(\theta)$, which is equal to the internal energy $\frac{1}{2}\Sigma S\,dL + \Sigma u\,dL$. Substituting $\frac{1}{2}P_1\Delta_1 + \frac{1}{2}P_2\Delta_2 + \frac{1}{2}P_3\Delta_3 = \frac{1}{2}\Sigma S\,dL$,

$$(1)(\theta) = \Sigma u\,dL$$

or

$$\theta = \sum u\,dL = \int_0^L \frac{Mm\,dx}{EI}$$

which is Eq. (10).

It is to be noted that, in finding deflections or rotations at points on statically determinate beams by the unit-load method, the direction of the deflection or rotation is the *same* as or *opposite* to that of the unit load or unit moment depending on whether the result is *positive* or *negative*.

Example 7. Find θ_B by the unit-load method (Fig. 18).

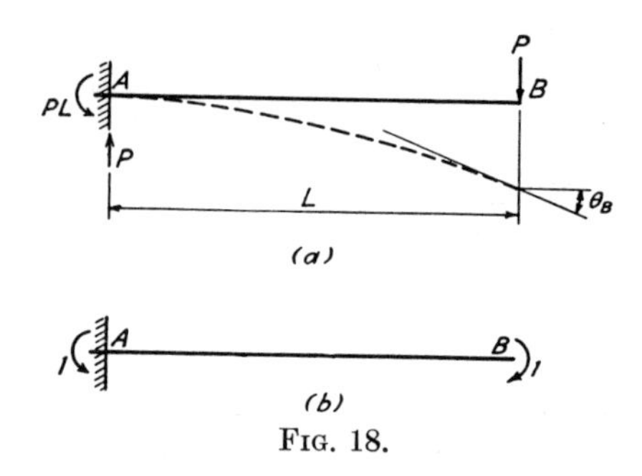

FIG. 18.

Solution

TABLE 7-1

Portion of beam	AB
Origin	B
Limits	$x = 0$ to $x = L$
M	$-Px$
m	-1

$$\theta_B = \int_0^L \frac{(-Px)(-1)\,dx}{EI} = \left[\frac{Px^2}{2EI}\right]_0^L = \frac{PL^2}{2EI} \text{ clockwise}$$

Example 8. Find θ_B by the unit-load method (Fig. 19).

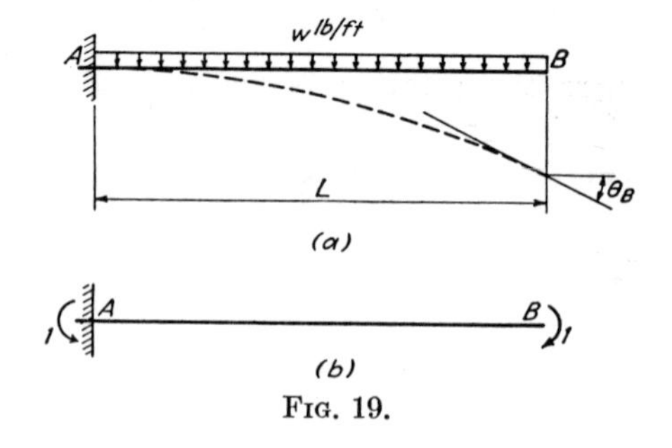

FIG. 19.

Solution

TABLE 8-1

Portion of beam	*AB*
Origin.	B
Limits.	$x = 0$ to $x = L$
M.	$-\frac{1}{2}wx^2$
m.	-1

$$\theta_B = \int_0^L \frac{(-wx^2/2)(-1)\,dx}{EI} = \left[\frac{wx^3}{6EI}\right]_0^L = \frac{wL^3}{6EI} \text{ clockwise}$$

Example 9. Find θ_A or θ_B by the unit-load method (Fig. 20).

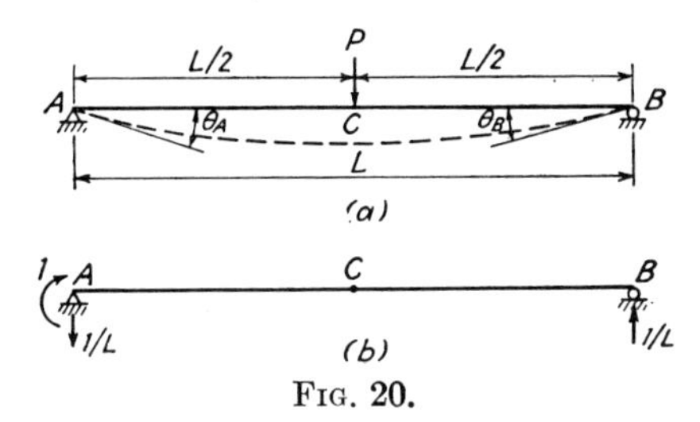

FIG. 20.

Solution

TABLE 9-1

Portion of beam	AC	CB
Origin.	A	B
Limits.	$x = 0$ to $x = \frac{1}{2}L$	$x = 0$ to $x = \frac{1}{2}L$
M.	$\frac{1}{2}Px$	$\frac{1}{2}Px$
m.	$1 - x/L$	x/L

$$\theta_A = \int_0^{L/2} \frac{\left(\frac{P}{2}x\right)\left(1 - \frac{x}{L}\right)dx}{EI} + \int_0^{L/2} \frac{\left(\frac{P}{2}x\right)\left(\frac{1}{L}x\right)dx}{EI}$$

$$= \frac{1}{EI}\left[\frac{Px^2}{4} - \frac{Px^3}{6L} + \frac{Px^3}{6L}\right]_0^{L/2} = \frac{PL^2}{16EI} \text{ clockwise}$$

$$\theta_B = \frac{PL^2}{16EI} \text{ counterclockwise}$$

Example 10. Find θ_A or θ_B by the unit-load method (Fig. 21).

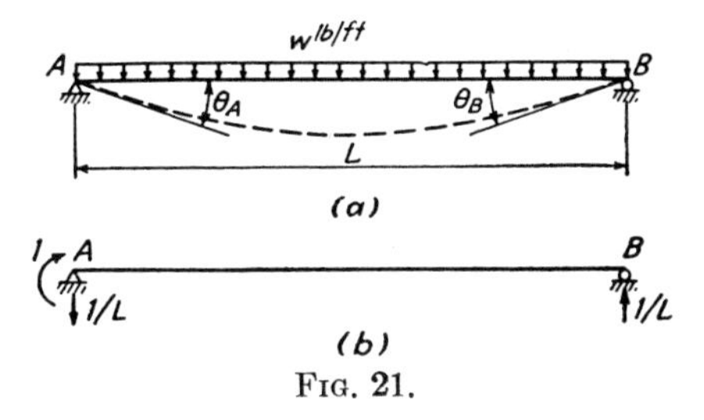

FIG. 21.

Solution

TABLE 10-1

Portion of beam	AB
Origin..........	B
Limits..........	$x = 0$ to $x = L$
M..............	$\frac{1}{2}wLx - \frac{1}{2}wx^2$
m..............	x/L

$$\theta_A = \int_0^L \frac{\left(\frac{wLx}{2} - \frac{wx^2}{2}\right)\left(\frac{1}{L}x\right) dx}{EI}$$

$$= \frac{1}{EI}\int_0^L \left(\frac{wx^2}{2} - \frac{wx^3}{2L}\right) dx = \frac{wL^3}{24EI} \text{ clockwise}$$

$$\theta_B = \frac{wL^3}{24EI} \text{ counterclockwise}$$

Example 11. Find θ_A in radians by the unit-load method (Fig. 22).

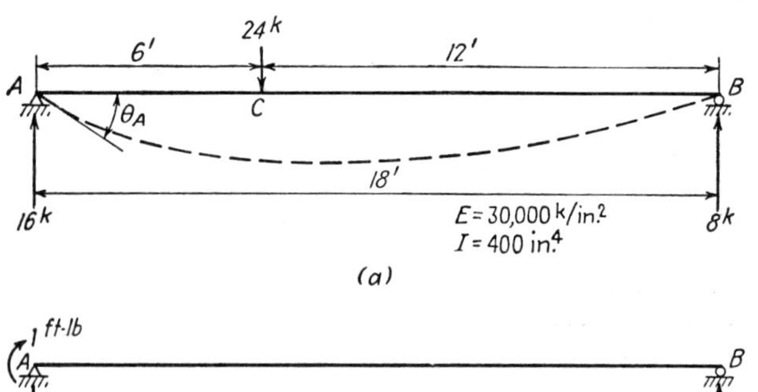

(b)

FIG. 22.

Solution

TABLE 11-1

Portion of beam	AC	CB
Origin..........	A	B
Limits..........	$x = 0$ to $x = 6$	$x = 0$ to $x = 12$
M..............	$16x$	$8x$
m..............	$1 - \frac{1}{18}x$	$\frac{1}{18}x$

$$EI\theta_A = \int_0^6 (16x)\left(1 - \frac{x}{18}\right) dx + \int_0^{12} (8x)\left(\frac{x}{18}\right) dx$$

$$= 224 + 256 = 480 \text{ kip-ft}^2$$

$$\theta_A = \frac{(480)(144)}{(30{,}000)(400)} = 0.00576 \text{ rad clockwise}$$

Example 12. Find θ_A or θ_B by the unit-load method (Fig. 23).

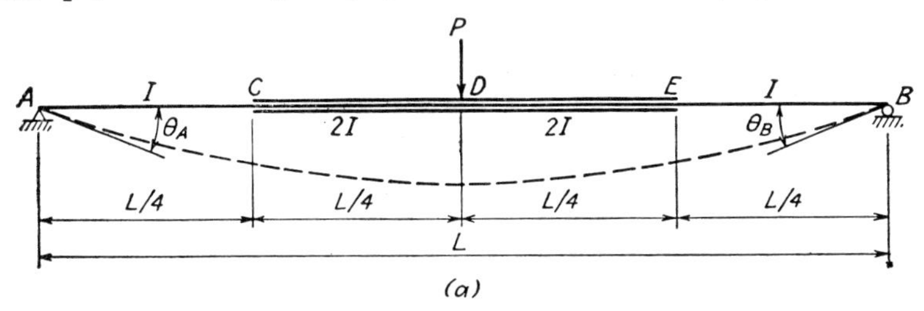

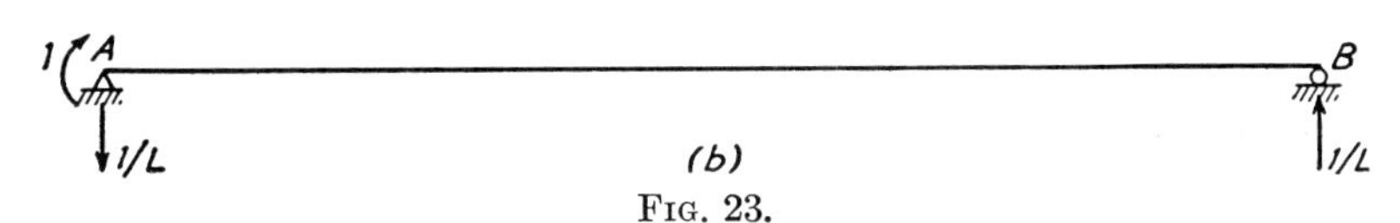

FIG. 23.

Solution

TABLE 12-1

Portion of beam	AC	CD	DE	EB
Origin	A	A	B	B
Limits	$x = 0$ to $x = \frac{L}{4}$	$x = \frac{L}{4}$ to $x = \frac{L}{2}$	$x = \frac{L}{4}$ to $x = \frac{L}{2}$	$x = 0$ to $x = \frac{L}{4}$
M	$\frac{1}{2}Px$	$\frac{1}{2}Px$	$\frac{1}{2}Px$	$\frac{1}{2}Px$
m	$1 - (1/L)x$	$1 - (1/L)x$	$(1/L)x$	$(1/L)x$
I	I	$2I$	$2I$	I

$$EI\theta_A = \int_0^{L/4} \left(\frac{P}{2}x\right)\left(1 - \frac{1}{L}x\right) dx + \frac{1}{2}\int_{L/4}^{L/2} \left(\frac{P}{2}x\right)\left(1 - \frac{1}{L}x\right) dx$$
$$+ \frac{1}{2}\int_{L/4}^{L/2} \left(\frac{Px}{2}\right)\left(\frac{1}{L}x\right) dx + \int_0^{L/4} \left(\frac{P}{2}x\right)\left(\frac{1}{L}x\right) dx = \frac{5PL^2}{128}$$

$$\theta_A = \frac{5PL^2}{128EI} \text{ clockwise}$$

$$\theta_B = \frac{5PL^2}{128EI} \text{ counterclockwise}$$

EXERCISES

8 to 10. Find θ_B by the unit-load method.

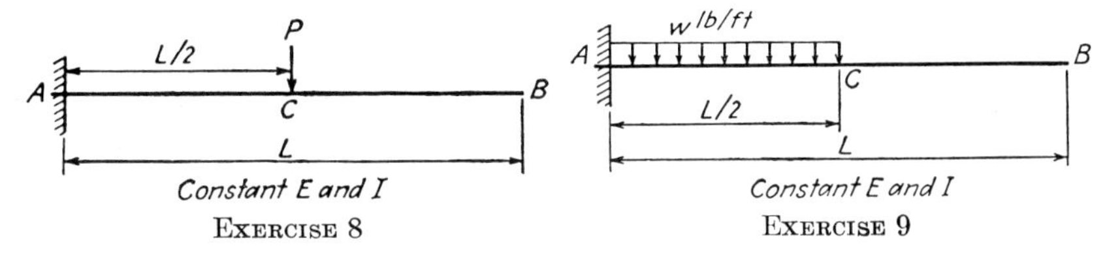

EXERCISE 8

EXERCISE 9

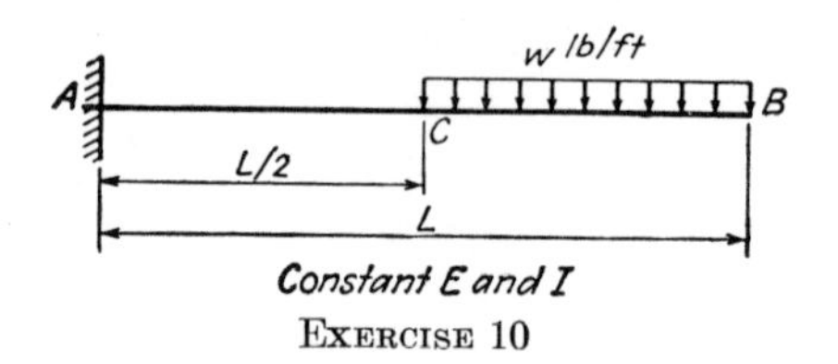

EXERCISE 10

11 and 12. Find θ_A and θ_B by the unit-load method.

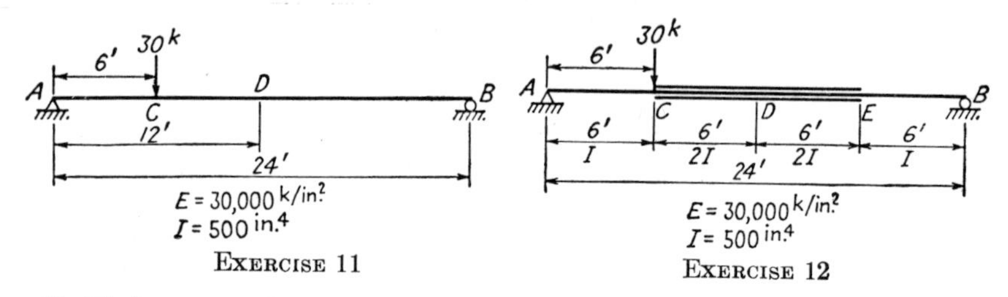

EXERCISE 12

13. Find θ_A, θ_B, and θ_D by the unit-load method.

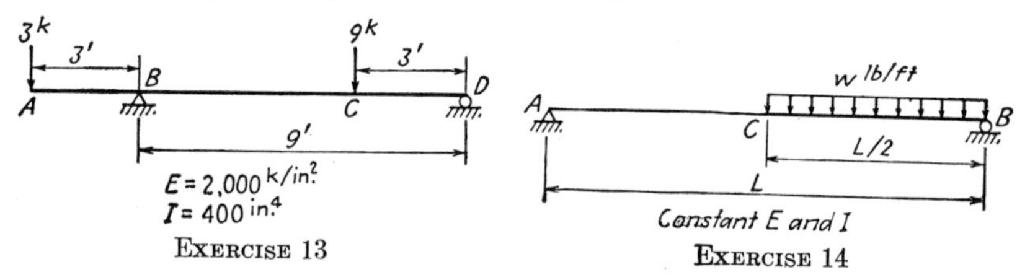

EXERCISE 14

14. Find θ_A and θ_B by the unit-load method.

9. Castigliano's Theorem—Derivation. Castigliano published two theorems in 1879 dealing with indeterminate structures. One of the two theorems is stated below, and the use of this theorem to find deflections (or rotations) of beams and frames actually involves processes identical with those of the unit-load method. However, the unit-load method, because of its simple approach, is preferred. The other theorem of Castigliano is known as the *theorem of least work*, which will be treated in Art. 25.

Castigliano's theorem can be stated as follows: The deflection in a certain direction at a certain point of a statically determinate structure is equal to the partial derivative of the total external work or the total internal energy with respect to a load applied at this point in the direction of the deflection.

The problem is to find the deflection Δ_1 at point 1 of the beam AB subjected to loads P_1, P_2, and P_3 (Fig. 24*a*). If the loads P_1, P_2, and P_3 are gradually applied to the beam, the external work done on the beam is

$$W = \tfrac{1}{2}P_1\Delta_1 + \tfrac{1}{2}P_2\Delta_2 + \tfrac{1}{2}P_3\Delta_3 \tag{11}$$

Now if a small load dP_1 is gradually added to the beam of Fig. 24a, the resulting deflections will be as shown in Fig. 24b. The work done additionally on the beam is

$$dW = \tfrac{1}{2}\,dP_1\,d\Delta_1 + P_1\,d\Delta_1 + P_2\,d\Delta_2 + P_3\,d\Delta_3 \tag{12}$$

The total work done on the beam owing to the gradually applied loads P_1, P_2, P_3, and then dP_1 is the sum of Eqs. (11) and (12).

$$W + dW = \tfrac{1}{2}P_1\Delta_1 + \tfrac{1}{2}P_2\Delta_2 + \tfrac{1}{2}P_3\Delta_3 + \tfrac{1}{2}\,dP_1\,d\Delta_1 + P_1\,d\Delta_1 + P_2\,d\Delta_2 + P_3\,d\Delta_3 \tag{13}$$

If the loads $(P_1 + dP_1)$, P_2, and P_3 are gradually applied to the beam

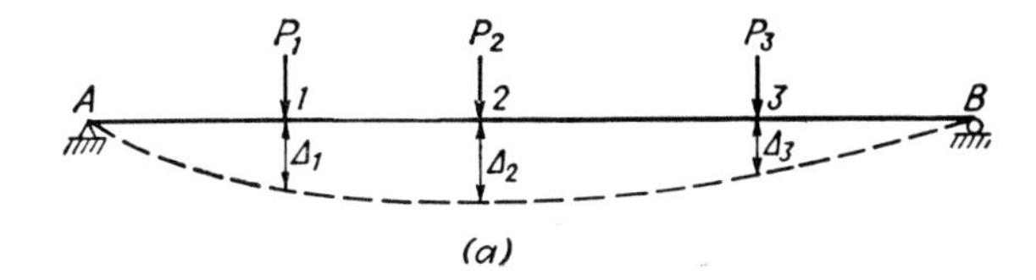

(a)

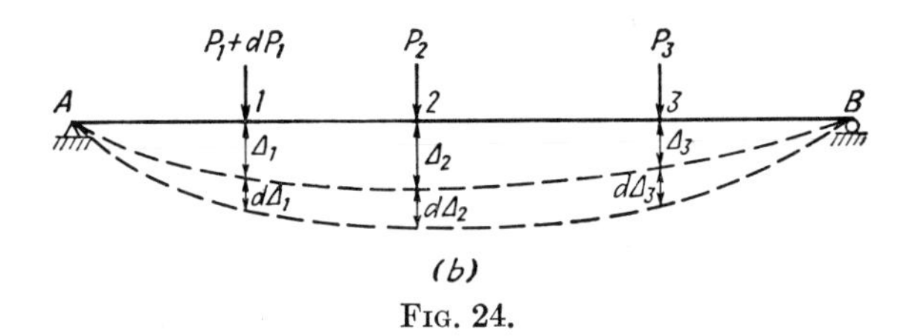

(b)

FIG. 24.

altogether, the total external work done on the beam should be (Fig. 24b)

$$W + dW = \tfrac{1}{2}(P_1 + dP_1)(\Delta_1 + d\Delta_1) + \tfrac{1}{2}P_2(\Delta_2 + d\Delta_2) + \tfrac{1}{2}P_3(\Delta_3 + d\Delta_3) \tag{14}$$

Since the total external work done on the beam must be the same regardless of whether the loads P_1, P_2, P_3 are first applied and then the load dP_1 is added or the loads $P_1 + dP_1$, P_2, P_3 are applied simultaneously, Eq. (13) can be equated to Eq. (14).

$$\begin{aligned} W + dW &= \tfrac{1}{2}P_1\Delta_1 + \tfrac{1}{2}P_2\Delta_2 + \tfrac{1}{2}P_3\Delta_3 + \tfrac{1}{2}\,dP_1\,d\Delta_1 + P_1\,d\Delta_1 \\ &\qquad + P_2\,d\Delta_2 + P_3\,d\Delta_3 \\ &= \tfrac{1}{2}(P_1 + dP_1)(\Delta_1 + d\Delta_1) + \tfrac{1}{2}P_2(\Delta_2 + d\Delta_2) \\ &\qquad \tfrac{1}{2}P_3(\Delta_3 + d\Delta_3) \end{aligned} \tag{15}$$

Simplifying Eq. (15),

$$\tfrac{1}{2}P_1\,d\Delta_1 + \tfrac{1}{2}P_2\,d\Delta_2 + \tfrac{1}{2}P_3\,d\Delta_3 = \tfrac{1}{2}\,dP_1\,\Delta_1 \tag{16}$$

Substituting Eq. (12) into the left side of Eq. (16) and neglecting the term $\frac{1}{2}\,dP_1\,d\Delta_1$,

$$\tfrac{1}{2}\,dW = \tfrac{1}{2}\,dP_1\,\Delta_1$$

or

$$\Delta_1 = \frac{dW}{dP_1} \tag{17a}$$

A similar derivation can be made for rotation, for which

$$\theta_1 = \frac{dW}{dM_1} \tag{17b}$$

Equation (17) is Castigliano's theorem.

10. Use of Castigliano's Theorem to Find Beam Deflections or Rotations. Castigliano's theorem, as represented by Eq. (17), can be used to find beam deflections or rotations, and since

$$\Delta_1 = \frac{dW}{dP_1} \qquad \text{or} \qquad \theta_1 = \frac{dW}{dM_1}$$

the method is sometimes called the *partial derivative* method. The external work W done on the beam can be equated to the internal energy $\frac{1}{2}\Sigma S\,dL$ stored in the beam, or

$$W = \tfrac{1}{2}\Sigma S\,dL \tag{18}$$

Substituting $S = \frac{My}{I}\,dA$ and $dL = \frac{My}{I}\frac{1}{E}\,dx$ into (18),

$$W = \frac{1}{2}\sum\left(\frac{My}{I}\,dA\right)\left(\frac{My}{I}\frac{1}{E}\,dx\right) = \frac{1}{2}\int_0^L\int_0^A y^2\,dA\,\frac{M^2}{EI^2}\,dx$$

$$= \frac{1}{2}\int_0^L \frac{M^2\,dx}{EI} \tag{19}$$

Substituting Eq. (19) into Eq. (17),

$$\Delta_1 = \frac{dW}{dP_1} = \frac{d\left(\dfrac{1}{2}\displaystyle\int_0^L \frac{M^2\,dx}{EI}\right)}{dP_1} = \int_0^L \frac{M\dfrac{dM}{dP_1}\,dx}{EI} \tag{20}$$

dM/dP_1 in Eq. (20) can be considered as the ratio of the increment dM in the bending moment at any section of the beam owing to an increment dP_1 in the load at the point 1, where the deflection is to be found; this ratio is obviously the bending moment at any section owing to a unit load at the point 1; or this ratio dM/dP_1 is equal to m as previously defined in the unit-load method. Equation (20) therefore reduces to Eq. (9). Thus the partial derivative method, though derived from a seemingly different conception, proves to be actually the same as the unit-load method. This fact will be further demonstrated by the following examples.

Example 13. Find the deflection Δ_D in inches by the partial derivative method (Fig. 25).

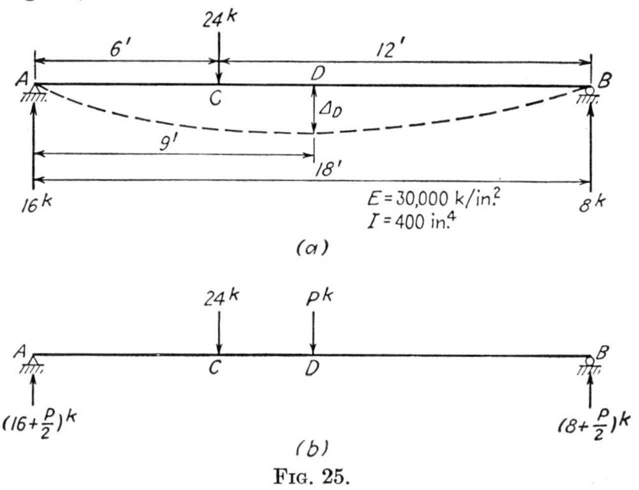

FIG. 25.

Solution. Although there is actually no load acting at point D, a load P at D must be added and later be made equal to zero after the partial differentiation. If there were actually a numerical load at D, it would be considered as P in deriving the expressions for M and this numerical value would be substituted for P after the partial differentiation is performed.

TABLE 13-1

Portion of beam	AC	CD	DB
Origin	A	A	B
Limits	$x = 0$ to $x = 6$	$x = 6$ to $x = 9$	$x = 0$ to $x = 9$
M	$(16 + \frac{1}{2}P)x$	$(16 + \frac{1}{2}P)x - 24(x - 6)$	$(8 + \frac{1}{2}P)x$

$$W = \frac{1}{2EI}\int_0^6 [(16 + \tfrac{1}{2}P)x]^2\,dx + \frac{1}{2EI}\int_6^9 [(16 + \tfrac{1}{2}P)x - 24(x - 6)]^2\,dx + \frac{1}{2EI}\int_0^9 [(8 + \tfrac{1}{2}P)x]^2\,dx$$

$$\Delta_D = \frac{dW}{dP} = \frac{1}{EI}\int_0^6 [(16 + \tfrac{1}{2}P)x](\tfrac{1}{2}x)\,dx + \frac{1}{EI}\int_6^9 [(16 + \tfrac{1}{2}P)x - 24(x - 6)](\tfrac{1}{2}x)\,dx + \frac{1}{EI}\int_0^9 [(8 + \tfrac{1}{2}P)x](\tfrac{1}{2}x)\,dx$$

Letting $P = 0$ in the above expression,

$$\Delta_D = \frac{dW}{dP} = \frac{1}{EI}\int_0^6 (16x)(\tfrac{1}{2}x)\,dx + \frac{1}{EI}\int_6^9 [16x - 24(x-6)](\tfrac{1}{2}x)\,dx + \frac{1}{EI}\int_0^9 (8x)(\tfrac{1}{2}x)\,dx$$

This expression for Δ_D is exactly the same as that obtained by the unit load method (see Example 5). So

$$\Delta_D = 0.358 \text{ in. downward}$$

Example 14. Find θ_A in radians by the partial derivative method (Fig. 26).

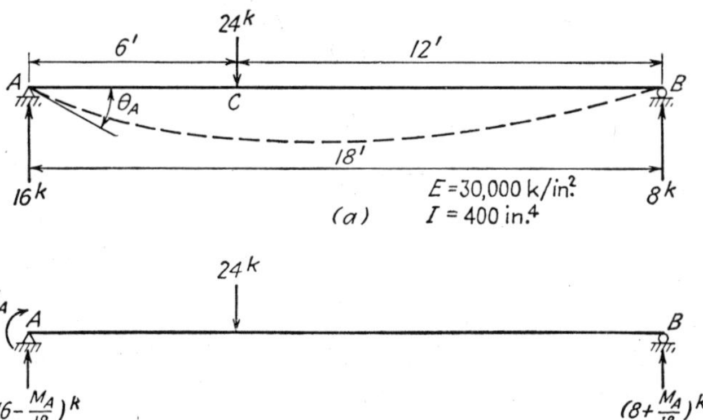

FIG. 26.

Solution

TABLE 14-1

Portion of beam	AC	CB
Origin	A	B
Limits	$x = 0$ to $x = 6$	$x = 0$ to $x = 12$
M	$M_A + \left(16 - \frac{M}{18}\right)x$	$\left(8 + \frac{M_A}{18}\right)x$

$$W = \frac{1}{2EI}\int_0^6 \left[M_A + \left(16 - \frac{M_A}{18}\right)x\right]^2 dx + \frac{1}{2EI}\int_0^{12}\left[\left(8 + \frac{M_A}{18}\right)x\right]^2 dx$$

$$\theta_A = \frac{dW}{dM_A} = \frac{1}{EI}\int_0^6 \left[M_A + \left(16 - \frac{M_A}{18}\right)x\right]\left[1 - \frac{x}{18}\right]dx + \frac{1}{EI}\int_0^{12}\left[\left(8 + \frac{M_A}{18}\right)x\right]\left[\frac{x}{18}\right]dx$$

Letting $M_A = 0$ in the above expression,

$$\theta_A = \frac{dW}{dM_A} = \frac{1}{EI}\int_0^6 (16x)\left(1 - \frac{x}{18}\right)dx + \frac{1}{EI}\int_0^{12}(8x)\left(\frac{x}{18}\right)dx$$

This expression for θ_A is exactly the same as that obtained by the unit-load method (see Example 11). Thus

$$\theta_A = 0.00576 \text{ rad clockwise}$$

EXERCISES

15. Find the deflection Δ_D by the partial derivative method.
16. Find θ_A and θ_B by the partial derivative method.

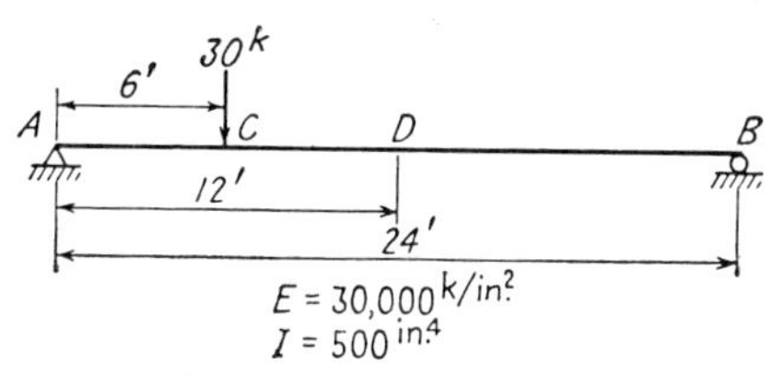

EXERCISES 15 and 16

11. The Unit-load Method—Application to Deflections and Rotations of Frames. The component in a certain direction of the total deflection or the rotation of the tangent at any point of a statically determinate frame with rigid joints can be found by the unit-load method. By one application of the unit load, only the deflection component in the direction of the unit load at its point of application can be determined. By determining the horizontal and vertical deflections of a point, the magnitude and direction of the total deflection can then be found. By one application of the unit moment, the rotation of the tangent at its point of application can be found. It is to be noted again that the direction of the deflection component or the rotation is the *same* as or the *opposite* to that of the unit load or the unit moment depending on whether the answer obtained is *positive* or *negative*. Attention needs also to be called to the assumption that, since the axial deformation of the members in a rigid frame owing to the direct axial stresses in them is always small, it can be neglected.

Example 15. Find θ_A, θ_B, θ_C, θ_D, Δ_H of B, Δ_H of C, and Δ_H of D in terms of EI by the unit-load method (Fig. 27).

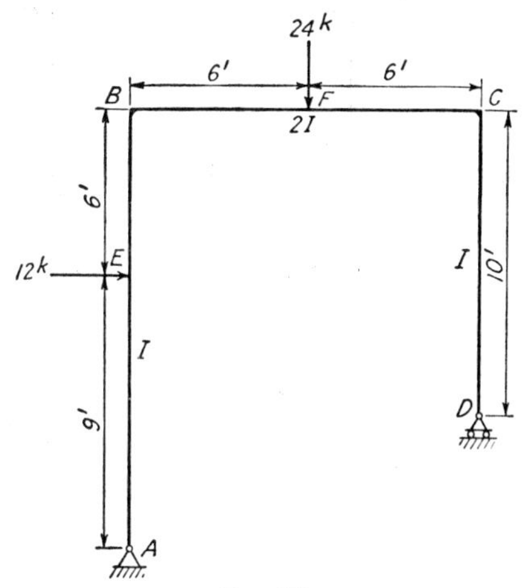

FIG. 27.

Solution. Refer to Fig. 28a to h, and see Table 15-1.

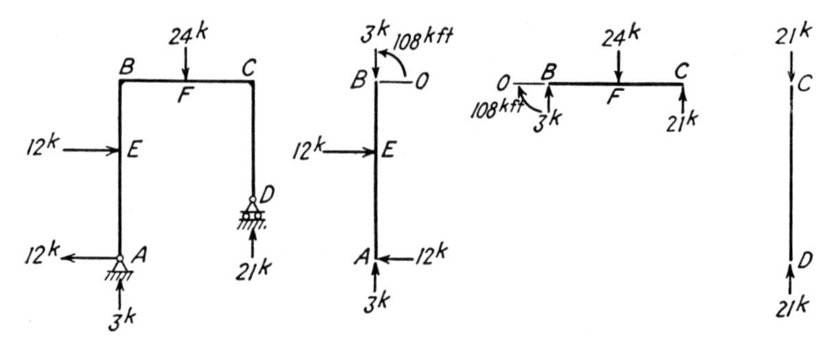

(a)-Free body diagrams from which values of M are obtained

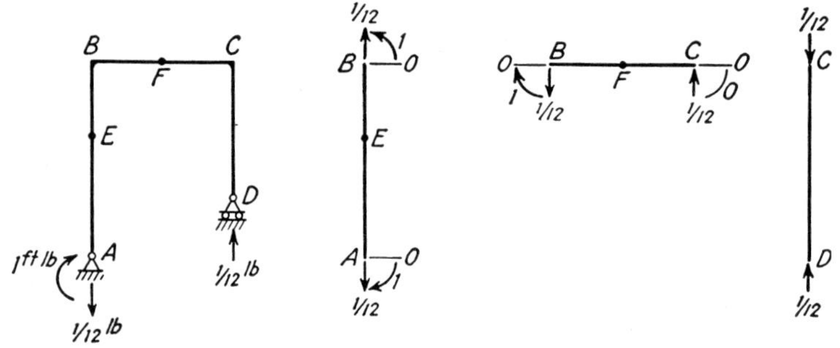

(b)-Free body diagrams from which values of m for θ_A are obtained

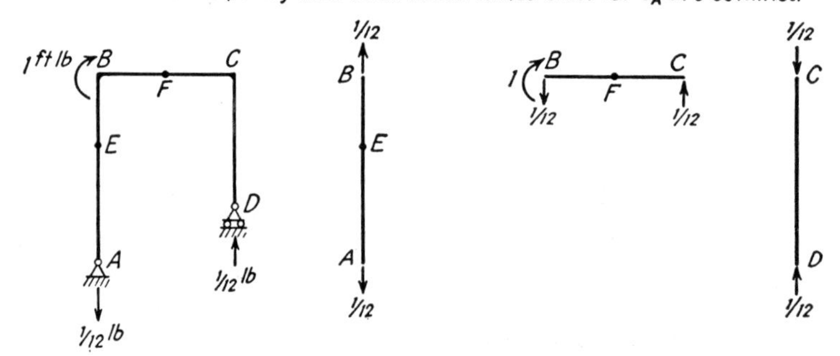

(c)-Free body diagrams from which values of m for θ_B are obtained

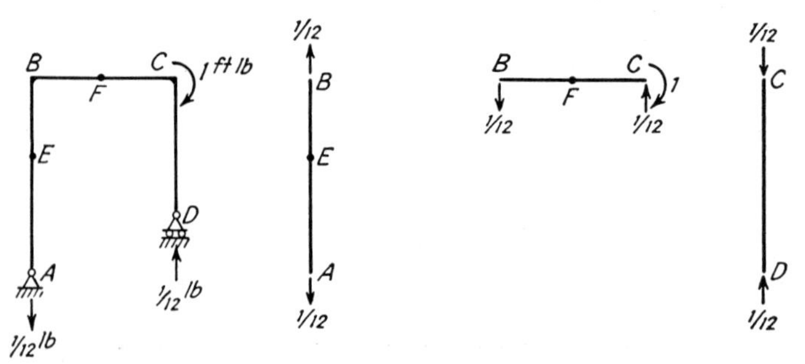

(d)-Free body diagrams from which values of m for θ_C are obtained

Fig. 28.

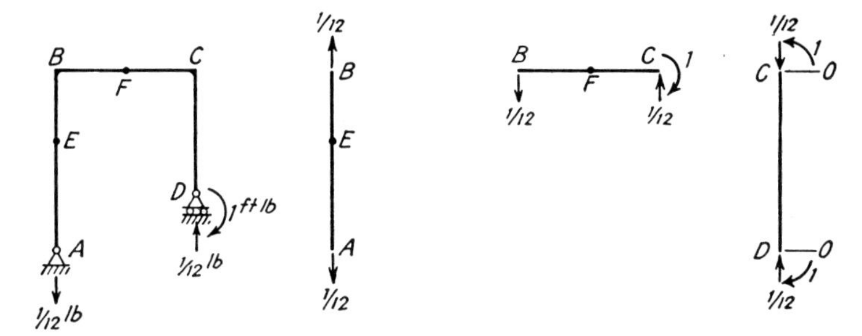

(e)-Free body diagrams from which values of m for θ_D are obtained

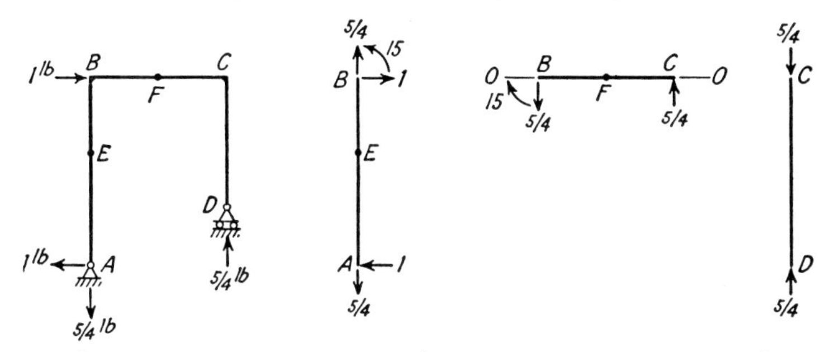

(f)-Free body diagrams from which values of m for Δ_H at B are obtained

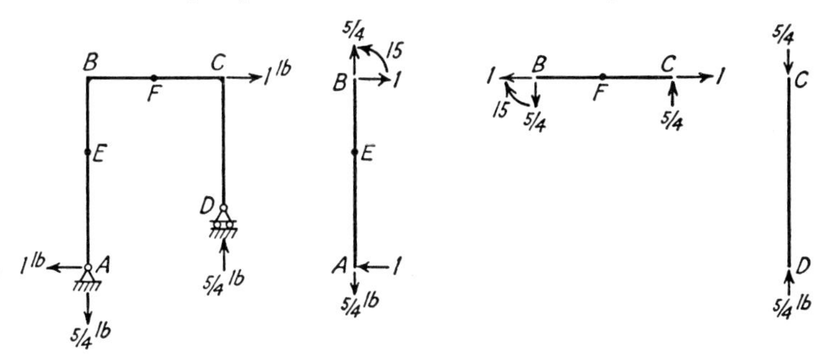

(g)-Free body diagrams from which values of m for Δ_H at C are obtained

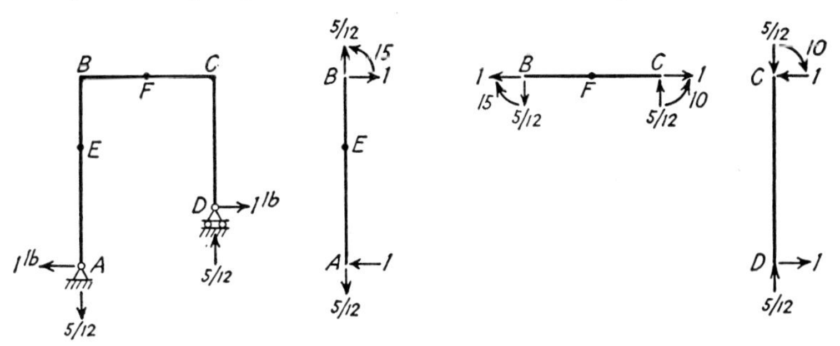

(h)-Free body diagrams from which values of m for Δ_H at D are obtained

FIG. 28 (*Continued*).

TABLE 15-1

Portion of beam	AE	EB	BF	FC	CD
Origin ...	A	B	B	C	D
Limits ...	$x = 0$ to $x = 9$	$x = 0$ to $x = 6$	$x = 0$ to $x = 6$	$x = 0$ to $x = 6$	$x = 0$ to $x = 10$
I	I	I	$2I$	$2I$	I
M	$12x$	108	$108 + 3x$	$21x$	0
m for θ_A	1	1	$1 - x/12$	$x/12$	0
m for θ_B	0	0	$1 - x/12$	$x/12$	0
m for θ_C	0	0	$-x/12$	$-1 + x/12$	0
m for θ_D	0	0	$-x/12$	$-1 + x/12$	-1
m for Δ_H of B...	x	$15 - x$	$15 - 5x/4$	$5x/4$	0
m for Δ_H of C...	x	$15 - x$	$15 - 5x/4$	$5x/4$	0
m for Δ_H of D ..	x	$15 - x$	$15 - 5x/12$	$10 + 5x/12$	x

Note: The algebraic sign of M or m is assumed positive if M or m causes compression on the outside of the frame.

$$EI\theta_A = \int_0^9 (12x)(1)\,dx + \int_0^6 (108)(1)\,dx + \frac{1}{2}\int_0^6 (108 + 3x)\left(1 - \frac{x}{12}\right)dx + \frac{1}{2}\int_0^6 (21x)\left(\frac{x}{12}\right)dx$$

$$= 486 + 648 + 261 + 63 = 1{,}458$$

$$\theta_A = 1{,}458\,\frac{\text{kip-ft}^2}{EI}\ \text{clockwise}$$

$$EI\theta_B = \frac{1}{2}\int_0^6 (108 + 3x)\left(1 - \frac{x}{12}\right)dx + \frac{1}{2}\int_0^6 (21x)\left(\frac{x}{12}\right)dx$$

$$= 261 + 63 = 324$$

$$\theta_B = 324\,\frac{\text{kip-ft}^2}{EI}\ \text{clockwise}$$

$$EI\theta_C = \frac{1}{2}\int_0^6 (108 + 3x)\left(-\frac{x}{12}\right)dx + \frac{1}{2}\int_0^6 (21x)\left(-1 + \frac{x}{12}\right)dx$$

$$= -90 - 126 = -216$$

$$\theta_C = 216\,\frac{\text{kip-ft}^2}{EI}\ \text{counterclockwise}$$

$EI\theta_D$ = same integrals as those for $EI\theta_C$

$$\theta_D = 216\,\frac{\text{kip-ft}^2}{EI}\ \text{counterclockwise}$$

$$EI\Delta_H \text{ of } B = \int_0^9 (12x)(x)\,dx + \int_0^6 (108)(15 - x)\,dx$$
$$+ \frac{1}{2}\int_0^6 (108 + 3x)(15 - \tfrac{5}{4}x)\,dx + \frac{1}{2}\int_0^6 (21x)(\tfrac{5}{4}x)\,dx$$
$$= 2{,}916 + 7{,}776 + 3{,}915 + 945 = 15{,}552$$

$$\Delta_H \text{ of } B = 15{,}552\,\frac{\text{kip-ft}^3}{EI} \text{ to the right}$$

$EI\Delta_H$ of C = same integrals as those for $EI\Delta_H$ of B

$$\Delta_H \text{ of } C = 15{,}552\,\frac{\text{kip-ft}^3}{EI} \text{ to the right}$$

$$EI\Delta_H \text{ of } D = \int_0^9 (12x)(x)\,dx + \int_0^6 (108)(15 - x)\,dx$$
$$+ \frac{1}{2}\int_0^6 (108 + 3x)(15 - \tfrac{5}{12}x)\,dx$$
$$+ \frac{1}{2}\int_0^6 (21x)(10 + \tfrac{5}{12}x)\,dx$$
$$= 2{,}916 + 7{,}776 + 4{,}815 + 2{,}205 = 17{,}712$$

$$\Delta_H \text{ of } D = 17{,}712\,\frac{\text{kip-ft}^3}{EI} \text{ to the right}$$

Example 16. Find θ_B, θ_C, θ_D, Δ_H of B, Δ_H and Δ_V of C, and Δ_H and Δ_V of D by the unit-load method (Fig. 29).

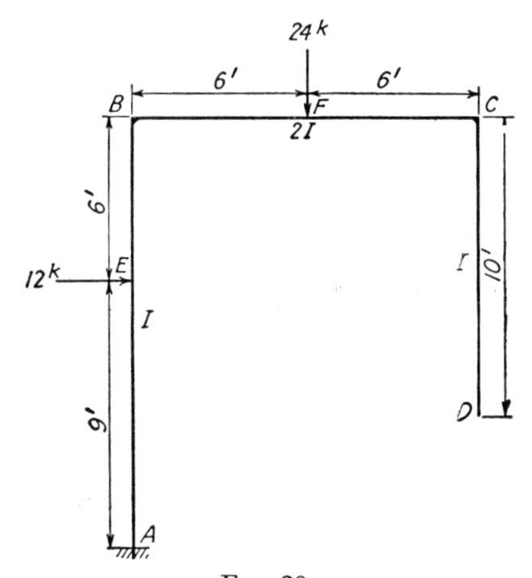

FIG. 29.

Solution. Refer to Fig. 30*a* to *i*, and see Table 16-1.

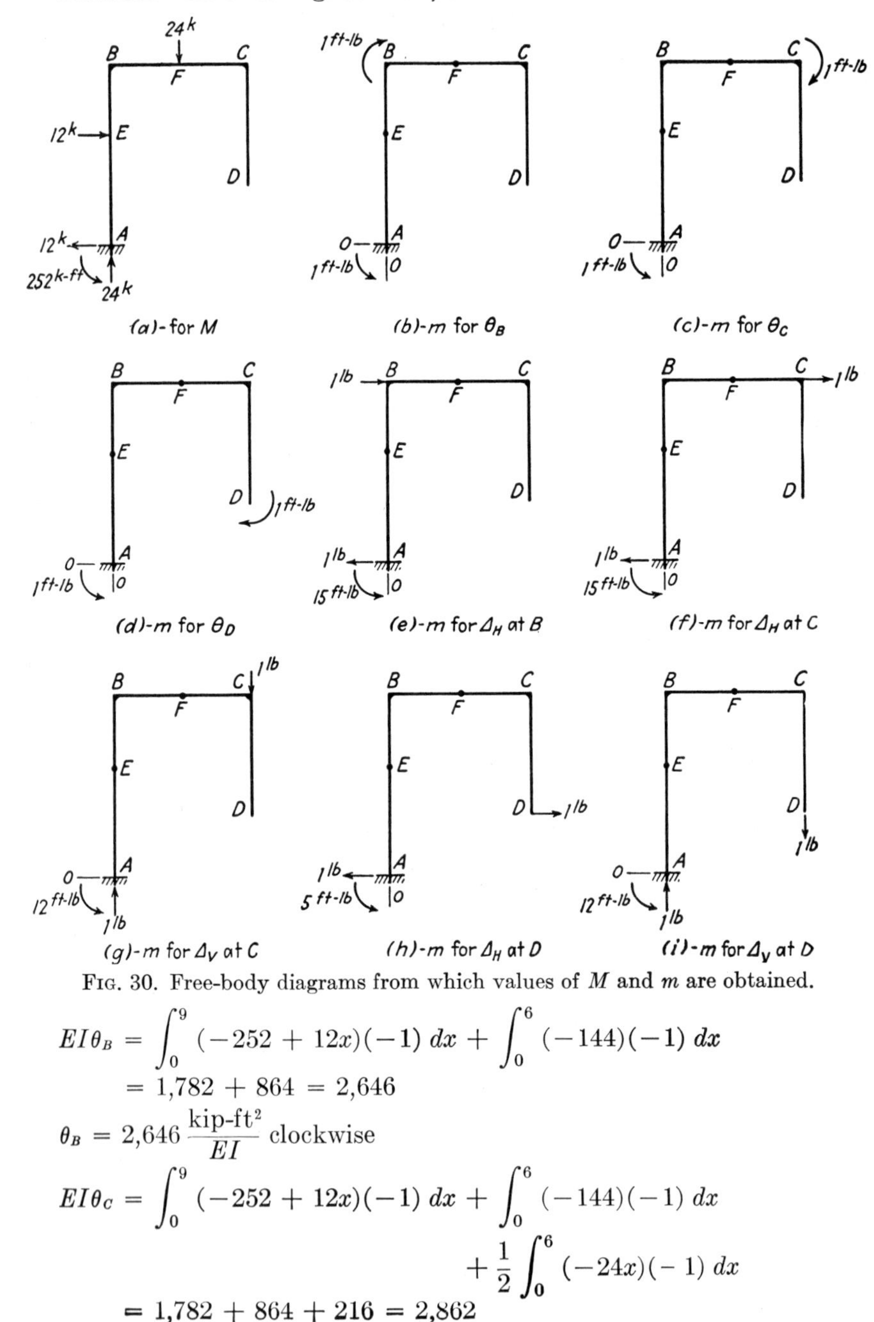

FIG. 30. Free-body diagrams from which values of M and m are obtained.

$$EI\theta_B = \int_0^9 (-252 + 12x)(-1)\,dx + \int_0^6 (-144)(-1)\,dx$$
$$= 1{,}782 + 864 = 2{,}646$$

$$\theta_B = 2{,}646 \frac{\text{kip-ft}^2}{EI} \text{ clockwise}$$

$$EI\theta_C = \int_0^9 (-252 + 12x)(-1)\,dx + \int_0^6 (-144)(-1)\,dx + \frac{1}{2}\int_0^6 (-24x)(-1)\,dx$$
$$= 1{,}782 + 864 + 216 = 2{,}862$$

TABLE 16-1

Portion of frame	AE	EB	BF	FC	CD
Origin...	A	B	F	C	D
Limits...	$x = 0$ to $x = 9$	$x = 0$ to $x = 6$	$x = 0$ to $x = 6$	$x = 0$ to $x = 6$	$x = 0$ to $x = 10$
I........	I	I	$2I$	$2I$	I
M.......	$-252 + 12x$	-144	$-24x$	0	0
m for θ_B..	-1	-1	0	0	0
m for θ_C..	-1	-1	-1	-1	0
m for θ_D..	-1	-1	-1	-1	-1
m for Δ_H of B...	$-15 + x$	$-x$	0	0	0
m for Δ_H of C...	$-15 + x$	$-x$	0	0	0
m for Δ_V of C...	-12	-12	$-(x + 6)$	$-x$	0
m for Δ_H of D...	$-5 + x$	$10 - x$	10	10	x
m for Δ_V of D...	-12	-12	$-(x + 6)$	$-x$	0

$$\theta_C = 2{,}862 \frac{\text{kip-ft}^2}{EI} \text{ clockwise}$$

$$EI\theta_D = \text{same integrals as those for } EI\theta_C = 2{,}862$$

$$\theta_D = 2{,}862 \frac{\text{kip-ft}^2}{EI} \text{ clockwise}$$

$$EI\Delta_H \text{ of } B = \int_0^9 (-252 + 12x)(-15 + x)\,dx + \int_0^6 (-144)(-x)\,dx$$

$$= 19{,}440 + 2{,}592 = 22{,}032$$

$$\Delta_H \text{ of } B = 22{,}032 \frac{\text{kip-ft}^3}{EI} \text{ to the right}$$

$$EI\Delta_H \text{ of } C = \text{same integrals as those for } EI\Delta_H \text{ of } B = 22{,}032$$

$$\Delta_H \text{ of } C = 22{,}032 \frac{\text{kip-ft}^3}{EI} \text{ to the right}$$

$$EI\Delta_V \text{ of } C = \int_0^9 (-252 + 12x)(-12)\,dx + \int_0^6 (-144)(-12)\,dx + \frac{1}{2}\int_0^6 (24x)(x + 6)\,dx$$

$$= 21{,}384 + 10{,}368 + 2160 = 33{,}912$$

$$\Delta_V \text{ of } C = 33{,}912 \frac{\text{kip-ft}^3}{EI} \text{ downward}$$

$$EI\Delta_H \text{ of } D = \int_0^9 (-252 + 12x)(-5 + x)\,dx + \int_0^6 (-144)(10 - x)\,dx + \frac{1}{2}\int_0^6 (-24x)(10)\,dx$$

$$= +1{,}620 - 6{,}048 - 2{,}160 = -6{,}588$$

$$\Delta_H \text{ of } D = 6{,}588\,\frac{\text{kip-ft}^3}{EI} \text{ to the left}$$

$$EI\Delta_V \text{ of } D = \text{same integrals as those for } EI\Delta_V \text{ of } C = 33{,}912$$

$$\Delta_V \text{ of } D = 33{,}912\,\frac{\text{kip-ft}^3}{EI} \text{ downward}$$

EXERCISES

17. Find θ_B, θ_D, Δ_H of B, and Δ_H of D in terms of EI by the unit-load method.

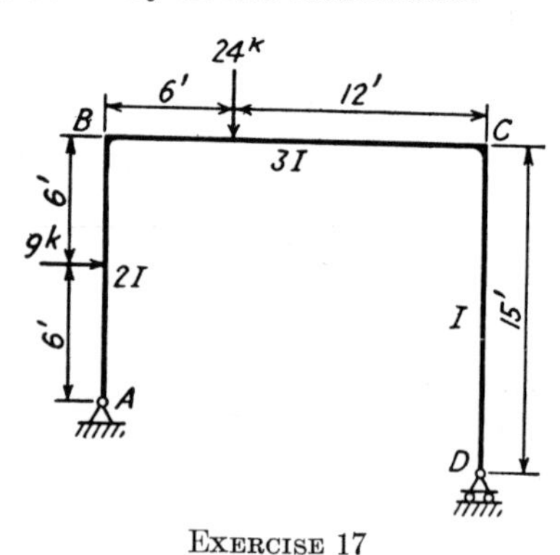

EXERCISE 17

18k
3'
A
B
C
6'
9'
D
E
6'
Constant E and I

EXERCISE 18

18. Find θ_A, θ_E, Δ_H and Δ_V of A, and Δ_H of D in terms of EI by the unit-load method.

19. Find θ_A, θ_C, Δ_H and Δ_V of A, and Δ_H of C in terms of EI by the unit-load method.

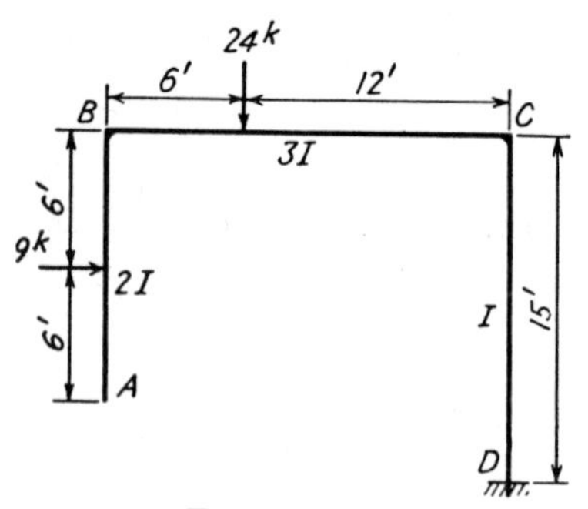

EXERCISE 19

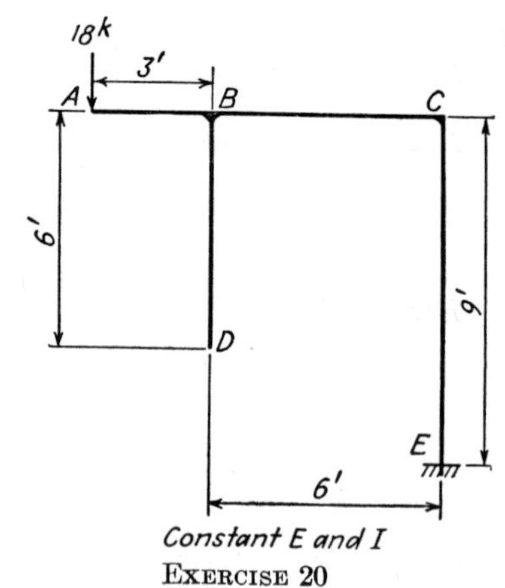

Constant E and I

EXERCISE 20

20. Find θ_A, θ_D, Δ_H and Δ_V of A, and Δ_H and Δ_V of D in terms of EI by the unit-load method.

12. Deflections and Rotations of Statically Determinate Frames Due to Movements of Supports. It can be shown that no internal stresses will be set up in statically *determinate* frames due to movements of supports. If the hinge A of rigid frame $ABCD$ yields 1 in. to the left (Fig. 31*a*), the joints B, C, and D will all move 1 in. to the left and all members remain straight as previously.

If the hinge yields 1 in. vertically downward (Fig. 31*b*), the unstressed frame $ABCD$ will take the position $A'B'C'D'$. Since all movements are small in comparison with the dimensions of the frame, it is assumed that

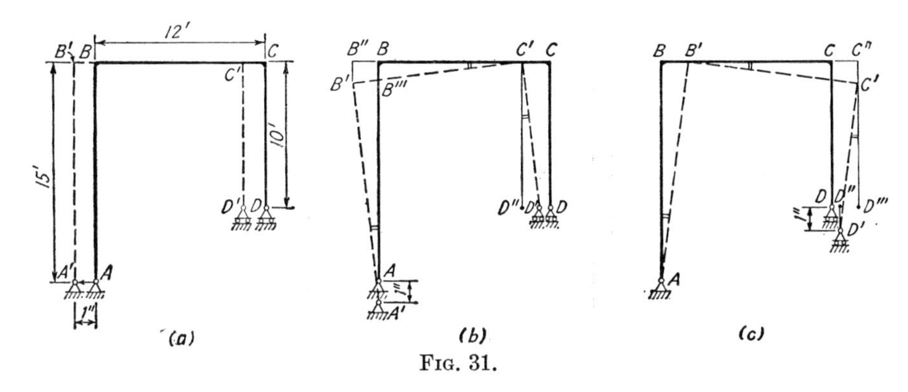

FIG. 31.

the length of a member does not change when its ends move in the perpendicular direction. Thus in Fig. 31*b*, $AA' = B''B' = 1$ in., and $BB'' = CC'$. Also, because angles $A'B'C'$ and $B'C'D'$ must remain right angles, the change in the directions of all three members must be the same; in other words, angle $B'A'B''' =$ angle $B''C'B' =$ angle $D''C'D'$.

$$\text{Angle } B''C'B' = \frac{B'B''}{B''C'} = \frac{1 \text{ in.}}{144 \text{ in.}} = \tfrac{1}{144} \text{ rad}$$

Then

$$B'B''' = (\tfrac{1}{144} \text{ rad})(A'B') = (\tfrac{1}{144})(180 \text{ in.}) = 1\tfrac{1}{4} \text{ in.} = CC'$$

and

$$DD' = DD'' - D'D'' = 1\tfrac{1}{4} \text{ in.} - (120 \text{ in.})(\tfrac{1}{144}) = 1\tfrac{1}{4} - \tfrac{5}{6} = \tfrac{5}{12} \text{ in.}$$

If the roller support at D yields 1 in. vertically downward (Fig. 31*c*), $C'C''$ must also be 1 in., or the angle $C''B'C' = 1 \text{ in.}/144 \text{ in.} = \tfrac{1}{144}$ rad. Then $BB' = \tfrac{1}{144} \times 180$ in. $= 1\tfrac{1}{4}$ in. $= CC''$, and

$$DD'' = DD''' - D''D''' = 1\tfrac{1}{4} \text{ in.} - \tfrac{1}{144} \times 120 \text{ in.} = 1\tfrac{1}{4} - \tfrac{5}{6} = \tfrac{5}{12} \text{ in.}$$

It is seen that, owing to each case of yielding of the external supports, the roller support must move a certain amount in the horizontal direction in order to keep the frame unstressed. If the support at D is a hinge, or if the frame is a statically *indeterminate* structure, the members of the frame will have to be bent to accommodate themselves to the new conditions. The problem of finding the stresses in a statically indeterminate frame due to movements of supports will be taken up in Chap. IV.

The unit-load method can be used to find the deflection or rotation at any point of a statically determinate frame owing to movements of supports. For instance, Fig. 32 is the free-body diagram of the frame shown in Fig. 31 subjected to a 1-lb load at D. If the hinge A moves 1 in. downward, the joints B, C, and D will also change position. Since the

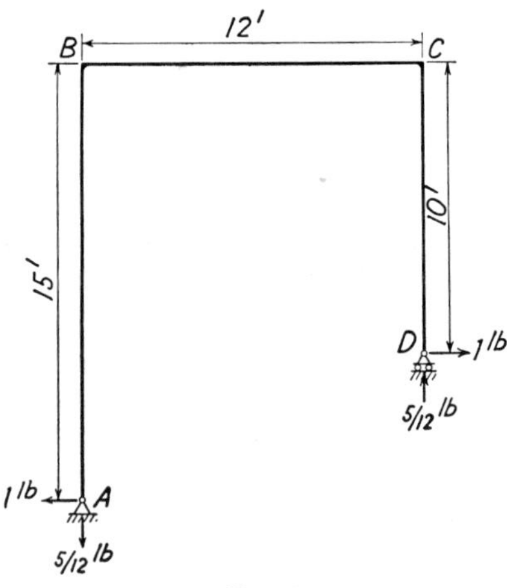

FIG. 32.

frame will not be stressed and thus there can be no internal energy stored in the frame, the total external work done by the four external forces or reactions on the frame must be zero. The horizontal reaction at A and the vertical reaction at D do no work because these forces are not displaced. Let the movement of D be Δ_H in. to the right. Then

$$\text{Total external work} = (\tfrac{5}{12}\text{ lb})(1\text{ in.}) + (1\text{ lb})(\Delta_H \text{ of } D) = 0$$
$$\Delta_H \text{ of } D = -\tfrac{5}{12}, \text{ or } \tfrac{5}{12} \text{ in. to the left}$$

This result checks with that of the purely geometrical consideration as discussed in the preceding paragraphs.

EXERCISES

21. Determine Δ_H and Δ_V of each joint if hinge A yields 2 in. to the left and 3 in. downward.

22. Determine Δ_H and Δ_V of each joint if the roller support at D yields 2 in. downward.

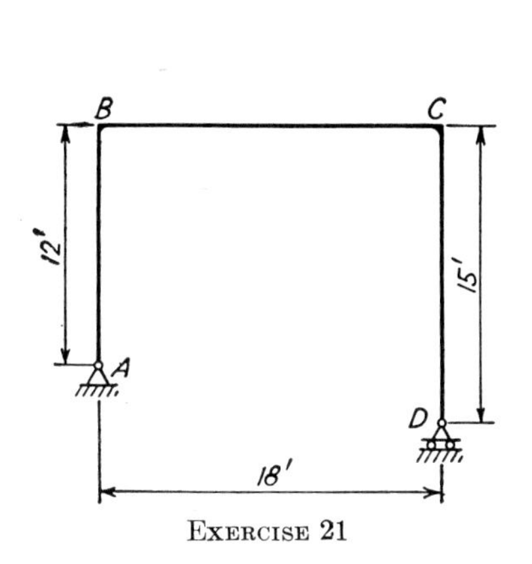

EXERCISE 21

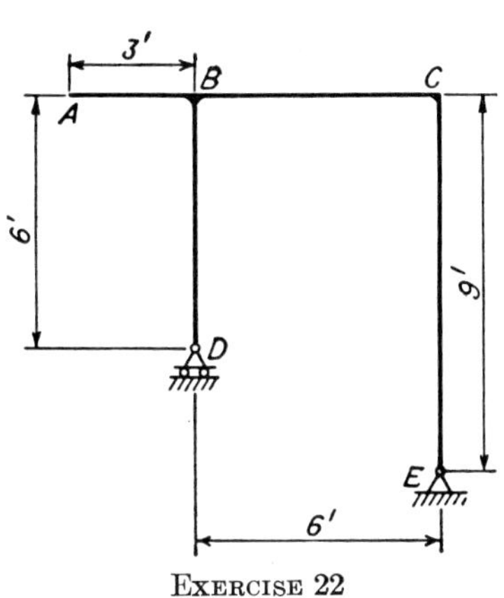

EXERCISE 22

13. The Moment-area Method—Derivation of Theorems. *Theorem* I. The angle in radians or the change in slope between the tangents to the elastic curve at two points on a straight member under bending is equal to the area of the M/EI diagram between those two points.

Theorem II. The deflection of a point on a straight member under bending in the direction perpendicular to the original straight axis of the member, measured from the tangent at another point on the member, is equal to the moment of the M/EI diagram between those two points about the point where this deflection occurs.

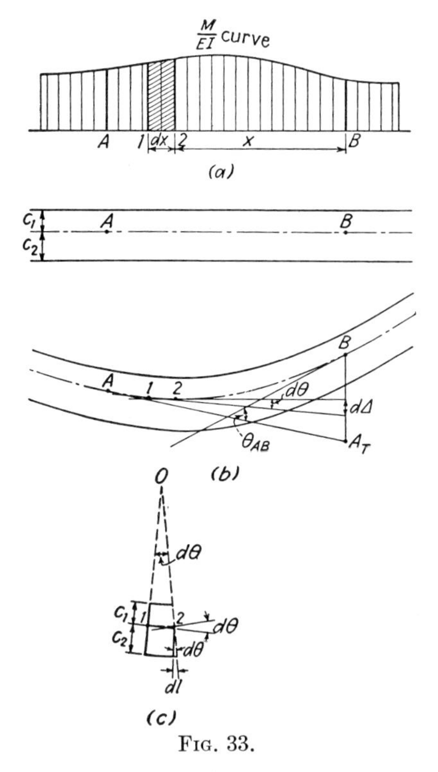

FIG. 33.

Derivation. Since the deflection of a member due to bending is usually small compared with the length of the original straight axis, it is generally assumed that the length of the elastic curve is equal to that of the original straight axis. Thus the length of curve AB (Fig. 33b) is equal to that of straight line AB. Theorem I states that

$$\theta_{AB} = \int_A^B \frac{M}{EI}\, dx \tag{21}$$

and Theorem II states that

$$A_T B = \int_A^B \frac{Mx}{EI}\, dx \tag{22}$$

Take any two points 1 and 2 at a distance dx apart on the elastic curve AB. Let $d\theta$ be the angle between the tangents at 1 and 2 (Fig. 33*b* and *c*). Since 1 and 2 are at an infinitesimal distance dx apart, curve 1-2 is a circular arc with center at point O (Fig. 33*c*). Draw a line through point 2 and parallel to line O-1. It can be seen, from Fig. 33*c*, that

$$d\theta = \frac{dl}{c_2} \tag{23}$$

Substituting $dl = (Mc_2/EI)\, dx$ into Eq. (23),

$$d\theta = \frac{(Mc_2/EI)\, dx}{c_2} = \frac{M}{EI}\, dx \tag{24}$$

Integrating Eq. (24) between the limits A and B,

$$\theta_{AB} = \int_A^B d\theta = \int_A^B \frac{M}{EI}\, dx$$

which is Eq. (21) or moment-area theorem I.

In Fig. 33*b*, prolong the tangents at 1 and 2 until they cut $d\Delta$ on $A_T B$. It can be seen that

$$d\Delta = x\, d\theta = x\left(\frac{M}{EI}\, dx\right) \tag{25}$$

Integrating Eq. (25) between the limits A and B,

$$A_T B = \int_A^B d\Delta = \int_A^B \frac{Mx}{EI}\, dx$$

which is Eq. (22) or moment-area theorem II.

14. The Moment-area Method—Application to Beam Deflections and Rotations. In applying the moment-area method to find beam deflections and rotations, it is necessary to plot or sketch the moment diagram (or the M/I diagram if I is variable) first and then to draw a qualitative picture of the elastic curve. The two theorems can be applied in succession to determine rotations and deflections at different points of the beam. All moment diagrams are plotted on the compression side.

Example 17. Find Δ_B and θ_B by the moment-area method (Fig. 34).

Solution

$$EI\theta_B = EI\theta_A \text{ plus the area of the } M \text{ diagram between } A \text{ and } B$$
$$= 0 + \frac{(PL)(L)}{2} = \frac{PL^2}{2}$$
$$\theta_B = \frac{PL^2}{2EI} \text{ clockwise}$$
$$EI\Delta_B = \text{deflection of } B \text{ from tangent at } A$$
$$= \left(\frac{PL^2}{2}\right)(\tfrac{2}{3}L) = \frac{PL^3}{3}$$
$$\Delta_B = \frac{PL^3}{3EI} \text{ downward}$$

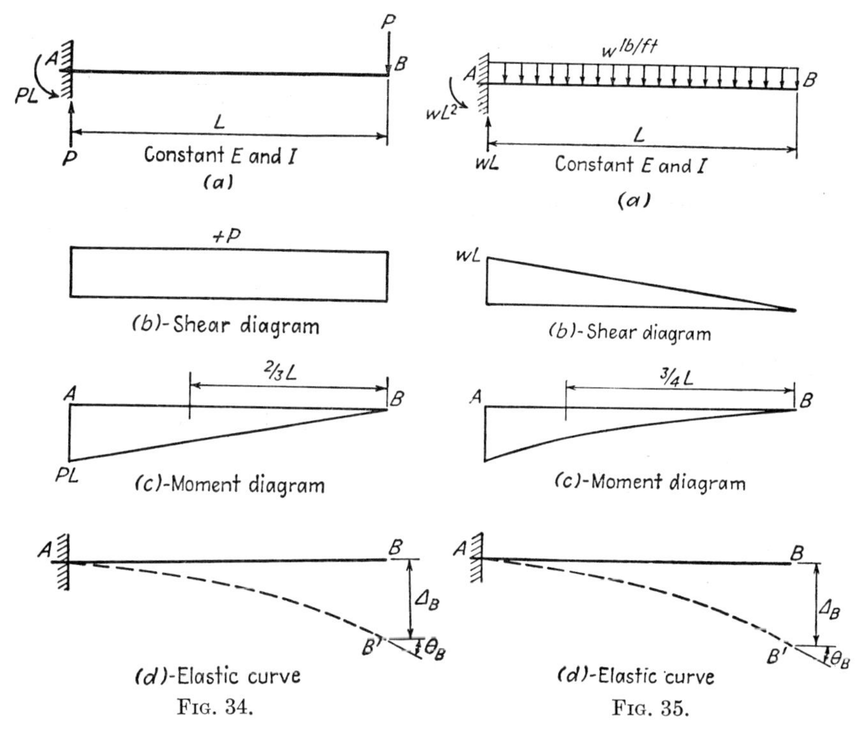

Fig. 34.　　　　Fig. 35.

Example 18. Find Δ_B and θ_B by the moment-area method (Fig. 35).

Solution

$$EI\theta_B = EI\theta_A \text{ plus the area of the } M \text{ diagram between } A \text{ and } B$$
$$= 0 + \frac{1}{3}\left(\frac{wL^2}{2}\right)(L) = \frac{wL^3}{6}$$

$$\theta_B = \frac{wL^3}{6EI} \text{ clockwise}$$

$$EI\Delta_B = \text{deflection of } B \text{ from tangent at } A$$

$$= \left(\frac{wL^3}{6}\right)(\tfrac{3}{4}L) = \frac{wL^4}{8}$$

$$\Delta_B = \frac{wL^4}{8EI} \text{ downward}$$

Example 19. Find θ_A or θ_B, and Δ_C by the moment-area method (Fig. 36).

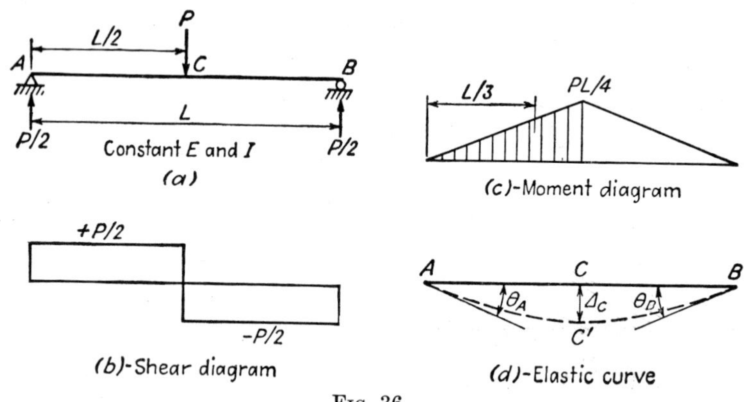

FIG. 36.

Solution

$$EI\theta_A = EI\theta_C \text{ plus the area of the } M \text{ diagram between } A \text{ and } C$$

$$= 0 + \frac{1}{2}\left(\frac{PL}{4}\right)\left(\frac{L}{2}\right) = \frac{PL^2}{16}$$

$$\theta_A = \frac{PL^2}{16EI} \text{ clockwise}$$

$$\theta_B = \frac{PL^2}{16EI} \text{ counterclockwise}$$

$$EI\Delta_C = \text{deflection of } A \text{ from tangent at } C$$

$$= \frac{PL^2}{16}\frac{L}{3} = \frac{PL^3}{48}$$

$$\Delta_C = \frac{PL^3}{48EI} \text{ downward}$$

Example 20. Find θ_A or θ_B, and Δ_C by the moment-area method (Fig. 37).

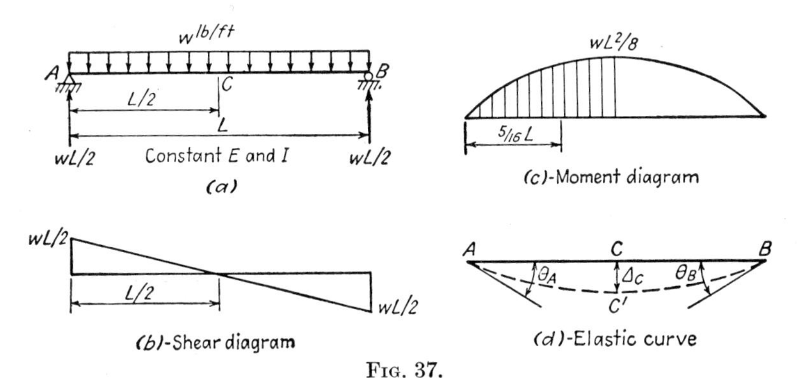

FIG. 37.

Solution

$$EI\theta_A = EI\theta_C \text{ plus the area of the } M \text{ diagram between } A \text{ and } C$$

$$= 0 + \frac{2}{3}\left(\frac{wL^2}{8}\right)\left(\frac{L}{2}\right) = \frac{wL^3}{24}$$

$$\theta_A = \frac{wL^3}{24EI} \text{ clockwise}$$

$$\theta_B = \frac{wL^3}{24EI} \text{ counterclockwise}$$

$$EI\Delta_C = \text{deflection of } A \text{ from tangent at } C$$

$$= \left(\frac{wL^3}{24}\right)\left(\frac{5L}{16}\right) = \frac{5wL^4}{384}$$

$$\Delta_C = \frac{5wL^4}{384EI} \text{ downward}$$

EXERCISES

23 to 25. Find θ_B and Δ_B by the moment-area method.

26. Find θ_A, θ_B, θ_D, Δ_A, and Δ_C by the moment-area method.

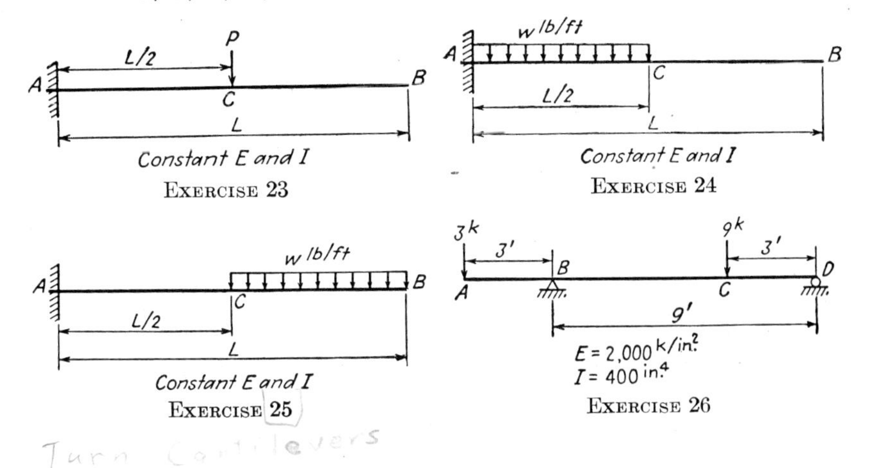

EXERCISE 23

EXERCISE 24

EXERCISE 25

EXERCISE 26

Turn Cantilevers Around

15. Special Case of Moment-area Method—the Conjugate-beam Method. In each of Examples 17 to 20, the elastic curve has a horizontal tangent at a certain point of the beam. Therefore the slope and deflection at any other point can be found by referring to this horizontal tangent as a basis when the two moment-area theorems are applied. In many cases, however, there may be no such horizontal tangent existing. Take, for example, the case of the beam shown in Fig. 38*a*. It is desired to find the slope θ_C and the deflection Δ_C at any point C on the beam. By the moment-area method, the procedure is as follows:

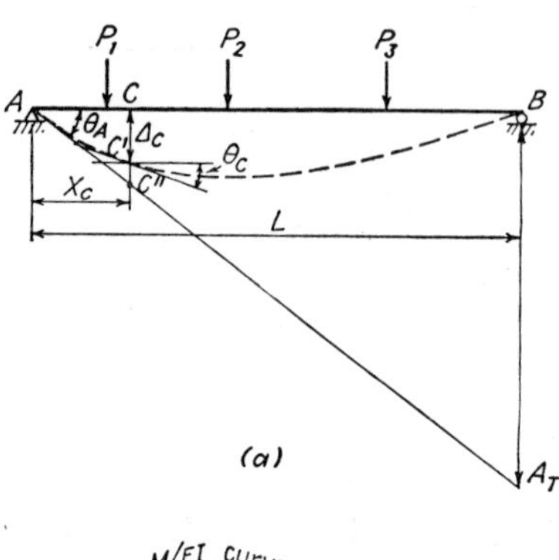

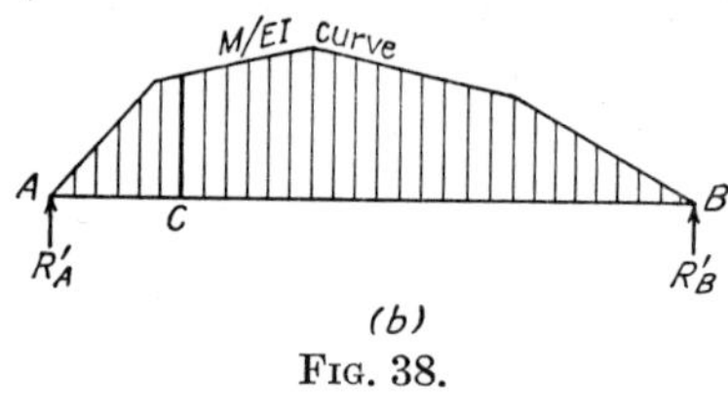

Fig. 38.

$$\theta_C = \theta_A - \left[\text{area of } \frac{M}{EI} \text{ diagram between } A \text{ and } C\right]$$

where

$$\theta_A = \frac{A_TB}{L} = \frac{1}{L}\left[\text{moment of } \frac{M}{EI} \text{ diagram between } A \text{ and } B \text{ about } B\right]$$

$$\Delta_C = CC'' - C'C''$$

where $CC'' = \theta_A x_C$ and $C'C''$ = deflection of C from tangent at A = moment of M/EI diagram between A and C about C. Thus

$$\theta_C = \left[\frac{\text{moment of } M/EI \text{ diagram between } A \text{ and } B \text{ about } B}{L}\right] - \left[\text{area of } \frac{M}{EI} \text{ diagram between } A \text{ and } C\right] \quad (26)$$

and

$$\Delta_C = \left[\frac{\text{moment of } M/EI \text{ diagram between } A \text{ and } B \text{ about } B}{L}\right] x_C - \left[\text{moment of } \frac{M}{EI} \text{ diagram between } A \text{ and } C \text{ about } C\right] \quad (27)$$

Now suppose that an imaginary, or an auxiliary, or a "conjugate" beam is defined as the original simple beam AB loaded by the M/EI diagram (Fig. 38*b*). Let R'_A and R'_B be the reactions to this conjugate beam and V'_C and M'_C be the shear and bending moment at C on this conjugate beam. The right side of Eq. (26) is obviously equal to V'_C, and the right side of Eq. (27) is obviously equal to M'_C. Thus

$$\theta_C = V'_C \tag{28}$$

and

$$\Delta_C = M'_C \tag{29}$$

It is to be noted that the above two equations can be applied between any two points A and B on the elastic curve, except that if the chord AB is not horizontal, θ_C is the angle between the tangent at C and the chord AB and Δ_C is the deflection of C measured from the chord AB. Equations (28) and (29) can be stated in words, thus:

Theorem I. *Conjugate-beam Method.* The angle between the tangent to the elastic curve at any point C between two points A and B on the elastic curve and the chord AB is equal to the shear at point C in a simple beam AB loaded with the M/EI diagram between A and B.

Theorem II. *Conjugate-beam Method.* The deflection of any point C between two points A and B on the elastic curve, measured from the chord AB, is equal to the bending moment at point C in a simple beam AB loaded with the M/EI diagram between A and B.

The conjugate-beam method is actually a special case of the moment-area method, or it can be considered as another way of describing the procedure in applying the moment-area theorems.

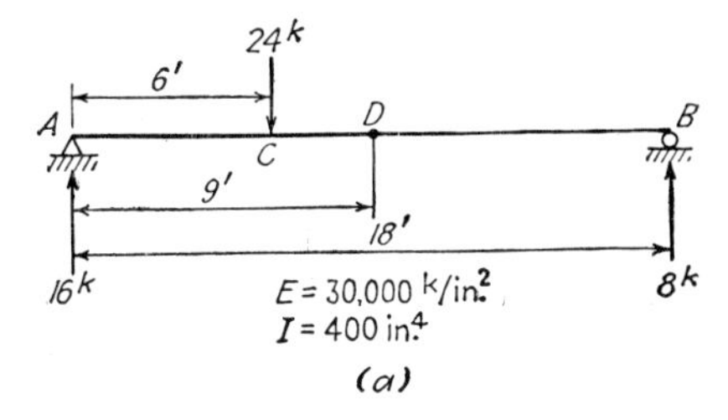

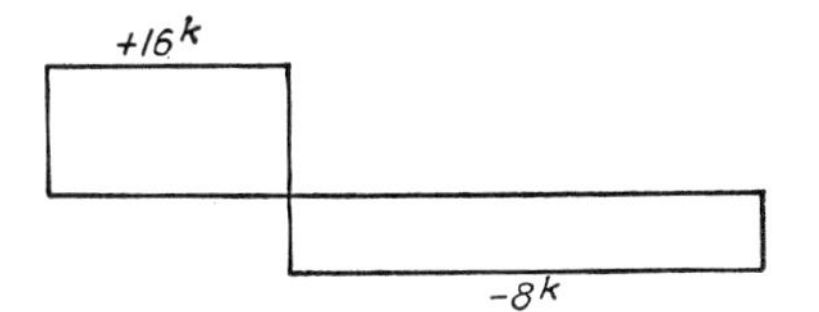

(b)-Shear diagram

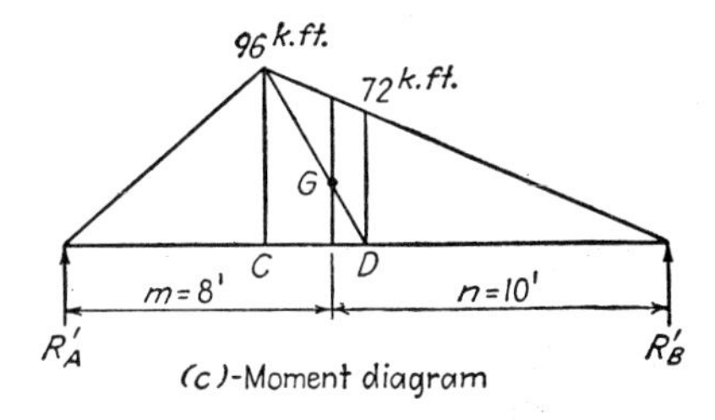

(c)-Moment diagram

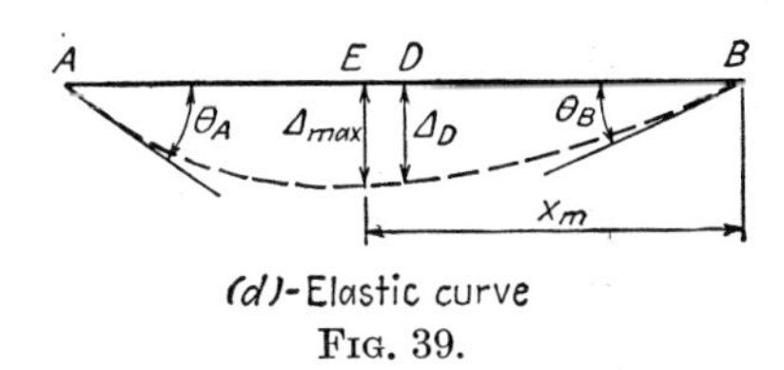

(d)-Elastic curve

FIG. 39.

Example 21. Find (1) θ_A and θ_B in radians, (2) Δ_D in inches, and (3) the position and amount of maximum deflection, by the conjugate-beam method (Fig. 39).

Solution. The conjugate-beam method will be used because the chord AB is the original straight axis of the beam. The conjugate beam AB is loaded with the M/EI diagram, but since EI is constant, it can be omitted from the calculations until the end. From Fig. 39c,

$$\text{Total area of } M \text{ diagram} = \tfrac{1}{2}(96)(18) = 864 \text{ kip-ft}^2$$

The distances m and n are, according to the simple formulas shown in Fig. 40,

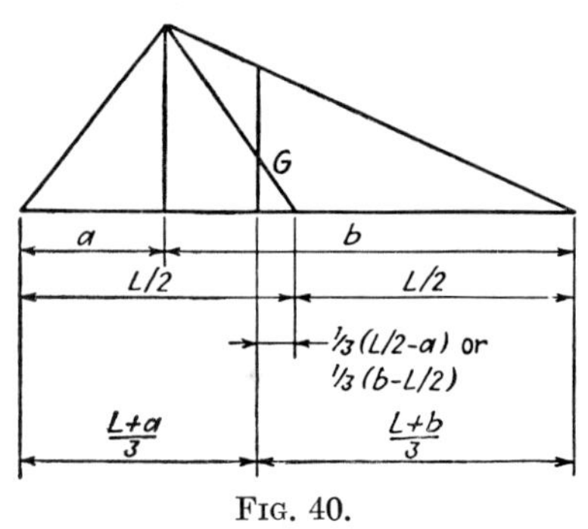

FIG. 40.

$$m = \frac{18 + 6}{3} = 8 \text{ ft} \qquad n = \frac{18 + 12}{3} = 10 \text{ ft}$$

$$R'_A = 864 \times \tfrac{10}{18} = 480 \text{ kip-ft}^2 = EI\theta_A$$

$$\theta_A = \frac{480 \text{ kip-ft}^2}{EI} = \frac{480 \times 144}{30{,}000 \times 400} = 0.00576 \text{ rad clockwise}$$

$$R'_B = 864 \times \tfrac{8}{18} = 384 \text{ kip-ft}^2 = EI\theta_B$$

$$\theta_B = \frac{384 \text{ kip-ft}^2}{EI} = \frac{384 \times 144}{30{,}000 \times 400} = 0.00461 \text{ rad counterclockwise}$$

$$M'_D = R'_B \times 9 - \tfrac{1}{2} \times 72 \times 9 \times 3 = 3{,}456 - 972 = 2{,}484 \text{ kip-ft}^3 = EI\Delta_D$$

$$\Delta_D = \frac{2{,}484 \times 1{,}728}{30{,}000 \times 400} = 0.358 \text{ in. downward}$$

Let the maximum deflection occur at E, which is at x_m ft from B. Then, since M'_E is maximum, $V'_E = 0$.

$$V'_E = R'_B - \tfrac{1}{2}(8x_m)(x_m) = 0 \qquad 384 - 4x_m^2 = 0$$

$$x_m = 4\sqrt{6} \text{ ft}$$

$$M'_E = R'_B x_m - \tfrac{1}{2}(8x_m)(x_m)\left(\frac{x_m}{3}\right) = (384)(4\sqrt{6}) - \tfrac{4}{3}(4\sqrt{6})^3$$

$$= 1{,}024\sqrt{6} = 2{,}508 \text{ kip-ft}^3 = EI\Delta_{\max}$$

$$\Delta_{\max} = \frac{2{,}508 \times 1{,}728}{30{,}000 \times 400} = 0.361 \text{ in. downward}$$

Example 22. Find θ_A or θ_B and Δ_D in terms of EI by the conjugate-beam method (Fig. 41).

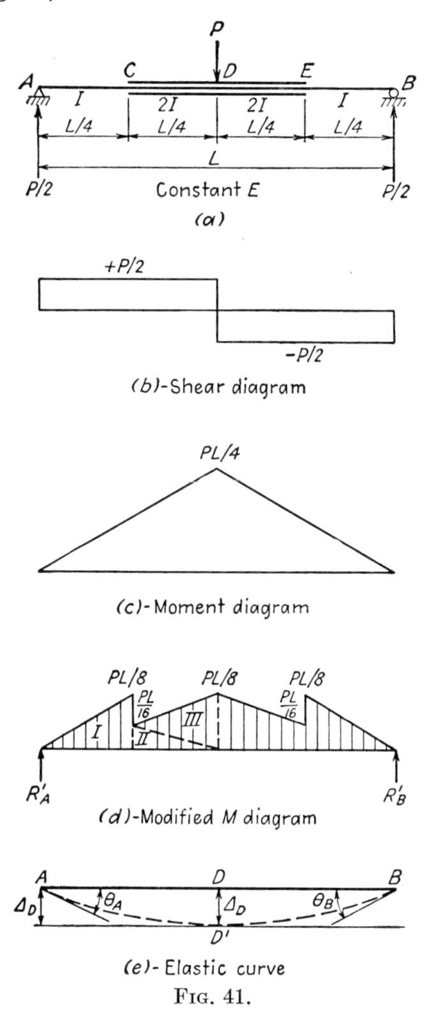

FIG. 41.

Solution. The moment diagram shown in Fig. 41*c* is modified to become Fig. 41*d* because the moment of inertia of the central portion is $2I$. The conjugate beam is as shown in Fig. 41*d*.

$$\begin{aligned} R'_A &= \text{area of one-half of the modified } M \text{ diagram} \\ &= \text{area I} + \text{area II} + \text{area III} \\ &= \frac{1}{2}\left(\frac{PL}{8}\right)\left(\frac{L}{4}\right) + \frac{1}{2}\left(\frac{PL}{16}\right)\left(\frac{L}{4}\right) + \frac{1}{2}\left(\frac{PL}{8}\right)\left(\frac{L}{4}\right) = \frac{5PL^2}{128} = EI\theta_A \end{aligned}$$

$$\theta_A = \frac{5PL^2}{128EI} \text{ clockwise}$$

$$\theta_B = \frac{5PL^2}{128EI} \text{ counterclockwise}$$

$$M'_D = (R'_A)\left(\frac{L}{2}\right) - (\text{area I})\left(\frac{L}{4} + \frac{L}{12}\right) - (\text{area II})\left(\frac{2}{3}\right)\left(\frac{L}{4}\right) - (\text{area III})\left(\frac{1}{3}\right)\left(\frac{L}{4}\right)$$

$$= \left(\frac{5PL^2}{128}\right)\left(\frac{L}{2}\right) - \left(\frac{PL^2}{64}\right)\left(\frac{L}{3}\right) - \left(\frac{PL^2}{128}\right)\left(\frac{L}{6}\right) - \left(\frac{PL^2}{64}\right)\left(\frac{L}{12}\right)$$

$$= \frac{3PL^3}{256} = EI\Delta_D$$

$$\Delta_D = \frac{3PL^3}{256EI} \text{ downward}$$

Note: The conventional moment-area method is just as easy as the conjugate-beam method by calling

$$\theta_A = \theta_D + \frac{\text{area I, II, III}}{EI} = 0 + \frac{\text{area I, II, III}}{EI}$$

$$\Delta_D = \text{deflection of } A \text{ from tangent at } D$$

$$= \frac{1}{EI} (\text{moment of areas I, II, and III about } A)$$

EXERCISES

27 and 28. Find (*a*) θ_A and θ_B in radians, (*b*) Δ_D in inches, and (*c*) the position and amount of maximum deflection, by the conjugate-beam method.

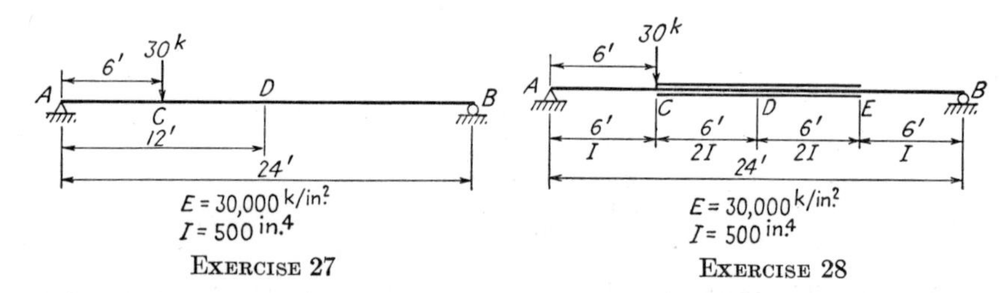

EXERCISE 27 EXERCISE 28

16. The Moment-area Method—Application to Deflections and Rotations of Frames. In applying the moment-area method to find rotations and deflections at points on a statically determinate rigid frame, it is important to sketch a correct moment diagram by plotting it on the compression side and a correct qualitative elastic curve for the whole frame. Then the two moment-area theorems are applied in succession to obtain the rotation and the horizontal and vertical deflection of any joint on the frame.

Example 23. Find θ, Δ_H, and Δ_V of joints A, B, C, and D by the moment-area method (Fig. 42a).

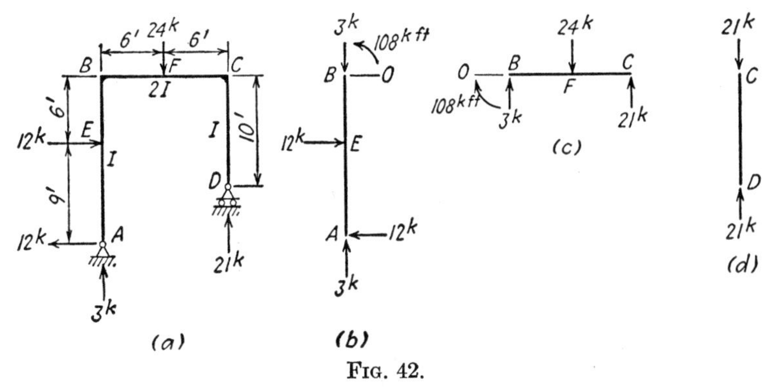

FIG. 42.

Solution. The free-body diagrams of the whole frame and of each member are shown in Fig. 42. The moment diagram (plotted on the compression side) shown in Fig. 43a is modified to become that of Fig. 43b by dividing all values along BC by 2 since the moment of inertia of BC is $2I$, while that of AB and CD is I. The elastic curve is plotted as Fig. 43c. In studying this curve, note that B' and C' stay at their original

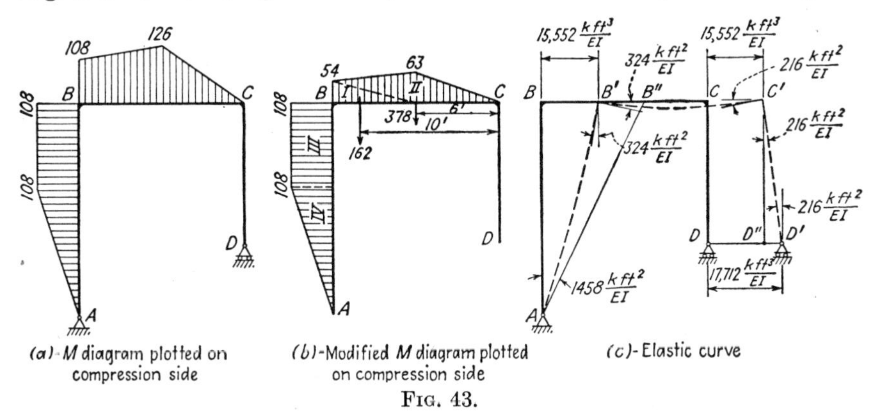

FIG. 43.

levels and both move the same Δ_H to the right. The conjugate-beam method, therefore, can be applied to member BC. The tangents at each joint are drawn as shown; the angle between the tangents at joints B' or C' must remain 90°. Note also that member CD remains straight. Applying the conjugate-beam method to member BC,

$$EI\theta_B = R'_B = (378)(\tfrac{6}{12}) + (162)(\tfrac{10}{12}) = 324 \text{ kip-ft}^2 \text{ clockwise}$$
$$EI\theta_C = R'_C = (162)(\tfrac{2}{12}) + (378)(\tfrac{6}{12}) = 216 \text{ kip-ft}^2 \text{ counterclockwise}$$
$$EI\theta_A = EI\theta_B + \text{area III} + \text{area IV} = 324 + (108 \times 6) + (\tfrac{1}{2})(108)(9)$$
$$= 1{,}458 \text{ kip-ft}^2 \text{ clockwise}$$

$$EI\theta_D = EI\theta_C = 216 \text{ kip-ft}^2 \text{ counterclockwise}$$

$$\begin{aligned} EI\Delta_H \text{ of } B &= BB' = BB'' - B'B'' \\ &= (\theta_A)(15) - [\text{deflection } B \text{ from tangent at } A] \\ &= (1{,}458)(15) - [\text{moment of areas III and IV about } B] \\ &= 21{,}870 - (648)(3) - (486)(9) \\ &= 15{,}552 \text{ kip-ft}^3 \text{ to the right} \\ EI\Delta_H \text{ of } D &= DD' = DD'' + D'D'' = 15{,}552 + \theta_C(10) \\ &= 15{,}552 + (216)(10) = 17{,}712 \text{ kip-ft}^3 \end{aligned}$$

$$\Delta_H \text{ of } D = 17{,}712 \frac{\text{kip-ft}^3}{EI} \text{ to the right}$$

The deflection components (θ, Δ_H, and Δ_V) of each joint are shown in Fig. 43*c*.

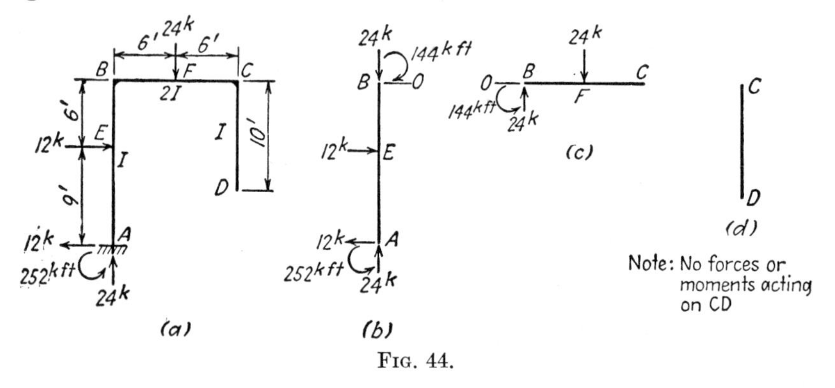

FIG. 44.

Example 24. Find θ, Δ_H, and Δ_V of joints A, B, C, and D by the moment-area method (Fig. 44*a*).

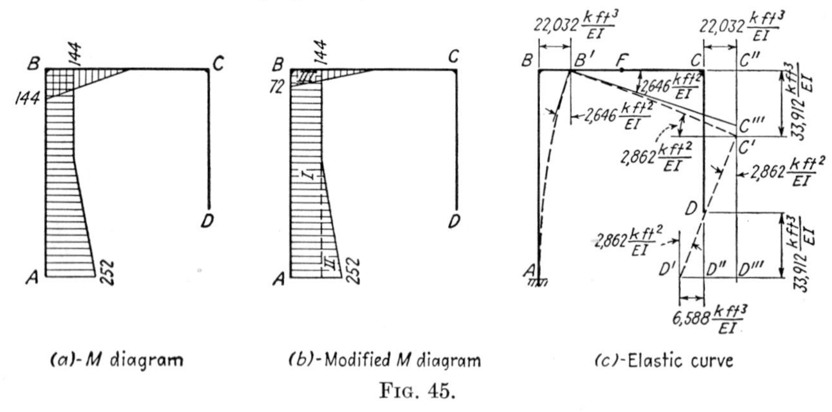

FIG. 45.

Solution. The free-body diagrams of the whole frame and of each separate member are shown in Fig. 44. In Fig. 45 are shown the moment diagram, the modified moment diagram, and the elastic curve. The portions $F'C'$ and $C'D'$ remain straight as they are not acted upon by any forces or moments. It can be seen that Δ_H of B is equal to Δ_H of C, Δ_V of B is zero, and Δ_V of C is equal to Δ_V of D.

$$EI\theta_B = EI\theta_A + \text{areas I and II}$$
$$= 0 + (144)(15) + (\tfrac{1}{2})(108)(9) = 2{,}646 \text{ kip-ft}^2$$

$$\theta_B = 2{,}646 \frac{\text{kip-ft}^2}{EI} \text{ clockwise}$$

$$EI\theta_C = EI\theta_B + \text{area III} = 2{,}646 + \frac{(72)(6)}{2} = 2{,}862 \text{ kip-ft}^2$$

$$\theta_C = 2{,}862 \frac{\text{kip-ft}^2}{EI} \text{ clockwise}$$

$$\theta_D = \theta_C = 2{,}862 \frac{\text{kip-ft}^2}{EI} \text{ clockwise}$$

$$EI\Delta_H \text{ of } B = BB' = \text{moment of areas I and II about } B$$
$$= (144)(15)(7.5) + (\tfrac{1}{2})(108)(9)(12) = 22{,}032 \text{ kip-ft}^3$$

$$\Delta_H \text{ of } B = \Delta_H \text{ of } C = 22{,}032 \frac{\text{kip-ft}^3}{EI} \text{ to the right}$$

$$EI\Delta_V \text{ of } C = C''C' = C''C''' + C'''C'$$
$$= (\theta_B)(12) + \text{moment of area III about } C$$
$$= (2{,}646)(12) + (\tfrac{1}{2})(72)(6)(10) = 33{,}912 \text{ kip-ft}^3$$

$$\Delta_V \text{ of } C = \Delta_V \text{ of } D = 33{,}912 \frac{\text{kip-ft}^3}{EI} \text{ downward}$$

$$EI\Delta_H \text{ of } D = D'D'' = D'D''' - D''D''' = (2{,}862)(10) - 22{,}032$$
$$= 6{,}588 \text{ kip-ft}^3$$

$$\Delta_H \text{ of } D = 6{,}588 \frac{\text{kip-ft}^3}{EI} \text{ to the left}$$

The deflection components (θ, Δ_H, and Δ_V) of each joint are shown in Fig. 45*c*.

EXERCISES

29. Find θ, Δ_H, and Δ_V at joints A, B, C, and D in terms of EI by the moment-area method.

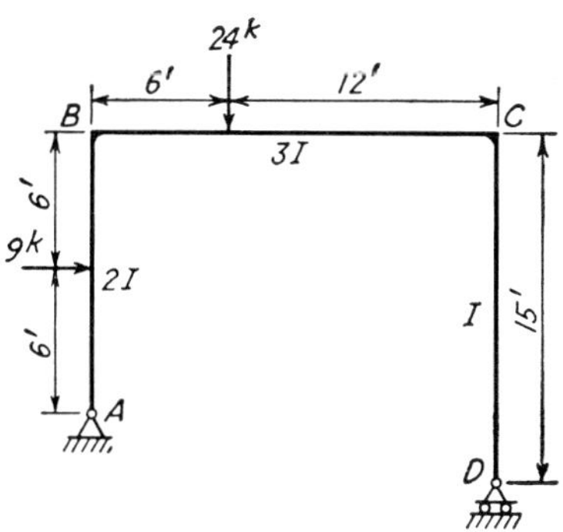

EXERCISE 29

30. Find θ, Δ_H, and Δ_V at joints A, B, C, D, and E in terms of EI by the moment-area method.

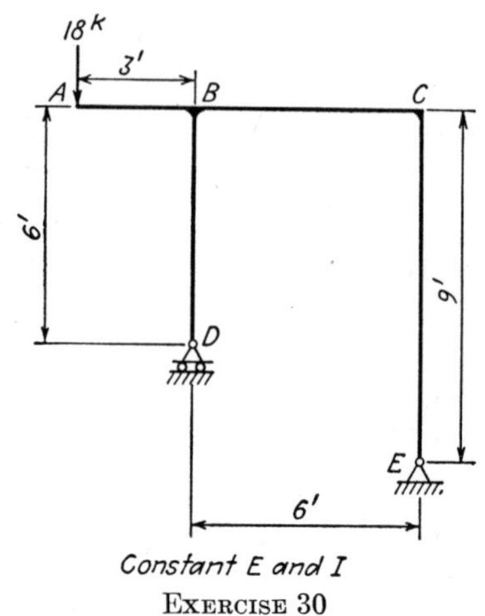

EXERCISE 30

31. Find θ, Δ_H, and Δ_V at joints A, B, C, and D in terms of EI by the moment-area method.

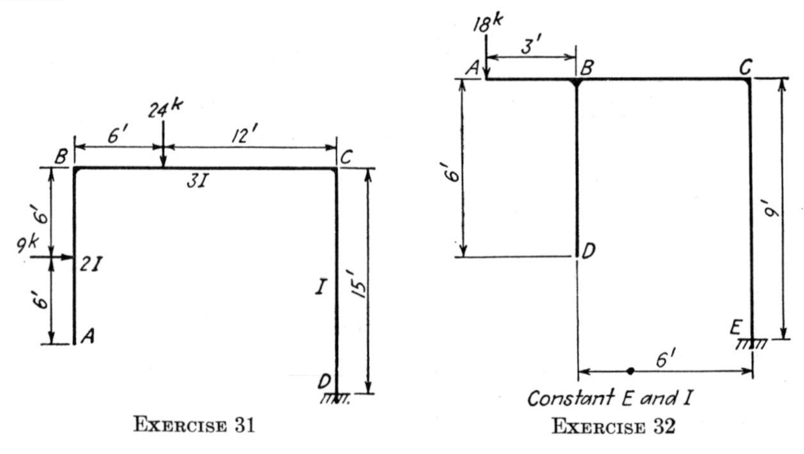

EXERCISE 31

EXERCISE 32

32. Find θ, Δ_H, and Δ_V at joints A, B, C, D, and E in terms of EI by the moment-area method.

CHAPTER III

DEFLECTION OF STATICALLY DETERMINATE TRUSSES

17. Names of Methods. The usual steel truss is a structure composed of individual members so joined together as to form a series of triangles. The joints are riveted or fastened together with pins but in either case are assumed to act as smooth hinges. It follows that the members are subjected to direct stress of tension or compression only and not subjected to bending since the ends are hinged and no loads are applied except at the joints themselves. They therefore remain straight. In addition to deformation due to direct stress caused by applied loadings, deformation may be due to other causes such as changes in temperature and errors of fabrication. When the lengths of members change, the joints or hinges must move accordingly to some new position so as to accommodate themselves to the new lengths. The amount and direction of this movement at each joint from its original position are called the deflection of the joint. With the changes in lengths of all members in a truss known, various methods as named below can be used to determine the deflection of any joint. The methods to be treated in order are (1) the unit-load method, (2) the angle-weights method, (3) the joint-displacements method, and (4) the graphical (Williot-Mohr) method. By the unit-load method, only one deflection component, *i.e.*, either the horizontal component or the vertical component of the total deflection, of one joint can be found in one application of the unit load. By the angle-weights method, the vertical deflection of all joints in one chord, *i.e.*, either the top chord or the bottom chord, can be found in one operation. By the joint-displacements method or the graphical method, the horizontal or vertical deflections of all joints can be determined at the same time.

18. The Unit-load Method. The basic formula of the unit-load method as derived in Art. 6 is

$$(1)(\Delta) = \Sigma u \, dL \tag{5}$$

In the case of a truss, there are a finite number of members; therefore Eq. (5) can be used as it stands if dL is replaced by ΔL of each member. The unit load and the u stresses are always expressed in terms of the same unit weight. Therefore Δ and dL are expressed in terms of the same unit length. The formula for computing truss deflection by the unit-load method is then

$$\Delta = \Sigma u \, \Delta L \tag{30}$$

where ΔL is the change in length of the member, which may be caused by the applied loading or other causes such as changes in temperature and errors of fabrication, and u is the stress in the same member due to the action of a unit load applied in the direction of the deflection at the point where deflection is being found. These values for all members may be entered in one table and the value of $\Sigma u \, \Delta L$ computed, but a diagrammatic representation as shown in Fig. 46 is considered advisable for the beginner.

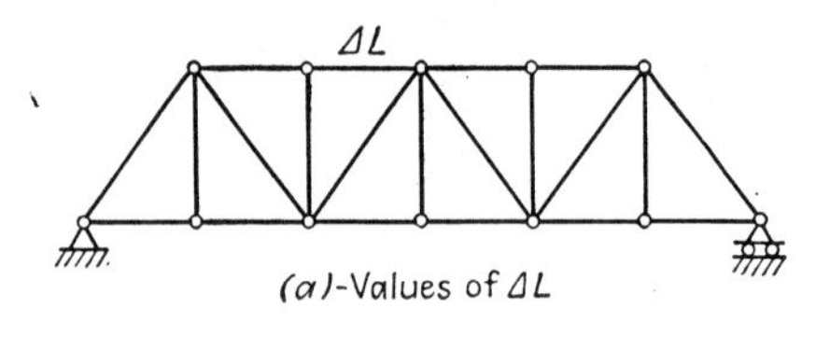

(a)-Values of ΔL

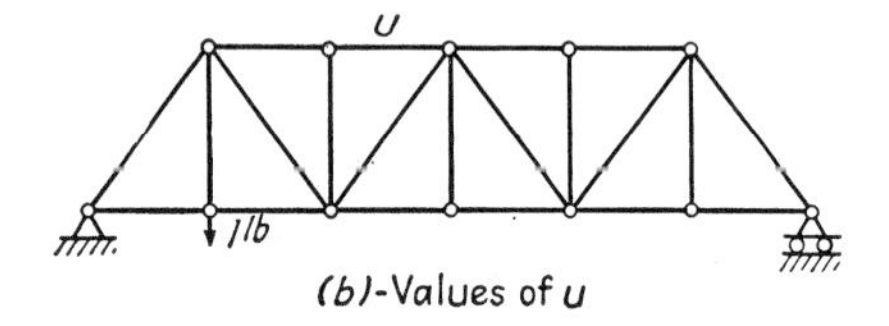

(b)-Values of u

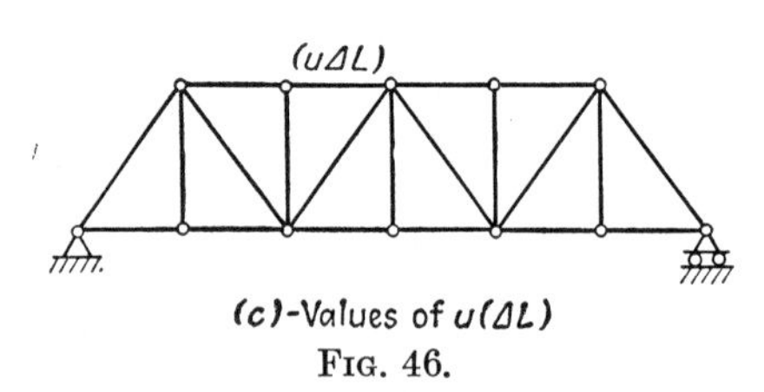

(c)-Values of u(ΔL)

FIG. 46.

Example 25. Determine the horizontal and vertical deflections of all lower chord joints due to the applied loading by the unit-load method (Fig. 47*a*).

Solution. Figure 47*a* shows the complete data with regard to the truss: dimensions, areas of members, loading. Figure 47*b* shows the reactions and stresses due to the applied loads; in solving for these the methods of joints or of sections are used freely, and the horizontal and vertical components of the stresses are shown so that the equilibrium of each joint can be easily checked by inspection. Figure 47*c* shows the changes in length of all members; a plus sign means lengthening, and a minus sign means shortening. Figure 47*d* shows the u stresses for Δ_H of L_1. The product of Fig. 47*c* and Fig. 47*d* is shown in Fig. 47*e*. $\Sigma(u) \, \Delta L$, or

TABLE 25-1

Joint	Δ_H	Δ_V
L_0	0	0
L_1	0.072 in. (right)	0.289 in. (down)
L_2	0.144 in. (right)	0.390 in. (down)
L_3	0.234 in. (right)	0.380 in. (down)
L_4	0.324 in. (right)	0

the algebraic sum of the values of $(u)\ \Delta L$ for each member in Fig. 47*e*, gives the Δ_H of L_1 and is written below Fig. 47*e*. Figure 47*f* and *g* shows the data with which Δ_V of L_1 is computed. Then Δ_H of L_2, Δ_V of L_2,

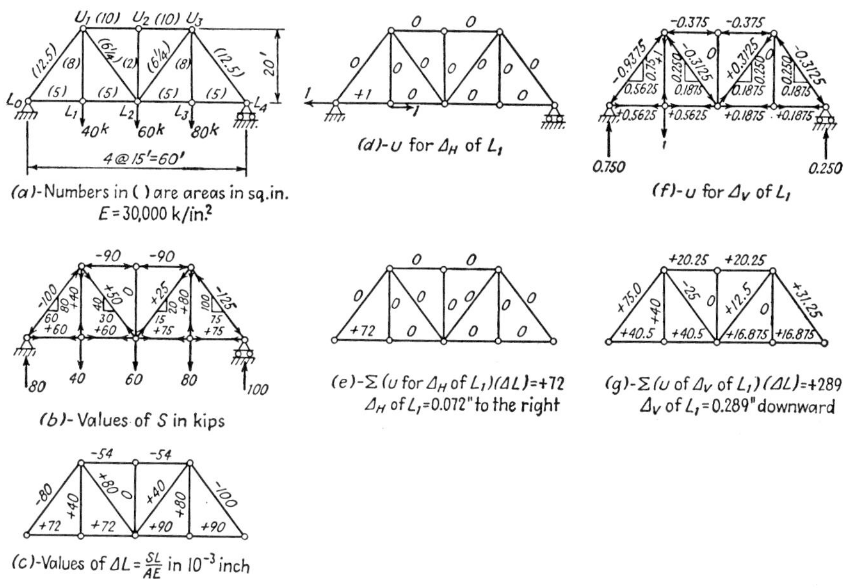

FIG. 47.

Δ_H of L_3, Δ_V of L_3, and Δ_H of L_4 are similarly computed. The required answers are summarized in Table 25-1.

Example 26. Determine the horizontal and vertical deflections of joint L_2 due to a temperature drop of 50°F in the lower chord only (Fig. 48*a*). Coefficient of expansion or contraction $= 6.5 \times 10^{-6}$ per degree Fahrenheit.

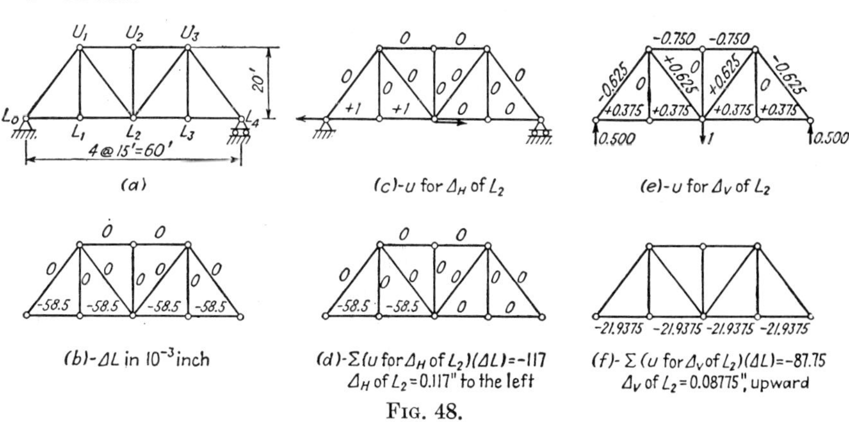

FIG. 48.

Solution

Total decrease in length in L_0L_1, L_1L_2, L_2L_3, or L_3L_4

$$= (180 \text{ in.})(6.5 \times 10^{-6})(50) = 58.5 \times 10^{-3} \text{ in.}$$

The balance of the solution is shown in Fig. 48*b* to *f*.

Example 27. Determine the horizontal and vertical deflections of joint L_2 if U_1L_2 were fabricated $\frac{1}{4}$ in. too short (Fig. 49*a*).

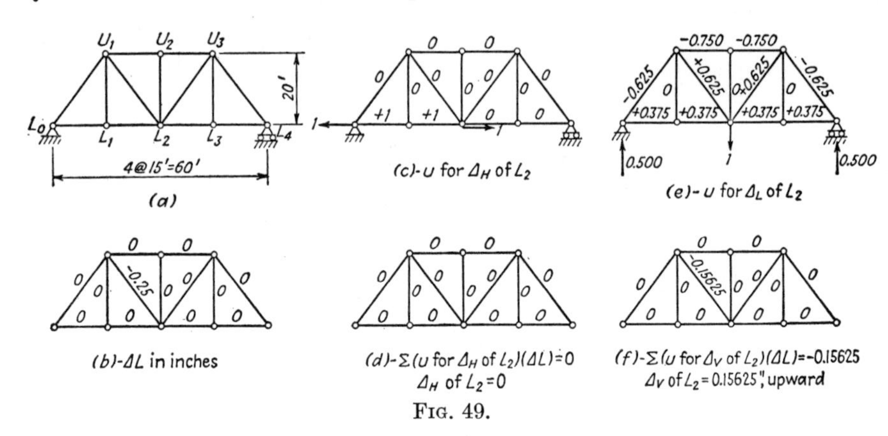

FIG. 49.

Solution. The solution of the problem is shown in Fig. 49*b* to *f*.

Example 28. Determine the relative movement between the joints L_1 and U_2 in the direction L_1U_2 (Fig. 50*a*).

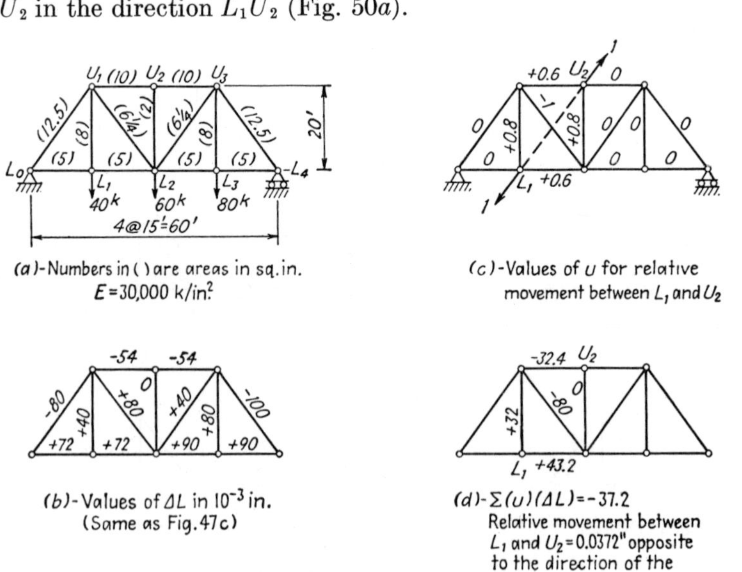

FIG. 50.

Solution. To find how much the two joints L_1 and U_2 move relatively in the direction joining L_1 and U_2, it is possible to find the absolute movements of U_2 and of L_1 in the direction L_1U_2 and add the two absolute movements to get the relative movement. However, this combined result can be more easily obtained by applying a pair of unit loads at L_1 and U_2 as shown in Fig. 50*c*. The u stresses are the combined stress, and the value of $\Sigma u \, \Delta L$ will be the same as the combined movement of the two absolute movements noted above. The balance of the solution is shown in Fig. 50.

Example 29. Determine the horizontal deflection of the roller support B due to the load of 12 kips at C (Fig. 51*a*).

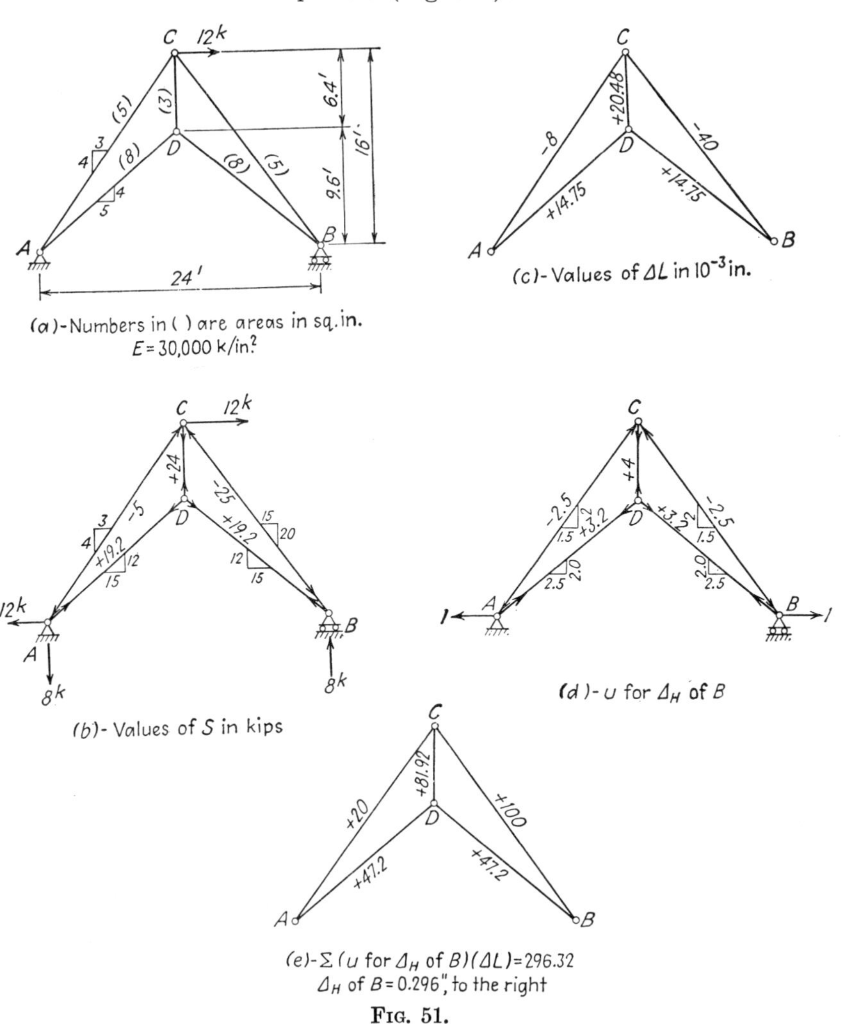

FIG. 51.

Solution. The solution of this problem is self-explanatory as outlined in the sketches of Fig. 51. It may be helpful, however, to derive the stress in AD. For this operation, the joint A is taken as the free body (see Fig. 52). Taking moments about C,

$$12 \times 16 - 8 \times 12 = (AD)^H \times 6.4$$
$$(AD)^H = 15 \text{ kips}$$
$$(AD)^V = 15 \times \tfrac{4}{5} = 12 \text{ kips}$$
$$(AD) = 15 \times \frac{6.4}{5} = 19.2 \text{ kips}$$

Fig. 52.

EXERCISES

33. Determine the horizontal and vertical deflections of all lower chord joints due to the applied loading by the unit-load method.

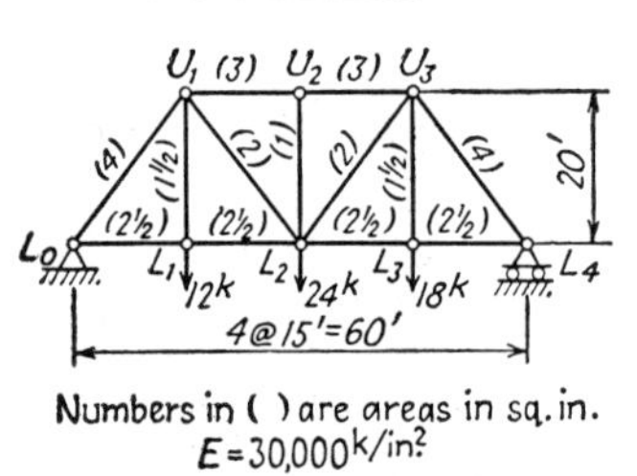

Exercises 33 and 36

34. Determine the horizontal and vertical deflections of joint L_4 due to a temperature drop of 60°F in the lower chord only. Coefficient of expansion or contraction $= 6.5 \times 10^{-6}$ per degree Fahrenheit.

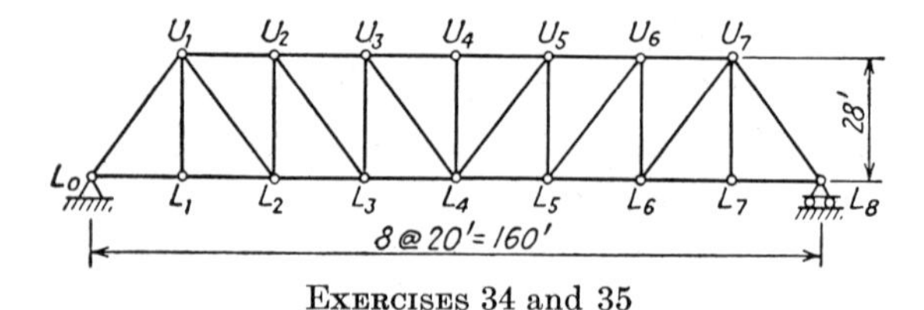

Exercises 34 and 35

35. Determine the horizontal and vertical deflections of joint L_4 if U_2U_3 were fabricated $\frac{1}{2}$ in. too long.

36. Determine the relative movement between the joints U_2 and L_3 in the direction U_2L_3.

37. Determine the horizontal deflection of the roller support B due to the load of 18 kips at C.

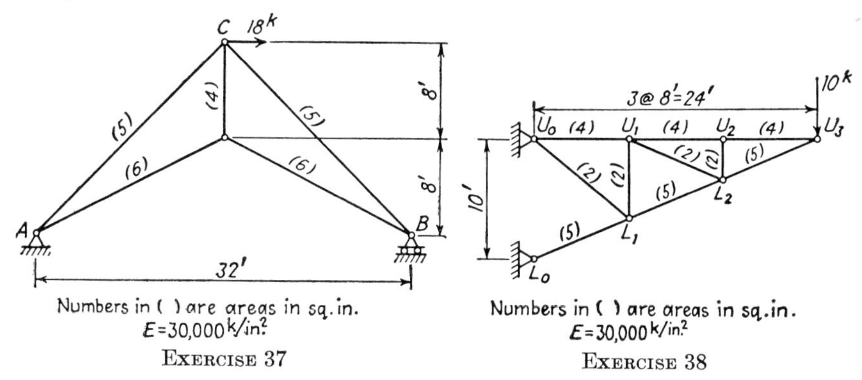

EXERCISE 37

EXERCISE 38

38. Determine the horizontal and vertical deflections of joint U_3 due to the applied loading by the unit-load method.

19. The Angle-weights Method. Before taking up the angle-weights method of solving truss deflections, it is necessary to derive the following proposition.

Proposition. If the three sides l_a, l_b, l_c opposite to angles A, B, C of triangle ABC have their lengths changed by unit elongations of e_a, e_b, e_c,

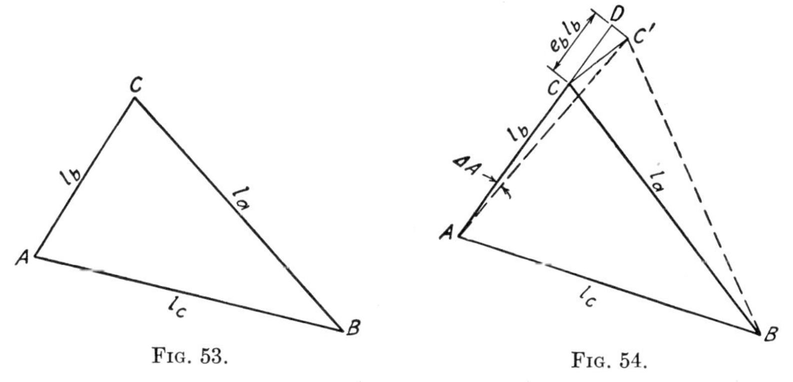

FIG. 53.

FIG. 54.

the changes in the angles ΔA, ΔB, ΔC, in terms of the angles A, B, C and the unit elongations e_a, e_b, e_c are as follows (Fig. 53):

$$\begin{aligned} \Delta A &= (e_a - e_b) \cot C + (e_a - e_c) \cot B \\ \Delta B &= (e_b - e_c) \cot A + (e_b - e_a) \cot C \\ \Delta C &= (e_c - e_a) \cot B + (e_c - e_b) \cot A \end{aligned} \tag{31}$$

The formula for ΔA will be derived first. Consider the effect of e_b only on ΔA (Fig. 54). Prolong AC to D, making $CD = e_b l_b$. Draw DC' perpendicular to AD and CC' perpendicular to BC. Triangle ABC' is the new shape of the deformed triangle due to e_b only. Since the sides of angle $DC'C$ are mutually perpendicular to the sides of angle C in the original triangle ABC, angle $DC'C$ = angle C.

$$\Delta A = -\frac{DC'}{AD \text{ or } AC} = -\frac{CD \cot DC'C}{AC} = -\frac{e_b l_b \cot C}{l_b} = -e_b \cot C \quad (32)$$

The minus sign is used in (32) because ΔA represents a decrease in the size of angle A.

Next consider the effect of e_c only on ΔA (Fig. 55). Prolong AB to D, making $BD = e_c l_c$. Draw DB' perpendicular to AD, and BB' perpendicular to BC.

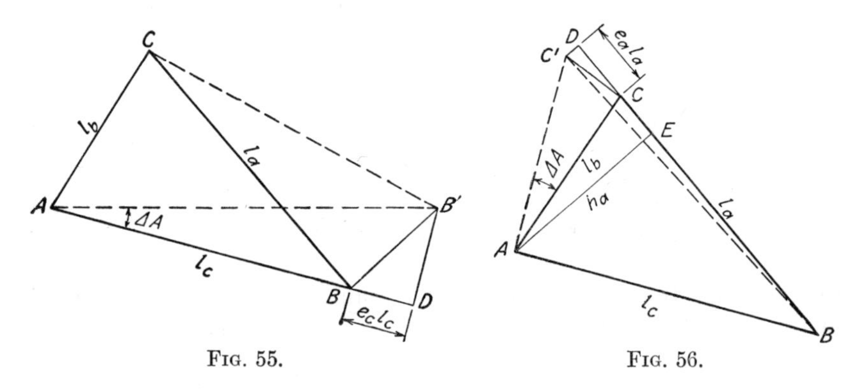

FIG. 55. FIG. 56.

Triangle $AB'C$ is the shape of the deformed triangle due to e_c only. Since the sides of angle $BB'D$ are mutually perpendicular to the sides of angle B in the original triangle ABC,

$$\text{angle } BB'D = \text{angle } B$$

$$\Delta A = -\frac{B'D}{AD \text{ or } AB} = -\frac{BD \cot BB'D}{AB} = -\frac{e_c l_c \cot B}{l_c}$$

$$= -e_c \cot B \quad (33)$$

Last consider the effect of e_a only on ΔA (Fig. 56). Prolong BC to D, making $CD = e_a l_a$. Draw DC' perpendicular to BD and CC' perpendicular to AC. Triangle ABC' is the shape of the deformed triangle. Since the sides of angle $DC'C$ are mutually perpendicular to the sides of angle C in the original triangle ABC, angle $DC'C$ = angle C.

$$\Delta A = +\frac{CC'}{AC} = +\frac{CD}{(\sin DC'C)AC} = +\frac{e_a l_a}{(\sin C)l_b} = +\frac{e_a l_a}{h_a} \quad (34)$$

In triangle ABC,

$$l_a = CE + BE = h_a \cot C + h_a \cot B \tag{35}$$

Substituting Eq. (35) into Eq. (34),

$$\Delta A = +\frac{e_a l_a}{h_a} = +\frac{e_a(h_a \cot C + h_a \cot B)}{h_a} = e_a(\cot B + \cot C) \tag{36}$$

When e_a, e_b, and e_c occur jointly in the lengths l_a, l_b, and l_c, the change ΔA in the angle A will be the sum of Eq. (32), (33), and (36), or

$$\Delta A = -e_b \cot C - e_c \cot B + e_a(\cot B + \cot C) = (e_a - e_b) \cot C + (e_a - e_c) \cot B$$

which is the first formula in Eqs. (31). The other two formulas in Eqs. (31) can be similarly derived.

Equations (31) can be stated in words as follows: The change in the angle is equal to the sum of two products, each obtained from multiplying the difference of the e of the opposite side minus the e of the adjacent side by the cotangent of the included angle.

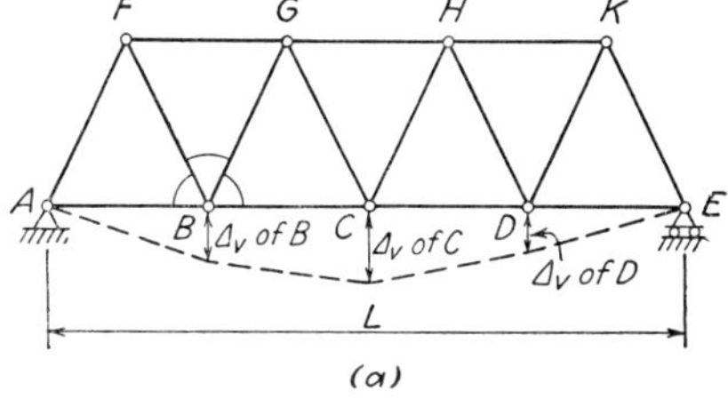

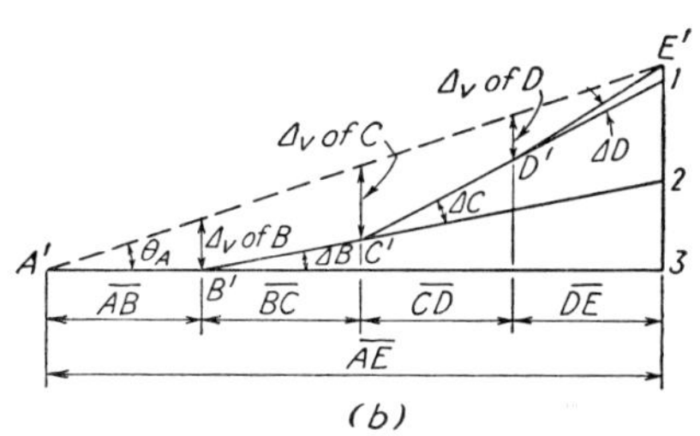

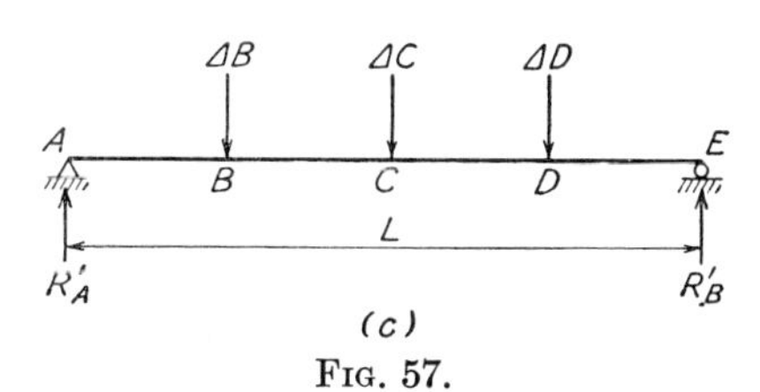

FIG. 57.

Now if it is required to find Δ_V of B, Δ_V of C, and Δ_V of D (Fig. 57*a*) at joints B, C, D of the truss owing to various values of *unit* elongations in all the members, the procedure as described below can be used. For joint B, determine changes in the sizes of angles ABF, FBG, and GBC, the total change ΔB in angle ABC being the sum of these three parts. Similarly ΔC and ΔD, the changes in angles C and D, are determined. Now, if ΔB, ΔC, and ΔD are all negative, the shape of the lower chord is the broken line $A'B'C'D'E'$ as shown in Fig. 57*b*. Actually the closing line $A'E'$ should stay horizontal; therefore Δ_V of B, Δ_V of C, and Δ_V of D are the vertical intercepts between this closing line $A'E'$ and the broken line $A'B'C'D'E'$. From Fig. 57*b*,

$$\theta_A = \frac{E'3}{A'3} = \frac{E'1 + 1\text{-}2 + 2\text{-}3}{L} = \frac{(DE)\,\Delta D + (CE)\,\Delta C + (BE)\,\Delta B}{L} \tag{37}$$

Note that since ΔB, ΔC, ΔD are small, $D'E' = D'\text{-}1$ can be considered equal to DE, $C'\text{-}1 = C'\text{-}2$ to CE, and $B'\text{-}2 = B'\text{-}3$ to BE. Referring to Fig. 57b again,

$$\begin{aligned} \Delta_V \text{ of } B &= (AB)\theta_A \\ \Delta_V \text{ of } C &= (AC)\theta_A - \Delta B(BC) \\ \Delta_V \text{ of } D &= (AD)\theta_A - \Delta B(BD) - \Delta C(CD) \end{aligned} \tag{38}$$

If a simple beam AE (Fig. 57c) is loaded by angle weights ΔB, ΔC, and ΔD at points B, C, and D,

$$R'_A = \frac{(\Delta D)(DE) + (\Delta C)(CE) + (\Delta B)(BE)}{L}$$

which is θ_A by comparing with Eq. (37). The bending moment at B, M'_B, on this simple beam AE is $R'_A(AB) = (AB)\theta_A$, which is Δ_V of B by comparing with Eq. (38). Also, the bending moments M'_C and M'_D at C' and D' are, respectively, Δ_V of C and Δ_V of D. Thus

$$\begin{aligned} M'_B &= \Delta_V \text{ of } B \\ M'_C &= \Delta_V \text{ of } C \\ M'_D &= \Delta_V \text{ of } D \end{aligned} \tag{39}$$

Equation (39) is the working formula for finding the vertical deflections at all joints on a straight chord of a truss by the angle-weights method. If the original shape of the truss chord is not straight, Eq. (39) must be modified but this modification will not be treated here.[1]

Example 30. Determine the vertical deflections of all lower chord joints due to the applied loading by the angle-weights method (Fig. 58a).

Solution. The stresses in the members are computed and shown in Fig. 58b. The unit strains are shown in Fig. 58c. By the use of

TABLE 30-1

Joint	Changes in angles (in terms of 10^{-3})	Summation of angle change
L_1	$L_0L_1U_1 = (-0.267 - 0.400)(\frac{3}{4}) + (-0.267 - 0.167)(\frac{4}{3}) = -1.077$ $U_1L_1L_2 = (+0.267 - 0.167)(\frac{4}{3}) + (+0.267 - 0.400)(\frac{3}{4}) = +0.033$	-1.044×10^{-3}
L_2	$L_1L_2U_1 = (+0.167 - 0.400)(0) + (+0.167 - 0.267)(\frac{4}{3}) = -0.133$ $U_1L_2U_2 = (-0.300 - 0)(0) + (-0.300 - 0.267)(\frac{3}{4}) = -0.425$ $U_2L_2U_3 = (-0.300 - 0)(0) + (-0.300 - 0.133)(\frac{3}{4}) = -0.325$ $U_3L_2L_3 = (+0.333 - 0.133)(\frac{4}{3}) + (+0.333 - 0.500)(0) = +0.267$	-0.616×10^{-3}
L_3	$L_2L_3U_3 = (+0.133 - 0.500)(\frac{3}{4}) + (+0.133 - 0.333)(\frac{4}{3}) = -0.542$ $U_3L_3L_4 = (-0.333 - 0.333)(\frac{4}{3}) + (-0.333 - 0.500)(\frac{3}{4}) = -1.514$	-2.056×10^{-3}

[1] Sutherland, Hale, and Harry L. Bowman, "Structural Theory," 4th ed., p. 203, John Wiley & Sons, Inc., New York, 1950.

Eq. (31) the angle changes in the angles around joints L_1, L_2, and L_3 are computed in Table 30-1 (refer to Fig. 58*d* for the cotangent function of angles). The answers are entered in Fig. 58*e*. The total angle change at each joint is applied to the conjugate truss as angle weights, as shown

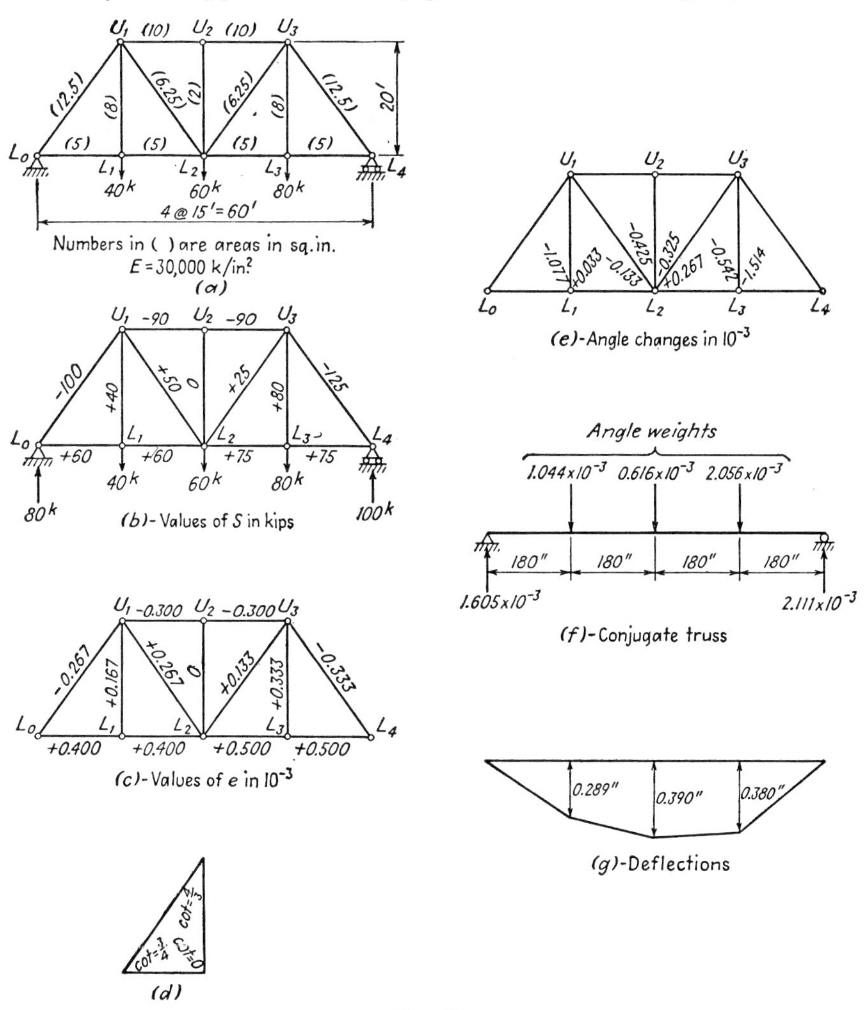

FIG. 58.

in Fig. 58*f*. The moment diagram of this conjugate truss is the deflected shape of the lower chord (Fig. 58*g*).

EXERCISE

39. Determine the vertical deflections of all lower chord joints due to the applied loading by the angle-weights method. (See drawing on p. 60.)

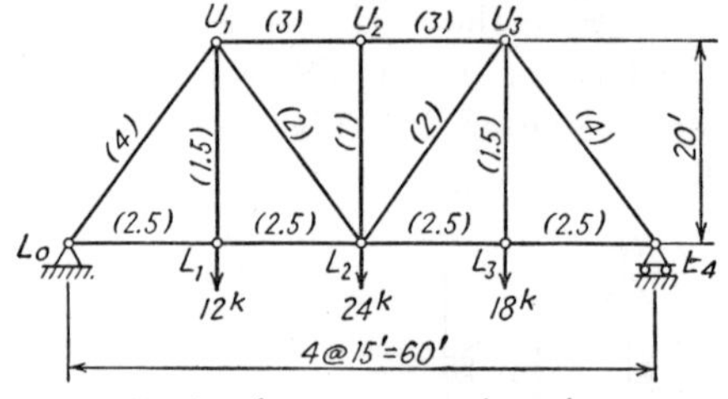

EXERCISE 39

20. The Joint-displacements Method. The joint-displacements method is an algebraic method by which the horizontal and vertical deflections of all joints of a truss due to the changes in the lengths of its members can be computed. The method is based on one equation, termed the *joint-displacements equation*, which is derived below.

Let AB be the member of an originally undeformed truss (Fig. 59). Owing to the changes in lengths of all members of the truss, the joint A

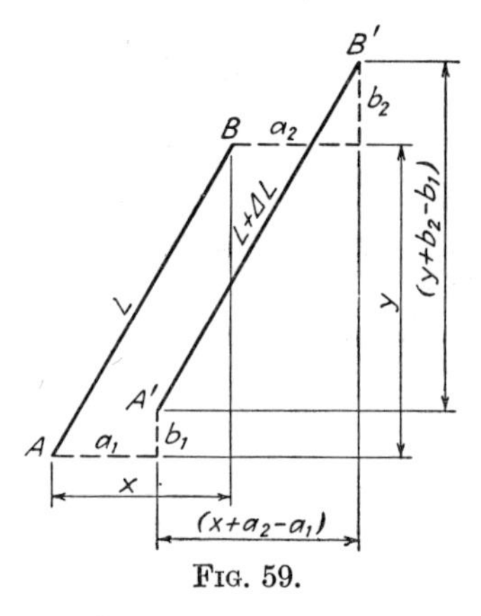

FIG. 59.

is displaced by a_1 units to the right and b_1 units up and the joint B by a_2 units to the right and b_2 units up. Let L be the original length of AB and ΔL be the increase in its length. Let x and y be the coordinates of B referred to A as origin. From Fig. 59, it is seen that

$$\overline{AB}^2 = L^2 = x^2 + y^2 \tag{40}$$

$$\overline{A'B'}^2 = (L + \Delta L)^2 = (x + a_2 - a_1)^2 + (y + b_2 - b_1)^2 \tag{41}$$

Subtracting Eq. (40) from Eq. (41),

$$2L\,\Delta L + (\Delta L)^2 = 2x(a_2 - a_1) + (a_2 - a_1)^2 + 2y(b_2 - b_1) + (b_2 - b_1)^2 \tag{42}$$

Neglecting the terms $(\Delta L)^2$, $(a_2 - a_1)^2$, and $(b_2 - b_1)^2$ in Eq. (42) and simplifying,

$$\Delta L = \frac{1}{L}[x(a_2 - a_1) + y(b_2 - b_1)] \tag{43}$$

Equation (43) expresses the change in the length of a member in terms of the displacements of its ends; hence it is called the joint-displacements equation. If, besides the quantities x, y, L, and ΔL, which are usually known, three of the four displacement components, a_1, b_1, a_2, and b_2, are known, the fourth can be determined. The use of this joint-displacements equation to determine the horizontal and vertical deflections of all joints in a truss will be illustrated by the following example.

Example 31. Determine the horizontal and vertical deflections of all joints due to the applied loading by the joint-displacements method (Fig. 60*a*).

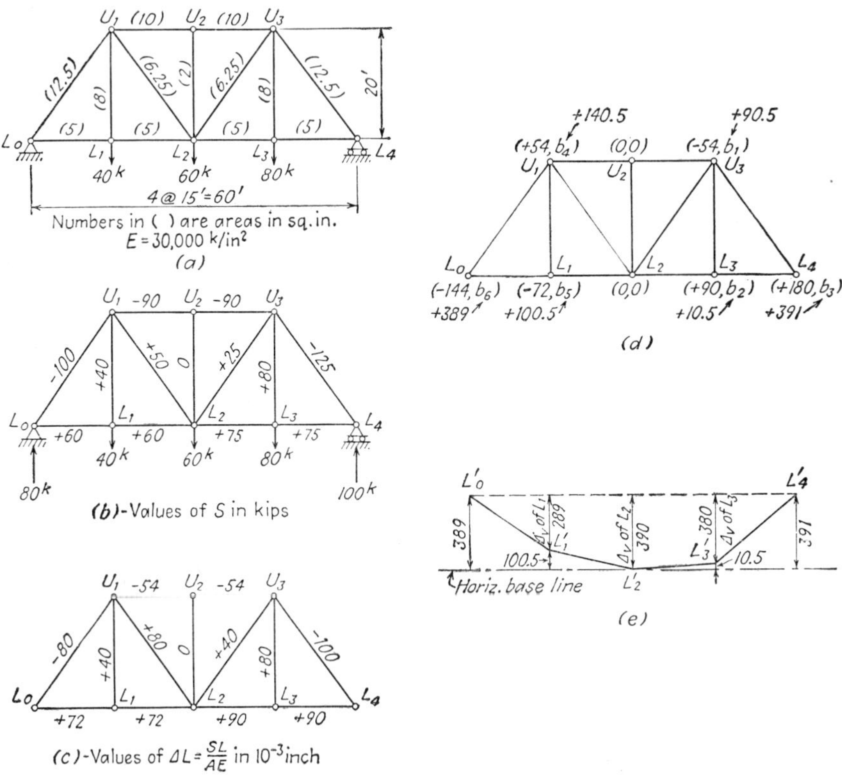

FIG. 60.

Solution. The location of joint L_0 does not change owing to the deformation of the truss, while the direction of every member does change.

In the first step of applying the joint-displacements method, however, any joint may be assumed to be fixed in location and any member through this joint to be fixed in direction. When the joint displacements of all joints are determined relative to this fixed joint and fixed direction, the whole deformed truss may be translated or rotated so as to fulfill certain physical requirements; for instance, the translation will bring the joint L_0' back to the hinge, and the rotation will bring the joint L_4' to the horizontal line through the hinge.

Assume joint L_2 to be the fixed point, or the *reference point.* Assume member L_2U_2 fixed in direction, or call member L_2U_2 the *reference member* (Fig. 60*d*). The horizontal and vertical deflections of the fixed point L_2 are (0,0). Because L_2U_2 is fixed in direction and its ΔL is zero, the deflections of joint U_2 are also (0,0). Since the horizontal deflection of L_2 is zero and the member L_2L_3 is lengthened by 90 units, it is obvious that joint L_3 moves 90 units to the right and its horizontal deflection is thus +90. Joint L_4 should move a distance equal to the sum of the lengthenings in members L_2L_3 and L_3L_4; its horizontal deflection, then, is $+90 + 90 = +180$. By the same reasoning, the horizontal deflections of joints L_1, L_0, U_3, and U_1 are as shown in Fig. 60*d*. The vertical deflections of joints U_3, L_3, and L_4 are unknowns and are noted as b_1, b_2, and b_3; so are the vertical deflections of joints U_1, L_1, and L_0, which are noted as b_4, b_5, and b_6. These unknown vertical deflections can be easily determined by the use of the joint-displacements equation (43). Applying Eq. (43) to the following members in succession,

Member L_2U_3:

$$+40 = (\tfrac{1}{25})[(+15)(-54 - 0) + (+20)(b_1 - 0)]$$
$$b_1 = +90.5$$

Member L_3U_3: If U_3 moves up 90.5 and U_3L_3 elongates by 80, L_3 must move only $90.5 - 80 = 10.5$ up.

$$b_2 = +10.5$$

Member U_3L_4:

$$-100 = (\tfrac{1}{25})[(+15)(+180 + 54) + (-20)(b_3 - 90.5)]$$
$$b_3 = +391$$

Member L_2U_1:

$$+80 = (\tfrac{1}{25})[(-15)(54 - 0) + (+20)(b_4 - 0)]$$
$$b_4 = +140.5$$

Member U_1L_1:

$$b_5 = 140.5 - 40 = +100.5$$

Member U_1L_0:

$$-80 = (\tfrac{1}{25})[(-15)(-144 - 54) + (-20)(b_6 - 140.5)]$$
$$b_6 = +389$$

With these values of b_1 to b_6 determined, and if only the vertical deflections of the lower chord joints are desired, a diagram such as that shown

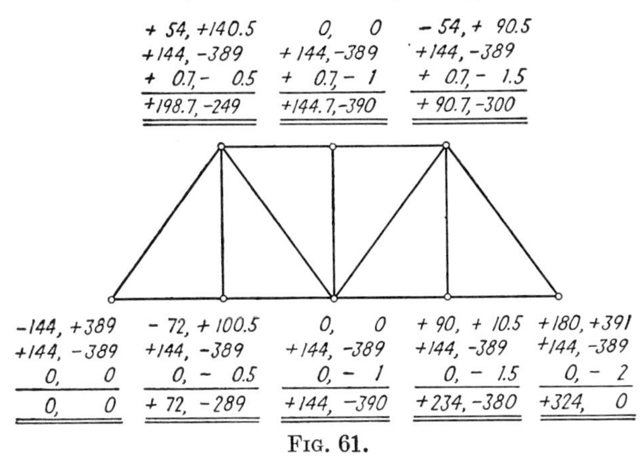

FIG. 61.

in Fig. 60*e* is all that is necessary. Draw a horizontal line through L'_2, plot L'_0 at $b_6 = 389$ units above the base line, L'_1 at $b_5 = 100.5$ units above, L'_3 at $b_2 = 10.5$ units above, and L'_4 at $b_3 = 391$ units above. Draw the closing line $L'_0L'_4$. The vertical intercepts between this closing line and the various points L'_1, L'_2, and L'_3 are the vertical deflections required, which are 0.289 in., 0.390 in., and 0.380 in., respectively.

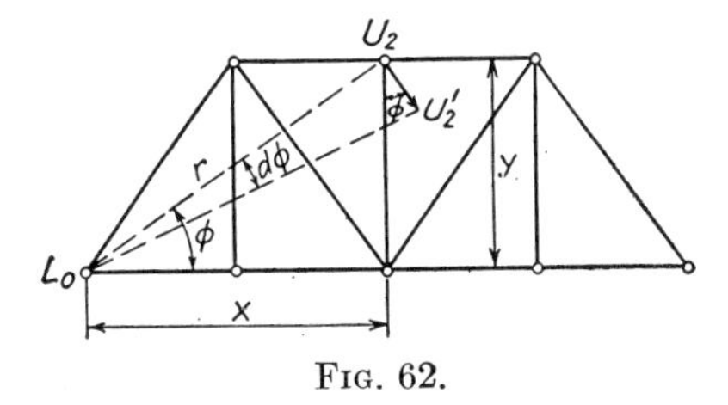

FIG. 62.

If the horizontal and vertical deflections of all joints are desired, they can be determined as described below. The numbers which appear at the first line (Fig. 61) are the horizontal and vertical displacements at all joints when joint L_2 is fixed in location and member L_2U_2 is fixed in direction. The displacements of L_0 are $(-144, +389)$. So the whole deformed truss should be translated 144 units to the right and 389 units downwards. These two displacements appear at the second line at all joints. Now L_4 is 2 units higher than L_0. So it is necessary to rotate the whole deformed truss through a clockwise angle $d\phi = 2$ units per 60 ft. Owing to this rotation, the horizontal and vertical deflections at each joint can be determined by the following procedure: Take, for example, joint U_2 (Fig. 62), the coordinates of which referred to the center of rotation L_0 as origin are (x,y).

$$\begin{aligned} \text{Horizontal displacement of } U_2 &= (U_2U_2')(\sin\phi) \\ &= (r)\,d\phi\,(\sin\phi) = y\,d\phi \\ \text{Vertical displacement of } U_2 &= (U_2U_2')(\cos\phi) \\ &= (r)\,d\phi\,(\cos\phi) = x\,d\phi \end{aligned} \tag{44}$$

Equations (44) are used to determine the horizontal and vertical deflections at all joints due to the clockwise rotation of $d\phi = 2$ units per 60 ft in the present problem. The horizontal movements of all lower chord joints are zero, because the y coordinates of these points referred to L_0 as origin are zero. The vertical movements of L_1, L_2, L_3, and L_4 are, respectively, $\frac{2}{60} \times 15 = 0.5$, $\frac{2}{60} \times 30 = 1$, $\frac{2}{60} \times 45 = 1.5$, and $\frac{2}{60} \times 60 = 2$. The vertical movements of U_1, U_2, and U_3 are the same as those of L_1, L_2, and L_3. The horizontal movements of U_1, U_2, and U_3 are $\frac{2}{60} \times 20 = 0.7$ unit because the y coordinates of these points referred to L_0 as origin are 20 ft. The signs of these displacements can best be determined by inspection. The displacements due to rotation appear at the third line of Fig. 61. The sums of all the three sets of movements, the first due to deformation when L_2 is the reference point and L_2U_2 is the reference member, the second due to translation, and the third due to rotation, give the horizontal and vertical deflections at all joints.

EXERCISE

40. Determine the horizontal and vertical deflections of all joints due to the applied loading by the joint-displacements method.

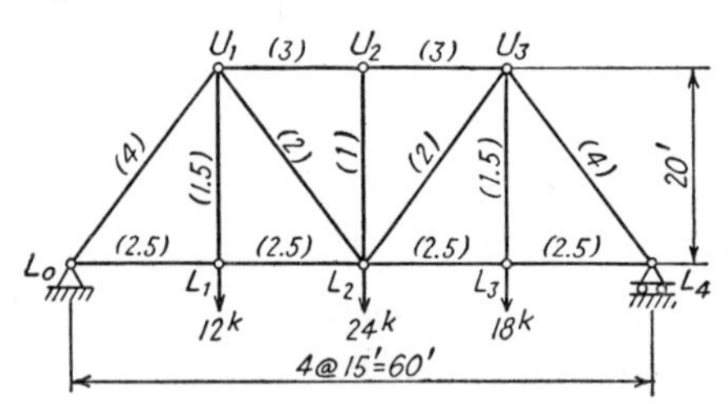

EXERCISE 40

21. The Graphical Method. The angle-weights method and the joint-displacements method are algebraic methods using only geometric relations. Thus it is easy to infer that the horizontal and vertical deflections of all joints can be found by a graphical solution. Theoretically it is possible to draw the shape of the deformed truss by using the new lengths of members as the sides of the component triangles. But the changed lengths are only a little longer or shorter than the original lengths, a fact which makes the deformed truss almost coincide with the original truss. This difficulty can be avoided by using two different scales in plotting the original lengths L and the changes in length, ΔL.

Take, for example, the Warren truss $ABCDE$ of Fig. 63*a*. Let $+7$ units be the lengthening in member AB; $+6$, $+3$, and $+9$ units the lengthenings in members BC, BD, and BE; -8, -5, and -4 units the shortenings in members AC, CD, and DE. The shape of the deformed truss can be found by drawing the new triangles $A'B'C'$, $B'C'D'$, and $B'D'E'$ in succession.

To start with triangle $A'B'C'$, let joint A' of the deformed truss coincide with joint A of the original truss (Fig. 63*a*), and let the direction of $A'B'$ coincide with that of AB. B', then, must fall at 7 units to the right of B in the direction AB. Joint C', the only unknown point on triangle $A'B'C'$, can be determined by the intersection of two arcs, using A' and B' as centers and the lengths of $A'C'$ and $B'C'$ as radii. This is performed by the following procedure: From C measure $CC_1 = 8$ units toward A or A' because the member AC shortens by 8 units. AC_1 is then the new length of member AC. Now, instead of drawing the arc with A' as center and $A'C_1$ as radius, which can be done only if CC_1 is plotted on the same scale as that for AC, the perpendicular to AC is drawn at C_1, an approximation which is permissible as long as the deformations are very small when compared with the original dimensions. This permits CC_1 to be plotted equal to 8 units and to a large scale. Draw $B'C_2$ parallel and equal to BC. Measure 6 units out from C_2 to C_3. $B'C_3$ is the new length of member BC because $B'C_2$ represents the original length and $C_2C_3 = 6$ units is the amount of lengthening. The perpendicular to C_3B' drawn at C_3 and the other one previously drawn at C_1 will intersect at C'. Triangle $A'B'C'$ is now completed.

The displacement of joint C can be scaled from C to C' in Fig. 63*a*. But the displacement polygon $CC_2C_3C'C_1$ can be isolated or drawn separately as shown in Fig. 63*d*, the scale of which is that of ΔL only and therefore can be made as large as desirable.

The next thing is to draw the triangle $B'C'D'$, for which B' and C' have already been located. Draw $C'D_1$ parallel and equal to CD. From D_1 measure $D_1D_2 = 5$ units. $C'D_2$ is the new length of member CD. Draw $B'D_3$ parallel and equal to BD. From D_3 measure $D_3D_4 = 3$ units. $B'D_4$ is the new length of member BD. The two perpendiculars to lines $C'D_2$ and $B'D_4$ erected, respectively, at points D_2 and D_4 intersect at

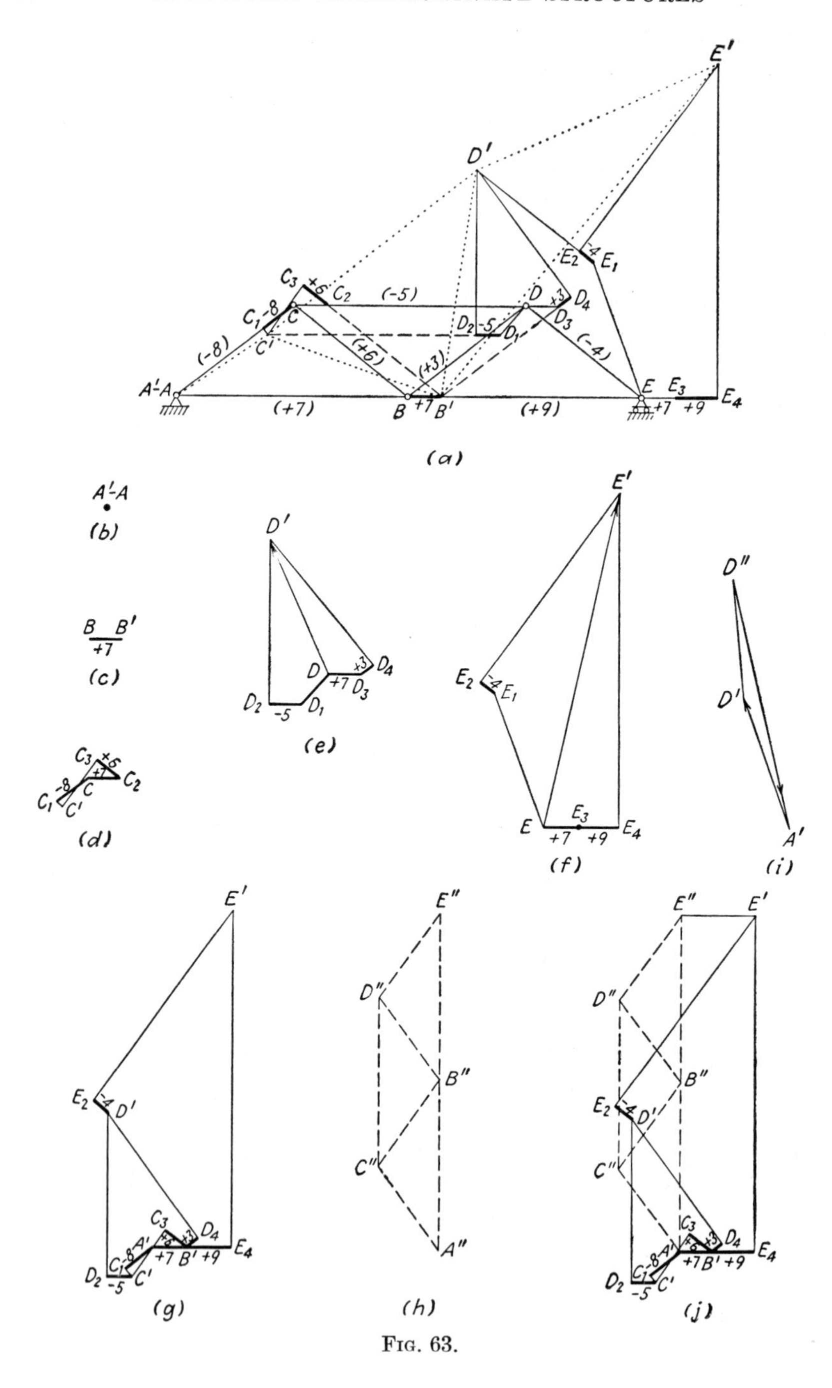
E'
D'
C3
+6
C2
(-5)
C1
-8
C
D
+3
D4
D2
-5
D1
D3
E2
-4
E1
(-8)
C'
(+6)
(+3)
(-4)
A'-A
(+7)
B
+7
B'
(+9)
E
E3
+7
+9
E4
(a)
(b)
(c)
(d)
(e)
(f)
D''
(i)
A'
E''
B''
C''
A''
(g)
(h)
(j)

Fig. 63.

D', which completes the triangle $B'C'D'$. Note that the displacement polygon $DD_3D_4D'D_2D_1$ can be drawn separately, as shown in Fig. 63*e*. The displacement of joint D is from D to D'.

Lastly, it is necessary to locate the joint E' for the triangle $B'D'E'$. Draw $D'E_1$ parallel and equal to DE. From E_1 measure $E_1E_2 = 4$ units. $D'E_2$ is the new length of member DE. Prolong $B'E$ to E_3, making $B'E_3 = BE$, or $EE_3 = 7$ units. From E_3 measure $E_3E_4 = 9$ units. $B'E_4$ is the new length of member BE. The two perpendiculars to lines $D'E_2$ and $B'E_4$ erected, respectively, at points E_2 and E_4 intersect at E', which is the new location of joint E. The displacement polygon $EE_4E'E_2E_1$ can be drawn separately, as shown in Fig. 63*f*.

Now if the displacement polygons, as shown separately in Fig. 63*b* to *f*, are superimposed upon one another, by keeping points A, B, C, D, and E coincident, the combined diagram of Fig. 63*g* is obtained. Then the deflections of joints A, B, C, D, and E can be scaled off from Fig. 63*g* from the common point A (B, C, D, or E) outward to points A' ($AA' = 0$), B', C', D', and E', respectively. This common point A is called the *reference point.* The combined diagram, shown as Fig. 63*g*, is called the deformation diagram or the Williot diagram.

Although the Williot diagram has been presented in a lengthy manner, it can be obtained directly as described below. After choosing joint A as the reference point and member AB as the reference member, locate point A' (Fig. 63*g*) and then point B' at 7 units to the right of A because the lengthening in AB is 7 units and joint B should move 7 units to the right relative to joint A. From A' measure $A'C_1 = 8$ units downward to the left in the direction of member AC because joint C should move 8 units downward to the left relative to joint A. From B' measure $B'C_3 = 6$ units upward to the left in the direction of member BC because joint C should move upward to the left relative to joint B. The two perpendiculars to the directions of members AC and BC, erected respectively at points C_1 and C_3, intersect at C'. From C' draw $C'D_2 = 5$ units horizontally to the left; and from B' measure $B'D_4 = 3$ units upward to the right in the direction of member BD. The two perpendiculars drawn at D_2 and D_4 intersect at D'. From D' measure $D'E_2 = 4$ units upward to the left in the direction of member DE, and from B' draw $B'E_4 = 9$ units horizontally to the right. The two perpendiculars drawn at E_2 and E_4 intersect at E'. The reader is advised to resketch the Williot diagram, as shown in Fig. 63*g*, while studying this paragraph.

The deformed truss now assumes the form $A'B'C'D'E'$ of Fig. 63*a*, and the movement of each joint can be measured in Fig. 63*g* from the reference point to the single prime point in question. Obviously it is necessary to rotate the deformed truss $A'B'C'D'E'$ of Fig. 63*a* through a clockwise angle of ($E'E_4$/span) rad so as to bring the point E' to the same level as the hinge or to coincide with point E_4. The additional displacements of all joints as caused by this rotation only, can be found as follows: In Fig. 63*g*, a vertical line through the hinge A (also called A' or A'') is made to intersect a horizontal line through E' at E'' (see Fig. 63*j*); then the vertical distance $E''A''$ is the movement of joint E due to the rotation above mentioned. With $A''E''$ as base (Fig. 63*h*), $A''B''C''D''E''$ is drawn similar to the original truss $ABCDE$. As to which side of the base $A''E''$ the rotation diagram $A''B''C''D''E''$ should be drawn on, the criterion is that $A''B''C''D''E''$ can be rotated by 90° in the plane to a position parallel and similar to the original truss, $ABCDE$.

It can be shown that the movements of joints B, C, D, and E due only to this rotation can be scaled in Fig. 63*h* from the double-prime point to the hinge A (A' or A''). For instance, it is required to prove that $D''A''$ in Fig. 63*h* equals line DA in Fig. 63*a* times the angle of rotation and that $D''A''$ in Fig. 63*h* is perpendicular to line DA in Fig. 63*a*.

Proof. Triangle $D''A''E''$ (Fig. 63*h*) is similar to triangle DAE (Fig. 63*a*). Since $E''A''$ is perpendicular to EA and

$$\text{angle } D''A''E'' = \text{angle } DAE,$$

$D''A''$ is perpendicular to DA. Also,

$$\frac{D''A''}{DA} = \frac{E''A''}{EA}$$

but,

$$\frac{E''A''}{EA} = \text{angle of rotation}$$

Thus

$$\frac{D''A''}{DA} = \text{angle of rotation}$$

or,

$$D''A'' = (DA) \text{ times the angle of rotation}$$

If Fig. 63*h* is superimposed on Fig. 63*g*, Fig. 63*j* is obtained. The total movement of each joint is the vector sum of the movement from the double-prime point to the hinge A (A' or A'') owing to rotation and from the reference point (in this case it happens that the hinge is chosen as the reference point) to the single-prime point owing to deformation, or the total movement can be measured directly from the double-prime point to the single-prime point in Fig. 63*j*. The vector-sum diagram for joint D only is shown as Fig. 63*i*. Thus in the graphical solution for the magnitude and direction of the deflection of each joint in a truss, a diagram like Fig. 63*j* is *all that is necessary*, and the deflection is measured from the double-prime point to the single-prime point. Figure 63*g* is known as the Williot diagram, Fig. 63*h* as the Mohr diagram, and Fig. 63*j* as the Williot-Mohr diagram.

If some joint other than the hinge had been chosen as the reference point, then owing to deformation only the hinge would have moved from the reference point to the single-prime point for the hinge. It is necessary therefore, in addition to the deformation and rotation effects,

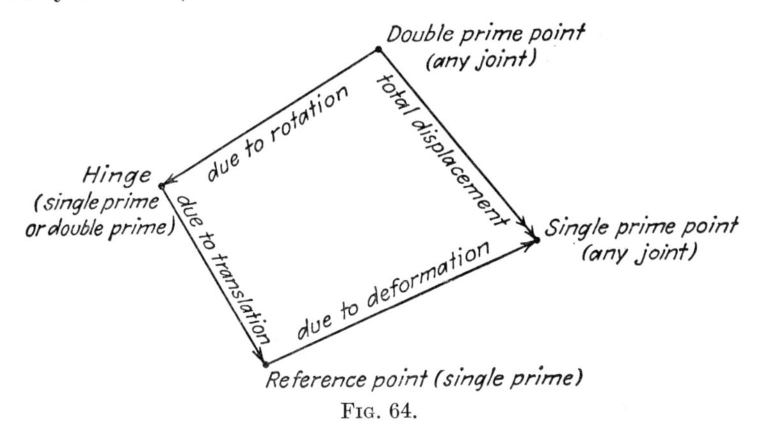

FIG. 64.

to translate the deformed truss through the displacement from the single-prime point for the hinge back to the reference point. Figure 64 shows the combination of all three vectors, and this discussion will be further illustrated in the latter part of Example 32.

Example 32. Determine the horizontal and vertical deflections of all joints due to the applied loading by the graphical method (Fig. 65*a*).

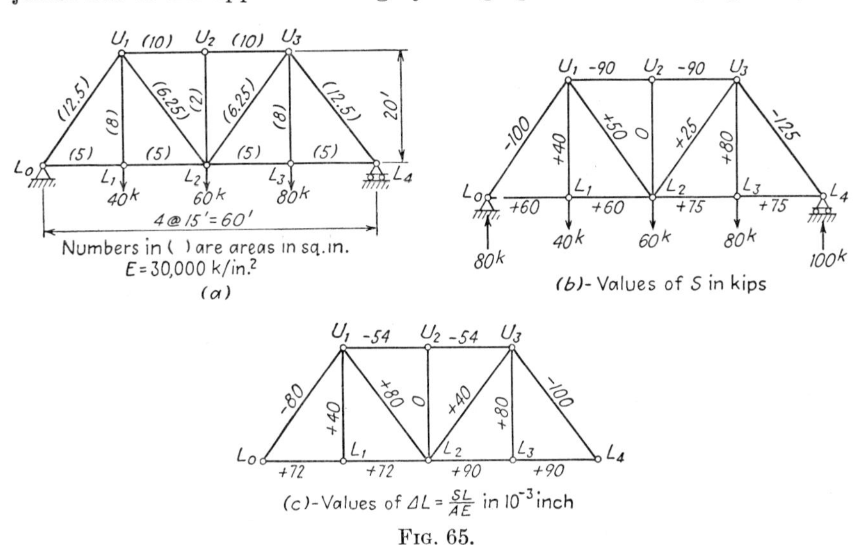

FIG. 65.

Solution. The stresses in all members of the truss due to the applied loading and the changes in the lengths of the members are shown in Fig. 65.

Two graphical solutions are shown in Figs. 66 and 67. It is to be noted that the solution shown in Fig. 67 takes much less space than that

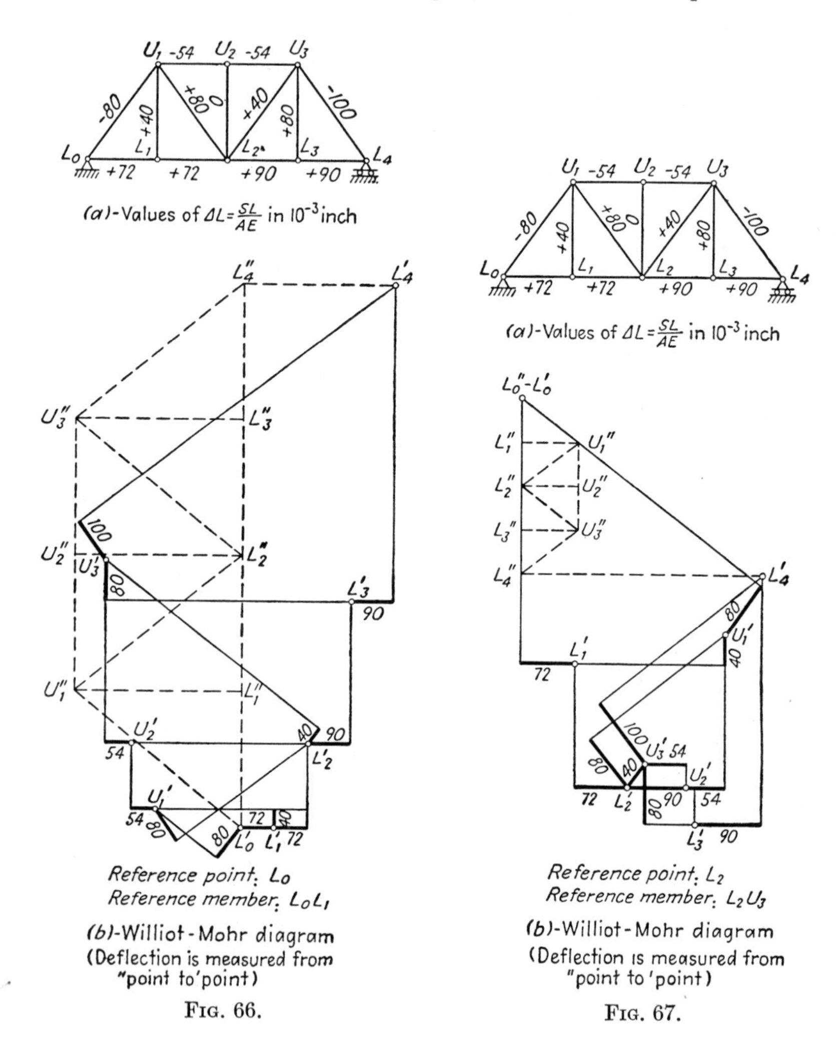

(a)-Values of $\Delta L = \frac{SL}{AE}$ in 10^{-3} inch

(b)-Williot-Mohr diagram (Deflection is measured from "point to 'point)

FIG. 66.

(a)-Values of $\Delta L = \frac{SL}{AE}$ in 10^{-3} inch

(b)-Williot-Mohr diagram (Deflection is measured from "point to 'point)

FIG. 67.

of Fig. 66 when the same scale is used, or in a certain available space a larger scale can be used if a good choice of reference point or reference member is made. Note also that the rotation diagram is on the left of $L_0'L_4''$ in Fig. 66 and on the right of $L_0'L_4''$ in Fig. 67, the criterion being that the rotation diagram can be rotated in the plane to a position parallel and similar to that of the original truss.

EXERCISES

41. Determine the horizontal and vertical deflections of all joints due to the applied loading by the graphical method. Check the solution by choosing a different reference point and reference member.

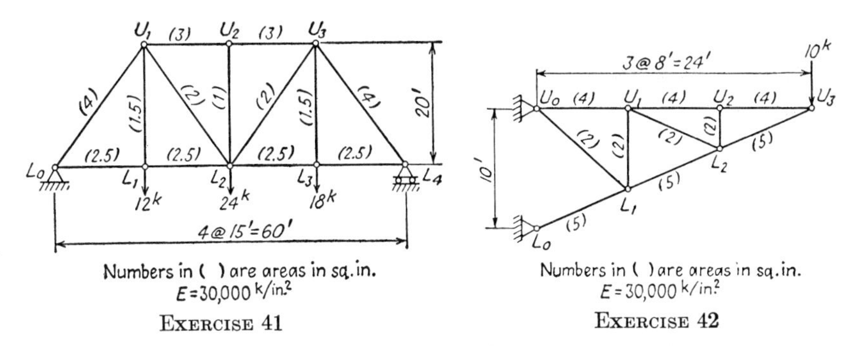

Numbers in ( ) are areas in sq.in.
$E = 30{,}000$ k/in.^{2}

EXERCISE 41

Numbers in ( ) are areas in sq.in.
$E = 30{,}000$ k/in.^{2}

EXERCISE 42

42. Determine the horizontal and vertical deflections of all joints due to the applied loading by the graphical method.

CHAPTER IV

ANALYSIS OF STATICALLY INDETERMINATE BEAMS AND FRAMES BY THE METHOD OF CONSISTENT DEFORMATION

22. The Method of Consistent Deformation. The method of consistent deformation is the most general method of analyzing statically indeterminate structures. Consider a structure which is statically indeterminate because it has more than adequate external supports for statical stability. The number of external reaction components (one for each roller support, two for each hinge support, and three for each fixed support) which are necessary and sufficient for statical stability is equal to the number of conditions for statical equilibrium. As explained earlier, the number of excess reaction components over those required for statical equilibrium is called the degree of indeterminacy of the structure. If the excess supports are removed and replaced by unknown force (or moment) reactions, a basic determinate structure under the action of the applied loading and these unknown reactions, or redundants, is obtained. The derived basic determinate structure must nevertheless still satisfy the physical requirements at the location of the excess supports now replaced by redundant reactions. If a roller support has been removed at a certain point, the requirement is that the deflection in the direction perpendicular to the supporting surface must be zero; if a hinge support has been removed, the two requirements are that the horizontal and vertical deflections at the point must both be zero; if a fixed support has been removed, the three requirements are that the rotation and horizontal and vertical deflections at the point must be zero. Thus there are always as many physical conditions of geometry as there are redundant reactions. After the redundant reaction components are found by using the conditions of geometry, the remaining two or three reaction components can be solved by the equations of statics. This method of consistent deformation will be now applied to the analysis of statically indeterminate beams and frames.

23. Analysis of Statically Indeterminate Beams by the Method of Consistent Deformation

Example 33. Determine all reaction components and draw shear and moment diagrams for the beam of Fig. 68*a*.

Solution. The beam as shown is fixed at A and simply supported at B; it has three reaction components V_A, V_B, and M_A. Statics offers only two conditions of equilibrium for a coplanar parallel-force system; so the given beam is statically indeterminate to the first degree. A basic determinate beam may now be set up by removing the simple support at B, thus obtaining a cantilever beam fixed at A and free at B. Acting on this cantilever beam are the applied loading P, which causes downward deflection Δ_B at the free end B (Fig. 68*b*), and the redundant reaction V_B, which causes the upward deflection $V_B\delta_B$ at the free end B (Fig. 68*c*), where δ_B is the upward deflection at the free end B caused by an upward load of 1 kip at B (Fig. 68*d*). The given beam shown in Fig. 68*a* is then obtained by superposing the beams shown in Fig. 68*b* and *c*. The condition of geometry here is that the deflection at B when both P and V_B are acting on the basic determinate cantilever beam must be zero. Thus

$$\Delta_B = V_B\delta_B \qquad \text{or} \qquad V_B = \frac{\Delta_B}{\delta_B}$$

Δ_B and δ_B can be found by any one of the methods treated in Chap. II. By the moment-area method

$$\Delta_B = \left(\frac{PL}{2EI}\right)\left(\frac{L}{2}\right)(\tfrac{1}{2})(\tfrac{5}{6}L) = \frac{5PL^3}{48EI}$$

$$\delta_B = \left(\frac{L}{EI}\right)(L)(\tfrac{1}{2})(\tfrac{2}{3}L) = \frac{L^3}{3EI}$$

$$V_B = \frac{\Delta_B}{\delta_B} = \tfrac{5}{16}P$$

By statics (Fig. 68*e*)

$$V_A = \tfrac{11}{16}P$$
$$M_A = \tfrac{3}{16}PL$$

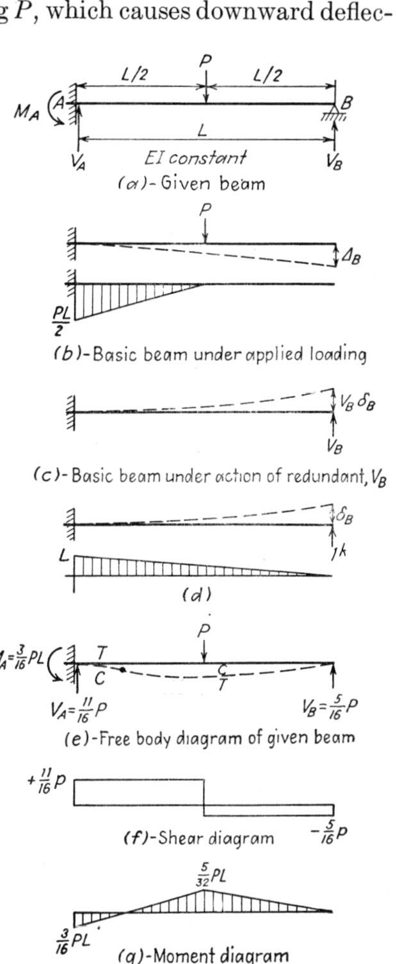

FIG. 68.

The shear and moment diagrams required are as shown in Fig. 68*f* and *g*.

Alternate Solution. When a simple beam AB is chosen as the basic determinate beam, the redundant is M_A and the condition required of geometry is that the rotation of the tangent at A should be zero. The given beam is the sum of the beams of Fig. 69*b* and *c*. The condition from which M_A can be determined is

$$\theta_A = M_A\phi_A \qquad \text{or} \qquad M_A = \frac{\theta_A}{\phi_A}$$

By the conjugate-beam method

$$\theta_A = \left(\frac{PL}{4EI}\right)\left(\frac{L}{2}\right)(\tfrac{1}{2}) = \frac{PL^2}{16EI}$$

$$\phi_A = \left(\frac{2}{3}\right)\left(\frac{1}{EI}\right)(L)(\tfrac{1}{2}) = \frac{L}{3EI}$$

$$M_A = \frac{\theta_A}{\phi_A} = \frac{3PL}{16}$$

By statics (Fig. 69*e*)

$$V_A = \tfrac{11}{16}P \qquad V_B = \tfrac{5}{16}P$$

The shear and moment diagrams can then be drawn as before.

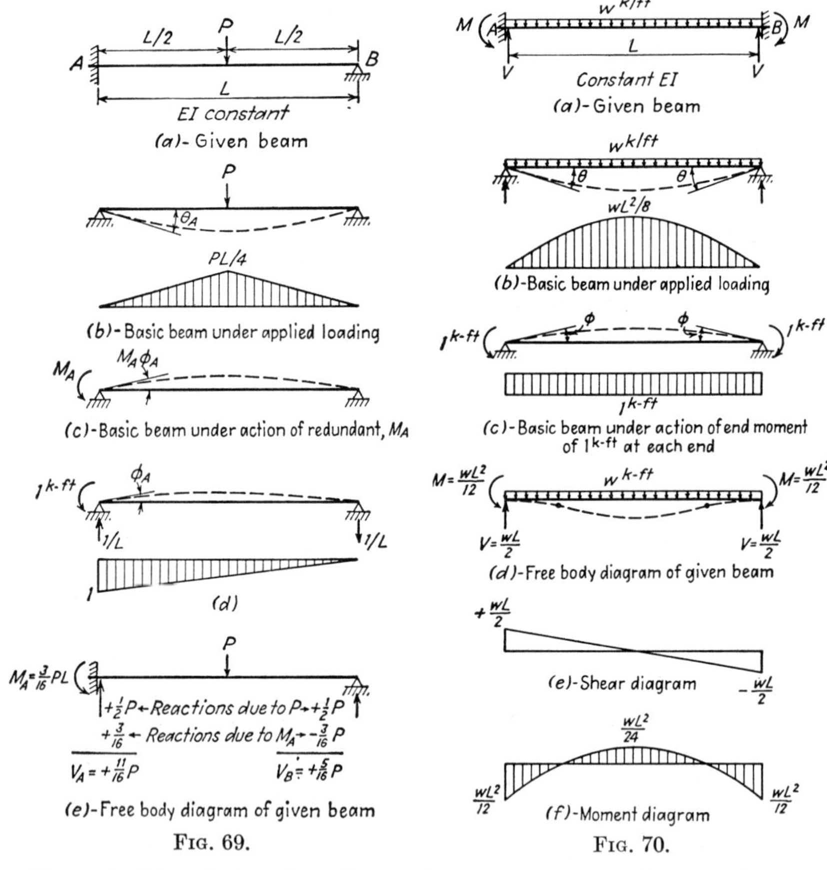

Fig. 69. Fig. 70.

Example 34. Determine all reaction components and draw shear and moment diagrams for the beam of Fig. 70*a*.

Solution. The beam is fixed at both ends, and it is statically indeterminate to the second degree. By symmetry, however, V at each end equals $wL/2$, and the only unknown is the fixed-end moment M at each end. A simple beam AB is chosen as the basic determinate beam, acting on which are the uniform load and the end moments M and M. The condition required of geometry is that the slope, θ, at either end caused by the uniform load (Fig. 70*b*) should be equal and opposite to the slope $(M)(\phi)$ caused by the end moments, where ϕ is the slope at either end due to an end moment of 1 kip-ft at each end (Fig. 70*c*). Thus

$$\theta = M\phi \qquad M = \frac{\theta}{\phi}$$

By the conjugate-beam method

$$EI\theta = \left(\frac{1}{2}\right)\left(\frac{2}{3}\right)\left(\frac{wL^2}{8}\right)(L) = \frac{wL^3}{24}$$

$$EI\phi = (\tfrac{1}{2})(1)(L) = \frac{L}{2}$$

$$M = \frac{EI\theta}{EI\phi} = \frac{wL^2}{12}$$

The shear and moment diagrams are as shown in Fig. 70*e* and *f*.

Alternate Solution. Suppose that the cantilever beam fixed at A and free at B is chosen as the basic determinate beam. Then the given beam

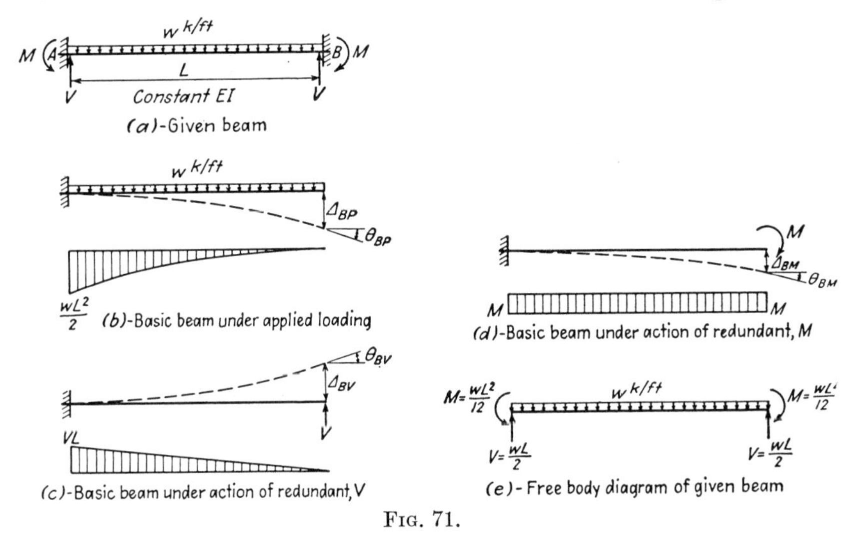

FIG. 71.

is equal to the sum of (1) the basic beam under applied loading (Fig. 71*b*), (2) the basic beam under action of redundant V (Fig. 71*c*), and (3) the basic beam under action of redundant M (Fig. 71*d*). Since the change in slope and the deflection at B of the given beam should both be zero, the conditions of geometry are

$$\Delta_{BP} + \Delta_{BM} = \Delta_{BV}$$

and

$$\theta_{BP} + \theta_{BM} = \theta_{BV}$$

Δ_{BP}, Δ_{BV}, and Δ_{BM} are the deflections at B due to applied loading, redundant V, and redundant M, respectively, while θ_{BP}, θ_{BV}, and θ_{BM} are the changes in slope due to the same causes.

Substituting the following quantities

$$\Delta_{BP} = \frac{wL^4}{8EI} \qquad \theta_{BP} = \frac{wL^3}{6EI}$$

$$\Delta_{BV} = \frac{VL^3}{3EI} \qquad \theta_{BV} = \frac{VL^2}{2EI}$$

$$\Delta_{BM} = \frac{ML^2}{2EI} \qquad \theta_{BM} = \frac{ML}{EI}$$

into the two conditions of geometry,

$$\frac{wL^4}{8} + \frac{ML^2}{2} = \frac{VL^3}{3}$$

and,

$$\frac{wL^3}{6} + ML = \frac{VL^2}{2}$$

Solving,

$$V = \frac{wL}{2} \qquad M = \tfrac{1}{12}wL^2$$

Example 35. Determine all reaction components for the beam of Fig. 72*a*.

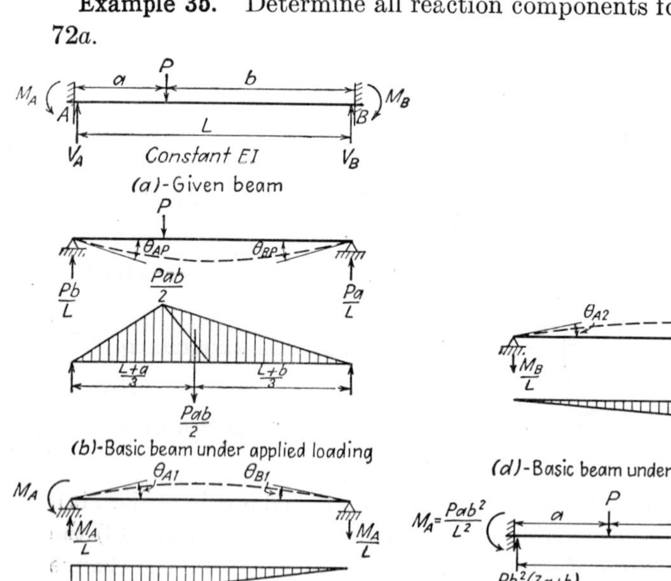

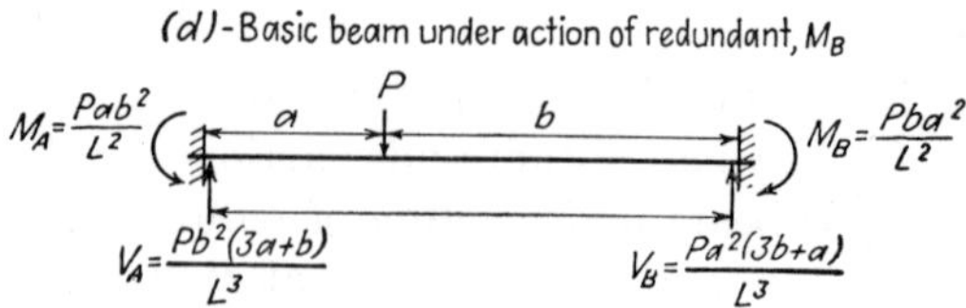

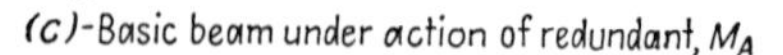

(c)-Basic beam under action of redundant, M_A

(e)-Free body diagram of given beam

Fig. 72.

Solution. The given beam is fixed at both ends and has four reaction components; so it is statically indeterminate to the second degree. The simple beam AB is chosen as the basic beam, where M_A and M_B are the redundants. The conditions of geometry require

$$\theta_{A1} + \theta_{A2} = \theta_{AP}$$

and

$$\theta_{B1} + \theta_{B2} = \theta_{BP}$$

By the conjugate-beam method

$$EI\theta_{AP} = \frac{Pab}{2}\frac{L+b}{3L} = \frac{Pab(L+b)}{6L}$$

$$EI\theta_{BP} = \frac{Pab}{2}\frac{L+a}{3L} = \frac{Pab(L+a)}{6L}$$

$$EI\theta_{A1} = \frac{2}{3}\frac{M_AL}{2} = \frac{M_AL}{3}$$

$$EI\theta_{B1} = \frac{1}{3}\frac{M_AL}{2} = \frac{M_AL}{6}$$

$$EI\theta_{A2} = \frac{1}{3}\frac{M_BL}{2} = \frac{M_BL}{6}$$

$$EI\theta_{B2} = \frac{2}{3}\frac{M_BL}{2} = \frac{M_BL}{3}$$

Substituting the above values into the two conditions,

$$\frac{M_AL}{3} + \frac{M_BL}{6} = \frac{Pab(L+b)}{6L}$$

$$\frac{M_AL}{6} + \frac{M_BL}{3} = \frac{Pab(L+a)}{6L}$$

Solving,

$$M_A = \frac{Pab^2}{L^2}$$

and

$$M_B = \frac{Pba^2}{L^2}$$

EXERCISES

43 and 44. Determine all reaction components by the method of consistent deformation, and draw shear and moment diagrams.

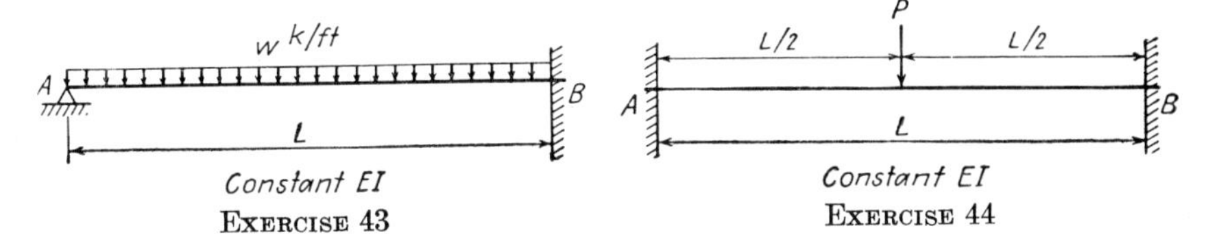

Constant EI
EXERCISE 43

Constant EI
EXERCISE 44

24. The Law of Reciprocal Deflections. It has been seen that much of the work in the solution of statically indeterminate structures by the method of consistent deformation involves the computation of deflections or rotations. This work can often be reduced or simplified by the application of the law of reciprocal deflections, which is derived below.

If, for any structure,

δ_{AB} = deflection at A in the direction AC (or rotation at A) due to a unit load applied at B in the direction BD

δ_{BA} = deflection at B in the direction BD due to a unit load applied at A in the direction AC (or due to a unit couple applied at A)

the law of reciprocal deflections can be stated as

$$\delta_{AB} = \delta_{BA} \tag{45}$$

The proof comes directly from the basic formula

$$(1)(\Delta) = \Sigma u\, dL \tag{5}$$

in the unit-load method.

Let u_A or $(dL)_A$ = total stress or total deformation in any one fiber due to a unit load applied at A in the direction AC (or due to a unit couple applied at A)

u_B or $(dL)_B$ = total stress or total deformation in any one fiber due to a unit load at B in the direction BD

Following the basic formula,

$$\delta_{AB} = \Sigma u_A (dL)_B \qquad \text{and} \qquad \delta_{BA} = \Sigma u_B (dL)_A \tag{46}$$

Since the total deformation must be directly proportional to the total stress,

$$(dL)_A = k(u_A) \qquad \text{and} \qquad (dL)_B = k(u_B) \tag{47}$$

in which k is a constant for each fiber. Substituting Eq. (47) into Eq. (46),

$$\delta_{AB} = \Sigma u_A (k u_B) = \Sigma k u_A u_B \qquad \text{and} \qquad \delta_{BA} = \Sigma u_B (k u_A) = \Sigma k u_A u_B$$

Thus

$$\delta_{AB} = \delta_{BA}$$

The application of the law of reciprocal deflections will be illustrated by the following examples, but the law will be used again and again in the subsequent part of this text.

Example 36. Determine all reactions and draw shear and moment diagrams for the beam of Fig. 73*a*.

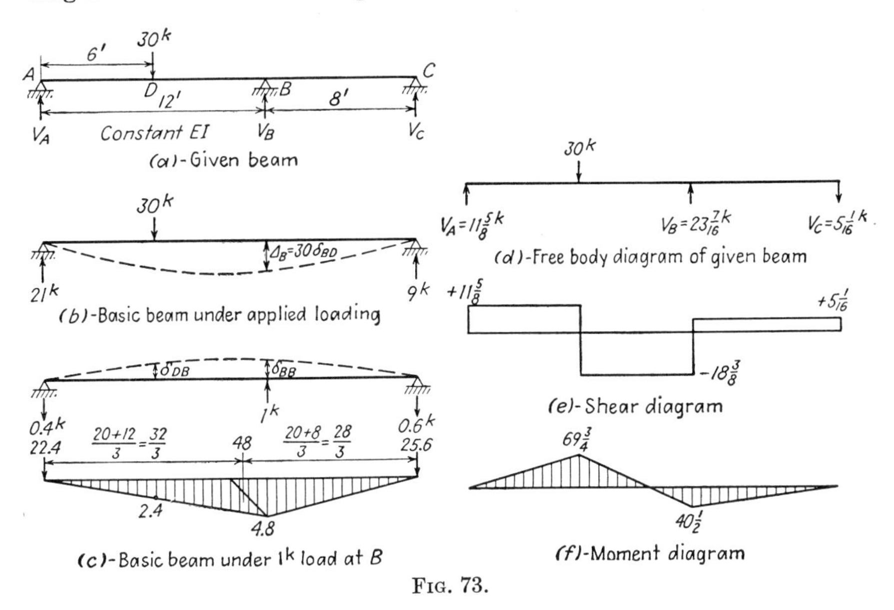

FIG. 73.

Solution. The beam as given has three supports, while two are enough for static equilibrium; so it is statically indeterminate to the first degree. The support at B is removed, and acting on the basic simple beam AC are the applied loading and the redundant reaction V_B. The condition required of geometry here is that the total deflection at B of the simple beam AC owing to the combined action of the 30-kip load and the redundant V_B is zero, or

$$\Delta_B = V_B \delta_{BB} \qquad V_B = \frac{\Delta_B}{\delta_{BB}}$$

By the conjugate-beam method (Fig. 73*c*)

$$EI\delta_{BB} = (22.4)(12) - (\tfrac{1}{2})(4.8)(12)(4) = 153.6 \text{ kip-ft}^3$$
$$EI\delta_{DB} = (22.4)(6) - (\tfrac{1}{2})(2.4)(6)(2) = 120 \text{ kip-ft}^3$$

By the law of reciprocal deflections,

$$EI\Delta_B = 30(EI\delta_{BD}) = 30(EI\delta_{DB}) = (30)(120) = 3{,}600 \text{ kip-ft}^3$$

Substituting,

$$V_B = \frac{EI\Delta_B}{EI\delta_{BB}} = \frac{3600}{153.6} = 23\tfrac{7}{16} \text{ kips upward}$$

By statics (Fig. 73*d*)

$$V_A = 21 - (0.4)(23\tfrac{7}{16}) = 11\tfrac{5}{8} \text{ kips upward}$$
$$V_C = 9 - (0.6)(23\tfrac{7}{16}) = -5\tfrac{1}{16} \text{ kips downward}$$

Alternate Solution. If the reaction at C is chosen as the redundant, the basic determinate beam is the overhanging beam ABC supported at A and B (Fig 74*a*). The condition required of geometry here is that the

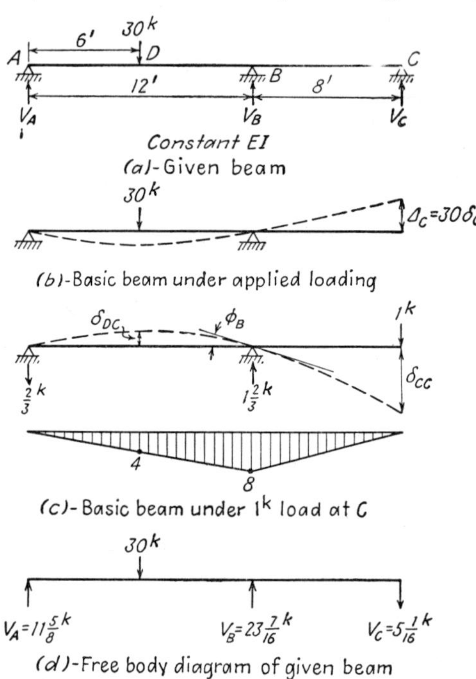

FIG. 74.

total deflection at C of the basic beam due to the applied load and the redundant V_C should equal zero, or

$$\Delta_C = V_C\delta_{CC} \qquad V_C = \frac{\Delta_C}{\delta_{CC}}$$

By the moment-area method (Fig. 74*c*)

$$EI\delta_{CC} = (8)(EI\phi_B) + \tfrac{1}{2}(8)(8)(\tfrac{16}{3})$$
$$= (8)(\tfrac{2}{3})\left(\frac{8 \times 12}{2}\right) + \tfrac{1}{2}(8)(8)(\tfrac{16}{3}) = \frac{1{,}280}{3} \text{ kip-ft}^3$$

By the conjugate-beam method (applied to AB of Fig. 74*c*)

$$EI\delta_{DC} = (\tfrac{1}{3})\left(\frac{8 \times 12}{2}\right)(6) - \left(\frac{4 \times 6}{2}\right)(2) = 72 \text{ kip-ft}^3$$

By the law of reciprocal deflections

$$EI\Delta_C = 30(EI\delta_{CD}) = 30(EI\delta_{DC}) = (30)(72) = 2{,}160 \text{ kip-ft}^3$$

$$V_C = \frac{\Delta_C}{\delta_C} = \frac{2{,}160}{1{,}280/3} = 5\tfrac{1}{16} \text{ kips downward}$$

By statics

$$V_A = 15 - (\tfrac{2}{3})(5\tfrac{1}{16}) = 11\tfrac{5}{8} \text{ kips upward}$$

$$V_B = 15 + (1\tfrac{2}{3})(5\tfrac{1}{16}) = 23\tfrac{7}{16} \text{ kips upward}$$

Example 37. Determine all reaction components and draw shear and moment diagrams for the beam of Fig. 75*a*.

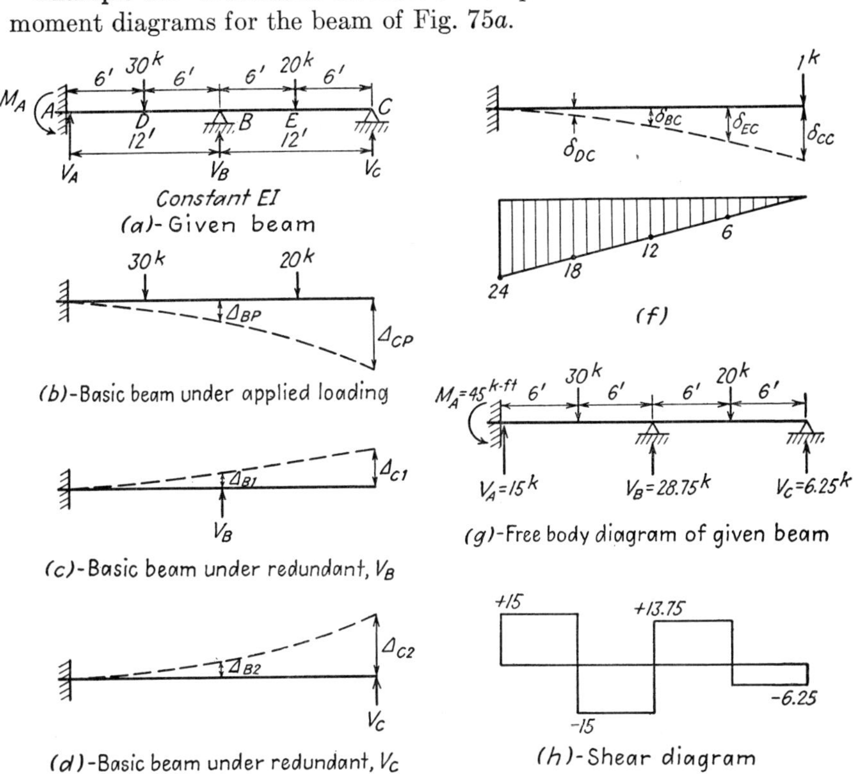

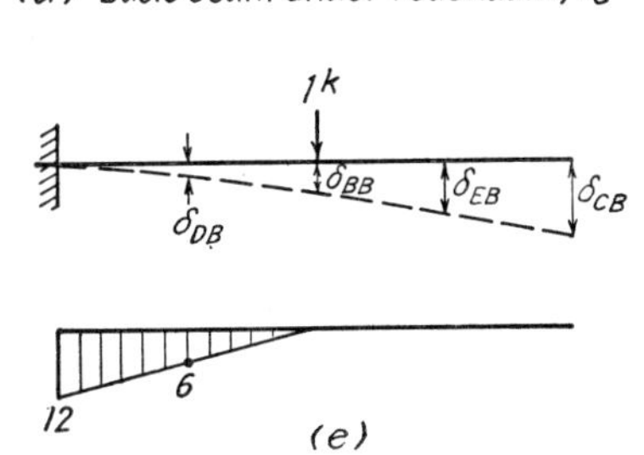

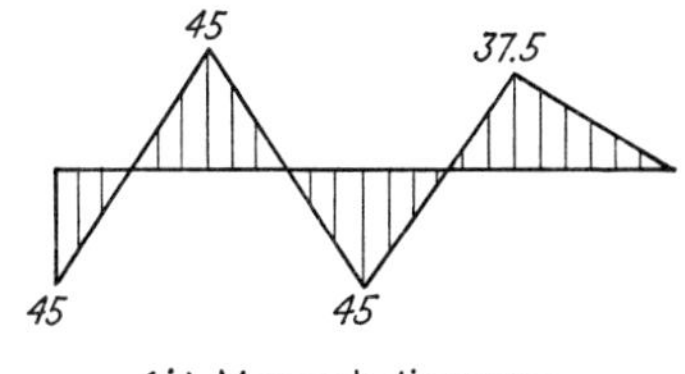

FIG. 75.

Solution. The given beam has four reaction components and is statically indeterminate to the second degree. By removing the supports at B and C the cantilever beam fixed at A and free at C is chosen as the basic determinate beam. Thus, from Fig. 75, it is seen that the sum of beams (b), (c), and (d) is the given beam. The conditions of geometry are

$$\Delta_{B1} + \Delta_{B2} = \Delta_{BP}$$
$$\Delta_{C1} + \Delta_{C2} = \Delta_{CP}$$

Before attempting to find the values of the Δ's in the above two equations, the following will first be done: The notation δ_{MN} will be defined as the deflection at M due to 1-kip load at N. If a 1-kip load acts at B of the basic beam (Fig. 75e), the deflections at the points D, B, E, and C are

$$\delta_{DB} = \left(\frac{12 \times 6}{2EI}\right)(4) + \left(\frac{6 \times 6}{2EI}\right)(2) = \frac{180}{EI}$$
$$\delta_{BB} = \left(\frac{12 \times 12}{2EI}\right)(8) = \frac{576}{EI}$$
$$\delta_{EB} = \left(\frac{12 \times 12}{2EI}\right)(14) = \frac{1{,}008}{EI}$$
$$\delta_{CB} = \left(\frac{12 \times 12}{2EI}\right)(20) = \frac{1{,}440}{EI}$$

If a 1-kip load acts at C of the basic beam (Fig. 75f), the deflections at the points D, B, E, and C are

$$\delta_{DC} = \left(\frac{24 \times 6}{2EI}\right)(4) + \left(\frac{18 \times 6}{2EI}\right)(2) = \frac{396}{EI}$$
$$\delta_{BC} = \left(\frac{24 \times 12}{2EI}\right)(8) + \left(\frac{12 \times 12}{2EI}\right)(4) = \frac{1{,}440}{EI}$$
$$\delta_{EC} = \left(\frac{24 \times 18}{2EI}\right)(12) + \left(\frac{6 \times 18}{2EI}\right)(6) = \frac{2{,}916}{EI}$$
$$\delta_{CC} = \left(\frac{24 \times 24}{2EI}\right)(16) = \frac{4{,}608}{EI}$$

By the law of reciprocal deflections

$$\Delta_{BP} = 30\delta_{BD} + 20\delta_{BE} = 30\delta_{DB} + 20\delta_{EB}$$
$$= 30\left(\frac{180}{EI}\right) + 20\left(\frac{1{,}008}{EI}\right) = \frac{25{,}560}{EI} \text{ downward}$$
$$\Delta_{CP} = 30\delta_{CD} + 20\delta_{CE} = 30\delta_{DC} + 20\delta_{EC}$$
$$= 30\left(\frac{396}{EI}\right) + 20\left(\frac{2{,}916}{EI}\right) = \frac{70{,}200}{EI} \text{ downward}$$

$$\Delta_{B1} = V_B\delta_{BB} = \frac{576}{EI} V_B \text{ upward}$$

$$\Delta_{B2} = V_C\delta_{BC} = \frac{1,440}{EI} V_C \text{ upward}$$

$$\Delta_{C1} = V_B\delta_{CB} = \frac{1,440}{EI} V_B \text{ upward}$$

$$\Delta_{C2} = V_C\delta_{CC} = \frac{4,608}{EI} V_C \text{ upward}$$

Substituting the above values into the two conditions of geometry,

$$576V_B + 1,440V_C = 25,560$$
$$1,440V_B + 4,608V_C = 70,200$$

Solving,

$$V_B = 28.75 \text{ kips upward}$$
$$V_C = 6.25 \text{ kips upward}$$

By statics

$$V_A = 15 \text{ kips upward}$$
$$M_A = 45 \text{ kip-ft counterclockwise}$$

The free-body, shear, and moment diagrams are as shown in Fig. 75*g* to *i*.

Alternate Solution. Suppose that the simple beam AC is chosen as the basic beam. The redundants are V_B and M_A. The conditions of

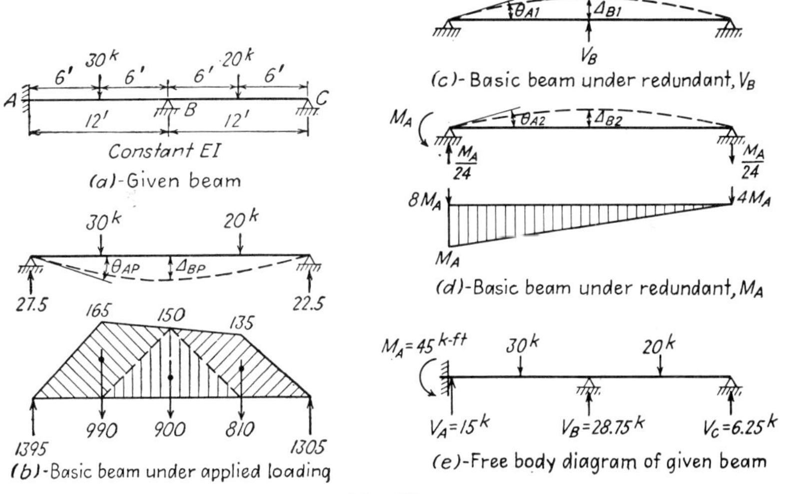

FIG. 76.

geometry are that the slope at A and the deflection at B should both be zero. Thus, referring to Fig. 76,

$$\theta_{A1} + \theta_{A2} = \theta_{AP}$$
$$\Delta_{B1} + \Delta_{B2} = \Delta_{BP}$$

By the conjugate-beam method

$$\theta_{AP} = \frac{810}{4EI} + \frac{900}{2EI} + \frac{3}{4}\frac{990}{EI} = \frac{1{,}395}{EI}$$

$$\Delta_{BP} = \frac{1}{EI}[(1{,}395)(12) - (990)(6) - (450)(2)] = \frac{9{,}900}{EI}$$

$$\Delta_{B1} = \frac{V_B(24)^3}{48EI} = \frac{288}{EI}V_B$$

$$\theta_{A1} = \frac{V_B(24)^2}{16EI} = \frac{36}{EI}V_B$$

$$\Delta_{B2} = \frac{1}{EI}\left[(4M_A)(12) - \frac{1}{2}\left(\frac{M_A}{2}\right)(12)(4)\right] = \frac{36}{EI}M_A$$

$$\theta_{A2} = \frac{8}{EI}M_A$$

Substituting the above values into the conditions of geometry,

$$36V_B + 8M_A = 1{,}395$$
$$288V_B + 36M_A = 9{,}900$$

Solving,

$$M_A = 45 \text{ kip-ft counterclockwise}$$
$$V_B = 28.75 \text{ kips upward}$$

By statics

$$V_A = 15 \text{ kips upward}$$
$$V_C = 6.25 \text{ kips upward}$$

EXERCISES

45 to 47. Determine all reactions by the method of consistent deformation, and draw shear and moment diagrams.

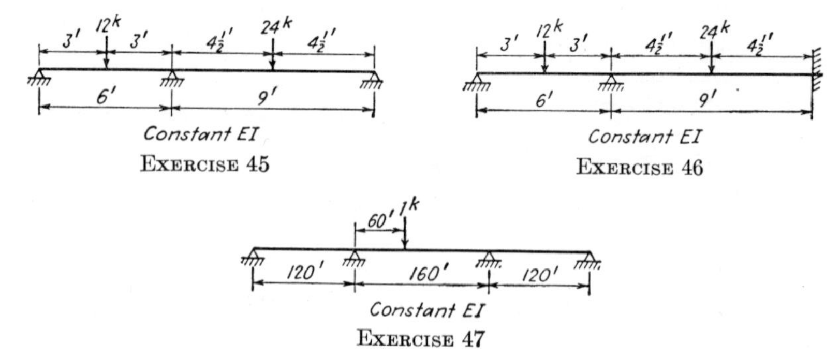

EXERCISE 45

EXERCISE 46

EXERCISE 47

25. The Theorem of Least Work. *Castigliano's Theorem of Least Work.* The redundant reaction components of a statically indeterminate structure are such as to make the total internal work a minimum.

The proof of the theorem of least work and its application can be explained first by the use of the example shown in Fig. 77. The beam shown in Fig. 77*a* is statically indeterminate to the first degree. It is obvious that the simple determinate beam of Fig. 77*b* subjected to the loads P_1 and P_2 and the redundant reaction V_B is equivalent to the given beam. The condition required of geometry with which V_B can be determined is that the deflection at B of the equivalent beam should be zero. This deflection, by Castigliano's theorem or by the partial derivative method, is $\Delta_B = \partial W/\partial V_B$; so the condition for determining V_B is

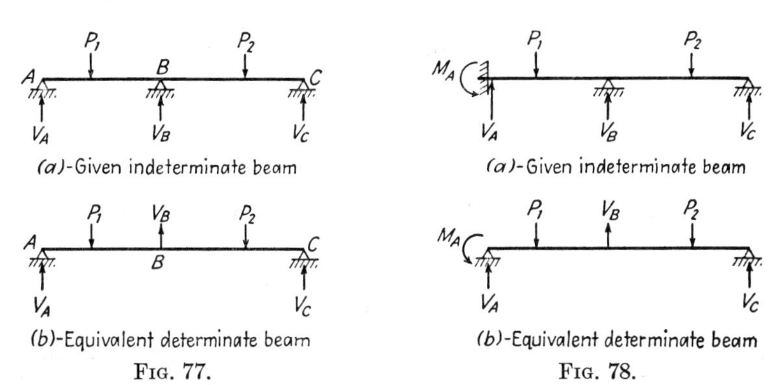

FIG. 77.

FIG. 78.

$\partial W/\partial V_B = 0$, or V_B is such as to make the total internal work a minimum. Strictly speaking, when the first derivative of a function at a certain value of the variable is zero, the function may be either a maximum or a minimum. A mathematical proof can be attempted for the present problem that the total work must be a minimum, but it seems unnecessary because it is physically inconceivable that the total work can be a maximum. In other words, when nature has its free choice, it will always tend to conserve work or energy. The same idea can be extended to the beam of Fig. 78*a*, the equivalent beam of which is shown in Fig. 78*b*. The conditions with which the redundants M_A and V_B can be determined are $\theta_A = \partial W/\partial M_A = 0$ and $\Delta_B = \partial W/\partial V_B = 0$; or the redundants M_A and V_B are such as to make the total internal work a minimum. The application of the theorem of least work will be illustrated by two examples. But it will be found that the solution is substantially identical with that of the method of consistent deformation. Thus in reality, the method of least work and the method of consistent deformation are identical methods, but the concept of consistent deformation may be considered preferable because it presents the physical picture rather than a philosophy.

Example 38. Determine all reaction components for the beam of Fig. 79*a* by the method of least work.

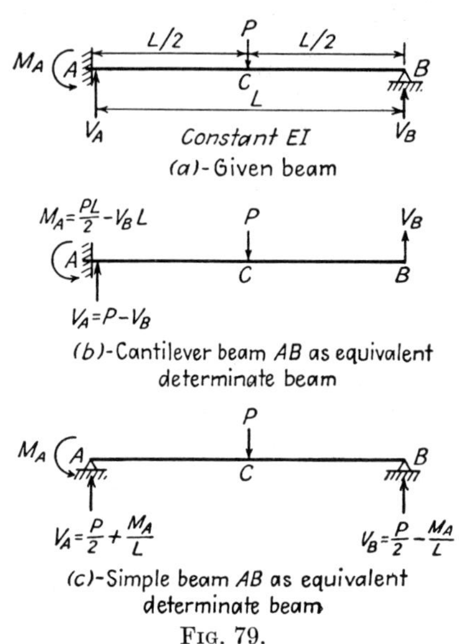

FIG. 79.

First Solution. Cantilever beam AB is chosen as the equivalent beam, with V_B as redundant (Fig. 79*b*).

TABLE 38-1

Portion of beam	AC	CB
Origin.	C	B
Limits.	$x = 0$ to $x = \frac{L}{2}$	$x = 0$ to $x = \frac{L}{2}$
M.	$V_B\left(x + \frac{L}{2}\right) - Px$	V_Bx

$$W = \frac{1}{2}\int \frac{M^2\,dx}{EI} = \frac{1}{2EI}\int_0^{L/2}\left[V_B\left(x+\frac{L}{2}\right) - Px\right]^2 dx + \frac{1}{2EI}\int_0^{L/2}(V_Bx)^2\,dx$$

$$\frac{\partial W}{\partial V_B} = \frac{1}{2EI}\int_0^{L/2} 2\left[V_B\left(x+\frac{L}{2}\right) - Px\right]\left(x+\frac{L}{2}\right) dx + \frac{1}{2EI}\int_0^{L/2} 2(V_Bx)(x)\,dx$$

$$= \frac{1}{EI}\left[\frac{V_B\left(x+\frac{L}{2}\right)^3}{3} - \frac{Px^3}{3} - \frac{PLx^2}{4}\right]_0^{L/2} + \frac{1}{EI}\left[\frac{V_Bx^3}{3}\right]_0^{L/2}$$

$$= \frac{1}{EI}\left[\frac{V_BL^3}{3} - \frac{5PL^3}{48}\right] = 0$$

or

$$V_B = \tfrac{5}{16}P$$

By statics

$$V_A = P - V_B = P - \tfrac{5}{16}P = \tfrac{11}{16}P$$

$$M_A = \frac{Pb}{2} - V_BL = \frac{PL}{2} - (\tfrac{5}{16}PL) = \tfrac{3}{16}PL$$

Second Solution. Simple beam AB is chosen as the equivalent beam, with M_A as redundant (Fig. 79c).

TABLE 38-2

Portion of beam	AC	CB
Origin..........	A	B
Limits..........	$x = 0$ to $x = \frac{L}{2}$	$x = 0$ to $x = \frac{L}{2}$
M..............	$-M_A + \left(\frac{P}{2} + \frac{M_A}{L}\right)x$	$\left(\frac{P}{2} - \frac{M_A}{L}\right)x$

$$W = \frac{1}{2}\int \frac{M^2\,dx}{EI} = \frac{1}{2EI}\int_0^{L/2}\left[-M_A + \left(\frac{P}{2} + \frac{M_A}{L}\right)x\right]^2 dx + \frac{1}{2EI}\int_0^{L/2}\left[\left(\frac{P}{2} - \frac{M_A}{L}\right)x\right]^2 dx$$

$$\frac{\partial W}{\partial M_A} = \frac{1}{2EI}\int_0^{L/2} 2\left[-M_A + \left(\frac{P}{2} + \frac{M_A}{L}\right)x\right]\left(-1 + \frac{x}{L}\right)dx + \frac{1}{2EI}\int_0^{L/2} 2\left[\left(\frac{P}{2} - \frac{M_A}{L}\right)x\right]\left(-\frac{x}{L}\right)dx$$

$$= \frac{1}{EI}\int_0^{L/2}\left[M_A - \frac{M_Ax}{L} - \frac{P}{2}x - \frac{M_A}{L}x + \left(\frac{P}{2} + \frac{M_A}{L}\right)\frac{x^2}{L}\right]dx + \frac{1}{EI}\int_0^{L/2} -\left(\frac{P}{2} - \frac{M_A}{L}\right)\frac{x^2}{L}\,dx$$

$$= \frac{1}{EI}\left[\frac{M_AL}{3} - \frac{PL^2}{16}\right] = 0$$

$$M_A = \frac{3PL}{16}$$

By statics

$$V_A = \frac{P}{2} + \frac{M_A}{L} = \frac{P}{2} + \tfrac{3}{16}P = \tfrac{11}{16}P$$

$$V_B = \frac{P}{2} - \frac{M_A}{L} = \frac{P}{2} - \tfrac{3}{16}P = \tfrac{5}{16}P$$

Example 39. Determine all reaction components for the beam of Fig. 80*a* by the method of least work.

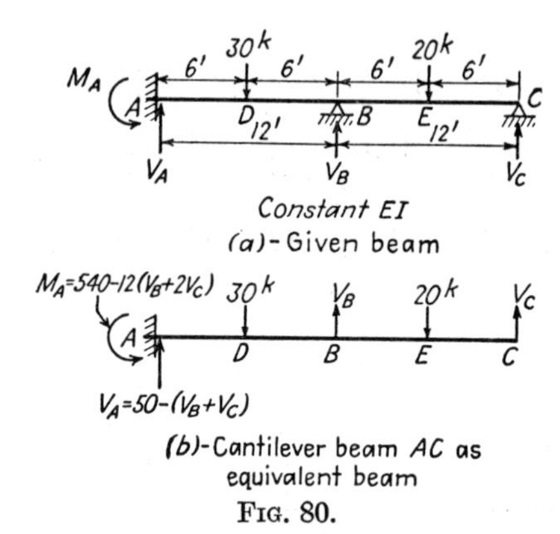

FIG. 80.

Solution. The cantilever beam fixed at A and free at C is chosen as the equivalent beam, with V_B and V_C as the redundants (Fig. 80*b*).

TABLE 39-1

Portion of beam	AD	DB	BE	EC
Origin	D	B	E	C
Limits	$x = 0$ to $x = 6$	$x = 0$ to $x = 6$	$x = 0$ to $x = 6$	$x = 0$ to $x = 6$
M	$V_C(x + 18) + V_B(x + 6) - 20(x + 12) - 30x$	$V_C(x + 12) - 20(x + 6) + V_Bx$	$V_C(x + 6) - 20x$	$+V_Cx$

$$W = \frac{1}{2EI}\int M^2\,dx$$

$$= \frac{1}{2EI}\int_0^6 \{[V_C(x + 18) + V_B(x + 6) - 20(x + 12) - 30x]^2$$

$$+ [V_C(x + 12) - 20(x + 6) + V_Bx]^2 + [V_C(x + 6) - 20x]^2 + (V_Cx)^2\}\,dx$$

$$\frac{\partial W}{\partial V_B} = \frac{2}{2EI}\int_0^6 \{[(V_B + V_C - 50)x + (6V_B + 18V_C - 240)](x + 6)$$

$$+ [(V_B + V_C - 20)x + (12V_C - 120)](x)\}\,dx$$

$$= \frac{1}{EI}\int_0^6 \{(2V_B + 2V_C - 70)x^2 + (12V_B + 36V_C - 660)x + (36V_B + 108V_C - 1{,}440)\}\, dx$$

$$= \frac{1}{EI}[576V_B + 1{,}440V_C - 25{,}560] = 0$$

or, on simplifying,

$$576V_B + 1{,}440V_C = 25{,}560$$

$$\frac{\partial W}{\partial V_C} = \frac{2}{2EI}\int_0^6 \{[(V_B + V_C - 50)x + (6V_B + 18V_C - 240)](x + 18) + [(V_B + V_C - 20)x + (12V_C - 120)](x + 12) + [(V_C - 20)x + 6V_C](x + 6) + [V_C x](x)\}\, dx$$

$$= \frac{1}{EI}\int_0^6 \{(2V_B + 4V_C - 90)x^2 + (36V_B + 72V_C - 1{,}620)x + (108V_B + 504V_C - 5{,}760)\}\, dx$$

$$= \frac{1}{EI}[1{,}440V_B + 4{,}608V_C - 70{,}200] = 0$$

or

$$1{,}440V_B + 4{,}608V_C = 70{,}200$$

Solving the two equations as derived from $\partial W/\partial V_B = 0$ and $\partial W/\partial V_C = 0$,

$$V_B = 28.75 \text{ kips upward}$$
$$V_C = 6.25 \text{ kips upward}$$

By statics

$$V_A = 50 - (V_B + V_C) = 15 \text{ kips upward}$$
$$M_A = 540 - 12(V_B + 2V_C) = 45 \text{ kip-ft counterclockwise}$$

EXERCISES

48. Determine all reaction components by the method of least work.
49. Determine all reactions by the method of least work.

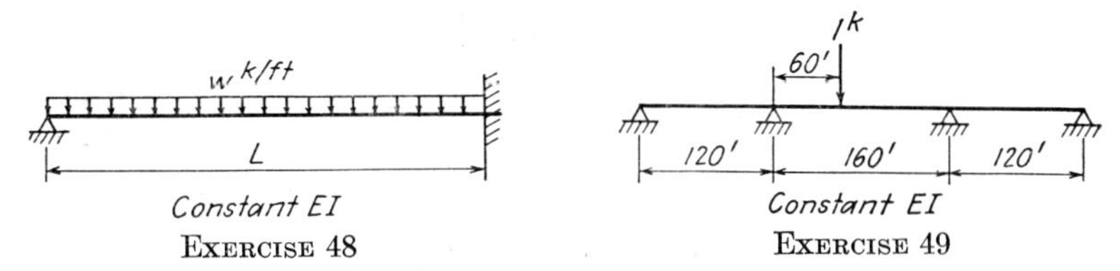

EXERCISE 48 EXERCISE 49

26. Induced Reactions on Statically Indeterminate Beams Due to Yielding of Supports. Either the vertical settlement of a simple support or the rotational slip of a fixed support of a statically indeterminate beam will alone induce reactions and, thereby, shears and moments in the beam in addition to those due to the loading. If more than one settlement or yielding is anticipated, it is better to determine the induced reactions due to one factor at a time and then combine the effects due to separate causes. The procedure of determining such induced reactions by the method of consistent deformation is as follows: First, a basic determinate structure is chosen by removing the excess supports *including* the yielding support and replacing them by redundant reaction components. The condition of geometry at the yielding support will be that the deflection or rotation there should be equal to the predicted amount of yielding, the conditions at all other redundants requiring zero deflection or rotation being the same as before. Thus the induced reactions and, thereby, the shears and moments in the beam can be determined.

Example 40. Determine all induced reaction components due to a vertical settlement of $\frac{1}{8}$ in. at the support B (Fig. 81a).

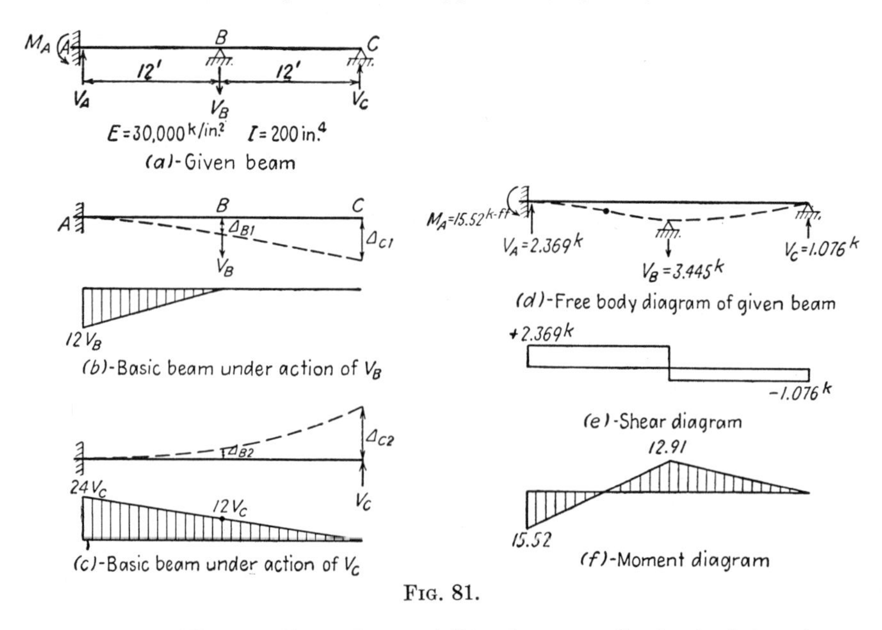

FIG. 81.

Solution. The cantilever beam AC is chosen as the basic determinate beam, with V_B and V_C as redundants. V_B and V_C are the only forces acting on this cantilever. The conditions of geometry are

$$\Delta_{B1} - \Delta_{B2} = \tfrac{1}{8} \text{ in.}$$
$$\Delta_{C1} - \Delta_{C2} = 0$$

Applying the moment-area method to Fig. 81b and c,

$$\Delta_{B1} = \frac{72V_B}{EI}(8) = 0.1659V_B \text{ in.}$$

$$\Delta_{C1} = \frac{72V_B}{EI}(20) = 0.4147V_B \text{ in.}$$

$$\Delta_{B2} = \frac{144V_C}{EI}(8) + \frac{72V_C}{EI}(4) = 0.4147V_C \text{ in.}$$

$$\Delta_{C2} = \frac{288V_C}{EI}(16) = 1.3271V_C \text{ in.}$$

Substituting the above values into the conditions of geometry,

$$0.1659V_B - 0.4147V_C = \tfrac{1}{8}$$
$$0.4147V_B - 1.3271V_C = 0$$

Solving,

$$V_C = 1.076 \text{ kips upward}$$
$$V_B = 3.445 \text{ kips downward}$$

By statics

$$V_A = 2.369 \text{ kips upward}$$
$$M_A = 15.52 \text{ kip-ft counterclockwise}$$

Alternate Solution. If the simple beam AB is chosen as the basic determinate beam, with V_B and M_A as redundants (Fig. 82), the condi-

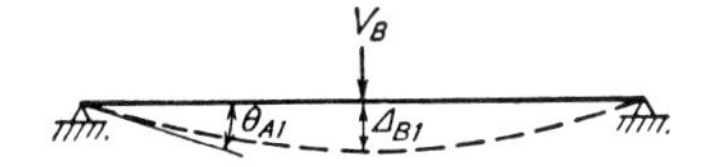

(a)-Basic beam under action of V_B

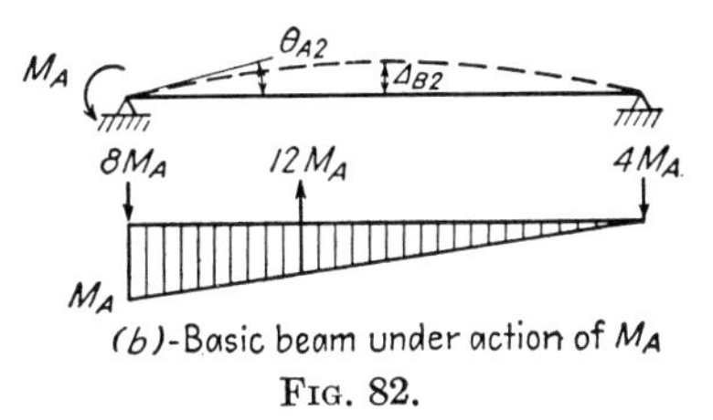

(b)-Basic beam under action of M_A

FIG. 82.

tions of geometry are

$$\theta_{A1} = \theta_{A2}$$
$$\Delta_{B1} - \Delta_{B2} = \tfrac{1}{8}$$

By the conjugate-beam method

$$\theta_{A1} = \frac{V_B L^2}{16EI} = \frac{36V_B}{EI}$$

$$\theta_{A2} = \frac{M_A L}{3EI} = \frac{8M_A}{EI}$$

$$\Delta_{B1} = \frac{V_B L^3}{48EI} = \frac{288V_B}{EI} = 0.082944V_B$$

$$\Delta_{B2} = \frac{4M_A}{EI}(12) - \frac{3M_A}{EI}(4) = \frac{36M_A}{EI} = 0.010368M_A$$

Substituting the above values into the conditions of geometry,

$$\frac{36V_B}{EI} = \frac{8M_A}{EI}$$

$$0.082944V_B - 0.010368M_A = \tfrac{1}{8} \text{ in.}$$

Solving,

$$V_B = 3.445 \text{ kips downward}$$
$$M_A = 15.52 \text{ kip-ft counterclockwise}$$

By statics

$$V_A = 2.369 \text{ kips upward}$$
$$V_C = 1.076 \text{ kips upward}$$

EXERCISE

50. Determine all induced reactions due to a vertical settlement of $\frac{1}{2}$ in. at the support B.

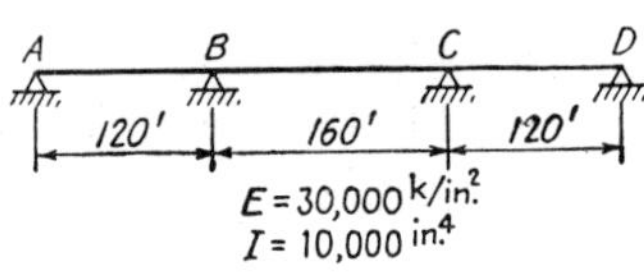

Exercise 50

27. Analysis of Statically Indeterminate Frames by the Method of Consistent Deformation. Since the direct stresses in the members of rigid frames are usually small when compared with bending stresses, the changes in the lengths of members owing to direct stresses can be and ordinarily are neglected. The conditions of consistent deformation for solving statically indeterminate frames therefore relate to the deformed frame resulting from bending stresses only. Although it has been found that there is little effect on the values of the redundants even if the changes in the length of the members were considered in making the equations of consistent deformation, this does not mean that there are no direct stresses and these direct stresses must be considered in the design of the members.

Example 41. Analyze the two-hinged rigid frame shown in Fig. 83*a* by the method of consistent deformation.

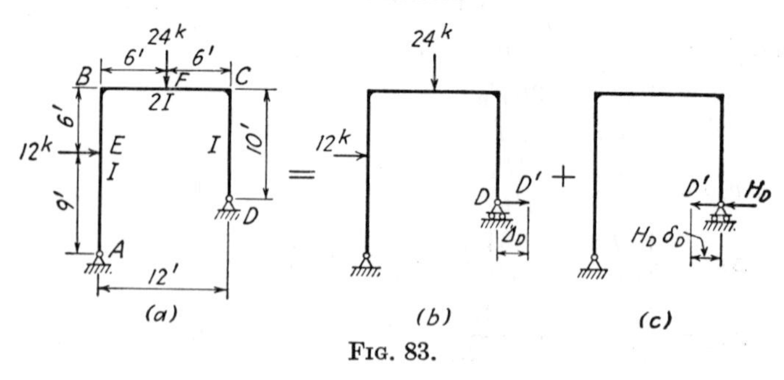

Fig. 83.

Solution. There are four reaction components to be determined; therefore the given frame is statically indeterminate to the first degree. Any of the four reaction components can be chosen as the redundant. In this example two solutions wherein H_D and then V_A are used as the redundant will be shown.

First Solution. When H_D is used as the redundant, the basic determinate structure is hinged at A and supported on horizontal rollers at D (Fig. 83b). Let δ_D be the horizontal deflection at D due to a 1-kip load applied at D. The condition of geometry with which H_D can be determined is that the deflection Δ_D to the right owing to the applied loading (Fig. 83b) must be equal to the deflection $H_D\delta_D$ to the left owing to the redundant H_D (Fig. 83c) since the total deflection at D is zero (Fig. 83a). Either the moment-area method or the unit-load method can be used to determine Δ_D and δ_D. The unit-load method is here shown. By referring to Fig. 84a and b, Table 41-1 is made for the computation of Δ_D.

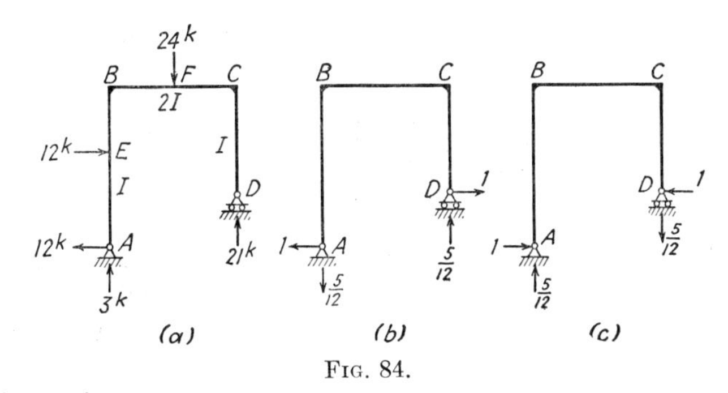

FIG. 84.

TABLE 41-1

Portion of frame	AE	EB	BF	FC	CD
Origin...	A	E	B	C	D
Limits...	$x = 0$ to $x = 9$	$x = 0$ to $x = 6$	$x = 0$ to $x = 6$	$x = 0$ to $x = 6$	$x = 0$ to $x = 10$
M (Fig. 84a)...	$12x$	108	$108 + 3x$	$21x$	0
m (Fig. 84b)...	x	$x + 9$	$15 - \frac{5}{12}x$	$10 + \frac{5}{12}x$	x
I........	I	I	$2I$	$2I$	I

$$EI\Delta_D = \int_0^9 (12x)(x)\,dx + \int_0^6 (108)(x+9)\,dx$$

$$+ \frac{1}{2}\int_0^6 (108 + 3x)(15 - \tfrac{5}{12}x)\,dx + \frac{1}{2}\int_0^6 (21x)(10 + \tfrac{5}{12}x)\,dx$$

$$= 2{,}916 + 7{,}776 + 4{,}815 + 2{,}205$$

$$= 17{,}712 \text{ kip-ft}^3$$

By referring to Fig. 84c Table 41-2 is made for the computation of δ_D.

TABLE 41-2

Portion of frame	AB	BC	CD
Origin	A	B	D
Limits	$x = 0$ to $x = 15$	$x = 0$ to $x = 12$	$x = 0$ to $x = 10$
M or m (Fig. 84c)	$-x$	$-15 + \frac{5}{12}x$	$-x$
I	I	$2I$	I

$$EI\delta_D = \int_0^{15} (-x)^2\,dx + \frac{1}{2}\int_0^{12} (-15 + \tfrac{5}{12}x)^2\,dx + \int_0^{10} (-x)^2\,dx$$
$$= 1{,}125 + 950 + \frac{1{,}000}{3} = \frac{7{,}225}{3}$$

From the condition of geometry

$$H_D = \frac{EI\Delta_D}{EI\delta_D} = \frac{17{,}712}{7{,}225/3} = 7.3545 \text{ kips to the left}$$

By statics

$$H_A = 4.6455 \text{ kips to the left}$$
$$V_A = 6.0643 \text{ kips upward}$$
$$V_D = 17.9357 \text{ kips upward}$$

The reaction components are here computed to an unusual degree of accuracy so that they may be checked with the results of the second solution. In practical problems the usual accuracy of three or four significant figures is all that is necessary.

Second Solution. When V_A is used as the redundant, the basic determinate structure is hinged at D and supported on vertical rollers at A

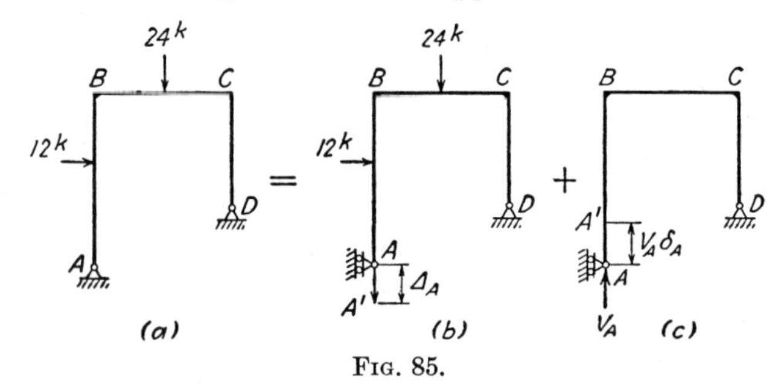

FIG. 85.

(Fig. 85b). It is assumed here that a roller support can exert either a push or a pull in a direction perpendicular to the supporting surface. The condition of geometry is that the deflection Δ_A in the downward direction due to the applied loading (Fig. 85b) must be equal to the

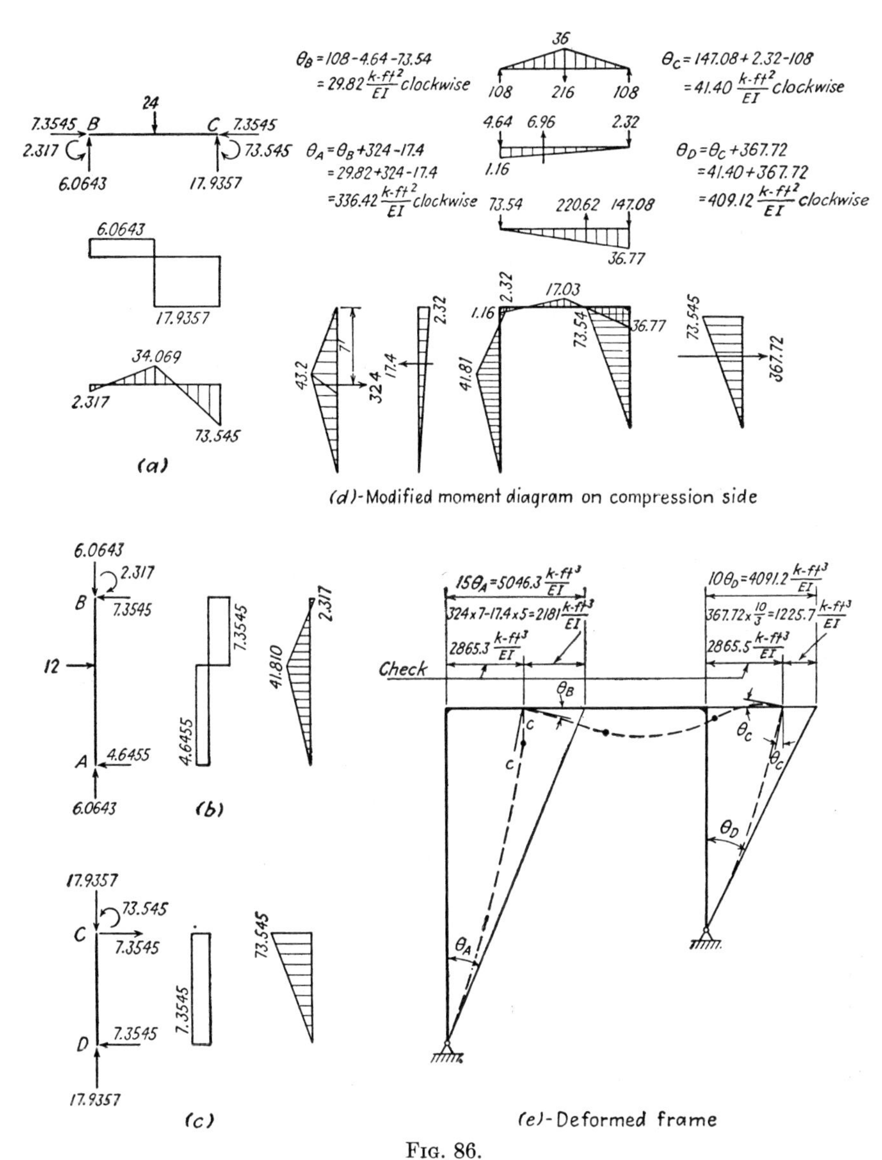

FIG. 86.

deflection $V_A\delta_A$ in the upward direction due to the action of V_A, δ_A being the vertical deflection at A due to a 1-kip vertical load at A (Fig. 85c). The unit-load method is used to determine Δ_A and δ_A, the values of which are

$$EI\Delta_A = 84{,}124.8 \text{ kip-ft}^3$$
$$EI\delta_A = 13{,}872 \text{ kip-ft}^3$$

From the condition of geometry

$$\Delta_A = V_A \delta_A$$

or,

$$V_A = \frac{EI\Delta_A}{EI\delta_A} = \frac{84{,}124.8}{13{,}872} = 6.0643 \text{ kips upward}$$

By statics

$$V_D = 17.9357 \text{ kips upward}$$
$$H_A = 4.6455 \text{ kips to the left}$$
$$V_A = 6.0643 \text{ kips upward}$$

Now all four reaction components on this two-hinged rigid frame are known. The free-body, shear, and moment diagrams for all members are shown in Fig. 86*a* to *c*. In plotting shear and moment diagrams, the usual sign convention for beams is observed, that is, shear is positive if the left side is being lifted relative to the right side, and moment is positive if it causes compression on the top of the section; vertical members are considered as horizontal members by looking at them from the right side.

It will be very interesting to review the geometry of the deformed frame by computing the horizontal deflection, vertical deflection, and rotation at every joint. After a qualitative sketch of the deformed frame is made by inspection, the moment-area method is best suited for determining all deflection components. This is shown diagrammatically in Fig. 86. It is to be noted that, in applying the moment-area method, the properties of the moment diagrams can be more easily found by separating the "final" moment diagram for each member into several component triangles (sometimes also parabolas), as shown in Fig. 86*d*. For instance, the moment diagram for member BC is the sum of three moment diagrams due to (1) the central load, (2) the left end moment, and (3) the right end moment. All moment diagrams are plotted on the compression side.

Example 42. Analyze the rigid frame with two fixed supports shown in Fig. 87*a* by the method of consistent deformation.

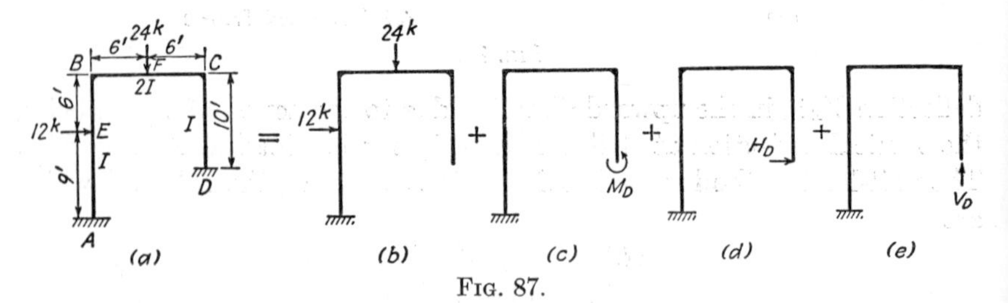

FIG. 87.

Solution. The given frame is statically indeterminate to the third degree. The frame fixed at A and free at D is chosen as the basic determinate structure, on which M_D, H_D, and V_D act as redundants (Fig. 87).

Because the rotation of the tangent and the horizontal and vertical deflections at D of the given structure (Fig. 87*a*) must all be zero, it follows that the sum of the rotations of tangents at D of the four structures shown in Fig. 87*b* to *e* must be zero; also, the sum of the horizontal deflections and the sum of the vertical deflections. The conditions of geometry are, in this case,

$$\Delta_{\theta P} + M_D\delta_{\theta M} + H_D\delta_{\theta H} + V_D\delta_{\theta V} = 0$$
$$\Delta_{HP} + M_D\delta_{HM} + H_D\delta_{HH} + V_D\delta_{HV} = 0$$
$$\Delta_{VP} + M_D\delta_{VM} + H_D\delta_{VH} + V_D\delta_{VV} = 0$$

where $\Delta_{\theta P}$, Δ_{HP}, and Δ_{VP} are the rotation, horizontal deflection, and vertical deflection at D of the basic structure due to the applied loading; $\delta_{\theta M}$, δ_{HM}, and δ_{VM} are the same quantities due to the action of a counterclockwise moment of 1 kip-ft at D; $\delta_{\theta H}$, δ_{HH}, and δ_{VH} are due to the action of a 1-kip load to the right at D; $\delta_{\theta V}$, δ_{HV}, and δ_{VV} are due to action of a 1-kip load upward at D. In inserting the numerical values of the Δ's or δ's in the above three equations, it is important to consider *counterclockwise* rotation of the tangent, horizontal deflection to the *right*, and vertical deflection *upward* as positive quantities and those in the opposite directions as negative quantities.

Before M_D, H_D, and V_D can be solved, it is necessary to evaluate the 12 values of the Δ's or δ's which appear in the three conditions of geometry. The following list of answers are obtained by the moment-area method (refer to Fig. 88):

$$\Delta_{\theta P} = -2{,}862 \frac{\text{kip-ft}^2}{EI} \text{ clockwise}$$

$$\Delta_{HP} = -6{,}588 \frac{\text{kip-ft}^3}{EI} \text{ to the left}$$

$$\Delta_{VP} = -33{,}912 \frac{\text{kip-ft}^3}{EI} \text{ downward}$$

$$\delta_{\theta M} = +31 \frac{\text{kip-ft}^2}{EI} \text{ counterclockwise}$$

$$\delta_{HM} = +147.5 \frac{\text{kip-ft}^3}{EI} \text{ to the right}$$

$$\delta_{VM} = +216 \frac{\text{kip-ft}^3}{EI} \text{ upward}$$

$$\delta_{\theta H} = +147.5 \frac{\text{kip-ft}^2}{EI} \text{ counterclockwise}$$

$$\delta_{HH} = +1{,}308.3 \frac{\text{kip-ft}^3}{EI} \text{ to the right}$$

$$\delta_{VH} = +810 \frac{\text{kip-ft}^3}{EI} \text{ upward}$$

$$\delta_{\theta V} = +216 \frac{\text{kip-ft}^2}{EI} \text{ counterclockwise}$$

$$\delta_{HV} = +810 \frac{\text{kip-ft}^3}{EI} \text{ to the right}$$

$$\delta_{VV} = +2{,}448 \frac{\text{kip-ft}^3}{EI} \text{ upward}$$

It is to be noted that by the law of reciprocal deflections

$$\delta_{\theta H} = \delta_{HM} = +147.5$$
$$\delta_{\theta V} = \delta_{VM} = +216$$
$$\delta_{HV} = \delta_{VH} = +810$$

Substituting the above values of Δ's and δ's into the conditions of geometry,

$$-2{,}862 + 31M_D + 147.5H_D + 216V_D = 0$$
$$-6{,}588 + 147.5M_D + 1{,}308.3H_D + 810V_D = 0$$
$$-33{,}912 + 216M_D + 810H_D + 2{,}448V_D = 0$$

Solving the three simultaneous equations,

$$M_D = +32.632 \text{ kip-ft counterclockwise}$$
$$H_D = -6.838 \text{ kips to the left}$$
$$V_D = +13.236 \text{ kips upward}$$

Putting the above values of M_D, H_D, and V_D back on the given frame and solving by the three equations of statics,

$$M_A = 26.342 \text{ kip-ft counterclockwise}$$
$$H_A = 5.162 \text{ kips to the left}$$
$$V_A = 10.764 \text{ kips upward}$$

The free-body, shear, and moment diagrams for all members in the given frame are drawn as Fig. 89*a* to *c*.

The deformed frame as shown in Fig. 89*e* is found by the moment-area method. The conditions of consistent deformation are obviously satisfied; so are the conditions of statics. When the analyzed structure satisfies both *statics* and *geometry*, the correctness of the solution is assured.

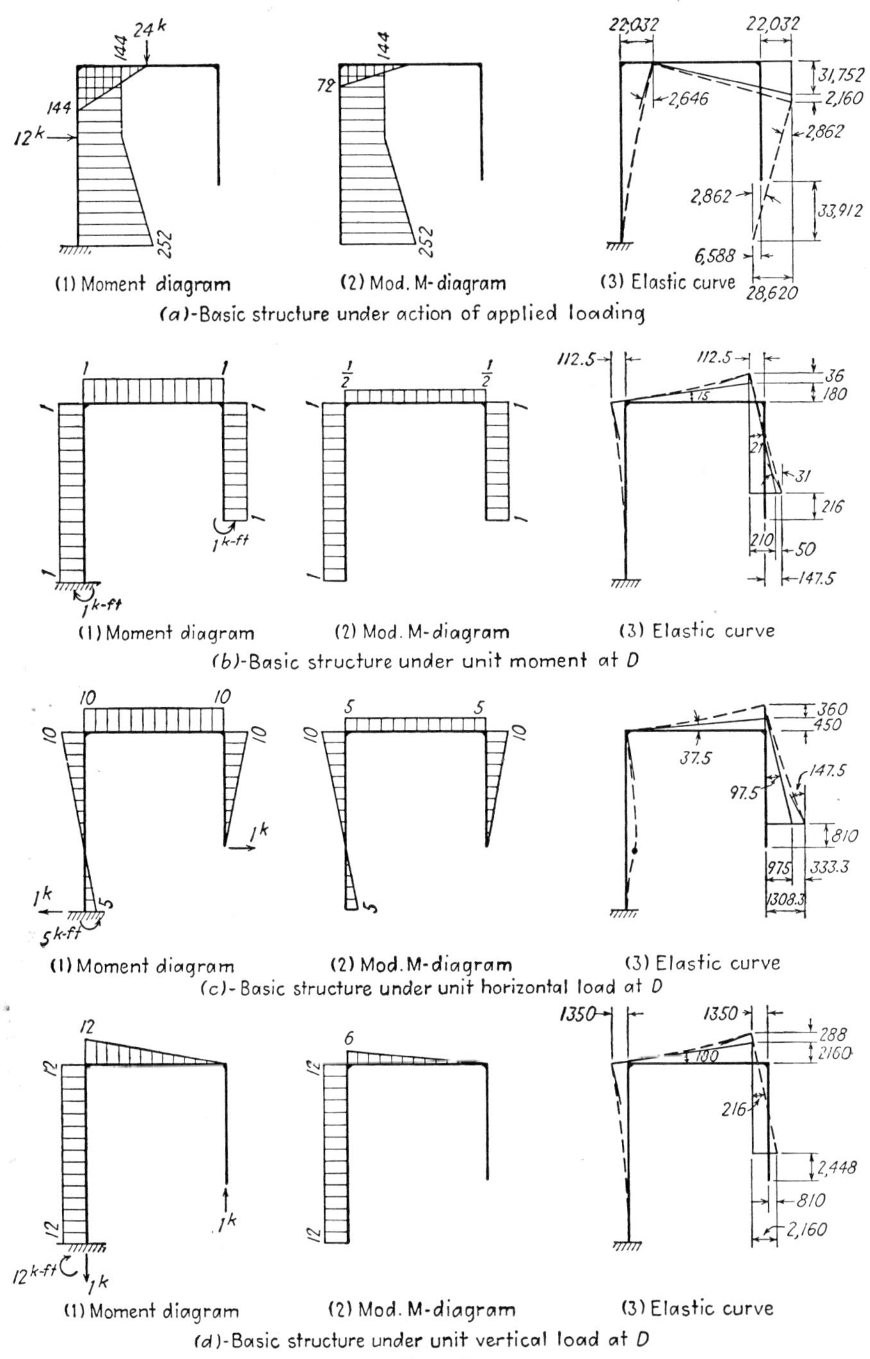
24^k
144
144
12^k
252
(1) Moment diagram
144
72
252
(2) Mod. M-diagram
22,032
22,032
31,752
2,646
2,160
2,862
2,862
33,912
6,588
(3) Elastic curve
28,620
(a)-Basic structure under action of applied loading
1^k-ft
1^k-ft
(1) Moment diagram
1/2
1/2
(2) Mod. M-diagram
112.5
112.5
36
180
15
21
31
216
210
50
147.5
(3) Elastic curve
(b)-Basic structure under unit moment at D
10
10
10
10
1^k
1^k
5
5^k-ft
(1) Moment diagram
5
5
10
10
5
(2) Mod. M-diagram
360
450
37.5
147.5
97.5
810
975
333.3
1308.3
(3) Elastic curve
(c)-Basic structure under unit horizontal load at D
12
12
12
1^k
12^k-ft
1^k
(1) Moment diagram
6
12
12
(2) Mod. M-diagram
1350
1350
288
2160
100
216
2,448
810
2,160
(3) Elastic curve
(d)-Basic structure under unit vertical load at D

Fig. 88.

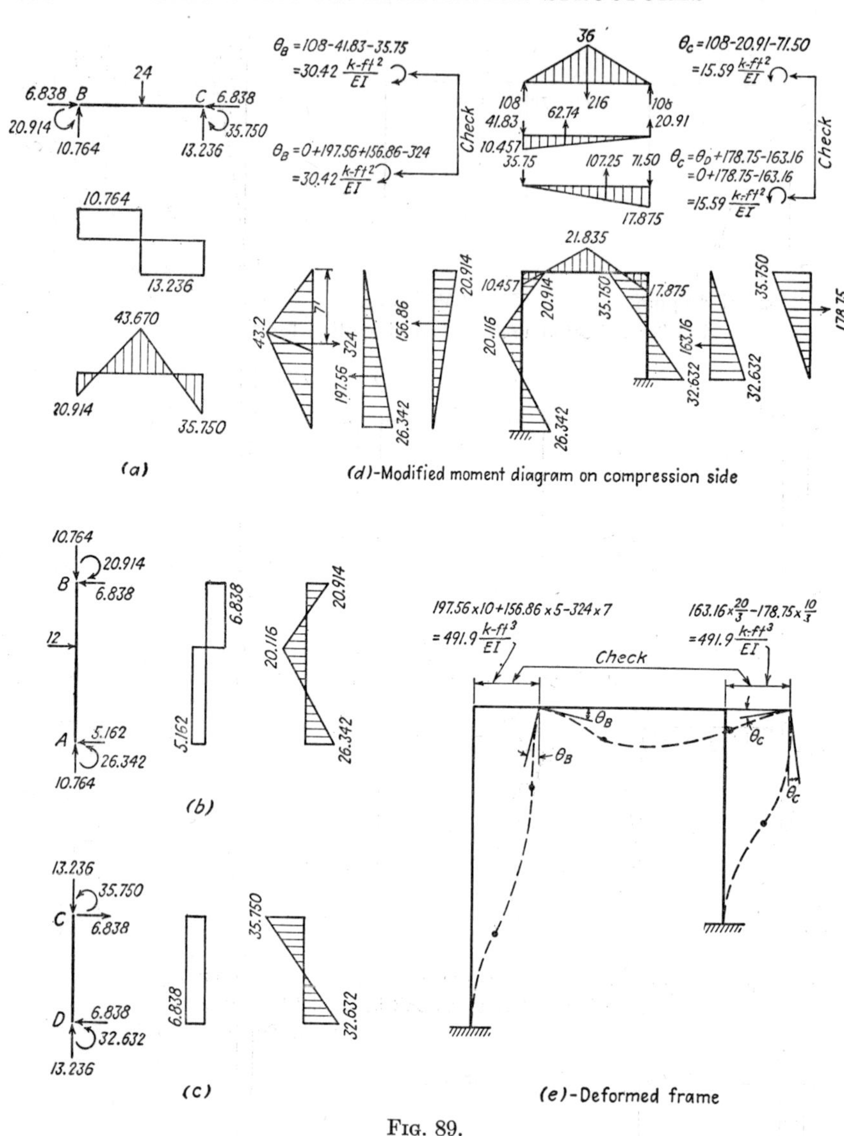
(a)
(b)
(c)
(d)-Modified moment diagram on compression side
(e)-Deformed frame
Check

FIG. 89.

EXERCISES

51 and 52. Analyze the rigid frame with two hinged supports by the method of consistent deformation. Draw shear and moment diagrams. Sketch the deformed structure.

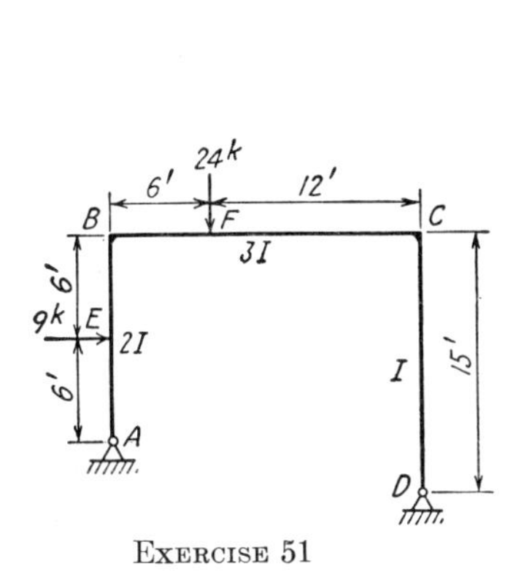

Exercise 51

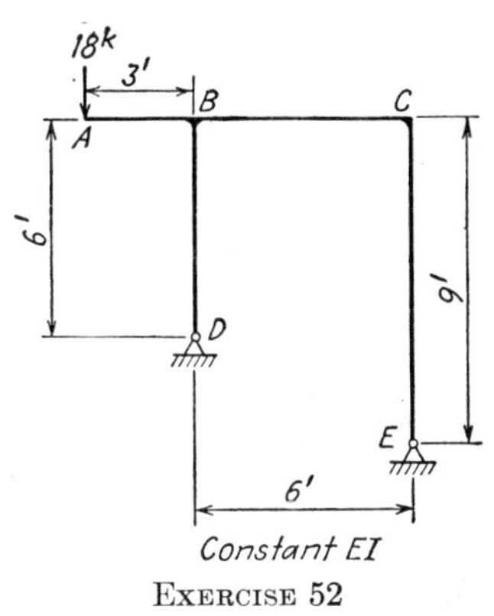

Exercise 52

53 and 54. Analyze the rigid frame with two fixed supports by the method of consistent deformation. Draw shear and moment diagrams. Sketch the deformed structure.

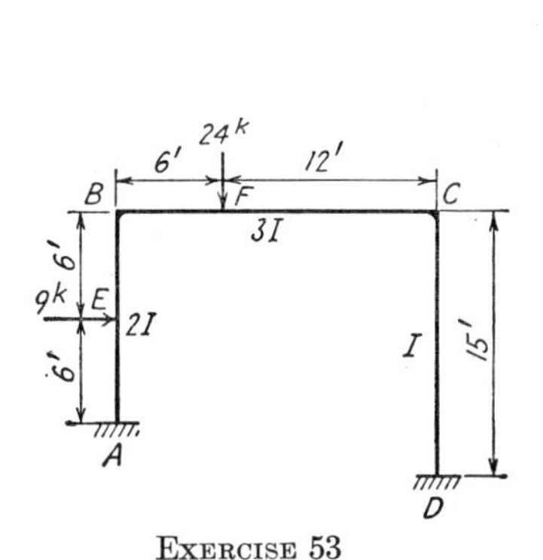

Exercise 53

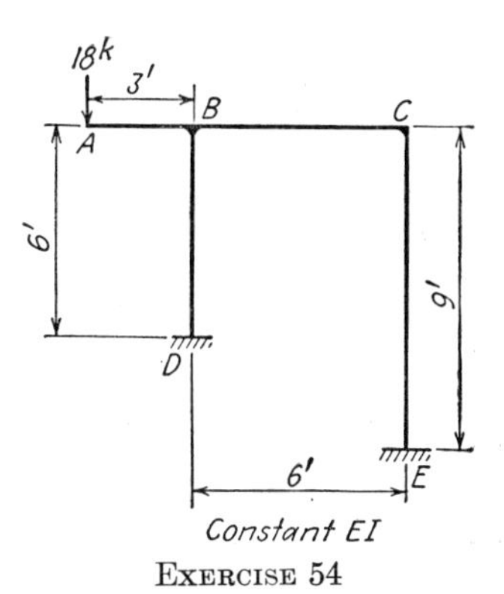

Exercise 54

28. Induced Reactions on Statically Indeterminate Frames Due to Yielding of Supports. The yielding of supports of statically indeterminate frames, either as a displacement or as a rotational slip (for fixed supports only), always induces reactions and thereby direct stresses, shears, and moments in all members of the frame. The procedure of finding the induced reactions is first to derive a basic determinate structure by removing the excess supports, or redundant reaction components including those which are subjected to yielding, and treating the redundant reaction components as loads on the basic structure. The conditions of geometry include those at the yielding supports where the deflection or rotation should be equal to the predicted amount of yielding and those at the other unyielding redundant supports where the deflection or rotation should still be zero. When the redundants are found, they are put back on the basic structure as loads and the remaining reaction components are solved by the equations of statics. The variation in direct stresses, shears, and moments in all members can then be found as usual.

Example 43. Determine all reaction components induced to act on the frames of Fig. 90*a* by the rotational slip of 0.002 rad clockwise of joint *D* and a vertical settlement of 0.45 in. at joint *D*. $E = 30{,}000$ kips/in.² $I = 800$ in.⁴

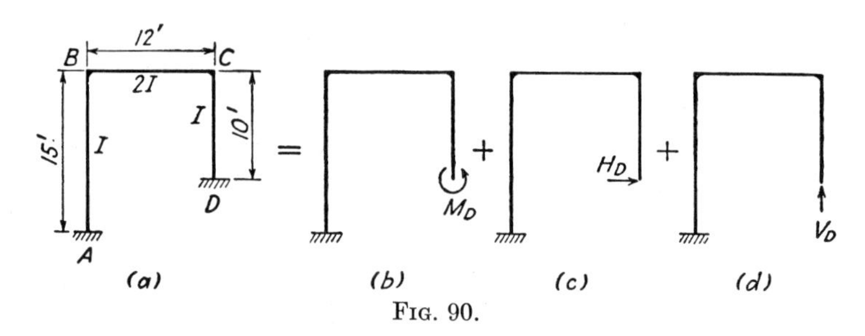

FIG. 90.

Solution. The frame fixed at *A* and free at *D* is chosen as the basic determinate structure on which only the redundant reaction components M_D, H_D, and V_D are acting (Fig. 90).

From Example 42, M_D causes

1. A rotation of $31M_D \dfrac{\text{kip-ft}^2}{EI}$ counterclockwise at *D*.

2. A horizontal deflection of $147.5M_D \dfrac{\text{kip-ft}^3}{EI}$ to the right at *D*.

3. A vertical deflection of $216M_D \dfrac{\text{kip-ft}^3}{EI}$ upward at *D*.

H_D causes

1. A rotation of $147.5H_D \dfrac{\text{kip-ft}^2}{EI}$ counterclockwise at *D*.

2. A horizontal deflection of $1{,}308.3H_D \dfrac{\text{kip-ft}^3}{EI}$ to the right at *D*.

3. A vertical deflection of $810H_D \dfrac{\text{kip-ft}^3}{EI}$ upward at *D*.

V_D causes:

1. A rotation of $216V_D \dfrac{\text{kip-ft}^2}{EI}$ counterclockwise at *D*.

2. A horizontal deflection of $810V_D \dfrac{\text{kip-ft}^3}{EI}$ to the right at D.

3. A vertical deflection of $2{,}448V_D \dfrac{\text{kip-ft}^3}{EI}$ upward at D.

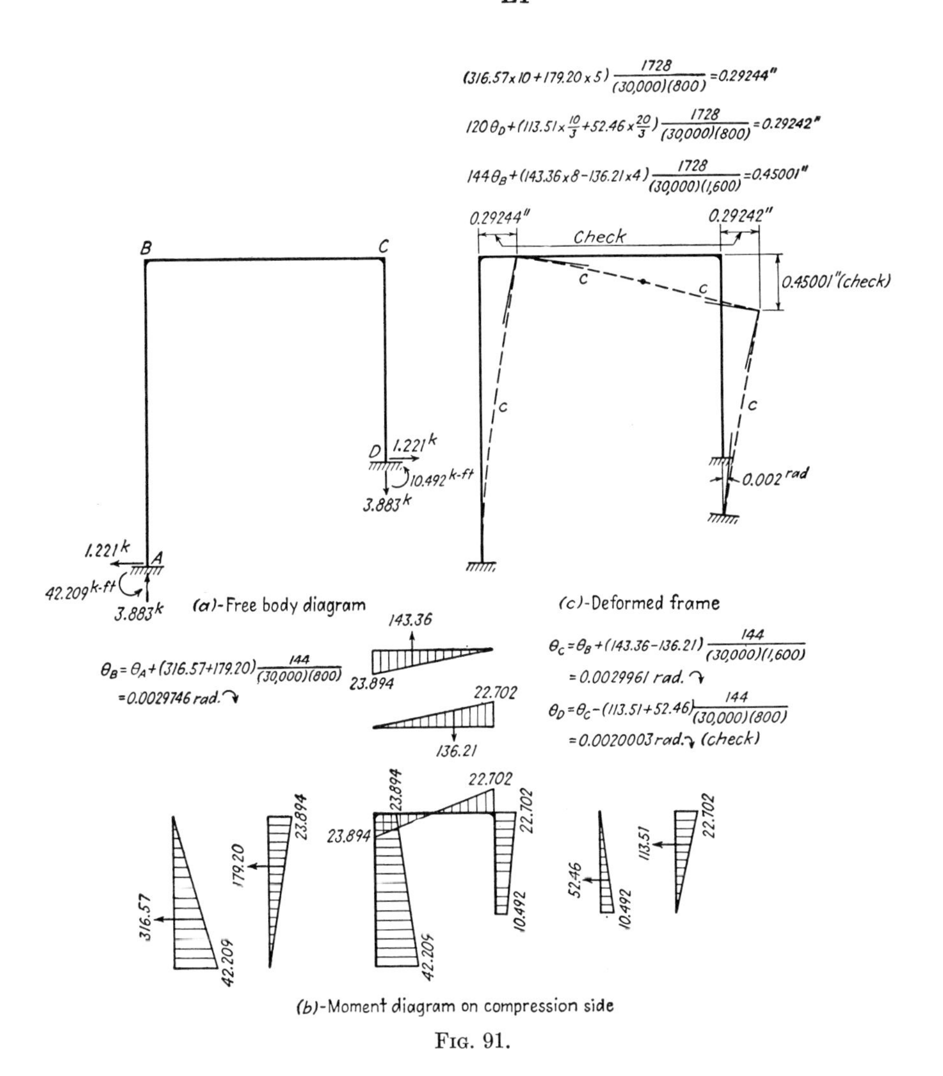

FIG. 91.

Now in the present case, M_D, H_D, and V_D should be such as to *cause*, or to be *caused* by, a rotation of 0.002 rad clockwise at D, a horizontal deflection of zero at D, and a vertical deflection of 0.45 in. downward at D. These are the conditions of geometry.

$$(+31M_D + 147.5H_D + 216V_D)\frac{144}{30{,}000 \times 800} = -0.002 \text{ rad}$$

$$(+147.5M_D + 1{,}308.3H_D + 810V_D)\frac{1{,}728}{30{,}000 \times 800} = 0$$

$$(+216M_D + 810H_D + 2{,}448V_D)\frac{1{,}728}{30{,}000 \times 800} = -0.45 \text{ in.}$$

Solving the above three equations,

$$M_D = +10.492 \text{ kip-ft counterclockwise}$$
$$H_D = +1.221 \text{ kips to the right}$$
$$V_D = -3.883 \text{ kips downward}$$

By statics

$$M_A = 42.209 \text{ kip-ft counterclockwise}$$
$$H_A = 1.221 \text{ kips to the left}$$
$$V_A = 3.883 \text{ kips upward}$$

The free-body diagram is shown in Fig. 91*a*. In Fig. 91*b* and *c*, the moment diagram is drawn and the deformed frame is found by the moment-area method. Note that, by starting from the fixed unyielding support at A and applying the moment-area theorems in succession, the rotation of the tangent at D is found to be 0.002 rad clockwise; the horizontal deflection is zero; and the vertical deflection is 0.45 in. in the downward direction.

EXERCISES

55. Determine all reaction components induced to act on the frame by a vertical settlement of ½ in. at the hinged support E. Draw shear and moment diagrams. Sketch the deformed structure.

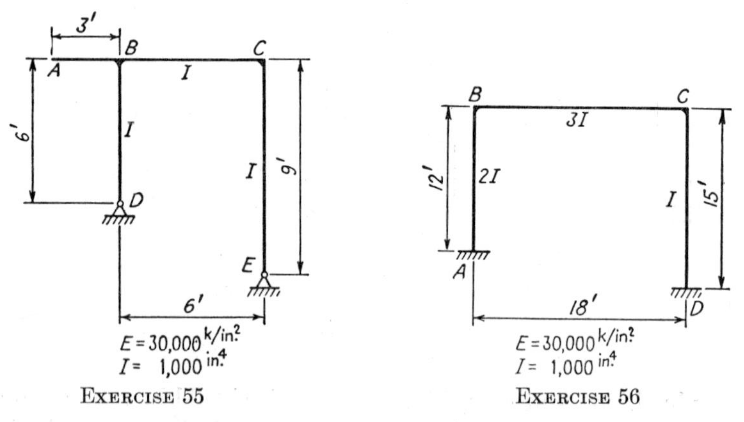

EXERCISE 55 EXERCISE 56

56. Determine all reaction components induced to act on the frame by a rotational slip of 0.002 rad clockwise of joint A and a vertical settlement of ½ in. at joint A. Draw shear and moment diagrams. Sketch the deformed structure.

CHAPTER V

ANALYSIS OF STATICALLY INDETERMINATE TRUSSES BY THE METHOD OF CONSISTENT DEFORMATION

29. Trusses Statically Indeterminate Because of External Redundant Reactions. The analysis of statically indeterminate trusses with external redundant reactions by the method of consistent deformation consists in first choosing a basic determinate truss on which the applied loading and the redundant reactions act and then applying the conditions of geometry requiring that the deflections in the direction of the redundant reactions must be zero. After the redundant reactions are found from the conditions of consistent deformation, they can be applied to the given truss and the remaining reactions found by the principles of statics.

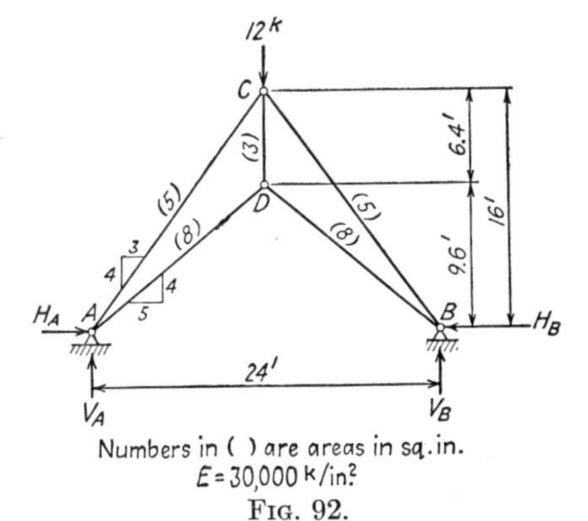

FIG. 92.

Example 44. Determine all reaction components and the stresses in all members of the two-hinged truss shown in Fig. 92.

Solution. The given truss is statically indeterminate to the first degree. The truss hinged at A and supported on rollers at B is chosen as the basic determinate truss (Fig. 93), with H_B as the redundant. The condition of geometry is $\Delta_B = H_B\delta_B$, where δ_B is the deflection to the left at B due to a 1-kip load applied to the left at B of the basic truss.

The unit-load method is used to evaluate Δ_B (Fig. 94) and δ_B (Fig. 95).

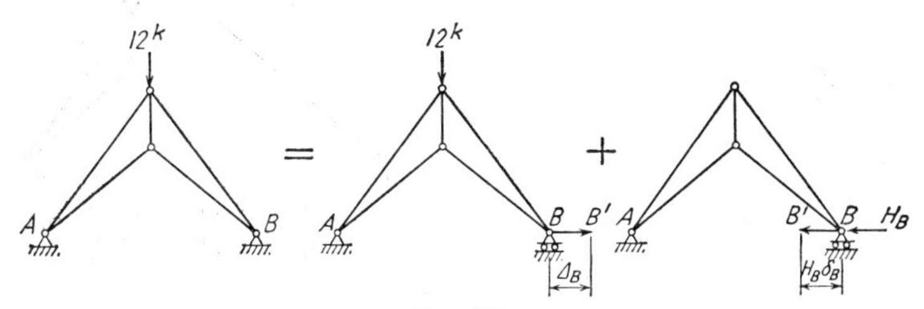

FIG. 93.

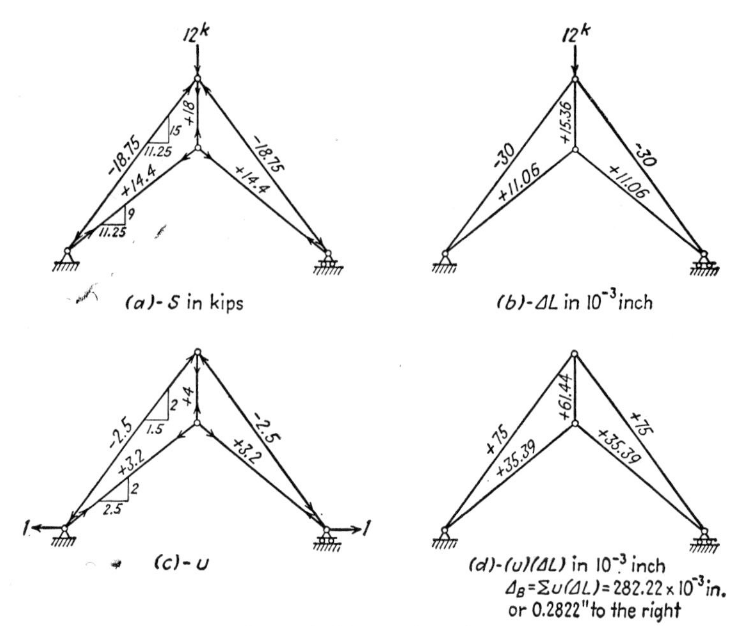

FIG. 94. Evaluation of Δ_B.

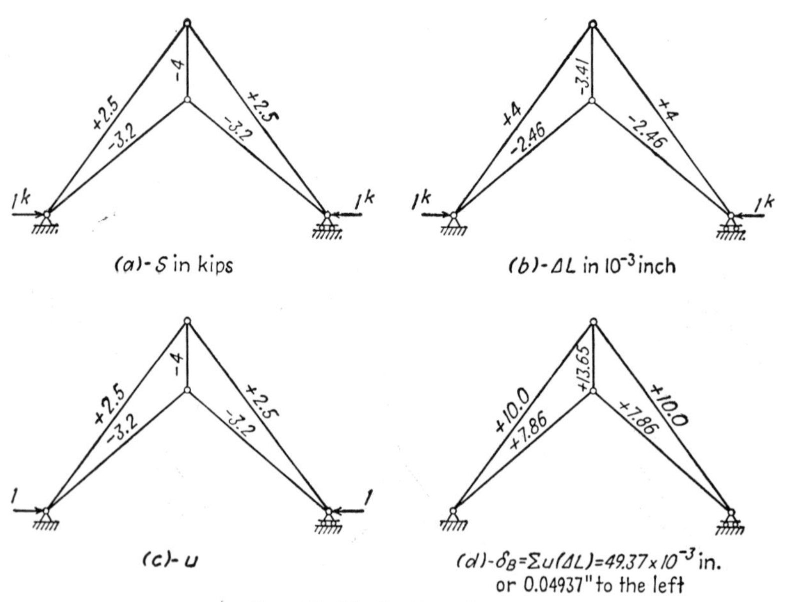

FIG. 95. Evaluation of δ_B.

Thus

$$\Delta_B = 0.2822 \text{ in. to the right}$$
$$\delta_B = 0.04937 \text{ in. to the left}$$

Then,

$$H_B = \frac{\Delta_B}{\delta_B} = \frac{0.2822}{0.04937} = 5.715 \text{ kips to the left}$$

By statics,

$$H_A = 5.715 \text{ kips to the right}$$
$$V_A = V_B = 6 \text{ kips upward}$$

The final *answer diagram* is shown in Fig. 96; the stresses in all members can be found by multiplying all values in Fig. 95*c* by 5.715 and adding the products to those in Fig. 94*a*.

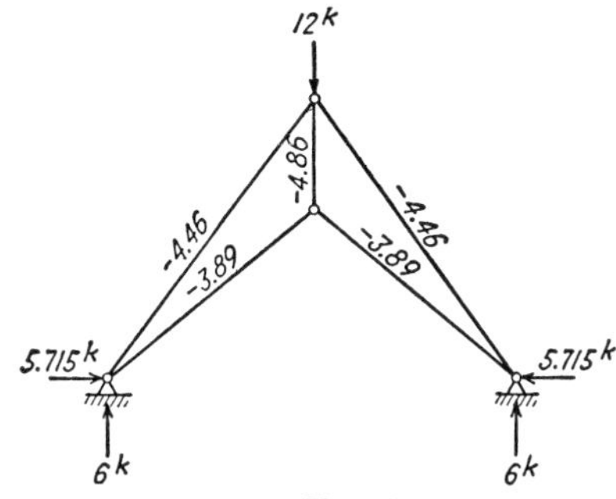

Fig. 96.

Example 45. Find the magnitude and direction of the vertical reactions V_0, V_2, V_6, and V_8 due to a load of 1 kip at joint L_3 (Fig. 97).

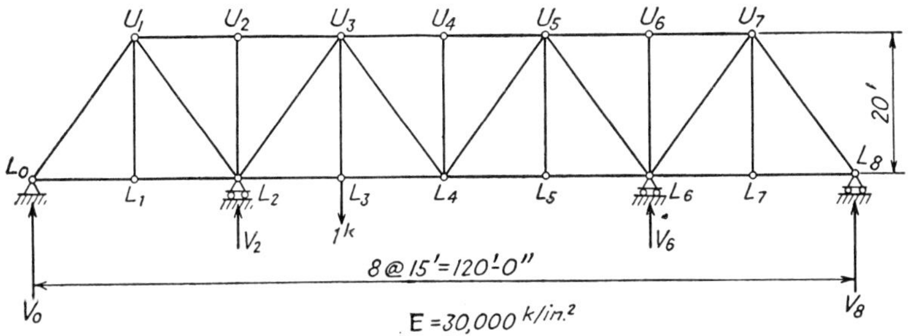

Fig. 97.

Solution. The given truss is statically indeterminate to the second degree. A simple truss supported only at joints L_0 and L_8 is chosen as the basic determinate truss on which the load of 1 kip and the redundant reactions V_2 and V_6 act simultaneously (Fig. 98). The conditions of geometry required from which V_2 and V_6 can be solved are

$$V_2\delta_{22} + V_6\delta_{26} = \delta_{23}$$

and

$$V_2\delta_{62} + V_6\delta_{66} = \delta_{63}$$

wherein the notation δ_{pq} means the vertical deflection at p due to a unit load applied at q.

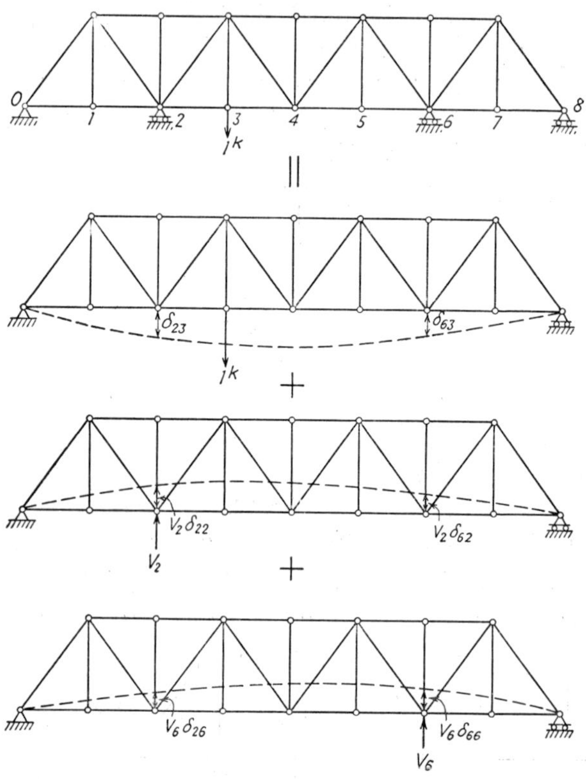

FIG. 98.

All the δ quantities in the above two equations can be found by determining the vertical deflections of all lower chord joints of the basic truss due to a load of 1 kip at joint L_2, if the law of reciprocal deflections is

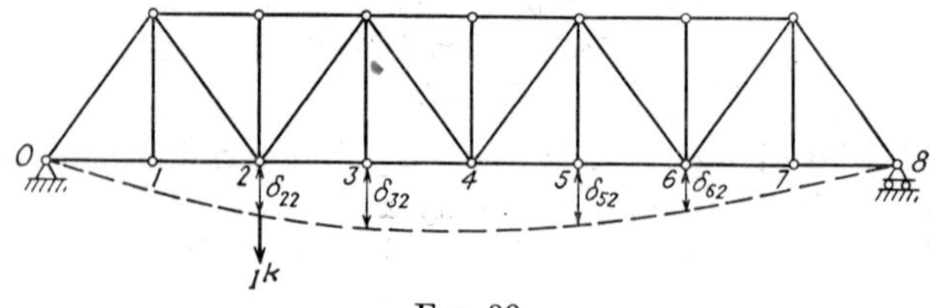

FIG. 99.

properly applied. Either the angle-weights method or the joint-displacements method can be used for this solution. The latter method has been used to obtain the following results (Fig. 99):

$$\delta_{22} = 4.4531 \times 10^{-3} \text{ in.}$$
$$\delta_{32} = 4.4844 \times 10^{-3} \text{ in.}$$
$$\delta_{52} = 3.3656 \times 10^{-3} \text{ in.}$$
$$\delta_{62} = 2.3844 \times 10^{-3} \text{ in.}$$

By symmetry

$$\delta_{66} = \delta_{22} = 4.4531 \times 10^{-3} \text{ in.}$$

and

$$\delta_{63} = \delta_{25}$$

But by the law of reciprocal deflections

$$\delta_{25} = \delta_{52}$$

or

$$\delta_{63} = \delta_{25} = \delta_{52} = 3.3656 \times 10^{-3} \text{ in.}$$

Also,

$$\delta_{26} = \delta_{62} = 2.3844 \times 10^{-3} \text{ in.}$$
$$\delta_{23} = \delta_{32} = 4.4844 \times 10^{-3} \text{ in.}$$

Substituting these values of δ into the two conditions of geometry,

$$4.4531V_2 + 2.3844V_6 = 4.4844$$
$$2.3844V_2 + 4.4531V_6 = 3.3656$$

Solving,

$$V_2 = 0.8444 \text{ kip upward}$$
$$V_6 = 0.3036 \text{ kip upward}$$

By statics

$$V_0 = 0.0842 \text{ kip downward}$$
$$V_8 = 0.0638 \text{ kip downward}$$

EXERCISES

57. Determine all reaction components and stresses in all members.

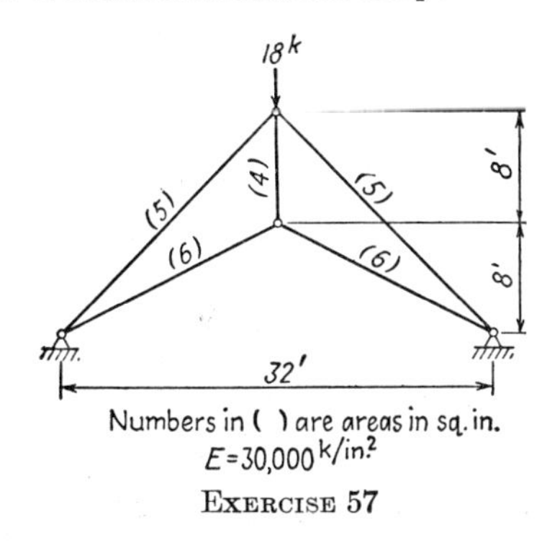

EXERCISE 57

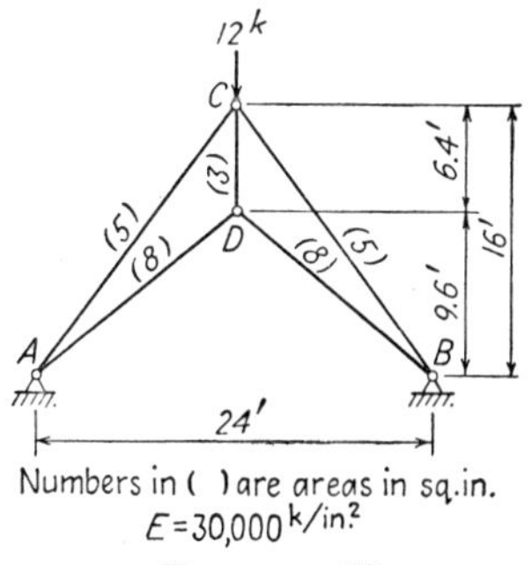

EXERCISE 58

58. Determine all reaction components and stresses in all members. Use member CD as the redundant member.

59. Determine all reactions. Use V_0 and V_8 as redundants.

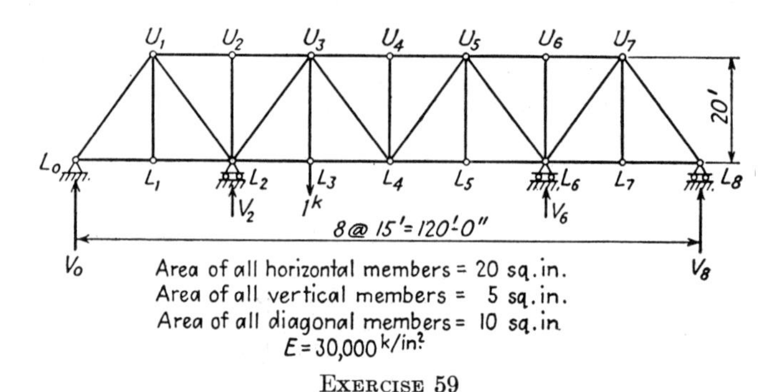

Exercise 59

60. Determine all reactions.

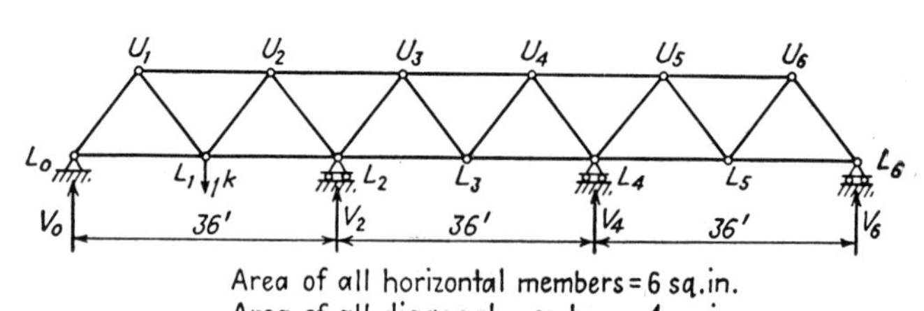

Exercise 60

30. Trusses Statically Indeterminate Because of Internal Redundant Members. The analysis of trusses statically indeterminate because of internal redundant members by the method of consistent deformation consists in first deriving a basic determinate truss by cutting the redundant members and replacing them by pairs of forces and then applying the conditions that the change in distance between the displaced end joints of each redundant member must be equal to the change in the length of the said redundant. Take, for example, the truss shown in Fig. 100*a*. This truss has one extra diagonal in the center panel and is statically indeterminate to the first degree. Or since the truss needs only $m = 2j - 3 = (2)(6) - 3 = 9$ members for static determinacy but has actually 10 members, it is internally indeterminate to the first degree. The member L_1U_2 is chosen as the redundant, and its action on the basic determinate truss is represented by the pair of forces x_a-x_a as shown in Fig. 100*c*. The total lengthening of the redundant member, with a tensile stress of x_a kips, is x_aL_a/A_aE. Now, applying the condition of geometry to the relative distance between the joints L_1 and U_2 (see Fig. 100),

$$\Delta - x_a\delta = \frac{x_a L_a}{A_a E}$$

in which δ is the relative movement (together) between joints L_1 and U_2 due to a pair of forces 1 kip-1 kip applied at joint L_1 and U_2. In other words, the pair of forces x_a-x_a must pull the joints L_1 and U_2 (which are

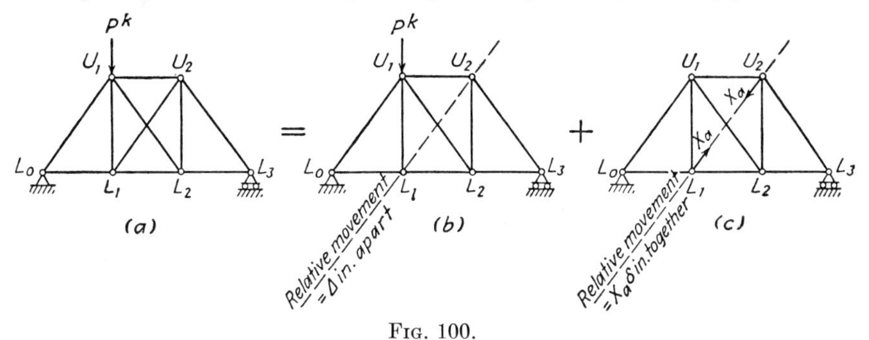

Fig. 100.

at Δ in. apart owing to the action of the applied loading) back together only part of the way, leaving the distance between L_1 and U_2 increased just by the *self-lengthening* of the redundant member which is $x_a L_a / A_a E$. It is to be noted that this *self-lengthening* (or self-shortening in other cases) is the chief characteristic which develops in applying the condition of geometry along the direction of the redundant member which is cut.

Example 46. Determine all reaction components and the stresses in all members (Fig. 101*a*).

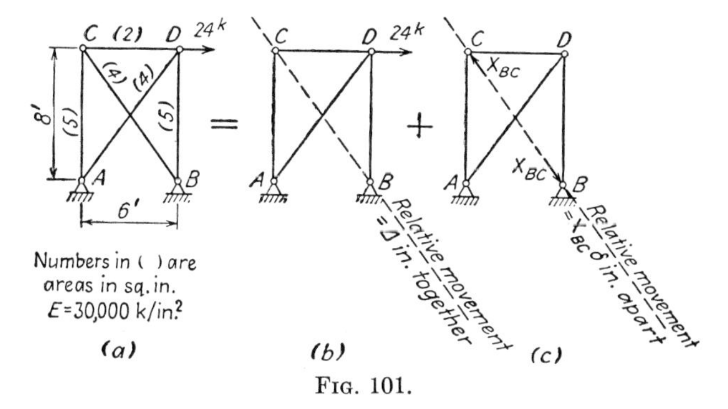

Fig. 101.

Solution. The given truss is statically indeterminate to the *first* degree. It can be considered to have either one external redundant reaction or one internal redundant member. Thus, if hinge B is replaced by a roller support, the stress in diagonal BC becomes zero and the member BC becomes superfluous. Or if diagonal BC is cut, the hinge reaction at B becomes known in direction; this direction must be vertical.

First Solution. The diagonal BC is chosen as the redundant, (Fig. 101). The condition of geometry is

$$\Delta \text{ in. (together)} - x_{BC}\delta \text{ (apart)} = \frac{x_{BC}L_{BC}}{A_{BC}E} \text{ (self-shortening)}$$

The unit-load method is used to evaluate Δ (Fig. 102) and δ (Fig. 103).

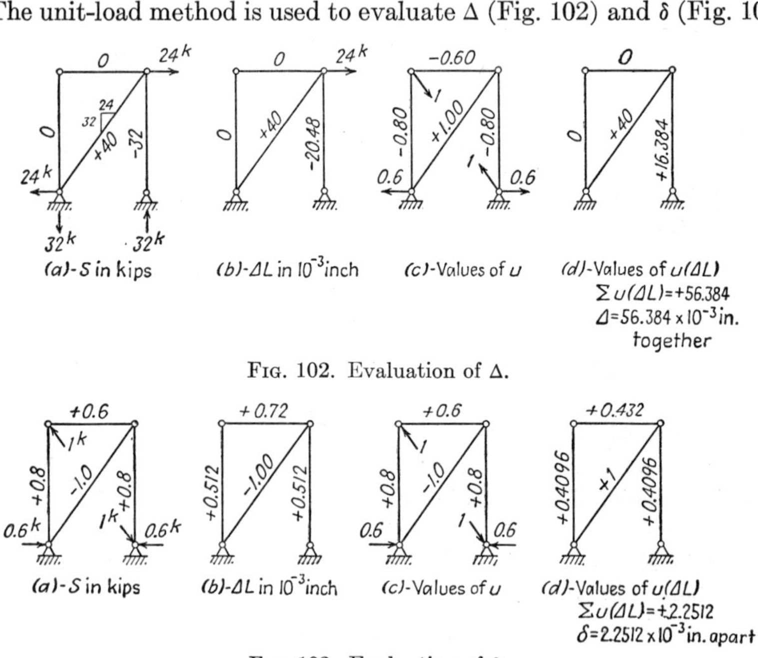

Fig. 102. Evaluation of Δ.

Fig. 103. Evaluation of δ.

Thus

$$\Delta = 56.384 \times 10^{-3} \text{ in. together}$$
$$\delta = 2.2512 \times 10^{-3} \text{ in. apart}$$

Substituting the values of Δ and δ into the equation of consistent deformation,

$$56.384 \times 10^{-3} - x_{BC}(2.2512 \times 10^{-3}) = \frac{x_{BC}(120)}{(4)(30{,}000)}$$

from which

$$x_{BC} = 17.34 \text{ kips compression}$$

The answer diagram is shown in Fig. 104. Values given can be found by adding 17.34 times the stresses shown in Fig. 103*a* to those shown in Fig. 102*a*.

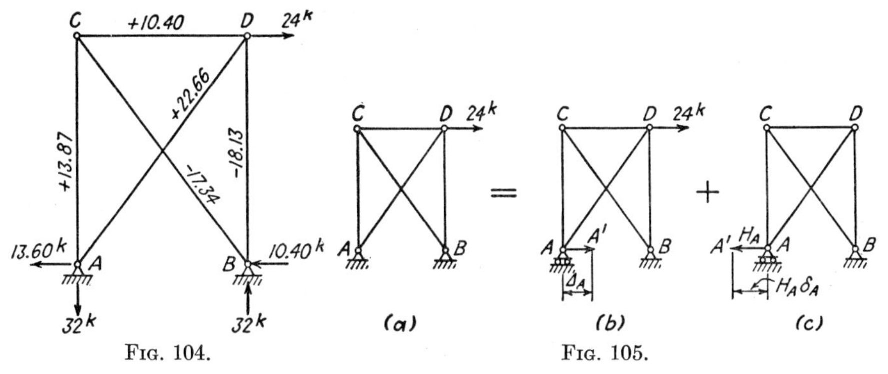

FIG. 104. FIG. 105.

Second Solution. The horizontal reaction at A is chosen as the redundant, Fig. 105. The condition of geometry is

$$\Delta_A = H_A\delta_A$$

The unit-load method has been used to determine the values of Δ_A and δ_A.

$$\Delta_A = 0.12278 \text{ in. to the right}$$
$$\delta_A = 9.032 \times 10^{-3} \text{ in. to the left}$$

Substituting the values of Δ_A and δ_A into the equation of consistent deformation,

$$\Delta_A = H_A\delta_A \qquad H_A = \frac{\Delta_A}{\delta_A} = \frac{0.12278}{0.009032} = 13.60 \text{ kips to the left}$$

The answer diagram is shown in Fig. 104.

Example 47. Find all reaction components and the stresses in all members of the truss shown in Fig. 106*a*.

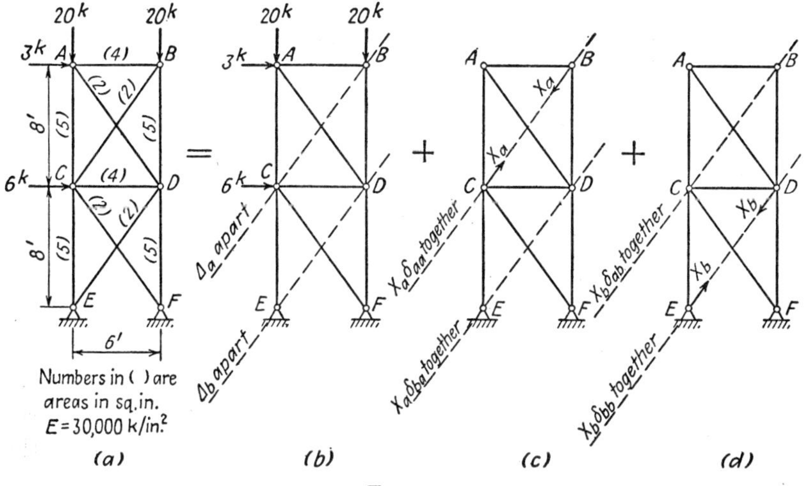

FIG. 106.

Solution. The given truss is statically indeterminate to the second degree. Note that it has one extra diagonal in the upper panel and either one extra diagonal or one extra reaction component in the lower panel.

In the solution below, the basic determinate structure is obtained by cutting the diagonals BC and DE and replacing them by pairs of forces x_a-x_a and x_b-x_b, which act on the basic truss in addition to the applied loading (Fig. 106).

The conditions for consistent deformation are

$$\Delta_a - x_A\delta_{aa} - x_b\delta_{ab} = \frac{x_a L_a}{A_a E}$$

$$\Delta_b - x_a\delta_{ba} - x_b\delta_{bb} = \frac{x_b L_b}{A_b E}$$

where x_a and x_b = tensile stresses in the redundant diagonals
δ_{pq} = relative movement in the p direction due to a pair of unit loads acting in the q direction

By the unit-load method, the following are computed (unit is the deflection in feet if E is considered to be 1 kip/in.^{2}):

Δ_a = 18.38 units together
Δ_b = 41.86 units apart
δ_{aa} = 8.128 units together
$\delta_{ab} = \delta_{ba}$ = 0.540 unit together
δ_{bb} = 7.588 units together

Substituting the values of Δ and δ found above into the equations of consistent deformation,

$$(-18.38) - x_a(8.128) - x_b(0.540) = \frac{x_a(10)}{(2)(1)}$$

$$(+41.86) - x_a(0.540) - x_b(7.588) = \frac{x_b(10)}{(2)(1)}$$

Solving,

x_a = stress in diagonal BC = -1.54 kips or 1.54 kips compression
x_b = stress in diagonal DE = $+3.39$ kips or 3.39 kips tension

The answer diagram is shown in Fig. 107*a*.

A check on the consistency of the geometry of deformation can be performed by cutting the diagonals AD and CF of the truss shown in Fig. 107*a* and showing that the relative movements along AD and CF are equal to the changes in the lengths of diagonals AD and CF. From Fig. 107 it is seen that the relative movement of 32.68 units together between joints A and D is consistent with the total shortening of 32.70

units of diagonal AD, and the same is true for diagonal CF. *Again*, when the results of analysis satisfy both *statics* and *geometry*, the correctness of the solution is certain.

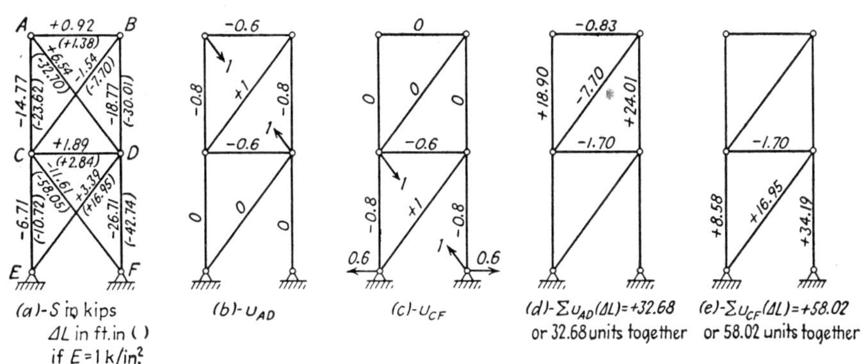

FIG. 107.

EXERCISES

61 and 62. Determine all reactions and stresses in all members.

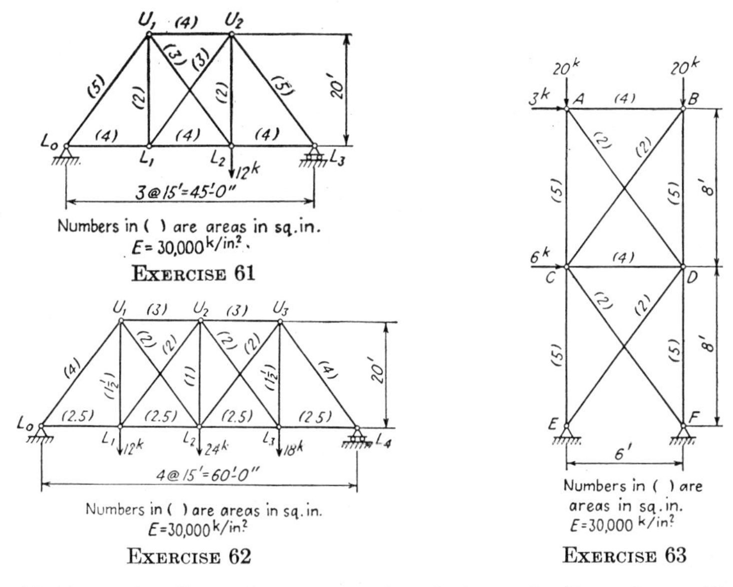

EXERCISE 61

EXERCISE 62

EXERCISE 63

63. Determine all reaction components and stresses in all members. Use members AD and CF as the redundant members.

31. Statically Indeterminate Trusses with Both External and Internal Redundants. The analysis of statically indeterminate trusses with both external and internal redundants by the method of consistent deformation is illustrated by the following example.

Example 48. Find all reactions and the stresses in all members of the truss shown in Fig. 108*a*.

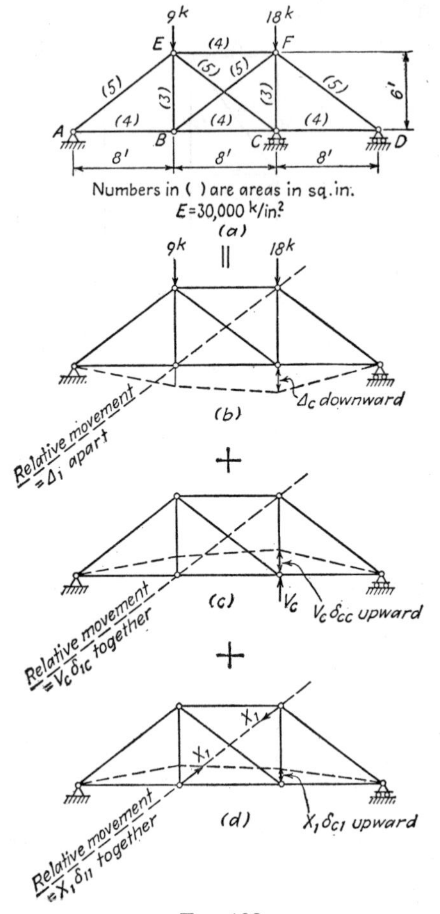

Fig. 108.

Solution. The given truss is statically indeterminate to the second degree; it has one external redundant reaction and one internal redundant member.

The basic determinate truss is obtained by removing the support at C and cutting the diagonal BF (Fig. 108). The conditions for consistent deformation are

$$\Delta_C(\text{downward}) - V_C\delta_{CC}(\text{upward}) - x_1\delta_{C1}(\text{upward}) = 0$$

$$\Delta_1(\text{apart}) - V_C\delta_{1C}(\text{together}) - x_1\delta_{11}(\text{together}) = \frac{x_1L_1}{A_1E}$$

in which x_1, L_1, and A_1 are, respectively, the total tensile stress, the length, and the area of member BF. The Δ and δ quantities are defined in Fig. 108.

The unit-load method is used to evaluate Δ_C and Δ_1 (Fig. 109), for which

$$\Delta_C = \tfrac{644}{9} \times 10^{-3} \text{ in. downward}$$
$$\Delta_1 = 8 \times 10^{-3} \text{ in. together}$$

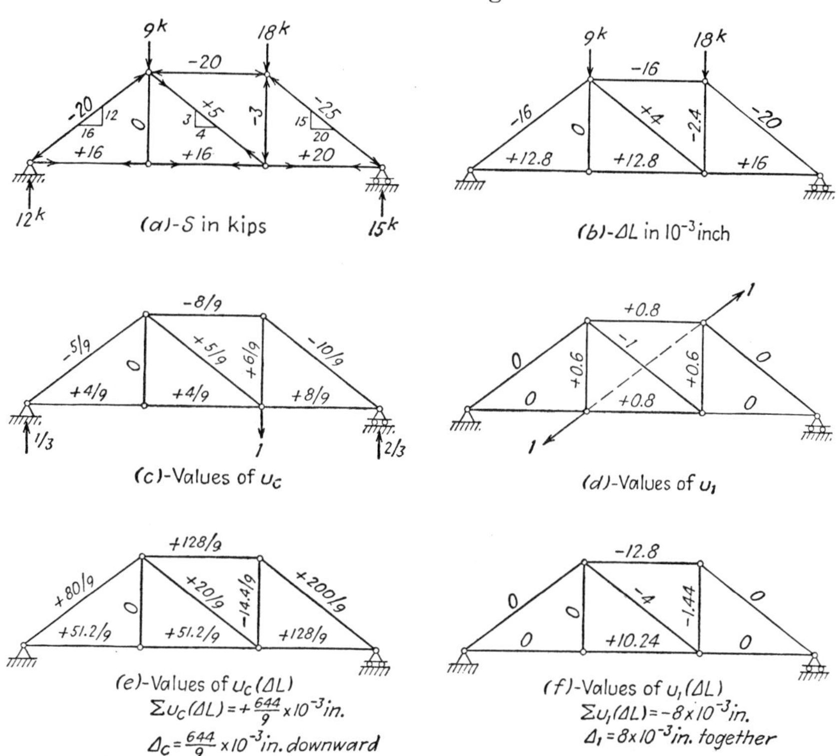

FIG. 109.

The unit-load method is also used to evaluate δ_{CC}, $\delta_{C1} = \delta_{1C}$, and δ_{11} (Fig. 110), for which

$$\delta_{CC} = \frac{276.8}{81} \times 10^{-3} \text{ in. upward}$$

$$\delta_{C1} = \frac{3.68}{9} \times 10^{-3} \text{ in. downward}$$

$$\delta_{1C} = \frac{3.68}{9} \times 10^{-3} \text{ in. apart}$$

$$\delta_{11} = 2.4 \times 10^{-3} \text{ in. together}$$

Substituting the Δ and δ values into the equations of consistent deformation (note the signs),

$$\frac{644}{9} - V_C\left(\frac{276.8}{81}\right) + x_1\left(\frac{3.68}{9}\right) = 0$$

$$-8 + V_C\left(\frac{3.68}{9}\right) - x_1(2.4) = \frac{x_1(120)}{(5)(30)}$$

Solving,

$$V_C = 20.96 \text{ kips upward}$$

$$x_1 = \text{stress in diagonal } EC = 0.18 \text{ kip tension}$$

The answer diagram is shown in Fig. 111.

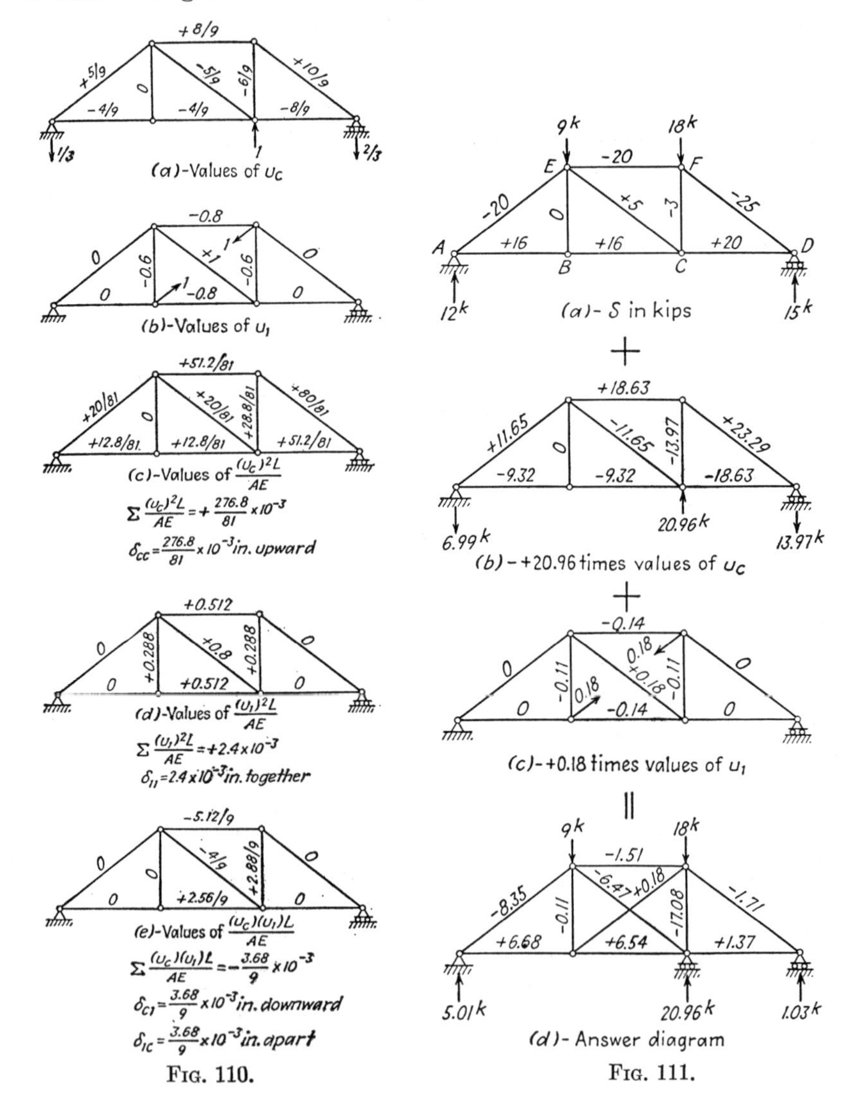

Fig. 110. Fig. 111.

EXERCISE

64. Determine all reactions and stresses in all members. Use reaction at D and stress in member EC as the redundants.

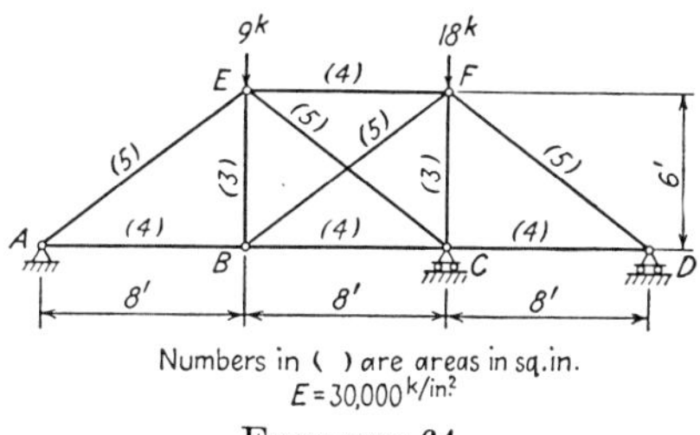

EXERCISE 64

32. Induced Reactions on Statically Indeterminate Trusses Due to Yielding of Supports. As discussed previously, the yielding of the supports of a statically indeterminate structure induces external reactions and corresponding internal stresses in the structure. The reactions may be determined by the method of consistent deformation as follows: (1) Choose a basic structure which is statically determinate by removing the redundant supports (the yielding support must be considered as redundant); (2) apply the conditions for consistent deformation to the points of application of all redundants.

Example 49. Determine V_0, V_2, V_6, and V_8 caused by a $\frac{1}{2}$-in. settlement of the support at L_2 (Fig. 112).

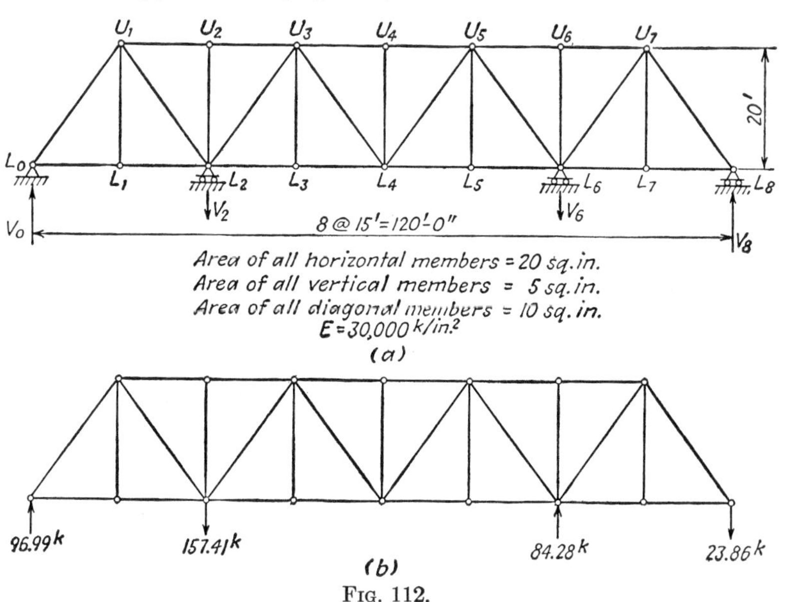

FIG. 112.

Solution. It has been computed in Example 45 that, of the simple truss supported at L_0 and L_8 only,

1. A downward load of 1 kip at L_2 causes
   a. A downward deflection of 4.4531×10^{-3} in. at L_2
   b. A downward deflection of 2.3844×10^{-3} in. at L_6
2. A downward load of 1 kip at L_6 causes
   a. A downward deflection of 2.3844×10^{-3} in. at L_2
   b. A downward deflection of 4.4531×10^{-3} in. at L_6

In the present problem, let the unknown reactions be V_2 kips downward at L_2 and V_6 kips downward at L_6. From the conditions of geometry it follows that

$$(4.4531 \times 10^{-3})(V_2) + (2.3844 \times 10^{-3})(V_6) = +0.500$$
$$(2.3844 \times 10^{-3})(V_2) + (4.4531 \times 10^{-3})(V_6) = 0$$

Solving,

$$V_2 = +157.41 \text{ kips} \quad \text{or} \quad V_2 = 157.41 \text{ kips downward}$$
$$V_6 = -84.28 \text{ kips} \quad \text{or} \quad V_6 = 84.28 \text{ kips upward}$$

By statics

$$V_0 = 96.99 \text{ kips upward}$$
$$V_8 = 23.86 \text{ kips downward}$$

The stresses in all members corresponding to the four reactions as shown in Fig. 112*b* can then be computed.

EXERCISE

65. Determine all reactions induced to act on the truss owing to a vertical settlement of $\frac{1}{2}$ in. at the second support.

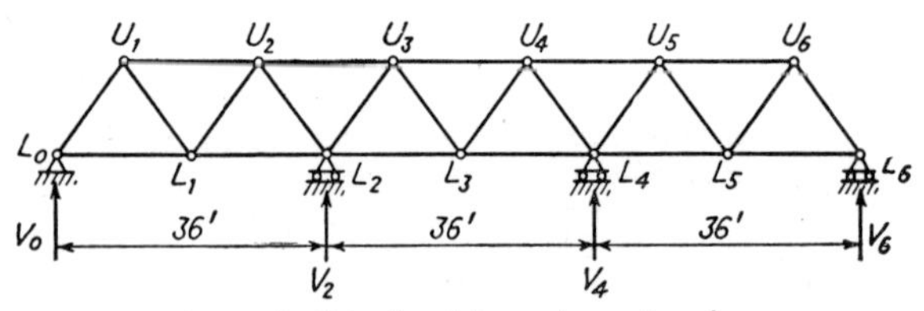

EXERCISE 65

CHAPTER VI

THE THREE-MOMENT EQUATION

33. The Three-moment Equation—Derivation. The three-moment equation expresses the relation between the bending moments at three successive supports of a continuous beam, subjected to a certain applied loading on the various spans, with or without unequal settlements of supports. This relation can be derived on the basis of the *continuity* of the elastic curve over the middle support; that is, the slope of the

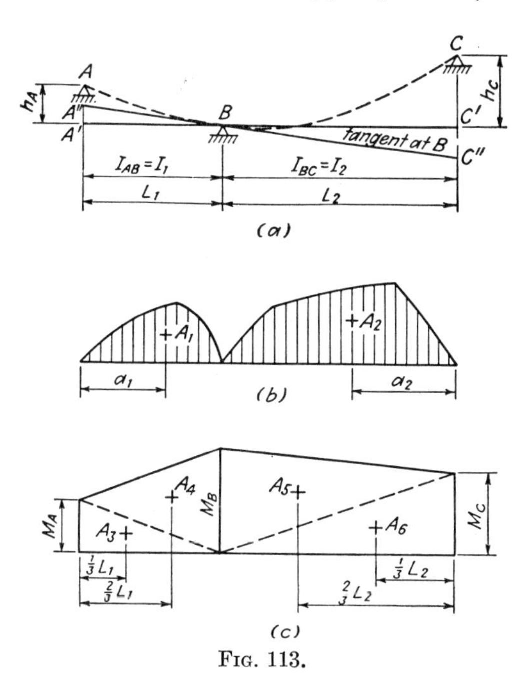

FIG. 113.

tangent at the right end of the left span must be equal to the slope of the tangent at the left end of the right span. The derivation follows:

Let AB and BC, Fig. 113, be two adjacent spans in a continuous beam in which owing to unequal settlements supports A and C are at higher elevations than support B by the amounts h_A and h_C, respectively. Let M_A, M_B, and M_C be the moments at A, B, and C where these moments are positive if they cause compression in the upper parts of the beam. Now consider Fig. 114, where the moment diagram on span AB is broken into two parts: Fig. 114*b* represents the moment diagram A due to the applied loading, with AB considered as a simple beam; Fig. 114*c* represents the moment diagram resulting from the moments at the supports. Figure 114*a* represents the entire moment diagram, or the sum of Fig. 114*b* and *c*. (For simplicity M_A and M_B are shown of the same sign as A, although usually they are negative and A is positive.)

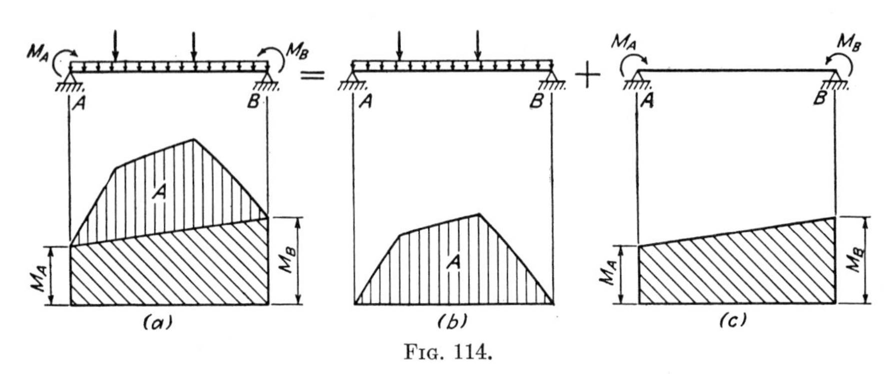

FIG. 114.

Figure 113 represents the moment diagram on the two spans AB and BC broken down into parts just as is the single span AB in Fig. 114. In the usual problems of analysis the simple beam moment diagrams A_1 and A_2 are known, and the object of the analysis is to find the moments M_A, M_B, and M_C at the supports.

A relation between M_A, M_B, and M_C can be derived from the condition that the beam is *continuous* at B, or the tangent at B to the elastic curve BA is on the same straight line as the tangent at B to the elastic curve BC (Fig. 113*a*). Expressed in another manner, the joint B can be considered as a rigid joint, and the tangents at B to the elastic curve on each side must remain at 180° to each other. In Fig. 113*a* a tangent at B is drawn, and from the condition of *continuity* discussed above, the following equation is obtained,

$$\frac{A'A''}{L_1} = \frac{C'C''}{L_2} \tag{48}$$

in which

$$A'A'' = h_A - AA'' = h_A - (\text{deflection of } A \text{ from the tangent at } B)$$

$$= h_A - \frac{1}{EI_1}\left[A_1a_1 + \frac{M_AL_1}{2}\left(\tfrac{1}{3}L_1\right) + \frac{M_BL_1}{2}\left(\tfrac{2}{3}L_1\right)\right] \tag{49}$$

and

$$C'C'' = CC'' - h_C = (\text{deflection of } C \text{ from the tangent at } B) - h_C$$

$$= \frac{1}{EI_2}\left[A_2a_2 + \frac{M_CL_2}{2}(\tfrac{1}{3}L_2) + \frac{M_BL_2}{2}(\tfrac{2}{3}L_2)\right] - h_C \tag{50}$$

Substituting Eqs. (49) and (50) into Eq. (48),

$$\frac{h_A}{L_1} - \frac{1}{EI_1L_1}[A_1a_1 + \tfrac{1}{6}M_AL_1^2 + \tfrac{1}{3}M_BL_1^2] = \frac{1}{EI_2L_2}[A_2a_2 + \tfrac{1}{6}M_CL_2^2 + \tfrac{1}{3}M_BL_2^2] - \frac{h_C}{L_2} \tag{51}$$

Multiplying every term in (51) by $6E$ and reducing,

$$M_A\left(\frac{L_1}{I_1}\right) + 2M_B\left(\frac{L_1}{I_1} + \frac{L_2}{I_2}\right) + M_C\left(\frac{L_2}{I_2}\right) = -\frac{6A_1a_1}{I_1L_1} - \frac{6A_2a_2}{I_2L_2} + \frac{6Eh_A}{L_1} + \frac{6Eh_C}{L_2} \tag{52}$$

Equation (52) is known as the *three-moment equation.*

34. Application of the Three-moment Equation to the Analysis of Statically Indeterminate Beams. The three-moment equation can be used to analyze statically indeterminate beams. For example, it is required to analyze the continuous beam of Fig. 115, which is subjected

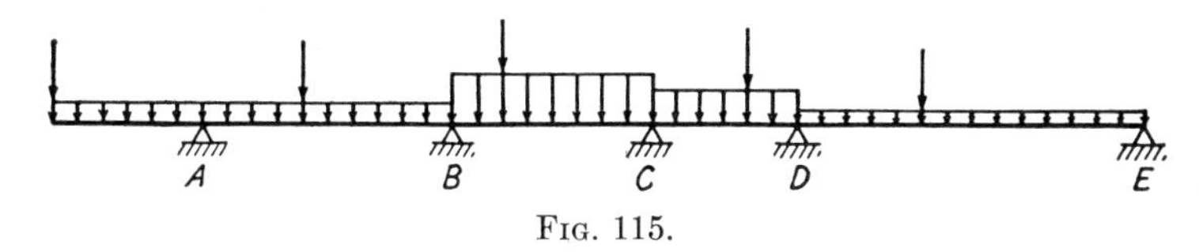

FIG. 115.

to the applied loading shown. The beam is statically indeterminate to the third degree but will be completely solved when the bending moments at all supports are known. The moments at supports A and E can be easily found by inspection. For determining the moments at supports B, C, and D, three three-moment equations can be written on the basis of *continuity* at supports B, C, and D. In other words, the bending moments M_B, M_C, and M_D are chosen as the redundants, and the conditions required of geometry are those of continuity, which can be expressed by the three-moment equation. Thus there are as many conditions of continuity as there are unknown moments at the intermediate supports. Once the bending moments at all supports are known, each span can be treated separately as being subjected to the applied loading on the span and the end moments. Shears at the ends of each span can be found by the laws of statics, and shear and moment diagrams can be drawn accordingly.

If the continuous beam has a fixed end, the bending moment at the fixed support is one of the unknowns (Fig. 116). This case may be treated as follows: The condition required of geometry which corresponds to the unknown fixed-end moment is that the slope of the tangent at A is zero. This condition can be met by adding an imaginary span A_0A

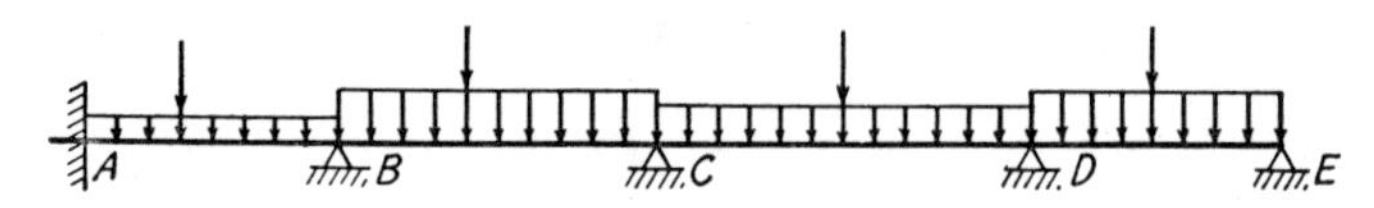

(a)-Continuous beam with fixed end

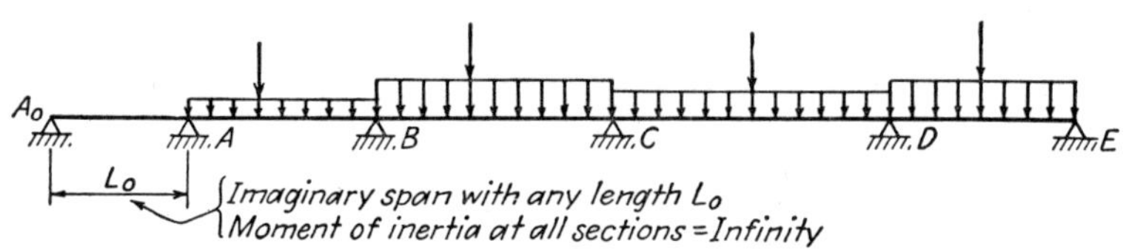

(b)-Equivalent continuous beam

Fig. 116.

of any length L_0 simply supported at A_0 and with an infinitely large moment of inertia at all sections. The three-moment equation, when applied to the spans A_0A and AB (Figs. 116 and 117) becomes

$$M_0\left(\frac{L_0}{\infty}\right) + 2M_A\left(\frac{L_0}{\infty} + \frac{L}{I}\right) + M_B\left(\frac{L}{I}\right) = -\frac{6Aa}{IL} + \frac{6Eh}{L} \tag{53}$$

Noting that $M_0 = 0$ and $L_0/\infty = 0$ in Eq. (53),

$$\frac{2M_AL}{I} + \frac{M_BL}{I} = -\frac{6Aa}{IL} + \frac{6Eh}{L} \tag{54}$$

Transforming Eq. (54),

$$\frac{A}{EI}\left(\frac{a}{L}\right) + \frac{M_AL}{2EI}\left(\tfrac{2}{3}\right) + \frac{M_BL}{2EI}\left(\tfrac{1}{3}\right) = \frac{h}{L} \tag{55}$$

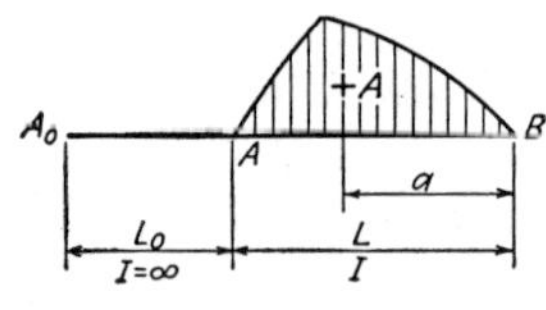

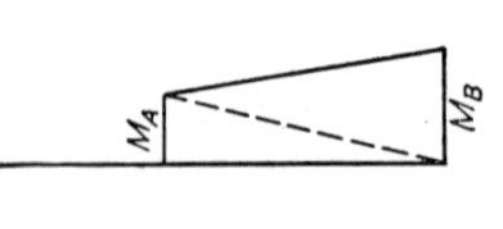

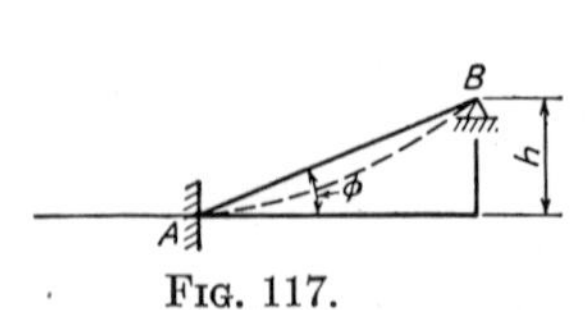

Fig. 117.

By the conjugate-beam method, the left side of Eq. (55) is the angle ϕ (Fig. 117) between the chord AB and the tangent at A, which in this case should be equal to h/L because the tangent at A must remain horizontal. Thus Eq. (53), instead of Eq. (55), can be used to express the condition of geometry required at the fixed support.

In deriving the three-moment equation, Eq. 52, the applied loading and unequal settlements of supports are considered simultaneously. In practice, it is found more convenient to consider the effect of each separately so that the designer may understand how much bending is caused by the applied loadings and how much by the predicted amount of unequal settlements.

Example 50. Analyze the continuous beam shown in Fig. 118*a* by using the three-moment equation. Draw shear and moment diagrams. Sketch the elastic curve.

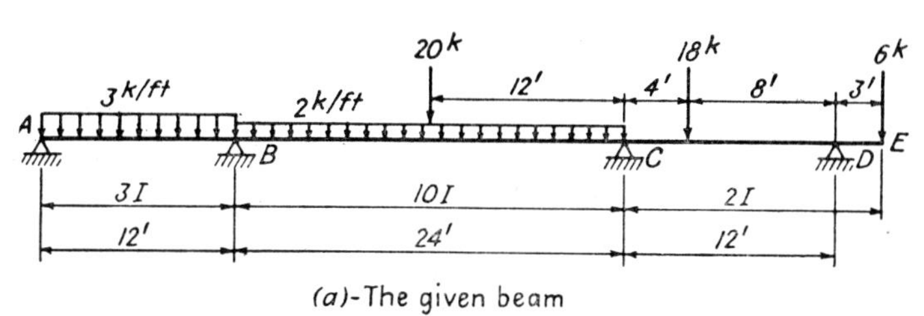

(a)-The given beam

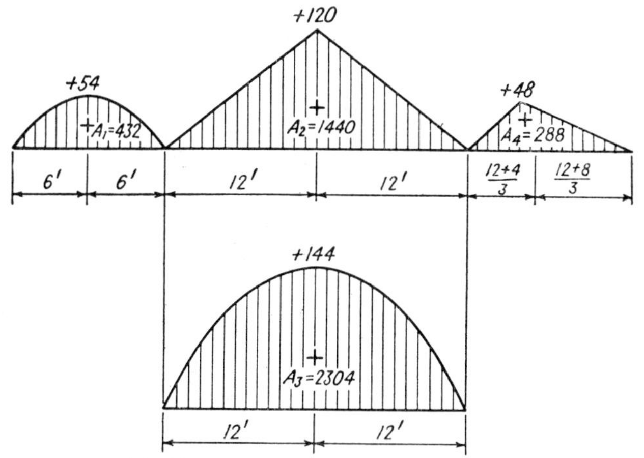

(b)-Moment diagrams on the simple spans resulting from the applied loading

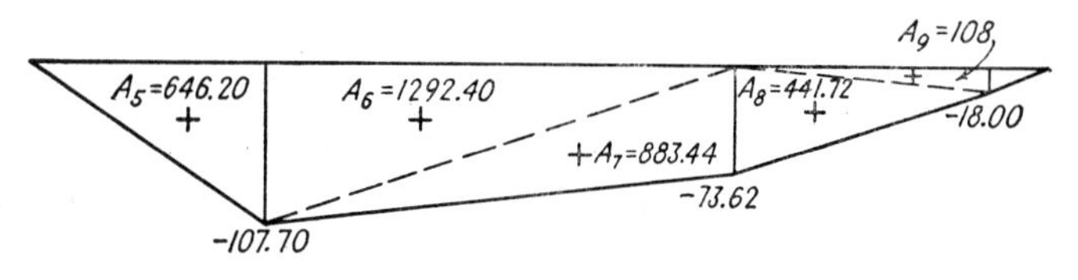

(c)-Moment diagrams on the simple spans due to the end moments

FIG. 118.

Solution. The moment diagrams on AB, BC, and CD by considering each span as a simple beam are shown in Fig. 118*b*. Note that, for span BC, separate moment diagrams are drawn for the uniform load and for the concentrated load.

By inspection, $M_A = 0$ and $M_D = -18$ kip-ft (negative because it causes compressive stress on the lower part of the beam at D).

Applying the three-moment equation to

Spans AB and BC:

$$M_A\left(\frac{12}{3I}\right) + 2M_B\left(\frac{12}{3I} + \frac{24}{10I}\right) + M_C\left(\frac{24}{10I}\right) = -\frac{(6)(432)(6)}{(3I)(12)} - \frac{(6)(3{,}744)(12)}{(10I)(24)}$$

Spans BC and CD:

$$M_B\left(\frac{24}{10I}\right) + 2M_C\left(\frac{24}{10I} + \frac{12}{2I}\right) + M_D\left(\frac{12}{2I}\right) = -\frac{(6)(3{,}744)(12)}{(10I)(24)} - \frac{(6)(288)(\frac{20}{3})}{(2I)(12)}$$

Simplifying,

$$12.8M_B + 2.4M_C = -1{,}555.2$$
$$2.4M_B + 16.8M_C = -1{,}495.2$$

Solving,

$$M_B = -107.70 \text{ kip-ft}$$
$$M_C = -73.62 \text{ kip-ft}$$

Before proceeding any further, it is well at this point to check the correctness of the values determined above for M_B and M_C. This can be done by checking θ_B in spans AB and BC and θ_C in spans BC and CD. Using the conjugate-beam method, it is obvious that the total moment diagram on each span is the sum of those in Fig. 118*b* and *c*.

Applying the conjugate beam method to

Span AB:

$$\theta_B = \frac{1}{E(3I)}\left[\tfrac{1}{2}A_1 - \tfrac{2}{3}A_5\right] = -71.60\,\frac{\text{kip-ft}^2}{EI} \text{ clockwise}$$

Span BC:

$$\theta_B = \frac{1}{E(10I)}\left[\tfrac{1}{2}(A_2 + A_3) - \tfrac{2}{3}A_6 - \tfrac{1}{3}A_7\right] = +71.59\,\frac{\text{kip-ft}^2}{EI} \text{ clockwise}$$

20k 18k 6k

3 k/ft 2 k/ft

107.70 73.62 18.00

End reactions due to applied loading	+18	+18	+10 +24	+ 10 + 24	+12	+6	+6
End reactions due to end moments	-8.975	+ 8.975	+ 1.420	- 1.420	+4.635	-4.635	0
Total end reaction	+9.025	+26.975	+35.420	+32.580	+16.635	+1.365	+6
	R_A=9.025^k	R_B=62.395^k		R_C=49.215^k		R_D=7.365^k	

(a)-Determination of reactions

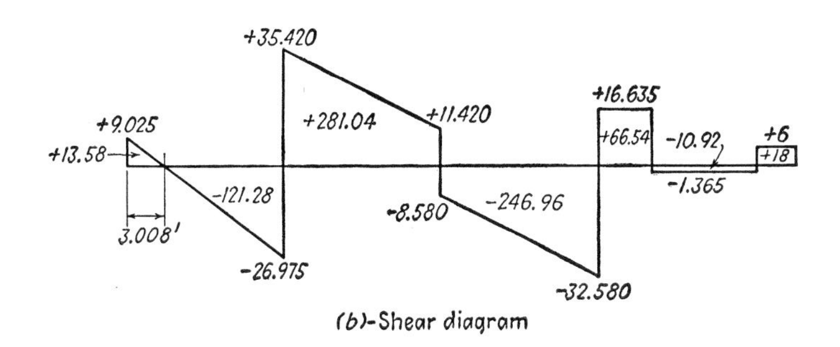

(b)-Shear diagram

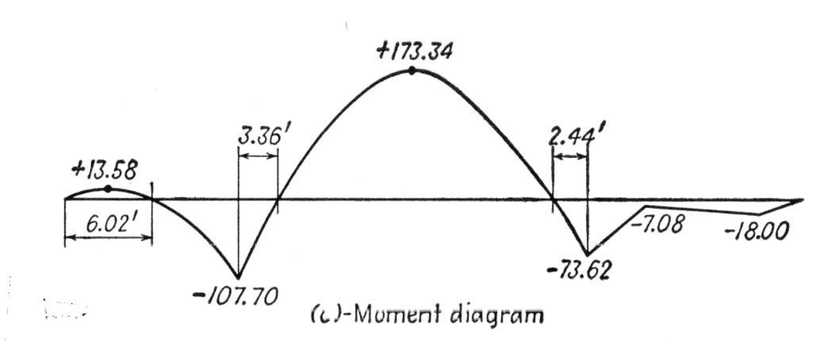

(c)-Moment diagram

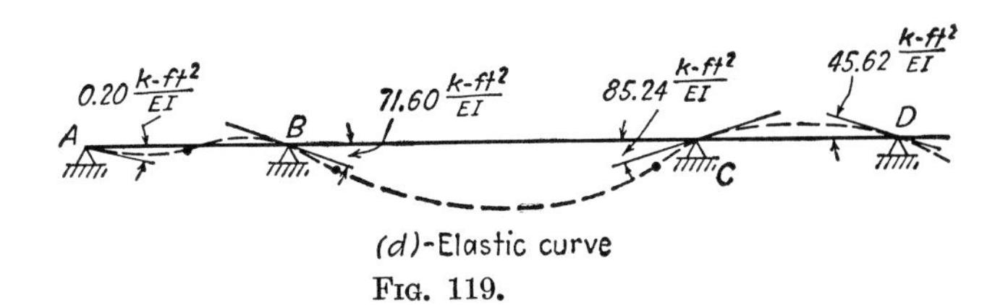

(d)-Elastic curve

Fig. 119.

Span BC:

$$\theta_C = \frac{1}{E(10I)} [\tfrac{1}{2}(A_2 + A_3) - \tfrac{1}{3}A_6 - \tfrac{2}{3}A_7]$$

$$= +85.22 \frac{\text{kip-ft}^2}{EI} \text{ counterclockwise}$$

Span CD:

$$\theta_B = \frac{1}{E(2I)} \left[\frac{20}{(3)(12)} A_4 - \tfrac{2}{3}A_8 - \tfrac{1}{3}A_9 \right]$$

$$= -85.24 \frac{\text{kip-ft}^2}{EI} \text{ counterclockwise}$$

It is to be noted that M_B and M_C were found on the basis of continuity at supports B and C, *i.e.*, by the three-moment equation. Their values must be checked accordingly, by the requirements for continuity at B and C: θ_B in span $AB = \theta_B$ in span BC, and θ_C in span $BC = \theta_C$ in span CD.

The reactions are determined as shown in Fig. 119*a*. The total end reaction at the end of each span is equal to the sum of the reaction due to the applied loading on the span (simply supported) and that due to the moments at the ends of the span. For instance, the sum of the end moments acting on span BC is 107.70 − 73.62 = 34.08 kip-ft *counterclockwise*, which requires a *clockwise* reaction couple, or an upward reaction of 34.08/24 = 1.420 kips at B, and a downward reaction of 1.420 kips at C. The total reaction to the continuous beam at support B is equal to the sum of the end reactions at B to spans BA and BC, or $R_B = 26.975 + 35.420 = 62.395$ kips. After all reactions are determined, the shear diagram is drawn as shown in Fig. 119*b*. The point of zero shear on span AB is at 9.025/3 = 3.008 ft from support A. The area of each branch of the shear diagram is computed and is indicated on the shear diagram. The moment diagram is plotted as shown in Fig. 119*c*, the relation that the change in moment between any two points is equal to the area of the shear diagram between those two points being used successively between convenient points. By so doing M_B and M_C are checked back to be −107.70 and −73.62, respectively, thus indicating that the reactions found above are correct. The *qualitative* elastic curve is shown in Fig. 119*d*.

Example 51. Analyze the continuous beam shown in Fig. 120*a* by using the three-moment equation. Draw shear and moment diagrams. Sketch the elastic curve.

Solution. The moment diagrams on AB, BC, and CD obtained by considering each span as a simple beam are shown in Fig. 120*b*. Since

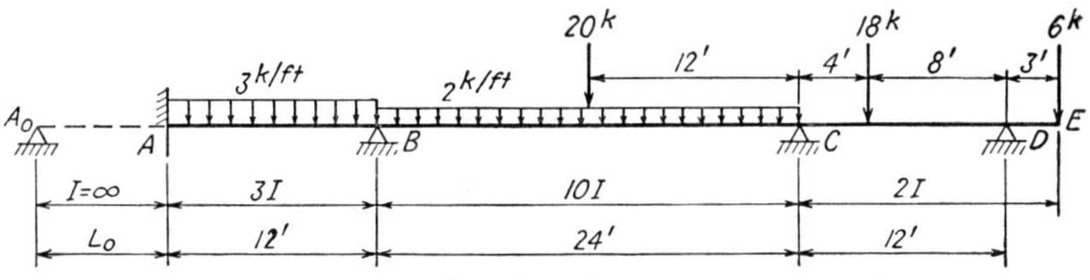

(a)-The given beam

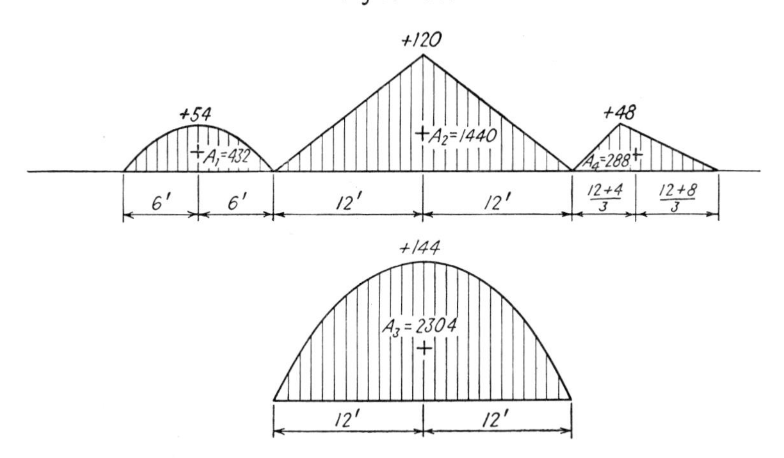

(b)-Moment diagrams on the simple spans resulting from the applied loading

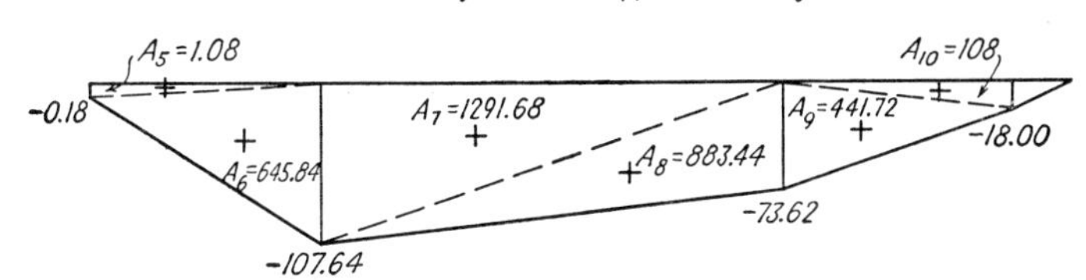

(c)-Moment diagrams on the simple spans due to the end moments

FIG. 120.

support A is fixed, an imaginary span, A_0A, of length L_0 and with $I = \infty$ at all sections, is added.

By inspection

$$M_{A_0} = 0$$
$$M_D = -18 \text{ kip-ft}$$

Applying the three-moment equations to

Spans A_0A and AB:

$$M_{A_0}\left(\frac{L_0}{\infty}\right) + 2M_A\left(\frac{L_0}{\infty} + \frac{12}{3I}\right) + M_B\left(\frac{12}{3I}\right) = -\frac{(6)(432)(6)}{(3I)(12)}$$

Spans AB and BC:

$$M_A\left(\frac{12}{3I}\right) + 2M_B\left(\frac{12}{3I} + \frac{24}{10I}\right) + M_C\left(\frac{24}{10I}\right) = -\frac{(6)(432)(6)}{(3I)(12)} - \frac{(6)(3{,}744)(12)}{(10I)(24)}$$

Spans BC and CD:

$$M_B\left(\frac{24}{10I}\right) + 2M_C\left(\frac{24}{10I} + \frac{12}{2I}\right) + M_D\left(\frac{12}{2I}\right) = -\frac{(6)(3{,}744)(12)}{(10I)(24)} - \frac{(6)(288)(\frac{20}{3})}{(2I)(12)}$$

Simplifying,

$$\begin{aligned} 8M_A + 4M_B \qquad\qquad &= -\ 432 \\ 4M_A + 12.8M_B + 2.4M_C &= -1{,}555.2 \\ +\ 2.4M_B + 16.8M_C &= -1{,}495.2 \end{aligned}$$

Solving,

$$\begin{aligned} M_A &= -0.18 \text{ kip-ft} \\ M_B &= -107.64 \text{ kip-ft} \\ M_C &= -73.62 \text{ kip-ft} \end{aligned}$$

It can be seen that the moment required to hold the tangent at A in the horizontal position is very small, because θ_A is only 0.20 kip-ft$^2/EI$ (Fig. 119*d*) when there is a simple support at A as in Example 50.

For checking the values determined above for M_A, M_B, and M_C the conjugate-beam method is again used to find θ_A, θ_B, and θ_C. The moment diagram on each span is the sum of those in Fig. 120*b* and *c*.

Applying the conjugate-beam method to

Span AB:

$$\theta_A = \frac{1}{E(3I)}\left[\tfrac{1}{2}A_1 - \tfrac{2}{3}A_5 - \tfrac{1}{3}A_6\right] = 0$$

$$\theta_B = \frac{1}{E(3I)}\left[\tfrac{1}{2}A_1 - \tfrac{1}{3}A_5 - \tfrac{2}{3}A_6\right] = -71.64\,\frac{\text{kip-ft}^2}{EI} \text{ clockwise}$$

Span BC:

$$\theta_B = \frac{1}{E(10I)}\left[\tfrac{1}{2}(A_2 + A_3) - \tfrac{2}{3}A_7 - \tfrac{1}{3}A_8\right] = +71.64\,\frac{\text{kip-ft}^2}{EI} \text{ clockwise}$$

$$\theta_C = \frac{1}{E(10I)}\left[\tfrac{1}{2}(A_2 + A_3) - \tfrac{1}{3}A_7 - \tfrac{2}{3}A_8\right] = +85.25\,\frac{\text{kip-ft}^2}{EI} \text{ counterclockwise}$$

Span CD:

$$\theta_C = \frac{1}{E(2I)}\left[\frac{(20)}{(3)(12)}A_4 - \tfrac{2}{3}A_9 - \tfrac{1}{3}A_{10}\right]$$

$$-85.24\,\frac{\text{kip-ft}^2}{EI} \text{ counterclockwise}$$

The reactions, shear and moment diagrams, and the sketch for the elastic curve can then be found in the same way as in Example 50.

Example 52. Analyze the continuous beam shown in Fig. 121*a* in the case of a $\frac{1}{2}$-in. settlement at support *B*, by using the three-moment equation. Draw shear and moment diagrams. Sketch the elastic curve.

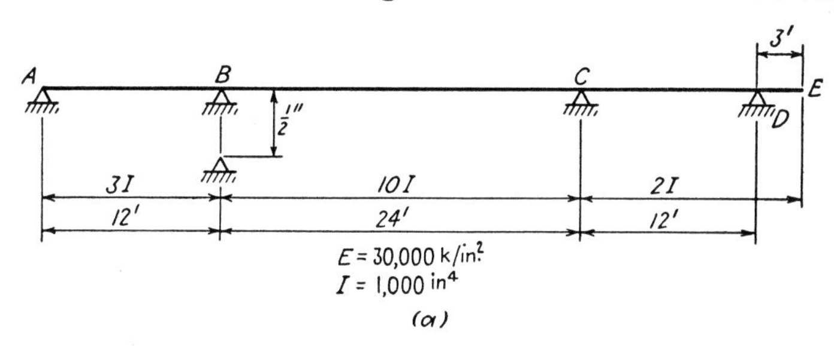

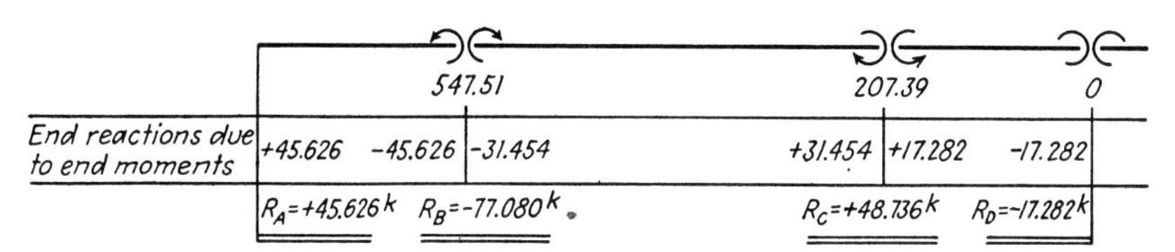

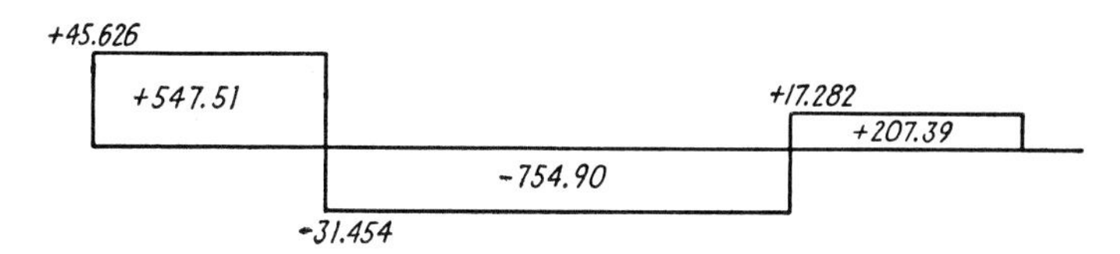

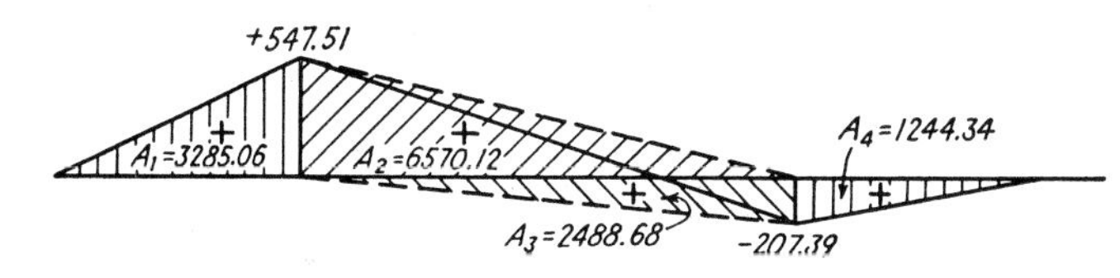

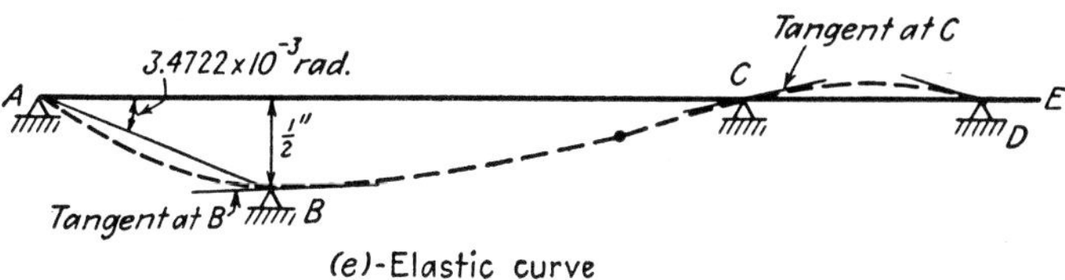

Fig. 121.

Solution. When there are unequal settlements of supports but no applied loading, the three-moment equation becomes

$$M_A\left(\frac{L_1}{I_1}\right) + 2M_B\left(\frac{L_1}{I_1} + \frac{L_2}{I_2}\right) + M_C\left(\frac{L_2}{I_2}\right) = +\frac{6Eh_A}{L_1} + \frac{6Eh_C}{L_2}$$

where the notation is as shown in Fig. 113*a*.

Applying the three-moment equation to

Spans AB and BC:

$$M_A\left(\frac{12}{3I}\right) + 2M_B\left(\frac{12}{3I} + \frac{24}{10I}\right) + M_C\left(\frac{24}{10I}\right) = +\frac{6E(+\frac{1}{2}\text{ in.})}{(12 \times 12)\text{ in.}} + \frac{6E(+\frac{1}{2}\text{ in.})}{(24 \times 12)\text{ in.}}$$

Spans BC and CD:

$$M_B\left(\frac{24}{10I}\right) + 2M_C\left(\frac{24}{10I} + \frac{12}{2I}\right) + M_D\left(\frac{12}{2I}\right) = +\frac{6E(-\frac{1}{2}\text{ in.})}{(24 \times 12)\text{ in.}}$$

Making $M_A = M_D = 0$ and simplifying,

$$(12.8M_B + 2.4M_C)\text{ kip-ft}^2 = 6{,}510.42\text{ kip-ft}^2$$
$$(2.4M_B + 16.8M_C)\text{ kip-ft}^2 = -2{,}170.14\text{ kip-ft}^2$$

Solving,

$$M_B = +547.51\text{ kip-ft}$$
$$M_C = -207.39\text{ kip-ft}$$

The reactions, together with shear and moment diagrams, are shown in Fig. 121*b* to *d*. The sketch of the elastic curve is shown in Fig. 121*e*.

As a check for the correctness of the value for M_B determined above, the slope of the tangent at *B*, as computed from span *AB* by the conjugate-beam method, should be equal to that from span *BC*. For a check on M_C the slope at *C* computed from span *BC* should be equal to that from span *CD*. See Fig. 121*e*.

Slope of tangent at *B* as computed from span *AB*

$$= 3.4722 \times 10^{-3} - \frac{1}{E(3I)}\left(\tfrac{2}{3}A_1\right)$$
$$= 3.4722 \times 10^{-3} - \frac{144}{(30{,}000)(3{,}000)}\left(\tfrac{2}{3}\right)(3{,}285.06)$$
$$= -0.0319 \times 10^{-3}\text{ rad counterclockwise}$$

Slope of tangent at *B* as computed from span *BC*

$$= 1.7361 \times 10^{-3} - \frac{1}{10EI}\left(\tfrac{2}{3}A_2\right) + \frac{1}{10EI}\left(\tfrac{1}{3}A_3\right)$$
$$= 1.7361 \times 10^{-3} - \frac{144}{(30{,}000)(10{,}000)}\left[\left(\tfrac{2}{3}\right)(6{,}570.12) - \left(\tfrac{1}{3}\right)(2{,}488.68)\right]$$
$$= +0.0319 \times 10^{-3}\text{ rad counterclockwise}$$

Slope of tangent at C as computed from span BC

$$= 1.7361 \times 10^{-3} + \frac{1}{10EI}\left[\tfrac{1}{3}A_2 - \tfrac{2}{3}A_3\right]$$
$$= 1.7361 \times 10^{-3} + \frac{144}{(30{,}000)(10{,}000)}\left[(\tfrac{1}{3})(6{,}570.12) - (\tfrac{2}{3})(2{,}488.68)\right]$$
$$= +1.9909 \times 10^{-3} \text{ rad counterclockwise}$$

Slope of tangent at C as computed from span CD

$$= \frac{1}{E(2I)}(\tfrac{2}{3}A_4) = \frac{144}{(30{,}000)(2{,}000)}(\tfrac{2}{3})(1{,}244.34)$$
$$= 1.9909 \times 10^{-3} \text{ rad counterclockwise}$$

Thus the two conditions for continuity at supports B and C are seen to be satisfied; and the correctness of the solution is assured.

Example 53. Analyze the continuous beam shown in Fig. 122*a* owing to the effect of a $\frac{1}{2}$-in. settlement at support B, by using the three-moment equation. Draw shear and moment diagrams. Sketch the elastic curve.

Solution. Applying the three-moment equation,

$$M_A\left(\frac{L_1}{I_1}\right) + 2M_B\left(\frac{L_1}{I_1} + \frac{L_2}{I_2}\right) + M_C\left(\frac{L_2}{I_2}\right) = +\frac{6Eh_A}{L_1} + \frac{6Eh_C}{L_2}$$

Spans A_0A and AB:

$$M_{A_0}\left(\frac{L_0}{\infty}\right) + 2M_A\left(\frac{L_0}{\infty} + \frac{12}{3I}\right) + M_B\left(\frac{12}{3I}\right) = +6E\left(\frac{-\frac{1}{2}}{144}\right)$$

Spans AB and BC:

$$M_A\left(\frac{12}{3I}\right) + 2M_B\left(\frac{12}{3I} + \frac{24}{10I}\right) + M_C\left(\frac{24}{10I}\right) = +6E\left(\frac{+\frac{1}{2}}{144}\right) + 6E\left(\frac{+\frac{1}{2}}{288}\right)$$

Spans BC and CD:

$$M_B\left(\frac{24}{10I}\right) + 2M_C\left(\frac{24}{10I} + \frac{12}{2I}\right) + M_D\left(\frac{12}{2I}\right) = +6E\left(\frac{-\frac{1}{2}}{288}\right)$$

Making $M_{A_0} = M_D = 0$ and simplifying,

$$\begin{aligned} 8M_A + 4M_B \qquad\qquad\quad &= -4{,}340.28 \text{ kip-ft}^2 \\ 4M_A + 12.8M_B + 2.4M_C &= +6{,}510.42 \text{ kip-ft}^2 \\ 2.4M_B + 16.8M_C &= -2{,}170.14 \text{ kip-ft}^2 \end{aligned}$$

Solving,

$$\begin{aligned} M_A &= -972.42 \text{ kip-ft} \\ M_B &= +859.76 \text{ kip-ft} \\ M_C &= -252.00 \text{ kip-ft} \end{aligned}$$

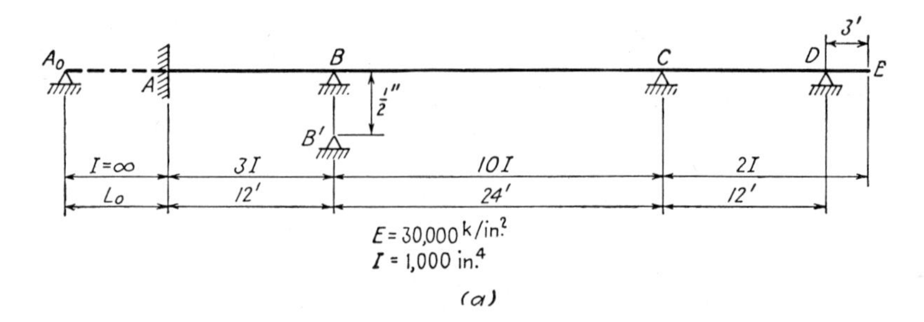

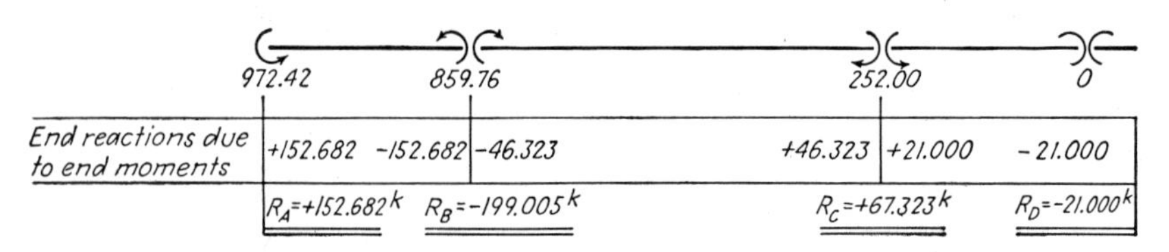

(b)-Determination of reactions

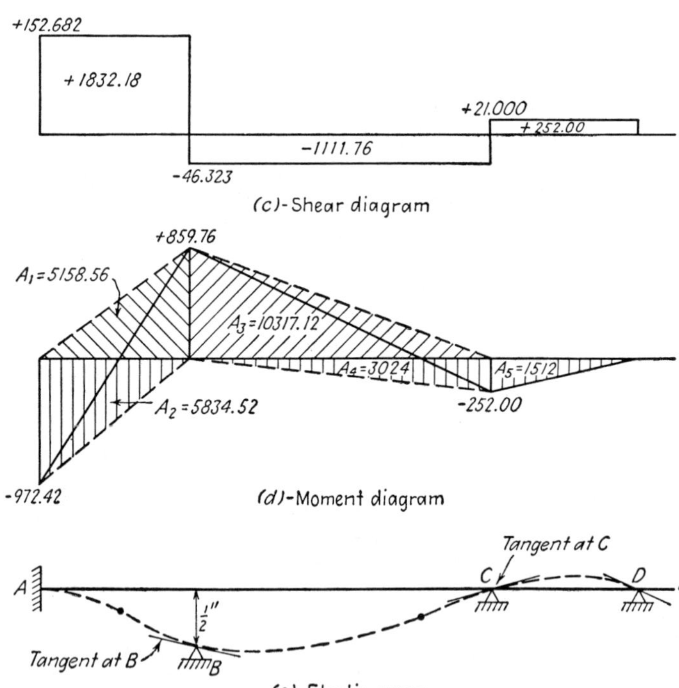

FIG. 122.

The reactions, shear and moment diagrams, and the elastic curve are shown in Fig. 122.

For checking the values of M_A, M_B, and M_C found above, the slopes of the tangents at A, B, and C are computed by the conjugate-beam method (see Fig. 122*d* and *e*).

Slope of tangent at A as computed from span AB

$$= 3.4722 \times 10^{-3} + \frac{1}{E(3I)} (\tfrac{1}{3}A_1 - \tfrac{2}{3}A_2) = 0$$

Slope of tangent at B as computed from span AB

$$= 3.4722 \times 10^{-3} + \frac{1}{E(3I)} (\tfrac{1}{3}A_2 - \tfrac{2}{3}A_1)$$
$$= +1.0815 \times 10^{-3} \text{ rad clockwise}$$

Slope of tangent at B as computed from span BC

$$= 1.7361 \times 10^{-3} + \frac{1}{E(10I)} (\tfrac{1}{3}A_4 - \tfrac{2}{3}A_3)$$
$$= -1.0815 \times 10^{-3} \text{ rad clockwise}$$

Slope of tangent at C as computed from span BC

$$= 1.7361 \times 10^{-3} + \frac{1}{E(10I)} (\tfrac{1}{3}A_3 - \tfrac{2}{3}A_4)$$
$$= +2.4192 \times 10^{-3} \text{ rad counterclockwise}$$

Slope of tangent at C as computed from span CD

$$= \frac{1}{E(2I)} (\tfrac{2}{3}A_5) = +2.4192 \times 10^{-3} \text{ rad counterclockwise}$$

Thus the three conditions of geometry at supports A, B, and C are satisfied, and the correctness of the solution is assured.

EXERCISES

66 and 67. Analyze the continuous beam shown by use of the three-moment equation. Draw shear and moment diagrams. Sketch the elastic curve.

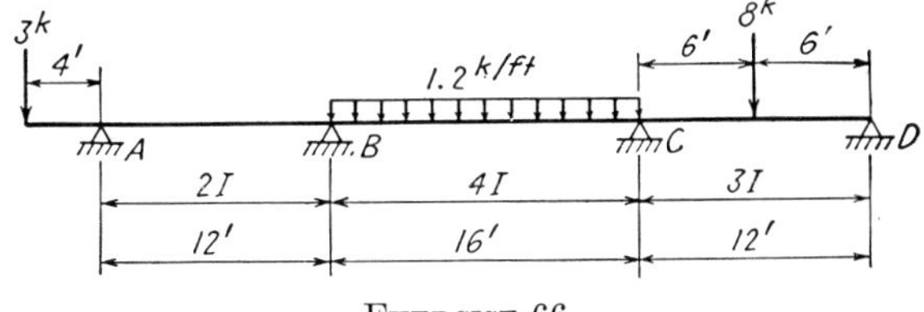

EXERCISE 66

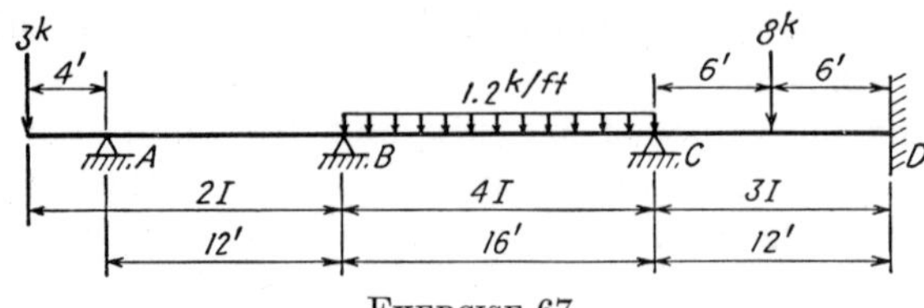

Exercise 67

68 and 69. By use of the three-moment equation analyze the continuous beam shown owing to a settlement of 0.125 in. at support B. Draw shear and moment diagrams. Sketch the elastic curve.

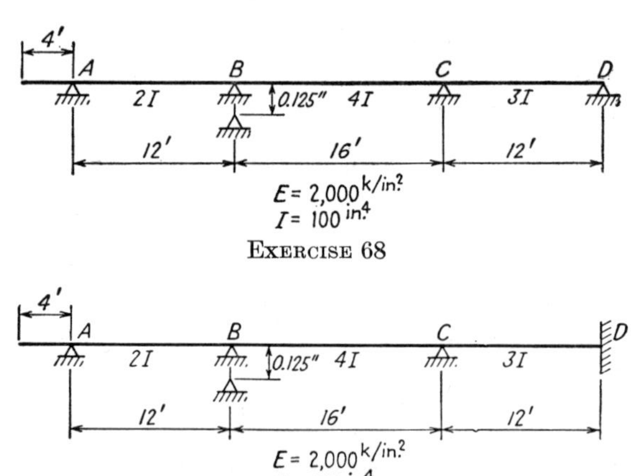

Exercise 68

Exercise 69

70 and 71. Analyze the continuous beam shown by use of the three-moment equation. Draw shear and moment diagrams. Sketch the elastic curve.

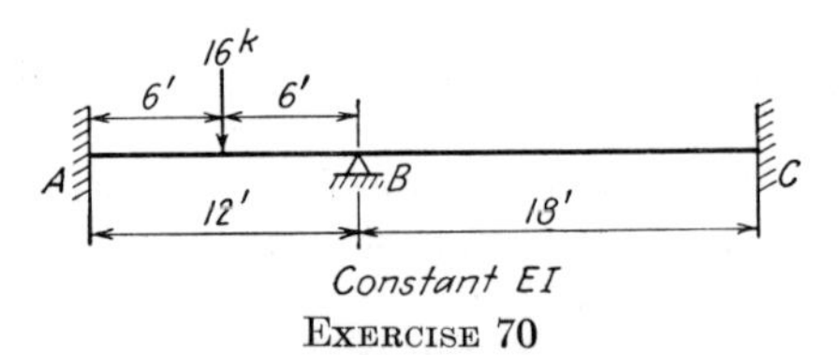

Exercise 70

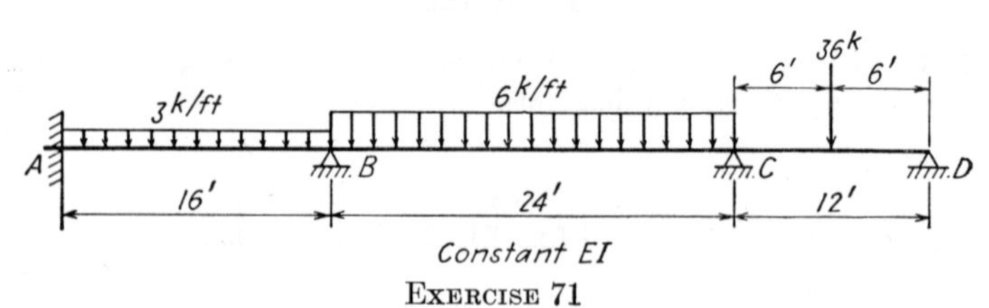

Exercise 71

CHAPTER VII

THE SLOPE-DEFLECTION METHOD

35. General Description of the Slope-deflection Method. The slope-deflection method can be used to analyze all types of statically indeterminate beams or rigid frames. In this method all joints are considered rigid; *i.e.*, the angles between members at the joints are considered not to change in value as loads are applied. Thus the joints at the interior supports of statically indeterminate beams can be considered 180° rigid joints; and ordinarily the joints in rigid frames are 90° rigid joints. When beams or frames are deformed, the rigid joints are considered to rotate only as a whole; in other terms, the angles between the tangents to the various branches of the elastic curve meeting at a joint remain the same as those in the original undeformed structure.

In the slope-deflection method the rotations of the joints are treated as unknowns. It will be shown later that for any one member bounded by two joints the end moments can be expressed in terms of the end rotations. But, to satisfy the condition of equilibrium, the sum of the end moments which any joint exerts on the ends of members meeting there must be zero, because the rigid joint in question is subjected to the sum of these end moments (only reversed in direction). This equation of equilibrium furnishes the necessary condition to cope with the unknown rotation of the joint, and when these unknown joint rotations are found, the end moments can be computed from the slope-deflection equations, which will be derived in the next article.

The theory in the preceding paragraph will be further clarified by the following example: Suppose that it is required to analyze the rigid frame of Fig. 123*a*, loaded as shown. This frame is statically indeterminate to the sixth degree. The method of consistent deformation could be used, but the amount of work involved would make that method too laborious. It is to be noted that, in this problem, because the frame is kept from horizontal movement by its connection at A and from vertical movement by the fixed bases at D and E, and since axial deformation of the members is usually neglected, all joints of this frame must remain in their original locations. (The cases in which some joints may change positions when the frame is deformed will be taken up later.) *Clockwise* joint rotations are considered to be *positive*, such as are shown in Fig.

123*a*. The free-body diagrams of all members are shown in Fig. 123*b*. At any one end of each member, there are three reaction components: direct pull or thrust, end shear, and end moment. The end moment which acts at end A of member AB is denoted as M_{AB}; that at end B of member AB as M_{BA}. *Counterclockwise* end moments acting on the members are considered to be *positive*, positive end moments being shown in Fig. 123*b*. It is possible, by the use of the slope-deflection equations to be derived in the next article, to express the end moments of each

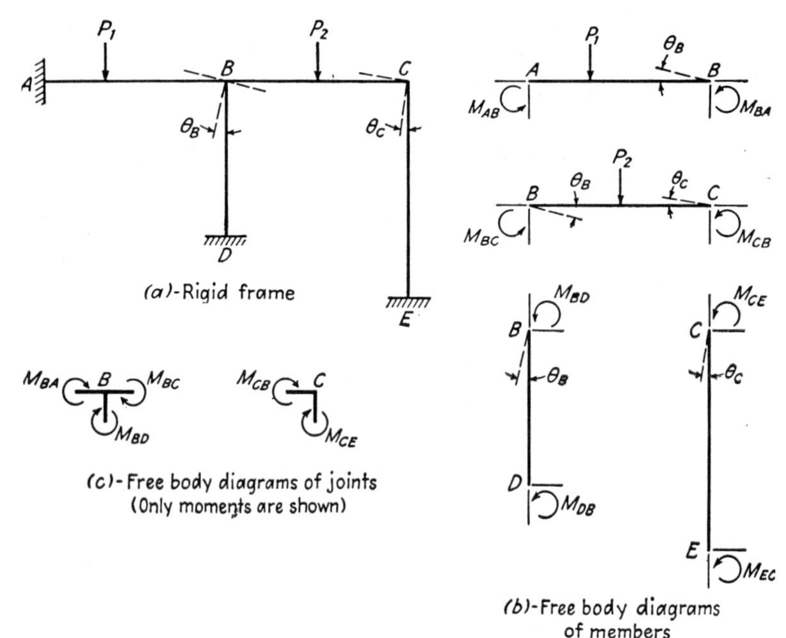

FIG. 123.

member in terms of the end rotations and the loading which acts on the member. Thus the eight end moments in this problem can be expressed in terms of the two unknown joint rotations. The free-body diagrams of all joints are shown in Fig. 123*c*. Of course, the action of the member on the joint consists of a force in the direction of the axis of the member, a force perpendicular to this axis, and a moment, each being the opposite of the action of the joint on the member. In Fig. 123*c*, only the moments are shown. These moments are drawn in their *positive* direction, which is *clockwise*. For equilibrium, summation of all moments acting on each joint must be zero. Thus

Joint condition at B:

$$M_{BA} + M_{BC} + M_{BD} = 0$$

Joint condition at C:

$$M_{CB} + M_{CE} = 0$$

The above two equations are necessary and sufficient to determine the values of θ_B and θ_C. All end moments can then be found by substituting the known joint rotations into the *slope deflection equations.* By the principles of statics the direct stress and shear and moment diagrams for each member can be found.

It has been repeatedly pointed out that the analysis of statically indeterminate structures must satisfy both *statics* and *geometry*. In the slope-deflection method of analyzing rigid frames the conditions required of the geometry of the deformed structure, which are those of the rigidity of the joints, are satisfied at the outset by calling the joint rotation one single unknown at each joint. Thus the conditions of statics, requiring that the sum of moments acting on each joint be zero, are used to solve for the joint rotations.

36. Derivation of the Slope-deflection Equations. In the slope-deflection equations the end moments acting at the ends of a member

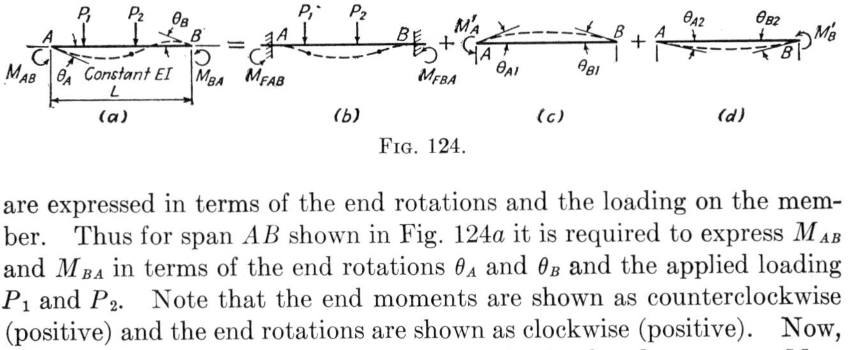

FIG. 124.

are expressed in terms of the end rotations and the loading on the member. Thus for span AB shown in Fig. 124*a* it is required to express M_{AB} and M_{BA} in terms of the end rotations θ_A and θ_B and the applied loading P_1 and P_2. Note that the end moments are shown as counterclockwise (positive) and the end rotations are shown as clockwise (positive). Now, with the applied loading on the member, the fixed end moments, M_{FAB} and M_{FBA} (both shown as counterclockwise) are required to hold the tangents at the ends fixed (Fig. 124*b*). The additional end moments, M'_A and M'_B, should be such as to cause rotations of θ_A and θ_B, respectively. If θ_{A1} and θ_{B1} are the end rotations caused by M'_A and θ_{A2} and θ_{B2} by M'_B (Fig. 124*c* and *d*), the conditions required of geometry are

$$\begin{aligned} \theta_A &= -\theta_{A1} + \theta_{A2} \\ \theta_B &= \theta_{B1} - \theta_{B2} \end{aligned} \tag{56}$$

By superposition

$$\begin{aligned} M_{AB} &= M_{FAB} + M'_A \\ M_{BA} &= M_{FBA} + M'_B \end{aligned} \tag{57}$$

By the conjugate-beam method

$$\theta_{A1} = \frac{M'_A L}{3EI} \qquad \theta_{B1} = \frac{M'_A L}{6EI}$$
$$\theta_{A2} = \frac{M'_B L}{6EI} \qquad \theta_{B2} = \frac{M'_B L}{3EI} \tag{58}$$

Substituting (58) into (56),

$$\theta_A = -\frac{M'_A L}{3EI} + \frac{M'_B L}{6EI}$$
$$\theta_B = \frac{M'_A L}{6EI} - \frac{M'_B L}{3EI} \tag{59}$$

Solving Eq. (59) for M'_A and M'_B,

$$M'_A = +\frac{2EI}{L}(-2\theta_A - \theta_B)$$
$$M'_B = +\frac{2EI}{L}(-2\theta_B - \theta_A) \tag{60}$$

Substituting Eq. (60) into Eq. (57),

$$M_{AB} = M_{FAB} + \frac{2EI}{L}(-2\theta_A - \theta_B)$$
$$M_{BA} = M_{FBA} + \frac{2EI}{L}(-2\theta_B - \theta_A) \tag{61}$$

Equations (61) are the slope-deflection equations which express the end moments in terms of the end rotations and the applied loading. Note again that counterclockwise moments acting at the ends of the member (M_{AB}, M_{BA}, M_{FAB}, and M_{FBA}) are positive and clockwise rotations (θ_A and θ_B) are positive.

The case in which joint B is deflected by an amount Δ in the direction perpendicular to the original axis of the member will be taken up later.

37. Application of the Slope-deflection Method to the Analysis of Statically Indeterminate Beams. The procedure of analyzing statically

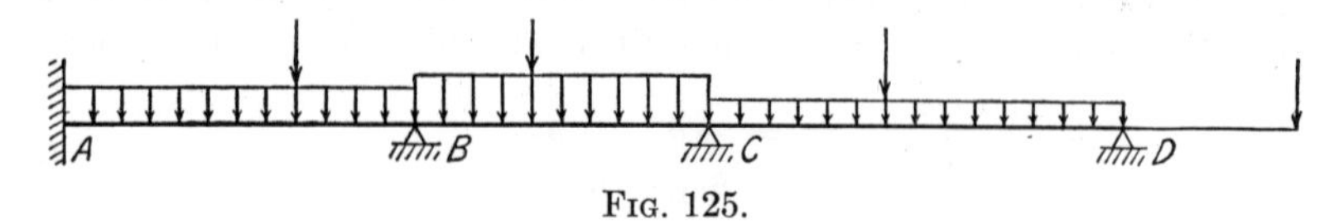

FIG. 125.

indeterminate beams by the slope-deflection method is as follows (see Fig. 125):

1. Determine the fixed-end moments at the ends of each span, using the formulas shown in Fig. 126.

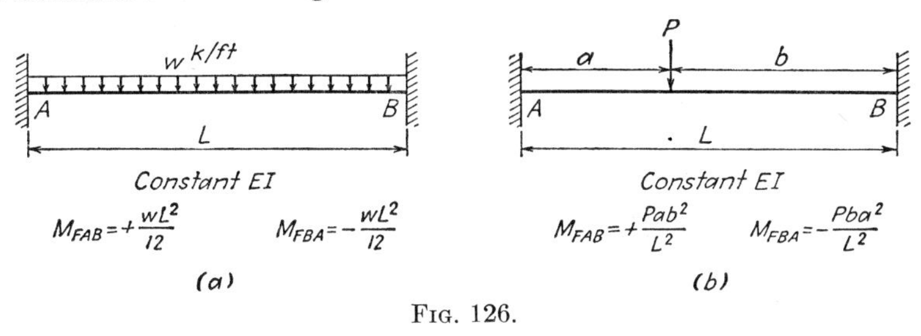

FIG. 126.

2. Express all end moments in terms of the fixed-end moments and the joint rotations by using the slope-deflection equations.

3. Establish simultaneous equations with the rotations at the supports as unknowns by applying the conditions that the sum of the end moments acting on the ends of the two members meeting at the support should be zero.

4. Solve for the rotations at all supports.

5. Substitute the rotations back into the slope-deflection equations, and compute the end moments.

6. Determine all reactions, draw shear and moment diagrams, and sketch the elastic curve.

Example 54. Analyze the continuous beam shown in Fig. 127 by the slope-deflection method. Draw shear and moment diagrams. Sketch the elastic curve.

Solution. In the slope-deflection equations

$$M_{AB} = M_{FAB} + \frac{2EI}{L}(-2\theta_A - \theta_B)$$

$$M_{BA} = M_{FBA} + \frac{2EI}{L}(-2\theta_B - \theta_A)$$

the coefficient $2EI/L$ before the parentheses is different for each span. If the $2EI/L$ values for all spans are made N times smaller, the effect will be only to make all θ values N times larger, while the products of the expressions $2EI/L$ and $(-2\theta_{\text{near}} - \theta_{\text{far}})$, or the values of the end moments, remain unchanged. If the absolute magnitudes of the θ values are not of direct interest, therefore, the relative values of $2EI/L$ can be used before the parentheses. If the relative values of I/L are called the relative stiffness, K, the slope-deflection equations become

$$\begin{aligned} M_{AB} &= M_{FAB} + K_{AB}(-2\theta_A - \theta_B) \\ M_{BA} &= M_{FBA} + K_{AB}(-2\theta_B - \theta_A) \end{aligned} \tag{62}$$

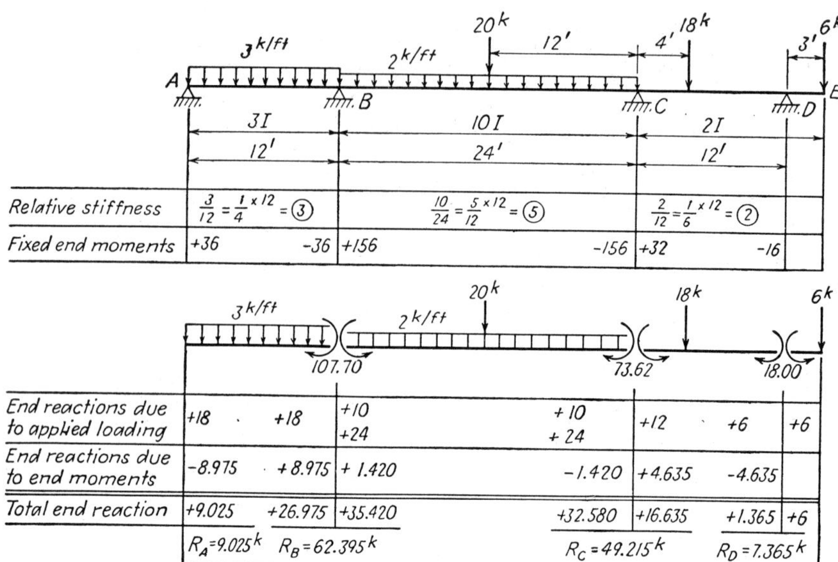

Relative stiffness	$\frac{3}{12}=\frac{1}{4}\times 12 = ③$		$\frac{10}{24}=\frac{5}{12}\times 12 = ⑤$		$\frac{2}{12}=\frac{1}{6}\times 12 = ②$		
Fixed end moments	+36	−36	+156	−156	+32	−16	

End reactions due to applied loading	+18	+18	+10 +24	+10 +24	+12	+6	+6
End reactions due to end moments	−8.975	+8.975	+1.420	−1.420	+4.635	−4.635	
Total end reaction	+9.025	+26.975	+35.420	+32.580	+16.635	+1.365	+6
	$R_A = 9.025^k$	$R_B = 62.395^k$		$R_C = 49.215^k$		$R_D = 7.365^k$	

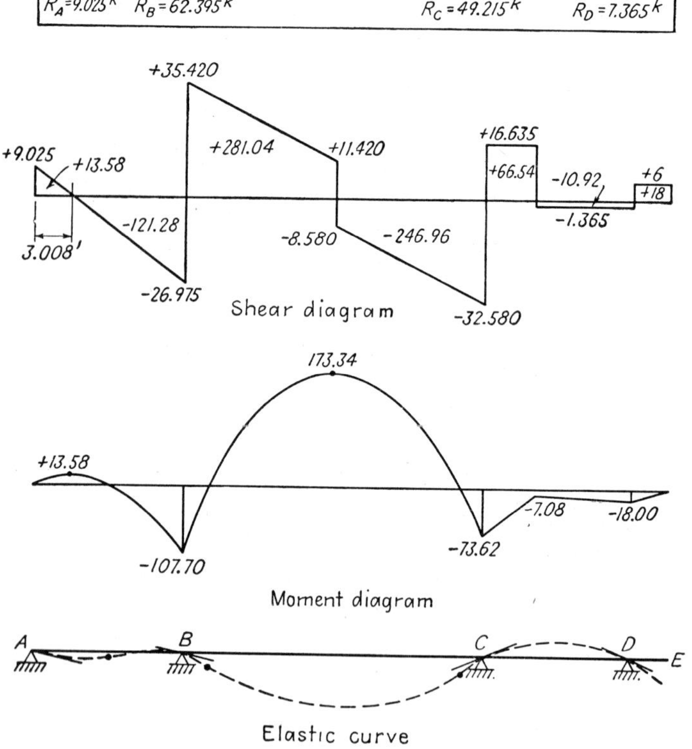

FIG. 127.

Relative Stiffness. In the present example the values of the relative stiffness are determined in Table 54-1.

TABLE 54-1

Span	AB	BC	CD
I	$3I$	$10I$	$2I$
L	12 ft	24 ft	12 ft
I/L	$3I/12$	$10I/24$	$2I/12$
K	3	5	2

Fixed-end Moments

$$
\begin{aligned}
M_{FAB} &= +\frac{(3)(12)^2}{12} = +36 \text{ kip-ft} \\
M_{FBA} &= -36 \text{ kip-ft} \\
M_{FBC} &= +\frac{(2)(24)^2}{12} + (\tfrac{1}{8})(20)(24) = +156 \text{ kip-ft} \\
M_{FCB} &= -156 \text{ kip-ft} \\
M_{FCD} &= +\frac{(18)(4)(8)^2}{(12)^2} = +32 \text{ kip-ft} \\
M_{FDC} &= -\frac{(18)(8)(4)^2}{(12)^2} = -16 \text{ kip-ft}
\end{aligned}
$$

Slope-deflection Equations. By using the modified slope-deflection equations (62), the following expressions for the end moments are found:

$$
\begin{aligned}
M_{AB} &= +36 + 3(-2\theta_A - \theta_B) = +36 - 6\theta_A - 3\theta_B \\
M_{BA} &= -36 + 3(-2\theta_B - \theta_A) = -36 - 6\theta_B - 3\theta_A \\
M_{BC} &= +156 + 5(-2\theta_B - \theta_C) = +156 - 10\theta_B - 5\theta_C \\
M_{CB} &= -156 + 5(-2\theta_C - \theta_B) = -156 - 10\theta_C - 5\theta_B \\
M_{CD} &= +32 + 2(-2\theta_C - \theta_D) = +32 - 4\theta_C - 2\theta_D \\
M_{DC} &= -16 + 2(-2\theta_D - \theta_C) = -16 - 4\theta_D - 2\theta_C
\end{aligned}
$$

Joint Conditions

Joint A:

$$M_{AB} = 0$$

Joint B:

$$M_{BA} + M_{BC} = 0$$

Joint C:

$$M_{CB} + M_{CD} = 0$$

Joint D:

$$M_{DC} = -18 \text{ kip-ft}$$

(*Note:* End moment acting on member CD at D is 18 kip-ft clockwise, or $M_{DC} = -18$ kip-ft.)

Simultaneous Equations. By substituting the slope-deflection equations into the joint conditions, the following simultaneous equations are established:

$$-6\theta_A - 3\theta_B = -36 \qquad (a)$$

$$-3\theta_A - 16\theta_B - 5\theta_C = -120 \qquad (b)$$

$$-5\theta_B - 14\theta_C - 2\theta_D = +124 \qquad (c)$$

$$-2\theta_C - 4\theta_D = -2 \qquad (d)$$

Note that if a diagonal is drawn downward to the right on the left side of Eqs. (*a*) to (*d*), not only are the coefficients on this diagonal predominant in their own equations, but the other coefficients are *symmetrical* with respect to this diagonal. This can be proved to be always true by the nature of the slope-deflection equations and the joint conditions. In order to observe this phenomenon, which sometimes serves as a check, it is important to arrange the unknowns in the order of θ_A, θ_B, θ_C, and θ_D along the horizontal direction and the joint conditions in the order of joints A, B, C, and D in the vertical direction.

Solutions for Relative Values of Joint Rotations

$$-6\theta_A - 3\theta_B = -36 \qquad (a)$$
$$-3\theta_A - 16\theta_B - 5\theta_C = -120 \qquad (b)$$
$$-5\theta_B - 14\theta_C - 2\theta_D = +124 \qquad (c)$$
$$-2\theta_C - 4\theta_D = -2 \qquad (d)$$

Subtracting twice (*b*) from (*a*),

$$+29\theta_B + 10\theta_C = +204 \qquad (e)$$

Subtracting (*d*) from twice (*c*),

$$-10\theta_B - 26\theta_C = +250 \qquad (f)$$

Multiplying (*e*) by 2.6,

$$+75.4\theta_B + 26\theta_C = +530.4 \qquad (g)$$

Adding (*f*) and (*g*),

$$+65.4\theta_B = +780.4$$
$$\theta_B = +11.9327 \qquad (h)$$

Substituting (*h*) into (*f*),

$$-119.327 - 26\theta_C = +250$$
$$\theta_C = -14.2049 \qquad (i)$$

Substituting (h) into (a),

$$-6\theta_A - 35.7981 = -36$$
$$\theta_A = +0.03365 \qquad (j)$$

Substituting (i) into (d),

$$+28.4098 - 4\theta_D = -2$$
$$\theta_D = +7.6025 \qquad (k)$$

The method of elimination is used. When the value of θ_B is found in (h), θ_C may be found from either (e) or (f), but (f) is preferred because the coefficient of θ_C in (f) is predominant; thus a greater accuracy is preserved. For the same reason, values of θ_A and θ_D are computed from (a) and (d), respectively.

Computation for End Moments

$$M_{AB} = +36 - 6\theta_A - 3\theta_B = +36 - (6)(+0.03365) - (3)(+11.9327) = 0$$

$$M_{BA} = -36 - 6\theta_B - 3\theta_A = -36 - (6)(+11.9327) - (3)(+0.03365) = -107.70 \text{ kip-ft}$$

$$M_{BC} = +156 - 10\theta_B - 5\theta_C = +156 - (10)(+11.9327) - (5)(-14.2049) = +107.70 \text{ kip-ft}$$

$$M_{CB} = -156 - 10\theta_C - 5\theta_B = -156 - (10)(-14.2049) - (5)(+11.9327) = -73.62 \text{ kip-ft}$$

$$M_{CD} = +32 - 4\theta_C - 2\theta_D = +32 - (4)(-14.2049) - (2)(+7.6025) = +73.62 \text{ kip-ft}$$

$$M_{DC} = -16 - 4\theta_D - 2\theta_C = -16 - (4)(+7.6025) - (2)(-14.2049) = -18.00 \text{ kip-ft}$$

Note that the four joint conditions $M_{AB} = 0$, $M_{BA} + M_{BC} = 0$, $M_{CB} + M_{CD} = 0$, and $M_{DC} = -18$ kip-ft are satisfied.

Reactions, Shear and Moment Diagrams, and Elastic Curve. The end moments determined above are shown to act at the ends of the spans (Fig. 127). Note again that positive end moments are counterclockwise when acting on the member. The shear and moment diagrams are drawn as usual. The sketch of the elastic curve, with relative slopes of tangents at the supports known, is also shown in Fig. 127.

Example 55. Analyze the continuous beam shown in Fig. 128 by the slope-deflection method. Draw shear and moment diagrams. Sketch the elastic curve.

Solution. The values of the relative stiffness and the fixed-end moments are computed and shown in Fig. 128. Extreme care must be exercised in determining these values because the subsequent computation, even though its own correctness can be checked and thus assured, depends nevertheless on these preliminary quantities.

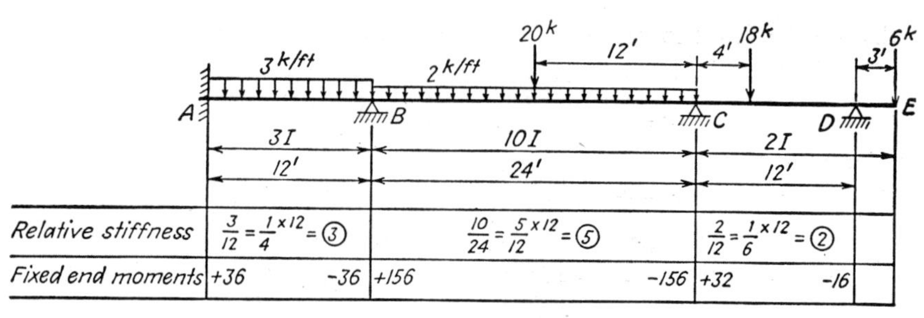

	AB		BC		CD		DE
Relative stiffness	$\frac{3}{12}=\frac{1\times 12}{4}=$ (3)		$\frac{10}{24}=\frac{5\times 12}{12}=$ (5)		$\frac{2}{12}=\frac{1\times 12}{6}=$ (2)		
Fixed end moments	+36	−36	+156	−156	+32	−16	

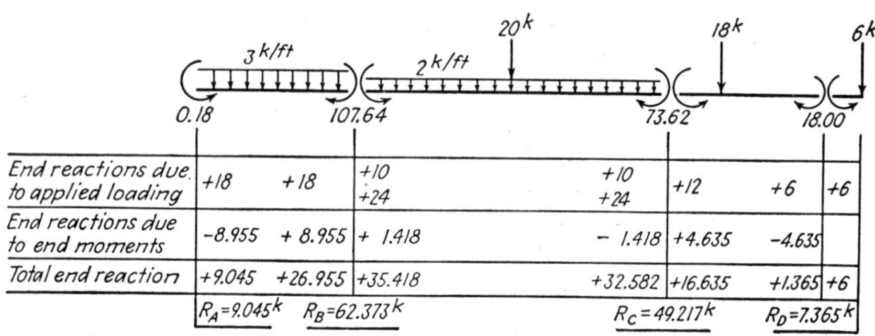

End reactions due to applied loading	+18	+18	+10 +24	+10 +24	+12	+6	+6
End reactions due to end moments	−8.955	+8.955	+1.418	−1.418	+4.635	−4.635	
Total end reaction	+9.045	+26.955	+35.418	+32.582	+16.635	+1.365	+6
	$R_A=9.045^k$	$R_B=62.373^k$		$R_C=49.217^k$		$R_D=7.365^k$	

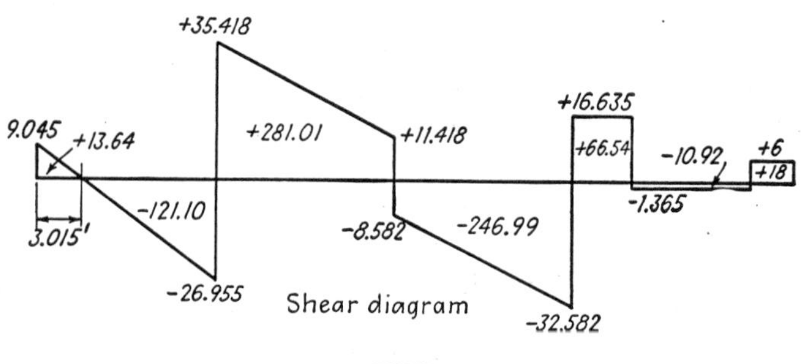

Shear diagram

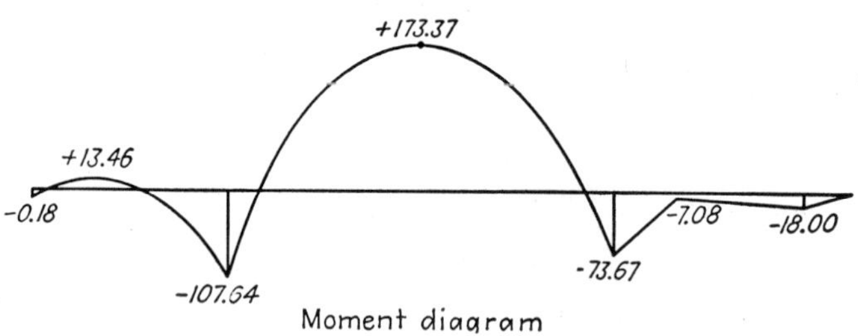

Moment diagram

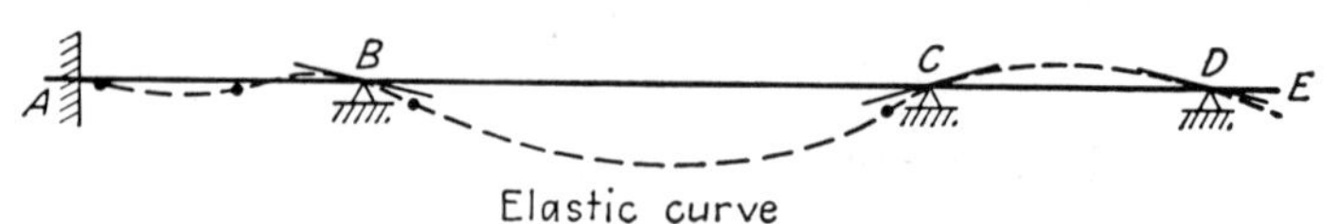

Elastic curve

FIG. 128.

Slope-deflection Equations. The modified slope-deflection equations

$$M_{AB} = M_{FAB} + K_{AB}(-2\theta_A - \theta_B)$$
$$M_{BA} = M_{FBA} + K_{AB}(-2\theta_B - \theta_A)$$

will be used.

In the present problem, θ_A is known to be zero.

$$\begin{aligned}
M_{AB} &= +36 + 3(-2\theta_A - \theta_B) = +36 - 3\theta_B \\
&= +36 - (3)(+11.9399) = +0.18 \text{ kip-ft} \\
M_{BA} &= -36 + 3(-2\theta_B - \theta_A) = -36 - 6\theta_B \\
&= -36 - (6)(+11.9399) = -107.64 \text{ kip-ft} \\
M_{BC} &= +156 + 5(-2\theta_B - \theta_C) = +156 - 10\theta_B - 5\theta_C \\
&= +156 - (10)(+11.9399) - (5)(-14.2076) = +107.64 \text{ kip-ft} \\
M_{CB} &= -156 + (5)(-2\theta_C - \theta_B) = -156 - 10\theta_C - 5\theta_B \\
&= -156 - (10)(-14.2076) - (5)(+11.9399) = -73.62 \text{ kip-ft} \\
M_{CD} &= +32 + 2(-2\theta_C - \theta_D) = +32 - 4\theta_C - 2\theta_D \\
&= +32 - (4)(-14.2076) - 2(+7.6038) = +73.62 \text{ kip-ft} \\
M_{DC} &= -16 + 2(-2\theta_D - \theta_C) = -16 - 4\theta_D - 2\theta_C \\
&= -16 - (4)(+7.6038) - (2)(-14.2076) = -18.00 \text{ kip-ft}
\end{aligned}$$

Joint Conditions

Joint B: $\quad M_{BA} + M_{BC} = 0 \qquad (a)$

Joint C: $\quad M_{CB} + M_{CD} = 0 \qquad (b)$

Joint D: $\quad M_{DC} = -18$ kip-ft $\qquad (c)$

$$-16\theta_B - 5\theta_C = -120 \qquad (a')$$

$$-5\theta_B - 14\theta_C - 2\theta_D = +124 \qquad (b')$$

$$-2\theta_C - 4\theta_D = -2 \qquad (c')$$

Subtracting (c') from twice (b'),

$$-10\theta_B - 26\theta_C = +250 \qquad (d)$$

Multiplying (d) by 1.6,

$$-16\theta_B - 41.6\theta_C = +400 \qquad (e)$$

Subtracting (e) from (a'),

$$+36.6\theta_C = -520$$
$$\theta_C = -14.2076 \qquad (f)$$

Substituting (f) into (a'),

$$-16\theta_B + 71.038 = -120$$
$$\theta_B = +11.9399 \qquad (g)$$

Substituting (f) into (c'),

$$+28.4152 - 4\theta_D = -2$$
$$\theta_D = +7.6038 \qquad (h)$$

These θ values are substituted into the slope-deflection equations to obtain the end moments. The reactions, shear and moment diagrams, and the elastic curve are shown in Fig. 128.

EXERCISES

72 to 75. Analyze the continuous beam shown by the slope-deflection method. Draw shear and moment diagrams. Sketch the elastic curve.

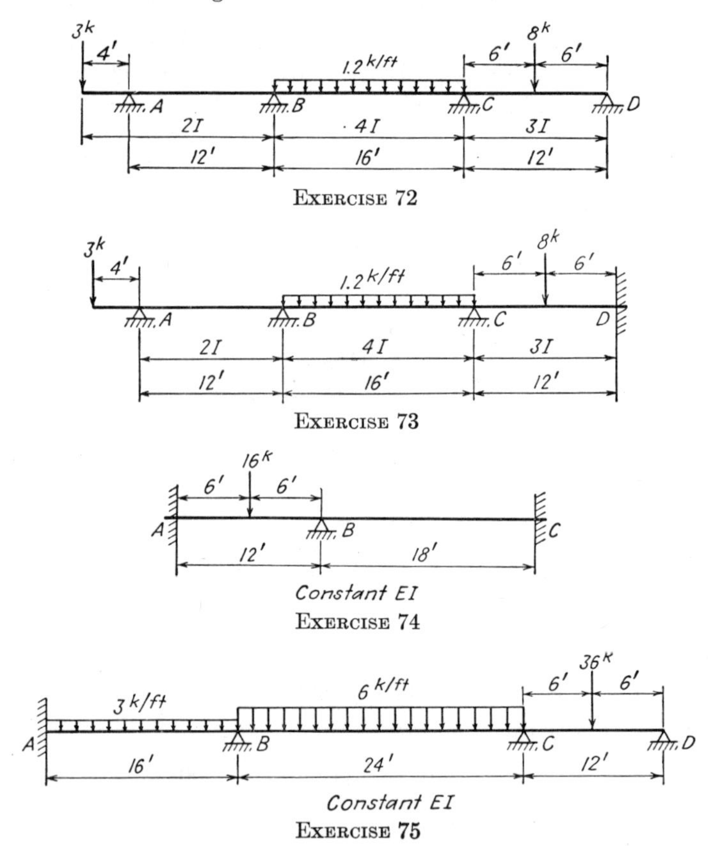

38. Slope-deflection Equations for Members Which Are Subjected to Applied Loadings and Unequal Movements of End Joints in the Direction Perpendicular to the Axis of the Member. The slope-deflection equations as shown in Eqs. (61) express the end moments in terms of the end rotations and the applied loading. If, in addition to the applied loading, the end joints are subjected to unequal movements in the direction perpendicular to the axis of the member, additional fixed end moments M'_{FAB} and M'_{FBA} (Fig 129c) are induced to act on the member to keep the tangents at the ends fixed. Then M'_A and M'_B

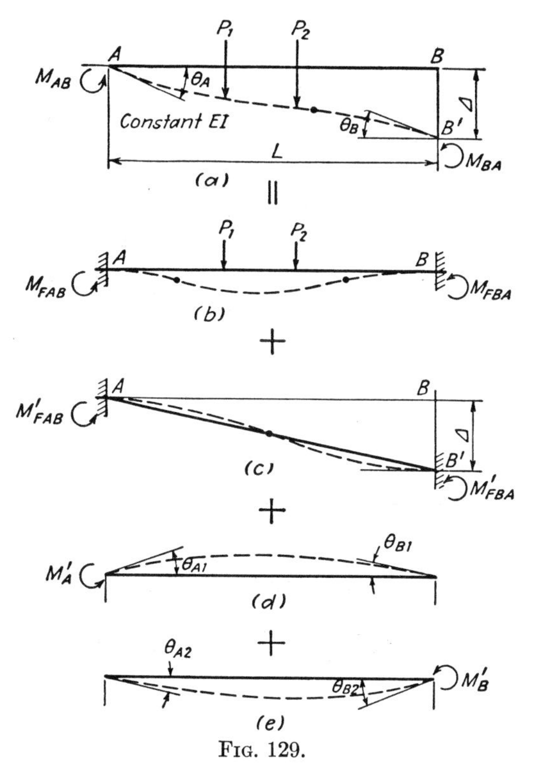

FIG. 129.

should be such as to cause rotations of θ_A and θ_B, respectively. The conditions required of geometry are

$$\begin{aligned} \theta_A &= -\theta_{A1} + \theta_{A2} \\ \theta_B &= \theta_{B1} - \theta_{B2} \end{aligned} \tag{63}$$

By superposition,

$$\begin{aligned} M_{AB} &= M_{FAB} + M'_{FAB} + M'_A \\ M_{BA} &= M_{FBA} + M'_{FBA} + M'_B \end{aligned} \tag{64}$$

From Eqs. (60),

$$\begin{aligned} M'_A &= +\frac{2EI}{L}(-2\theta_A - \theta_B) \\ M'_B &= +\frac{2EI}{L}(-2\theta_B - \theta_A) \end{aligned} \tag{60}$$

For determining M'_{FAB} and M'_{FBA}, the moment-area method will be used (Fig. 130). Note that R, the angle measured from the original direction of member AB to the line joining the displaced joints, is *positive* when *clockwise*. Note also that

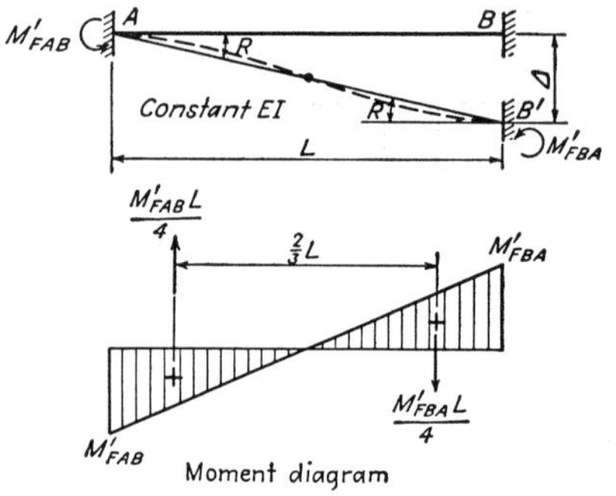

$$R = \frac{\Delta}{L}$$

By the first moment-area theorem,

Change of slope between tangents at A and B' = area of $\frac{M}{EI}$ diagram between A and B' = 0

or,

$$M'_{FAB} = M'_{FBA}$$

FIG. 130.

Deflection of B' from tangent at A $= \frac{1}{EI}\left[\frac{M'_{FAB}(L)}{4}\right][\tfrac{2}{3}L] = \frac{M'_{FAB}L^2}{6EI} = \Delta$

or,

$$M'_{FAB} = M'_{FBA} = \frac{6EI\Delta}{L^2} = \frac{6EIR}{L} \tag{65}$$

Substituting Eqs. (60) and (65) into Eqs. (64),

$$\begin{aligned} M_{AB} &= M_{FAB} + \frac{2EI}{L}(-2\theta_A - \theta_B + 3R) \\ M_{BA} &= M_{FBA} + \frac{2EI}{L}(-2\theta_B - \theta_A + 3R) \end{aligned} \tag{66}$$

Equations (66) are the slope-deflection equations which express the end moments in terms of the end rotations, the applied loading, and the angle R between the line joining the deflected end joints and the original direction of the member. If the angle R is zero, Eqs. (66) reduce to Eqs. (61).

39. Analysis of Statically Indeterminate Beams Due to Yielding of Supports. The general slope-deflection equations (66) can be used to analyze statically indeterminate beams due to the combined action of applied loadings and unequal settlement of supports. But as discussed previously, usually the effect of the yielding of one support only is investigated, and the results may be combined with those of applied loadings or settlement at other supports. In the following two examples, the use of the slope-deflection method to analyze statically indeterminate beams due to the yielding of one support is illustrated.

Example 56. Analyze the continuous beam shown in Fig. 131 owing to the effect of a $\frac{1}{2}$-in. settlement at support B by the slope-deflection method. Draw shear and moment diagrams. Sketch the elastic curve.

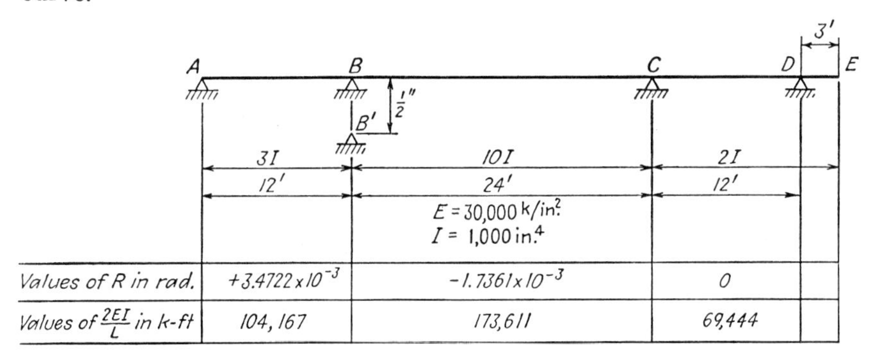

Values of R in rad.	$+3.4722 \times 10^{-3}$	-1.7361×10^{-3}	0	
Values of $\frac{2EI}{L}$ in k-ft	104,167	173,611	69,444	

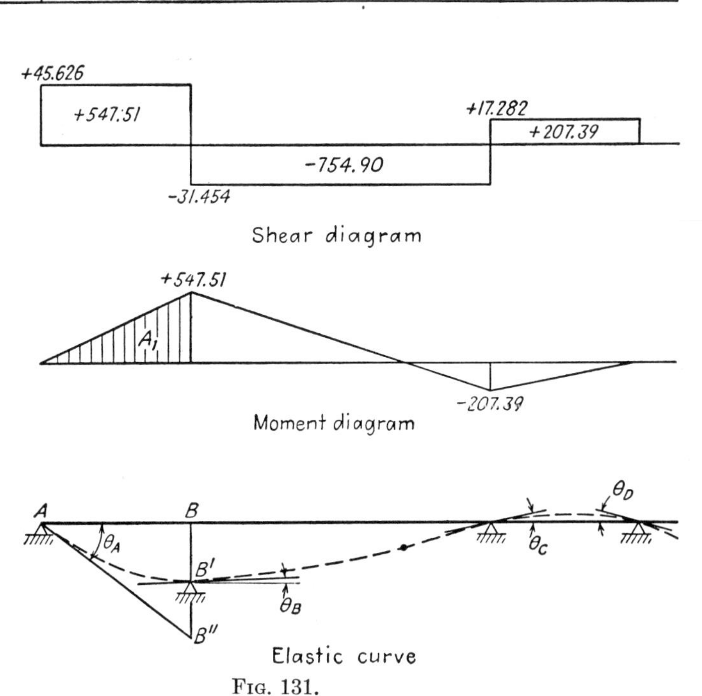

End reactions due to end moments	+45.626 -45.626	-31.454 ... +31.454	+17.282 -17.282	
Reactions	$R_A = +45.626^k$ $R_B = -77.080^k$		$R_C = +48.736^k$ $R_D = -17.282^k$	

Fig. 131.

Solution. The slope-deflection equations

$$M_{AB} = M_{FAB} + \frac{2EI}{L}(-2\theta_A - \theta_B + 3R)$$
$$M_{BA} = M_{FBA} + \frac{2EI}{L}(-2\theta_B - \theta_A + 3R)$$

are to be used; in this example there are no fixed-end moments due to applied loadings, but there are R values for spans AB and BC. Inasmuch as the absolute values of R in radians are known the absolute values of $2EI/L$ in kip-feet *must* be used.

Span AB:
$$R = +\frac{0.500 \text{ in.}}{144 \text{ in.}} = +3.4722 \times 10^{-3} \text{ rad}$$
$$\frac{2EI}{L} = \frac{(2)(30{,}000)(3{,}000)}{(144)(12)} = 104{,}167 \text{ kip-ft}$$

Span BC:
$$R = -\frac{0.500}{288} = -1.7361 \times 10^{-3} \text{ rad}$$
$$\frac{2EI}{L} = \frac{(2)(30{,}000)(10{,}000)}{(144)(24)} = 173{,}611 \text{ kip-ft}$$

Span CD:
$$R = 0$$
$$\frac{2EI}{L} = \frac{(2)(30{,}000)(2{,}000)}{(144)(12)} = 69{,}444 \text{ kip-ft}$$

Slope-deflection Equations

$$M_{AB} = 104{,}167\,[-2\theta_A - \theta_B + (3)(+3.4722 \times 10^{-3})]$$
$$= -208{,}333\theta_A - 104{,}167\theta_B + 1{,}085.06$$
$$M_{BA} = 104{,}167\,[-2\theta_B - \theta_A + (3)(+3.4722 \times 10^{-3})]$$
$$= -208{,}333\theta_B - 104{,}167\theta_A + 1{,}085.06$$
$$M_{BC} = 173{,}611\,[-2\theta_B - \theta_C + (3)(-1.7361 \times 10^{-3})]$$
$$= -347{,}222\theta_B - 173{,}611\theta_C - 904.22$$
$$M_{CB} = 173{,}611\,[-2\theta_C - \theta_B + (3)(-1.7361 \times 10^{-3})]$$
$$= -347{,}222\theta_C - 173{,}611\theta_B - 904.22$$
$$M_{CD} = 69{,}444\,[-2\theta_C - \theta_D + (3)(0)] = -138{,}889\theta_C - 69{,}444\theta_D$$
$$M_{DC} = 69{,}444\,[-2\theta_D - \theta_C + (3)(0)] = -138{,}889\theta_D - 69{,}444\theta_C$$

Joint Conditions

Joint A: $\quad M_{AB} = 0 \quad$ (a)

Joint B: $\quad M_{BA} + M_{BC} = 0 \quad$ (b)

Joint C: $\quad M_{CB} + M_{CD} = 0 \quad$ (c)

Joint D: $\quad M_{DC} = 0 \quad$ (d)

$$-208{,}333\theta_A - 104{,}167\theta_B = -1{,}085.06 \qquad (a')$$

$$-104{,}167\theta_A - 555{,}555\theta_B - 173{,}611\theta_C = -180.84 \qquad (b')$$

$$- 173{,}611\theta_B - 486{,}111\theta_C - 69{,}444\theta_D = +904.22 \qquad (c')$$

$$-69{,}444\theta_C - 138{,}889\theta_D = 0 \qquad (d')$$

Subtracting twice (b') from (a'),

$$+1{,}006{,}943\theta_B + 347{,}222\theta_C = -723.38 \qquad (e)$$

From (d'),

$$\theta_D = -\tfrac{1}{2}\theta_C \qquad (f)$$

Substituting (f) into (c'),

$$-173{,}611\theta_B - 451{,}389\theta_C = +904.22 \qquad (g)$$

Dividing (e) by 1,006,943,

$$\theta_B + 0.344830\theta_C = -0.71839 \times 10^{-3} \qquad (h)$$

Dividing (g) by 173,611,

$$-\theta_B - 2.600000\theta_C = +5.20831 \times 10^{-3} \qquad (i)$$

Adding (h) and (i),

$$-2.255170\theta_C = +4.48992 \times 10^{-3}$$

$$\theta_C = -1.99095 \times 10^{-3} \qquad (j)$$

Substituting (j) into (f),

$$\theta_D = -\tfrac{1}{2}\theta_C = +0.99547 \times 10^{-3} \qquad (k)$$

Substituting (j) into (h),

$$\theta_B = -0.71839 \times 10^{-3} + 0.68654 \times 10^{-3} = -0.03185 \times 10^{-3} \qquad (l)$$

Substituting (l) into (a'),

$$-\theta_A - \tfrac{1}{2}\theta_B = -5.20830 \times 10^{-3}$$

$$\theta_A = +5.22423 \times 10^{-3} \qquad (m)$$

Computations for End Moments

$$M_{AB} = (-208{,}333)(+5.22423 \times 10^{-3}) - (104{,}167)(-0.03185 \times 10^{-3}) + 1{,}085.06 = 0$$

$$M_{BA} = (-208{,}333)(-0.03185 \times 10^{-3}) - (104{,}167)(+5.22423 \times 10^{-3}) + 1{,}085.06 = +547.51 \text{ kip-ft}$$

$$M_{BC} = (-347{,}222)(-0.03185 \times 10^{-3}) - (173{,}611)(-1.99095 \times 10^{-3}) - 904.22 = -547.51 \text{ kip-ft}$$

$$M_{CB} = (-347{,}222)(-1.99095 \times 10^{-3}) - (173{,}611)(-0.03185 \times 10^{-3}) - 904.22 = -207.39 \text{ kip-ft}$$

$$M_{CD} = (-138{,}889)(-1.99095 \times 10^{-3}) - (69{,}444)(+0.99547 \times 10^{-3}) = +207.39 \text{ kip-ft}$$

$$M_{DC} = (-138{,}889)(+0.99547 \times 10^{-3}) - (69{,}444)(-1.99095 \times 10^{-3}) = 0$$

The reactions, shear and moment diagrams, and the elastic curve are shown in Fig. 131.

A good independent check can be made by finding BB' (Fig. 131) by the moment-area method.

$$BB' = BB'' - B'B''$$

$$= (144 \text{ in.})\theta_A - \frac{1}{E(3I)} (\text{moment of area } A_1 \text{ about } B)$$

$$= (144)(5.22423 \times 10^{-3}) - \frac{1{,}728}{(30{,}000)(3{,}000)} \left[\frac{(547.51)(12)}{2} (4) \right]$$

$$= 0.75229 \text{ in.} - 0.25229 \text{ in.} = 0.5000 \text{ in. } (\textit{check})$$

Example 57. Analyze the continuous beam shown in Fig. 132 owing to the effect of a $\frac{1}{2}$-in. settlement at support B by the slope-deflection method. Draw shear and moment diagrams. Sketch the elastic curve.

Solution. *Values of R and $2EI/L$*

Span AB:
$$R = + \frac{0.500}{144} = +3.4722 \times 10^{-3} \text{ rad}$$
$$\frac{2EI}{L} = \frac{(2)(30{,}000)(3{,}000)}{(144)(12)} = 104{,}167 \text{ kip-ft}$$

Span BC:
$$R = - \frac{0.500}{288} = -1.7361 \times 10^{-3} \text{ rad}$$
$$\frac{2EI}{L} = \frac{(2)(30{,}000)(10{,}000)}{(144)(24)} = 173{,}611 \text{ kip-ft}$$

Span CD:
$$R = 0$$
$$\frac{2EI}{L} = \frac{(2)(30{,}000)(2{,}000)}{(144)(12)} = 69{,}444 \text{ kip-ft}$$

Slope-deflection Equations. In the present example, θ_A is known to be zero.

$$M_{AB} = 104{,}167[-2\theta_A - \theta_B + 3(+3.4722 \times 10^{-3})] = -104{,}167\theta_B + 1{,}085.06$$
$$M_{BA} = 104{,}167[-2\theta_B - \theta_A + 3(+3.4722 \times 10^{-3})] = -208{,}333\theta_B + 1{,}085.06$$
$$M_{BC} = 173{,}611[-2\theta_B - \theta_C + 3(-1.7361 \times 10^{-3})] = -347{,}222\theta_B - 173{,}611\theta_C - 904.22$$
$$M_{CB} = 173{,}611[-2\theta_C - \theta_B + 3(-1.7361 \times 10^{-3})] = -347.222\theta_C - 173{,}611\theta_B - 904.22$$
$$M_{CD} = 69{,}444[-2\theta_C - \theta_D + 3(0)] = -138{,}889\theta_C - 69{,}444\theta_D$$
$$M_{DC} = 69{,}444[-2\theta_D - \theta_C + 3(0)] = -138{,}889\theta_D - 69{,}444\theta_C$$

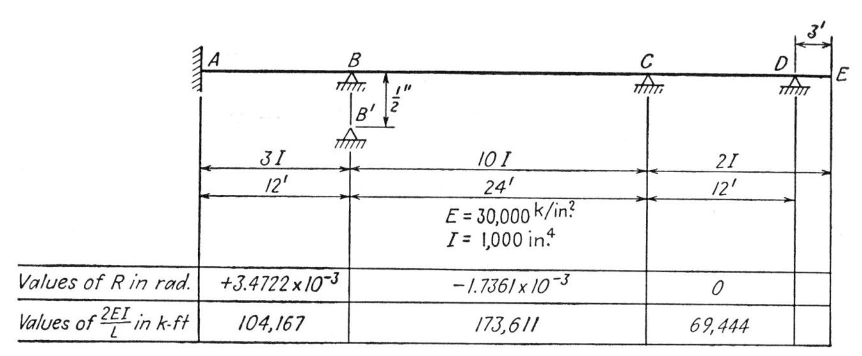

	A–B	B–C	C–D	
Values of R in rad.	$+3.4722 \times 10^{-3}$	-1.7361×10^{-3}	0	
Values of $\frac{2EI}{L}$ in k-ft	104,167	173,611	69,444	

972.42   859.76   252.00   0

End reactions due to end moments	+152.682	−152.682	−46.323	+46.323	+21.000	−21.000
Reactions	$R_A = +152.682^k$	$R_B = -199.005^k$		$R_C = +67.323^k$		$R_D = -21.000^k$

Shear diagram

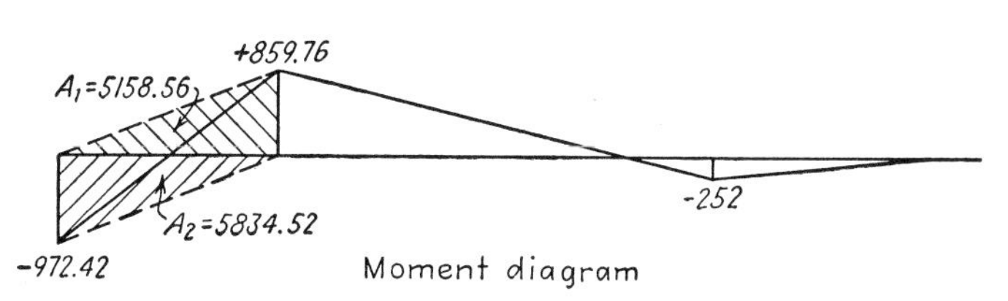

Moment diagram

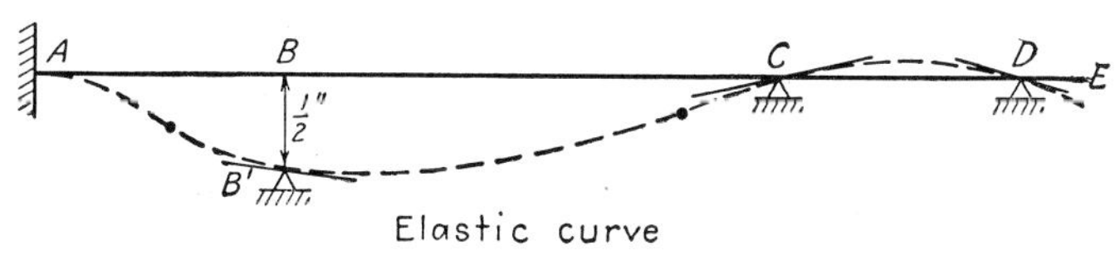

Elastic curve

FIG. 132.

Joint Conditions

Joint B: $\quad M_{BA} + M_{BC} = 0 \quad (a)$

Joint C: $\quad M_{CB} + M_{CD} = 0 \quad (b)$

Joint D: $\quad M_{DC} = 0 \quad (c)$

$$-555{,}555\theta_B - 173{,}611\theta_C = -180.84 \qquad (a')$$

$$-173{,}611\theta_B - 486{,}111\theta_C - 69{,}444\theta_D = +904.22 \qquad (b')$$

$$-69{,}444\theta_C - 138{,}889\theta_D = 0 \qquad (c')$$

Solving,

$$\theta_B = +1.08150 \times 10^{-3}$$
$$\theta_C = -2.41916 \times 10^{-3}$$
$$\theta_D = +1.20958 \times 10^{-3}$$

Computations for End Moments. The end moments are computed by substituting the θ values into the slope-deflection equations.

$$M_{AB} = +972.42 \text{ kip-ft}$$
$$M_{BA} = +859.76 \text{ kip-ft}$$
$$M_{BC} = -859.76 \text{ kip-ft}$$
$$M_{CB} = -252.00 \text{ kip-ft}$$
$$M_{CD} = +252.00 \text{ kip-ft}$$
$$M_{DC} = 0$$

The reactions, shear and moment diagrams, and the elastic curve are shown in Fig. 132.

A good independent check can be made by finding BB' (Fig. 132) by the moment-area method.

$$BB' = \frac{1}{E(3I)} [(\text{moment of area } A_2 \text{ about } B) - (\text{moment of area } A_1 \text{ about } B)]$$

$$= \frac{1{,}728}{(30{,}000)(3{,}000)} [(5{,}834.52)(8) - (5{,}158.56)(4)]$$

$$= 0.50000 \text{ in. } (check)$$

EXERCISES

76 and 77. Analyze the continuous beam shown owing to a settlement of 0.125 in. at support B by the slope-deflection method. Draw shear and moment diagrams. Sketch the elastic curve.

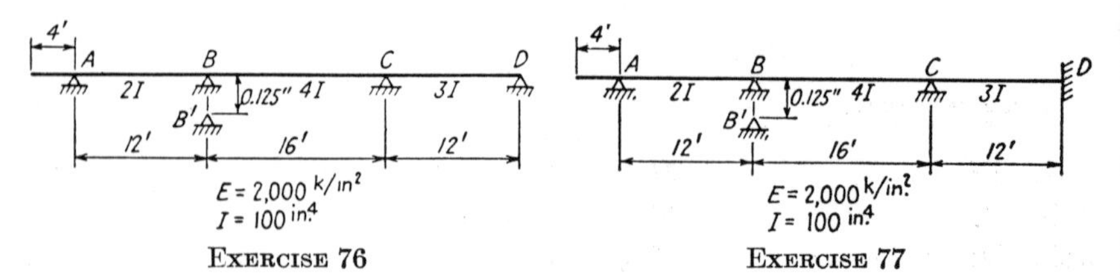

EXERCISE 76 EXERCISE 77

40. Application of the Slope-deflection Method to the Analysis of Statically Indeterminate Frames. *Case* I—*without Joint Movements.* The modified slope deflection equations

$$M_{AB} = M_{FAB} + K_{AB}(-2\theta_A - \theta_B)$$
$$M_{BA} = M_{FBA} + K_{AB}(-2\theta_B - \theta_A)$$

can be used to analyze statically indeterminate frames wherein all joints remain fixed in location during deformation. Again the axial deformation in the members owing to direct stress is neglected in applying the conditions of consistent deformation to the analysis; but direct stresses, together with shears and moments, must be considered in the design of sections. The conditions for consistent deformation are those of the rigidity of joints, or the angle between any two tangents to the elastic curves meeting at one joint must remain the same as that in the original undeformed structure. In the slope-deflection method the rotation at each joint is considered as the unknown, while the condition corresponding to this unknown is one of statics; that is, the sum of the moments, as expressed by the slope-deflection equations, acting on the joint is equal to zero. Thus there are always as many conditions of statics as unknown rotations. After the latter are solved, all end moments can be found from the slope-deflection equations. With all end moments known, the direct stresses, shears, and moments in all members can be found by applying the principles of statics to the individual members.

Example 58. Analyze the rigid frame shown in Fig. 133 by the slope-deflection method. Find the direct stresses, shears, and moments in all members. Sketch the deformed structure.

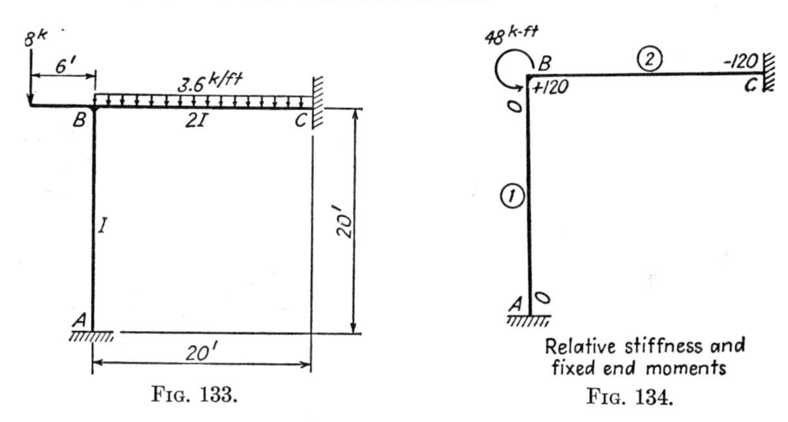

FIG. 133.

FIG. 134.

Solution. Joints A and C are fixed; joint B must be at 20 ft to the left of C and at 20 ft above A, and therefore it remains stationary. Thus all three joints cannot move, and R is zero for all members.

Relative Stiffness and Fixed-end Moments (Fig. 134)

TABLE 58-1

Span AB	$I/20$	1
Span BC	$2I/20$	2

$$M_{FBC} = + \frac{(3.6)(20)^2}{12} = +120 \text{ kip-ft}$$
$$M_{FCB} = -120 \text{ kip-ft}$$

Slope-deflection Equations. θ_A and θ_C are both zero.

$$M_{AB} = 0 + 1(-2\theta_A - \theta_B) = -\theta_B = -12 \text{ kip-ft}$$
$$M_{BA} = 0 + 1(-2\theta_B - \theta_A) = -2\theta_B = (-2)(+12) = -24 \text{ kip-ft}$$
$$M_{BC} = +120 + 2(-2\theta_B - \theta_C) = +120 - 4\theta_B = (+120) - (4)(+12) = +72 \text{ kip-ft}$$
$$M_{CB} = -120 + 2(-2\theta_C - \theta_B) = -120 - 2\theta_B = (-120) - (2)(+12) = -144 \text{ kip-ft}$$

Joint Condition

$$M_{BA} + M_{BC} - 48 = 0$$
$$-2\theta_B + 120 - 4\theta_B - 48 = 0$$
$$\theta_B = +12$$

The free-body, shear, and moment diagrams of the individual members are shown in Fig. 135. The free-body diagram, moment diagram, and the elastic curve are shown in Fig. 136.

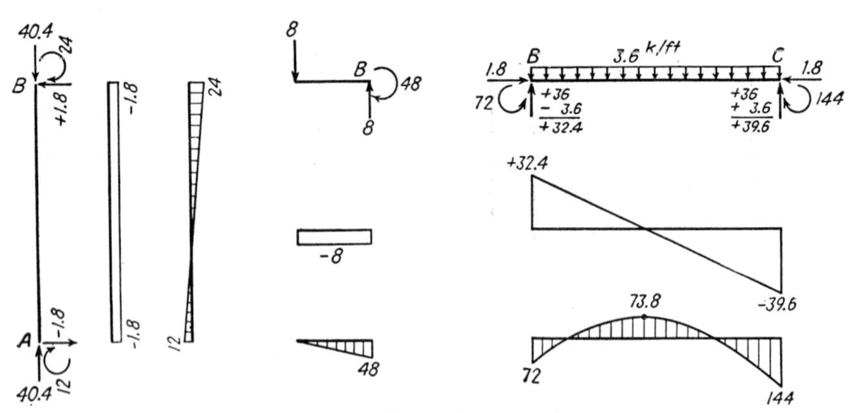

FIG. 135.

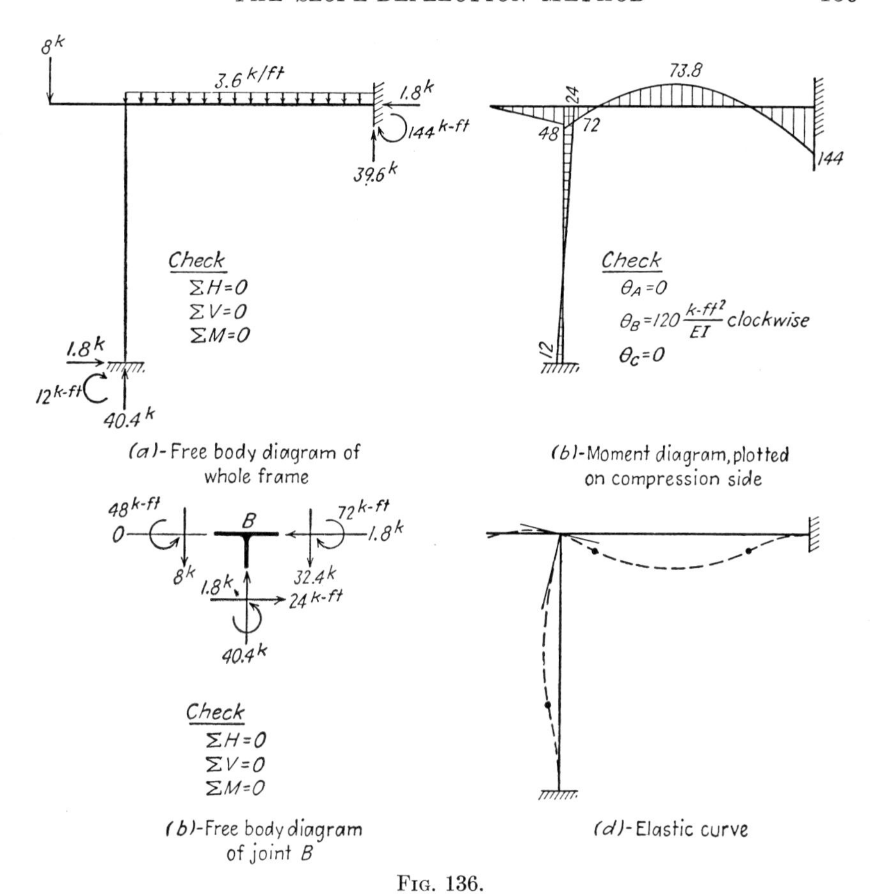

FIG. 136.

Example 59. Analyze the rigid frame shown in Fig. 137 by the slope-deflection method. Find the direct stresses, shears, and moments in all members. Sketch the deformed structure.

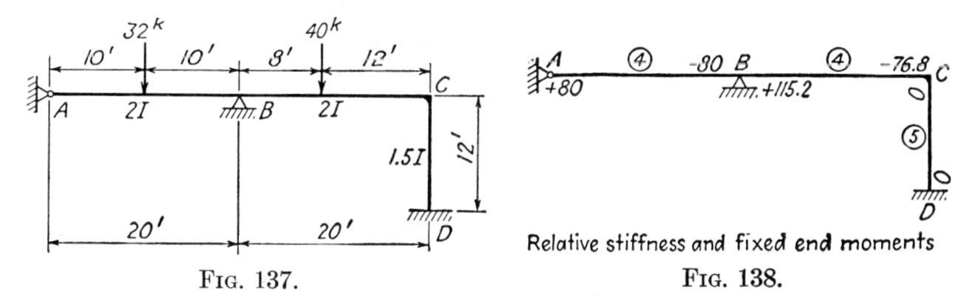

FIG. 137. FIG. 138.

Solution. Joints A, B, C, and D must all remain fixed in location; thus, R is zero for all members.

Relative Stiffness and Fixed-end Moments (Fig. 138)

TABLE 59-1

AB...........	$\left(\frac{2I}{20} = \frac{I}{10}\right) \times 40$	4
BC...........	$\left(\frac{2I}{20} = \frac{I}{10}\right) \times 40$	4
CD...........	$\left(\frac{1.5I}{12} = \frac{I}{8}\right) \times 40$	5

Span AB:

$$M_{FAB} = +\frac{(32)(20)}{8} = +80 \text{ kip-ft}$$

$$M_{FBA} = -80 \text{ kip-ft}$$

Span BC:

$$M_{FBC} = +\frac{(40)(8)(12)^2}{(20)^2} = +115.2 \text{ kip-ft}$$

$$M_{FCB} = -\frac{(40)(12)(8)^2}{(20)^2} = -76.8 \text{ kip-ft}$$

Slope-deflection Equations. θ_D is known to be zero.

$$M_{AB} = +80 + 4(-2\theta_A - \theta_B) = +80 - 8\theta_A - 4\theta_B$$
$$= +80 - (8)(+9.5322) - (4)(+0.9356) = 0$$
$$M_{BA} = -80 + 4(-2\theta_B - \theta_A) = -80 - 8\theta_B - 4\theta_A$$
$$= -80 - (8)(+0.9356) - (4)(+9.5322) = -125.61 \text{ kip-ft}$$
$$M_{BC} = +115.2 + 4(-2\theta_B - \theta_C) = +115.2 - 8\theta_B - 4\theta_C$$
$$= +115.2 - (8)(+0.9356) - (4)(-4.4746) = +125.61 \text{ kip-ft}$$
$$M_{CB} = -76.8 + 4(-2\theta_C - \theta_B) = -76.8 - 8\theta_C - 4\theta_B$$
$$= -76.8 - (8)(-4.4746) - (4)(+0.9356) = -44.75 \text{ kip-ft}$$
$$M_{CD} = 0 + 5(-2\theta_C - \theta_D) = -10\theta_C = +44.75 \text{ kip-ft}$$
$$M_{DC} = 0 + 5(-2\theta_D - \theta_C) = -5\theta_C = +22.37 \text{ kip-ft}$$

Joint Conditions

$$M_{AB} = 0 \qquad +80 - 8\theta_A - 4\theta_B = 0 \qquad (a)$$
$$M_{BA} + M_{BC} = 0 \qquad -80 - 8\theta_B - 4\theta_A + 115.2 - 8\theta_B - 4\theta_C = 0 \qquad (b)$$
$$M_{CB} + M_{CD} = 0 \qquad -76.8 - 8\theta_C - 4\theta_B - 10\theta_C = 0 \qquad (c)$$

$$-8\theta_A - 4\theta_B = -80 \qquad (a')$$
$$-4\theta_A - 16\theta_B - 4\theta_C = -35.2 \qquad (b')$$
$$-4\theta_B - 18\theta_C = +76.8 \qquad (c')$$

Solving,

$$\theta_A = +9.5322$$
$$\theta_B = +0.9356$$
$$\theta_C = -4.4746$$

The free-body, shear, and moment diagrams of the individual members are shown in Fig. 139. The free-body diagram, moment diagram, and the elastic curve are shown in Fig. 140.

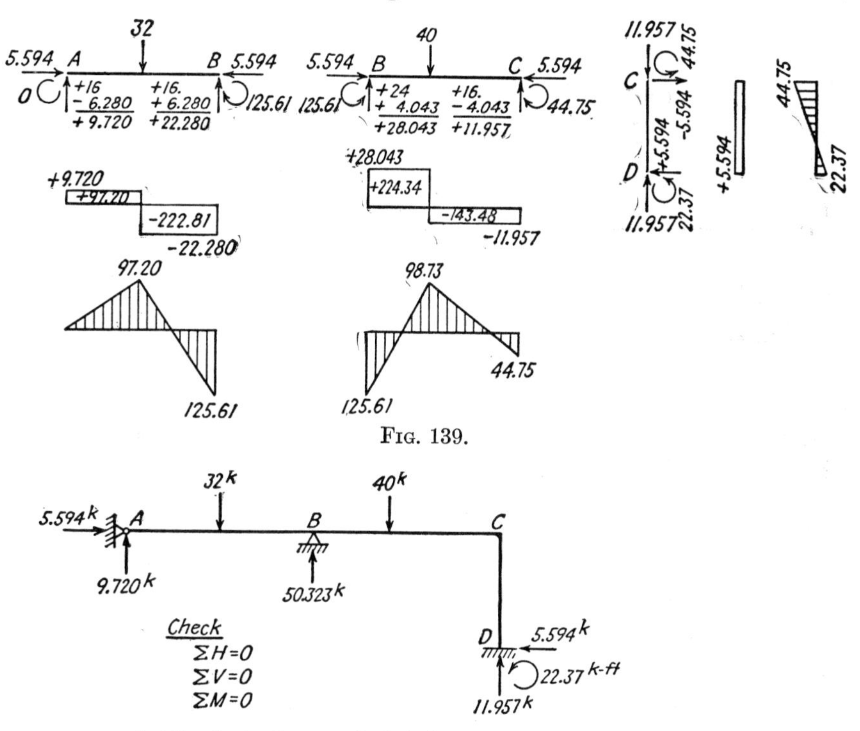

FIG. 139.

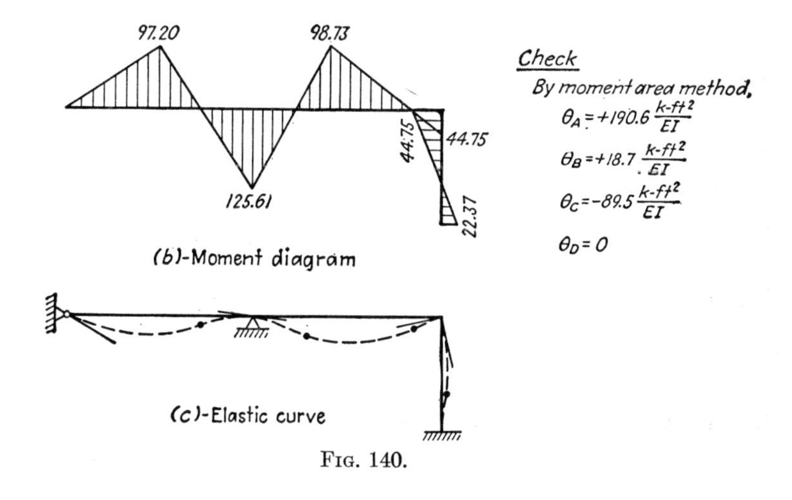

FIG. 140.

Example 60. Analyze the rigid frame shown in Fig. 141 by the slope-deflection method. Draw shear and moment diagrams. Sketch the deformed structure.

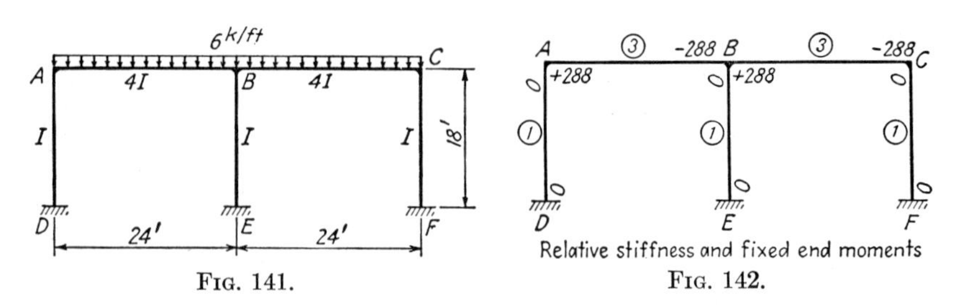

FIG. 141.

FIG. 142.

Solution. Joints D, E, and F are fixed. Joints A, B, and C cannot move in the vertical direction, but each may shift the same distance in the horizontal direction. In the present example, however, on account of the symmetry both in the properties of the frame and in the applied loading, joints A, B, and C will not have any horizontal displacement. Thus R is equal to zero for all members.

Relative Stiffness and Fixed-end Moments (Fig. 142)

TABLE 60-1

AB, BC	$\left(\frac{4I}{24} = \frac{I}{6}\right) \times 18$	3
AD, BE, CF	$\left(\frac{I}{18}\right) \times 18$	1

$$M_{FAB} = M_{FBC} = + \frac{(6)(24)^2}{12} = +288 \text{ kip-ft}$$
$$M_{FBA} = M_{FCB} = -288 \text{ kip-ft}$$

Slope-deflection Equations. θ_D, θ_E, and θ_F are known to be zero. By symmetry: (1) $\theta_C = -\theta_A$ and (2) $\theta_B = 0$.

$$M_{AB} = +288 + 3(-2\theta_A - \theta_B) = +288 - 6\theta_A = +288 - (6)(+36) = +72 \text{ kip-ft}$$
$$M_{BA} = -288 + 3(-2\theta_B - \theta_A) = -288 - 3\theta_A = -288 - (3)(+36) = -396 \text{ kip-ft}$$
$$M_{BC} = +288 + 3(-2\theta_B - \theta_C) = +288 + 3\theta_A = +288 + (3)(+36) = +396 \text{ kip-ft}$$
$$M_{CB} = -288 + 3(-2\theta_C - \theta_B) = -288 + 6\theta_A = -288 + (6)(+36) = -72 \text{ kip-ft}$$

$$M_{AD} = 0 + 1(-2\theta_A - \theta_D) = -2\theta_A = (-2)(+36) = -72 \text{ kip-ft}$$
$$M_{DA} = 0 + 1(-2\theta_D - \theta_A) = -\theta_A = -(+36) = -36 \text{ kip-ft}$$
$$M_{BE} = 0 + 1(-2\theta_B - \theta_E) = 0$$
$$M_{EB} = 0 + 1(-2\theta_E - \theta_B) = 0$$
$$M_{CF} = 0 + 1(-2\theta_C - \theta_F) = +2\theta_A = (+2)(+36) = +72 \text{ kip-ft}$$
$$M_{FC} = 0 + 1(-2\theta_F - \theta_C) = +\theta_A = +(+36) = +36 \text{ kip-ft}$$

Joint Conditions

Joint A: $\quad M_{AB} + M_{AD} = 0 \qquad +288 - 6\theta_A - 2\theta_A = 0$

$$\theta_A = +36$$

The joint conditions at B and C are satisfied on account of the known facts, (1) $\theta_C = -\theta_A$, and (2) $\theta_B = 0$.

The free-body, shear, and moment diagrams of the individual members are shown in Fig. 143. The free-body diagram, moment diagram, and elastic curve are shown in Fig. 144.

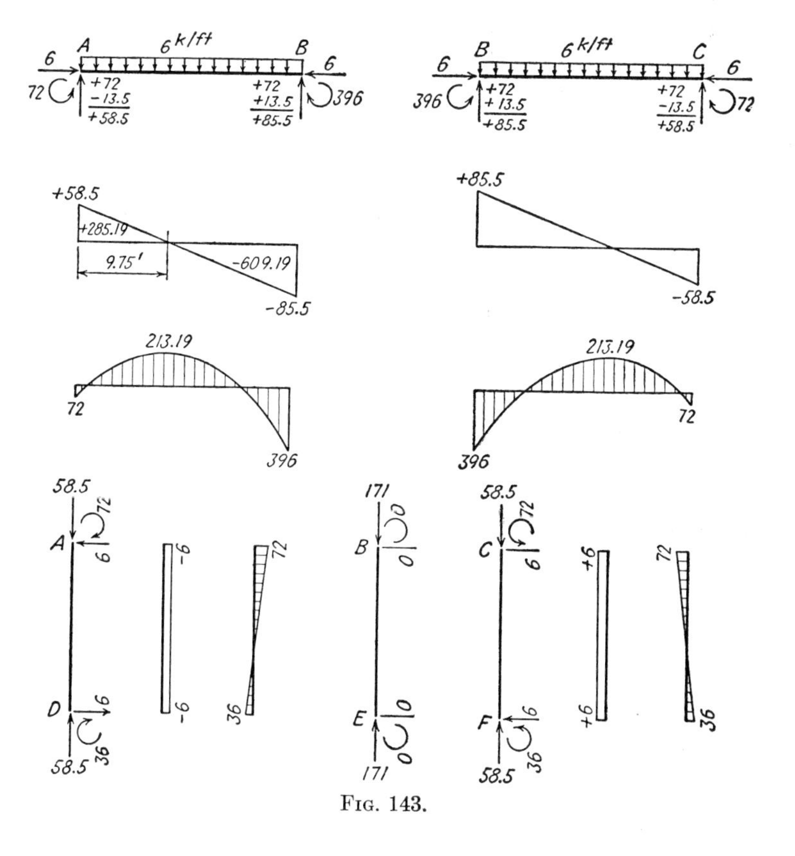

Fig. 143.

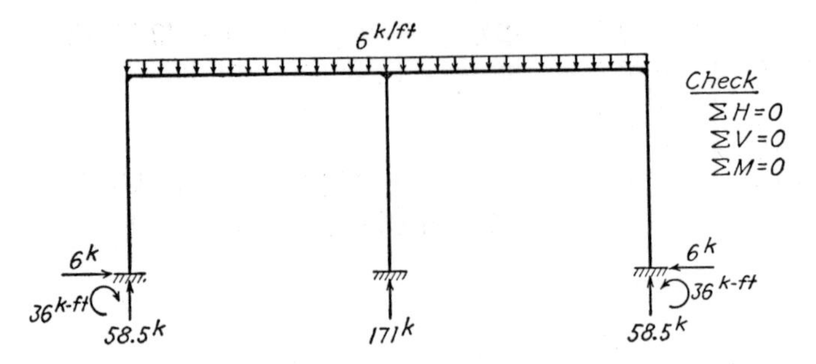

(a)-Free body diagram of whole frame

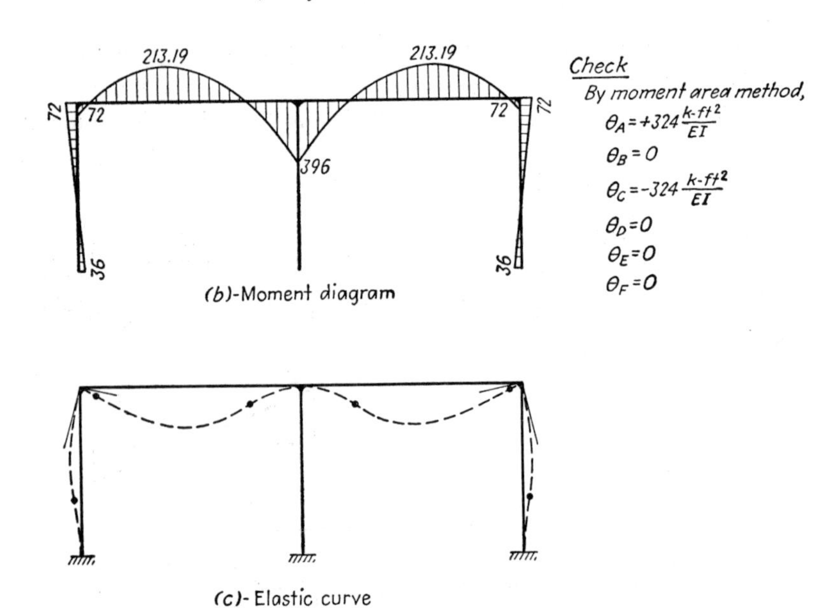

(b)-Moment diagram

(c)-Elastic curve

FIG. 144.

EXERCISES

78 to 84. Analyze the rigid frame shown by the slope-deflection method. Draw shear and moment diagrams. Sketch the deformed structure.

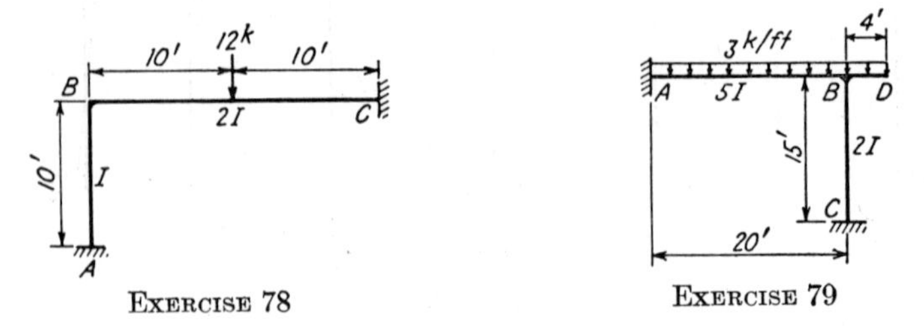

EXERCISE 78

EXERCISE 79

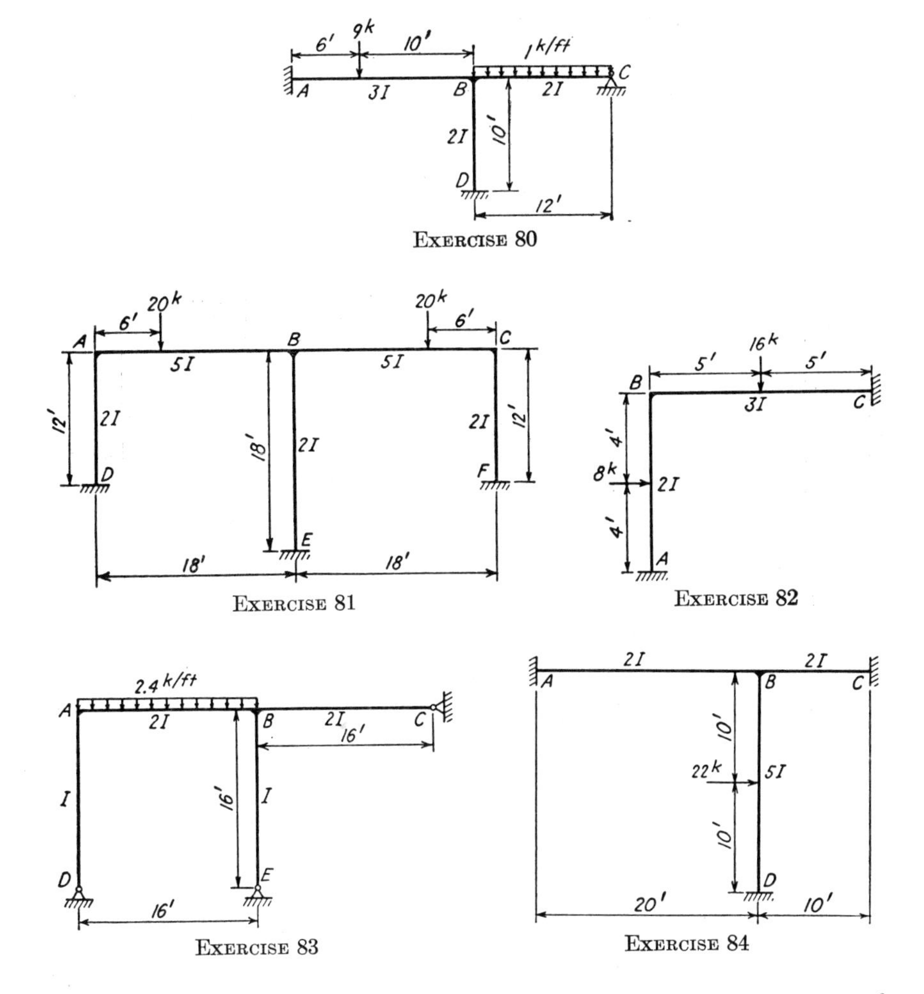

Exercise 80

Exercise 81

Exercise 82

Exercise 83

Exercise 84

41. Application of the Slope-deflection Method to the Analysis of Statically Indeterminate Frames. *Case* II—*with Joint Movements.* When loads are applied to statically indeterminate frames, there are cases in which some joints move unknown distances, although usually in known directions. Take, for instance, the rigid frame of Fig. 145*a*; the joints *D*, *E*, and *F* are fixed; but the joints *A*, *B*, and *C* may move equal distances in the horizontal direction. This horizontal movement is generally called *sidesway*. Assume that the amount of sidesway is Δ to the right; then

$$R_{AD} = \frac{\Delta}{H_1} \qquad R_{BE} = \frac{\Delta}{H_2} \qquad R_{CF} = \frac{\Delta}{H_3}$$

Thus the slope-deflection equations (66)

$$M_{AB} = M_{FAB} + \frac{2EI}{L}(-2\theta_A - \theta_B + 3R)$$
$$M_{BA} = M_{FBA} + \frac{2EI}{L}(-2\theta_B - \theta_A + 3R) \quad (66)$$

must be used for members AD, BE, and CF. It is necessary, then, to seek another condition to cope with the unknown amount of sidesway, Δ.

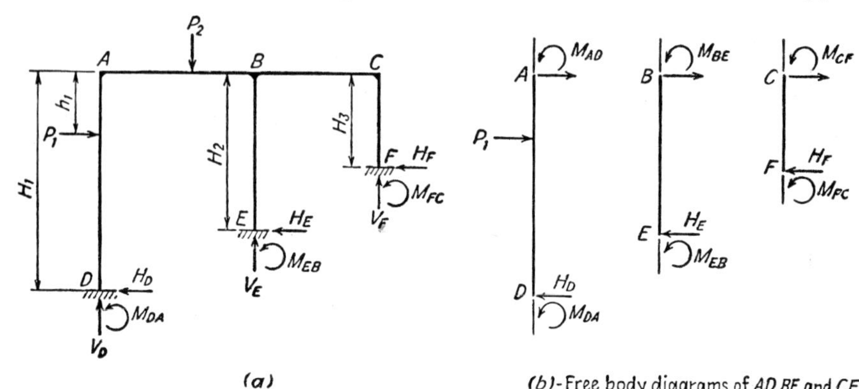

FIG. 145.

By applying the equations of statics to the free bodies of members AD, BE, and CF (Fig. 145b),

$$H_D = +\frac{P_1 h_1}{H_1} + \frac{M_{AD} + M_{DA}}{H_1}$$
$$H_E = \frac{M_{BE} + M_{EB}}{H_2}$$
$$H_F = \frac{M_{CF} + M_{FC}}{H_3}$$

By applying the equation $\Sigma H = 0$, of statics to the whole frame shown in Fig. 145a,

$$+P_1 - H_D - H_E - H_F = 0 \quad (67)$$

Equation (67) is generally called the *shear* equation or the *bent* equation, which furnishes the extra condition corresponding to the extra unknown, Δ.

Next, take the rigid frame of Fig. 146a. The joints G, H, and K are fixed; but the joints A, B, and C may move an amount, Δ_1, horizontally to the right, and the joints D, E, and F an amount, Δ_2. Obviously, Δ_1 must be greater than Δ_2. By definition

$$R_{AD} = R_{BE} = R_{CF} = \frac{\Delta_1 - \Delta_2}{H_1}$$

and

$$R_{DG} = R_{EH} = \frac{\Delta_2}{H_2} \qquad R_{FK} = \frac{\Delta_2}{H_3}$$

The slope-deflection equations in the form of Eqs. (66) are to be used for all vertical members. It is necessary to have two conditions correspond-

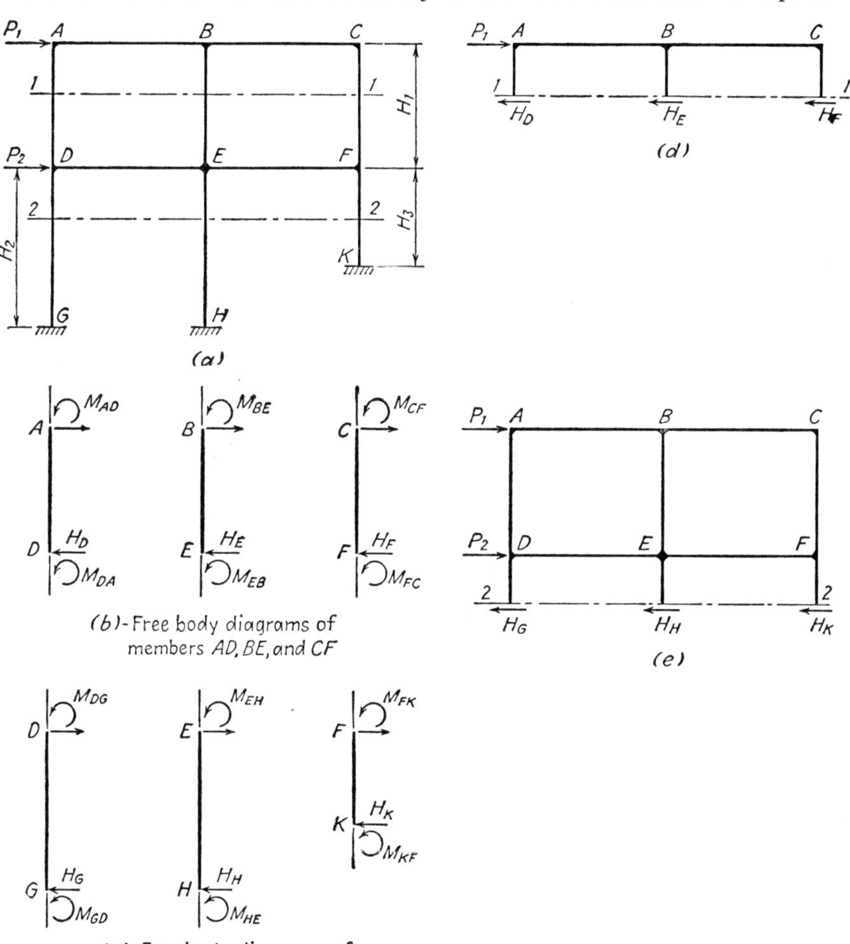

FIG. 146.

ing to the unknowns Δ_1 and Δ_2. From the free-body diagrams of AD, BE, and CF (Fig. 146*b*)

$$H_D = \frac{M_{AD} + M_{DA}}{H_1} \qquad H_E = \frac{M_{BE} + M_{EB}}{H_1} \qquad H_F = \frac{M_{CF} + M_{FC}}{H_1}$$

The shear condition for the upper story, by referring to the free body shown in Fig. 146*d*, is

$$+P_1 - H_D - H_E - H_F = 0 \tag{68}$$

From the free-body diagrams of *DG*, *EH*, and *FK* (Fig. 146*c*)

$$H_G = \frac{M_{DG} + M_{GD}}{H_2} \qquad H_H = \frac{M_{EH} + M_{HE}}{H_2} \qquad H_K = \frac{M_{FK} + M_{KF}}{H_3}$$

The shear condition for the lower story, by referring to the free body shown in Fig. 146*e*, is

$$+(P_1 + P_2) - H_G - H_H - H_K = 0 \tag{69}$$

Equations (68) and (69) furnish the two extra conditions, corresponding to the two extra unknowns Δ_1 and Δ_2.

The application of the slope-deflection method to the analysis of statically indeterminate frames wherein some joints are displaced during deformation will be illustrated by the following examples.

Example 61. Analyze the rigid frame shown in Fig. 147 by the slope-deflection method. Draw shear and moment diagrams. Sketch the deformed structure.

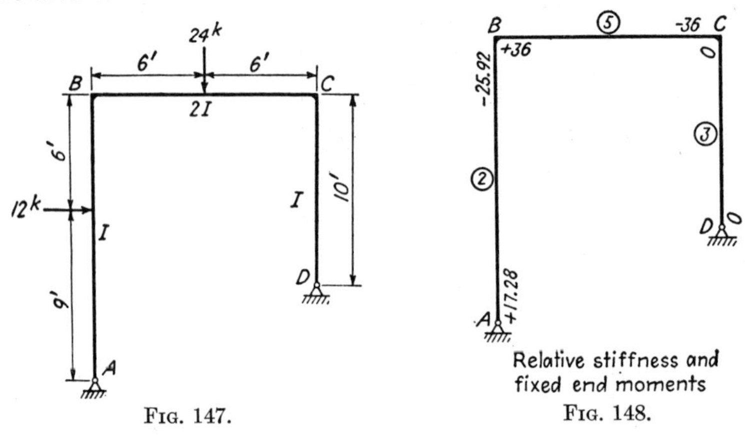

FIG. 147. FIG. 148.

Solution. *Relative Stiffness and Fixed-end Moments* (Fig. 148)

TABLE 61-1

AB.	$\frac{I}{15} \times 30$	2
BC.	$\left(\frac{2I}{12} = \frac{I}{6}\right) \times 30$	5
CD.	$\frac{I}{10} \times 30$	3

Span AB:

$$M_{FAB} = +\frac{(12)(9)(6)^2}{(15)^2} = +17.28 \text{ kip-ft}$$

$$M_{FBA} = -\frac{(12)(6)(9)^2}{(15)^2} = -25.92 \text{ kip-ft}$$

Span BC:

$$M_{FBC} = +\frac{(24)(12)}{8} = +36 \text{ kip-ft}$$

$$M_{FCB} = -36 \text{ kip-ft}$$

Relative Values of R (Fig. 149)

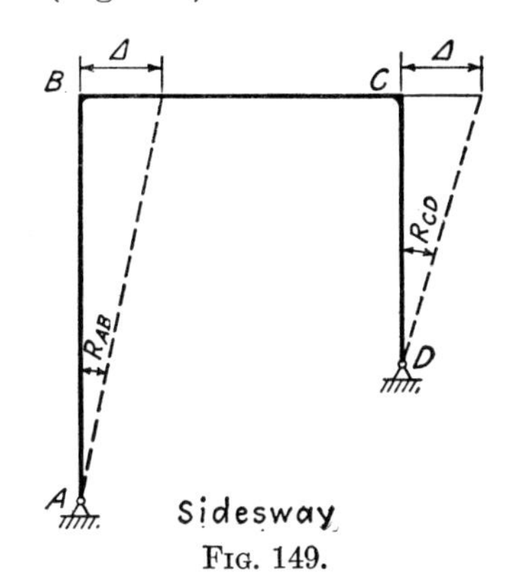

FIG. 149.

TABLE 61-2

R_{AB}...........	$\frac{\Delta}{15} \times 30$	$2R$
R_{BC}...........	0	0
R_{CD}...........	$\frac{\Delta}{10} \times 30$	$3R$

If $R_{AB} = 2R$, $R_{CD} = 3R$.

Slope-deflection Equations. The slope-deflection equations

$$\begin{aligned} M_{AB} &= M_{FAB} + \frac{2EI}{L}(-2\theta_A - \theta_B + 3R) \\ M_{BA} &= M_{FBA} + \frac{2EI}{L}(-2\theta_B - \theta_A + 3R) \end{aligned} \tag{66}$$

can be modified to take the following form,

$$\begin{aligned} M_{AB} &= M_{FAB} + K_{AB}(-2\theta_A - \theta_B + R_{\text{rel}}) \\ M_{BA} &= M_{FBA} + K_{AB}(-2\theta_B - \theta_A + R_{\text{rel}}) \end{aligned} \tag{70}$$

wherein K_{AB} is the relative stiffness and R_{rel} is the relative size of the angle between the line joining the deflected ends and the original axis.

Note that $2EI/L$ is replaced by K_{AB} and $3R$ is replaced by R_{rel}. These changes do not affect the values of the end moments at all. Equations (70) will be used in the present example.

$$
\begin{aligned}
M_{AB} &= +17.28 + 2(-2\theta_A - \theta_B + 2R) = +17.28 - 4\theta_A - 2\theta_B + 4R \\
M_{BA} &= -25.92 + 2(-2\theta_B - \theta_A + 2R) = -25.92 - 4\theta_B - 2\theta_A + 4R \\
M_{BC} &= +36 + 5(-2\theta_B - \theta_C) = +36 - 10\theta_B - 5\theta_C \\
M_{CB} &= -36 + 5(-2\theta_C - \theta_B) = -36 - 10\theta_C - 5\theta_B \\
M_{CD} &= 0 + 3(-2\theta_C - \theta_D + 3R) = -6\theta_C - 3\theta_D + 9R \\
M_{DC} &= 0 + 3(-2\theta_D - \theta_C + 3R) = -6\theta_D - 3\theta_C + 9R
\end{aligned}
$$

Joint Conditions

$$M_{AB} = 0 \qquad (a)$$

$$M_{BA} + M_{BC} = 0 \qquad (b)$$

$$M_{CB} + M_{CD} = 0 \qquad (c)$$

$$M_{DC} = 0 \qquad (d)$$

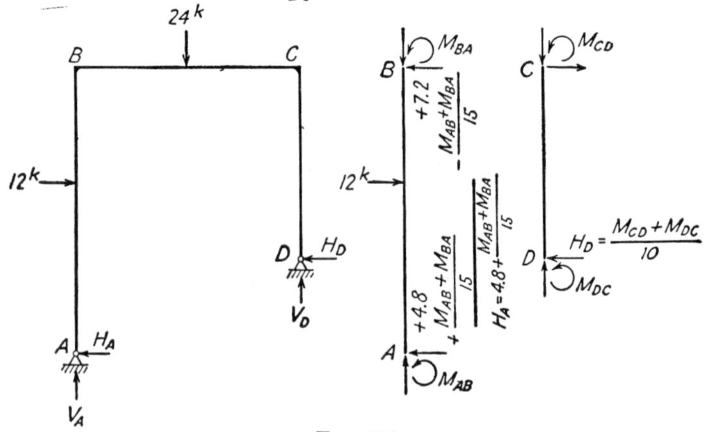

FIG. 150.

Shear Condition (Fig. 150). Substituting

$$H_A = 4.8 + \frac{M_{AB} + M_{BA}}{15}$$

and

$$H_D = \frac{M_{CD} + M_{DC}}{10}$$

into the shear condition $12 - H_A - H_D = 0$

$$+12 - \left(4.8 + \frac{M_{AB} + M_{BA}}{15}\right) - \left(\frac{M_{CD} + M_{DC}}{10}\right) = 0$$

$$+360 - 144 - 2(M_{AB} + M_{BA}) - 3(M_{CD} + M_{DC}) = 0$$

$$12\theta_A + 12\theta_B + 27\theta_C + 27\theta_D - 70R = -233.28$$

or

$$4\theta_A + 4\theta_B + 9\theta_C + 9\theta_D - \tfrac{70}{3}R = -77.76 \qquad (e)$$

$$
\begin{aligned}
-4\theta_A - 2\theta_B \qquad\qquad\qquad + 4R &= -17.28 \\
-2\theta_A - 14\theta_B - 5\theta_C \qquad\quad + 4R &= -10.08 \\
- 5\theta_B - 16\theta_C - 3\theta_D + 9R &= +36 \\
- 3\theta_C - 6\theta_D + 9R &= 0 \\
+4\theta_A + 4\theta_B + 9\theta_C + 9\theta_D - \tfrac{70}{3}R &= -77.76
\end{aligned}
$$

Solving the five simultaneous equations,

$$
\begin{aligned}
\theta_A &= +22.430 \\
\theta_B &= +1.9882 \\
\theta_C &= +2.7604 \\
\theta_D &= +27.276 \\
R &= +19.104
\end{aligned}
$$

In deriving the shear equation (*e*), M_{AB} and M_{DC} could have been called zero, but the slope-deflection equations for them are substituted in order that the symmetry of the coefficients with respect to the diagonal drawn across the simultaneous equations (*a*) to (*e*) may be shown.

Computations for End Moments

$$M_{AB} = +17.28 - (4)(+22.430) - (2)(+1.9882) + (4)(+19.104) = 0$$
$$M_{BA} = -25.92 - (4)(+1.9882) - (2)(+22.430) + (4)(+19.104) = -2.317 \text{ kip-ft}$$
$$M_{BC} = +36 - (10)(+1.9882) - (5)(+2.7604) = +2.316 \text{ kip-ft}$$
$$M_{CB} = -36 - (10)(+2.7604) - (5)(+1.9882) = -73.545 \text{ kip-ft}$$
$$M_{CD} = -(6)(+2.7604) - (3)(+27.276) + (9)(+19.104) = +73.545 \text{ kip-ft}$$
$$M_{DC} = -(6)(+27.276) - (3)(+2.7604) + (9)(+19.104) = 0$$

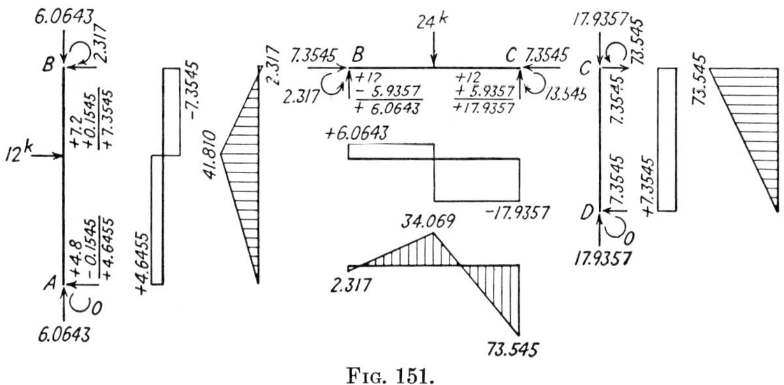

FIG. 151.

The free-body, shear, and moment diagrams of the individual members are shown in Fig. 151. The free-body diagram, moment diagram, and elastic curve of the whole frame are shown in Fig. 152.

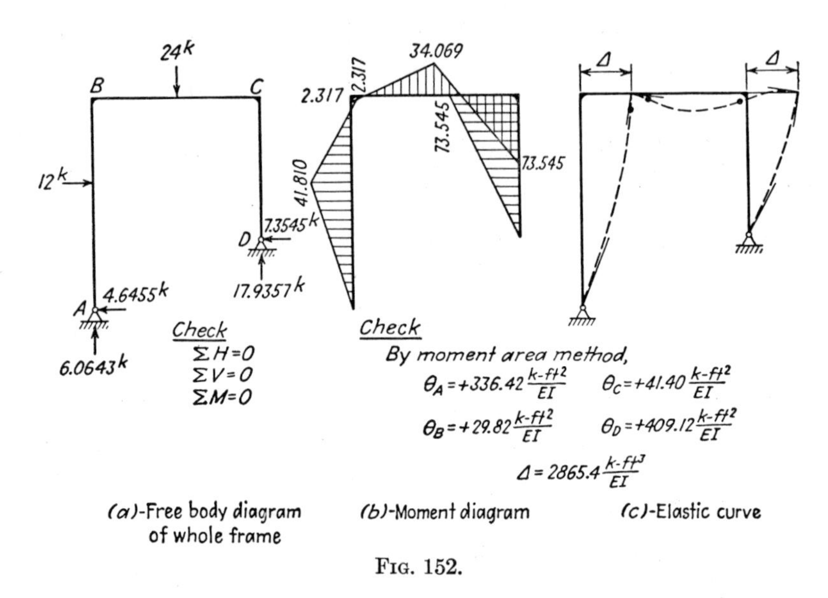

Fig. 152.

Example 62. Analyze the rigid frame shown in Fig. 153 by the slope-deflection method. Draw shear and moment diagrams. Sketch the deformed structure.

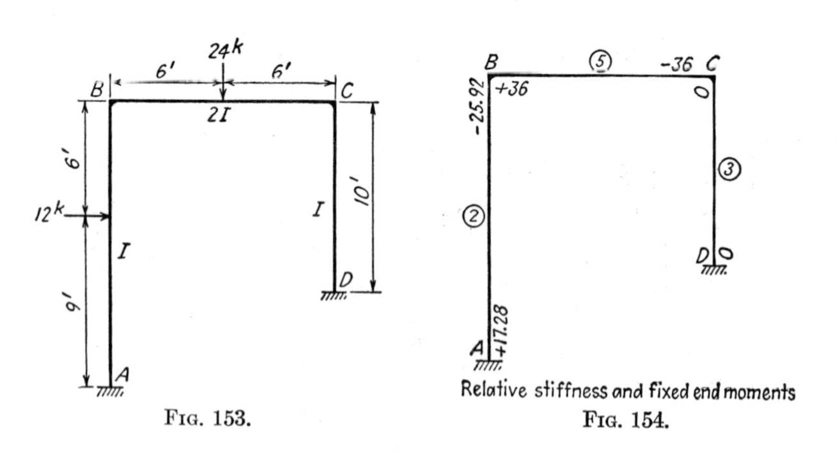

Fig. 153.

Fig. 154.

Solution. *Relative Stiffness and Fixed-end Moments* (Fig. 154)

TABLE 62-1

AB............	$\frac{I}{15} \times 30$	2
BC............	$\left(\frac{2I}{12} = \frac{I}{6}\right) \times 30$	5
CD............	$\frac{I}{10} \times 30$	3

Span AB:

$$M_{FAB} = +\frac{(12)(9)(6)^2}{(15)^2} = +17.28 \text{ kip-ft}$$

$$M_{FBA} = -\frac{(12)(6)(9)^2}{(15)^2} = -25.92 \text{ kip-ft}$$

Span BC:

$$M_{FBC} = +\frac{(24)(12)}{8} = +36 \text{ kip-ft}$$

$$M_{FCB} = -36 \text{ kip-ft}$$

Relative Values of R (Fig. 155)

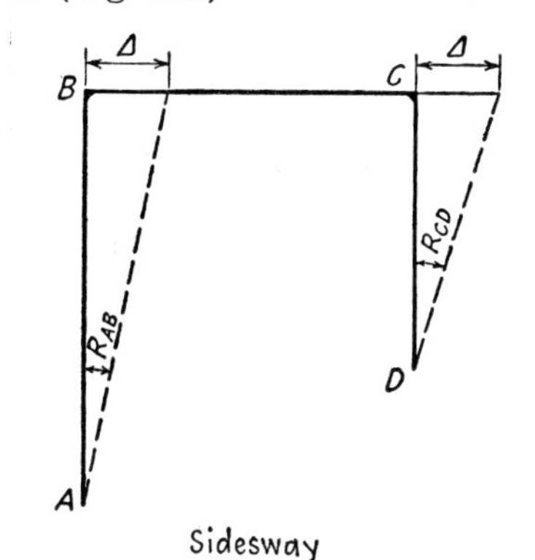

FIG. 155.

TABLE 62-2

R_{AB}...........	$\frac{\Delta}{15} \times 30$	$2R$
R_{BC}...........	0	0
R_{CD}...........	$\frac{\Delta}{10} \times 30$	$3R$

If $R_{AB} = 2R$, $R_{CD} = 3R$.

Slope-deflection Equations. The slope-deflection equations

$$M_{AB} = M_{FAB} + K_{AB}(-2\theta_A - \theta_B + R_{\text{rel}})$$
$$M_{BA} = M_{FBA} + K_{AB}(-2\theta_B - \theta_A + R_{\text{rel}})$$

will be used. θ_A and θ_D are known to be zero.

$$M_{AB} = +17.28 + 2(-2\theta_A - \theta_B + 2R) = +17.28 - 2\theta_B + 4R$$
$$M_{BA} = -25.92 + 2(-2\theta_B - \theta_A + 2R) = -25.92 - 4\theta_B + 4R$$
$$M_{BC} = +36 + 5(-2\theta_B - \theta_C) = +36 - 10\theta_B - 5\theta_C$$
$$M_{CB} = -36 + 5(-2\theta_C - \theta_B) = -36 - 10\theta_C - 5\theta_B$$
$$M_{CD} = 0 + 3(-2\theta_C - \theta_D + 3R) = -6\theta_C + 9R$$
$$M_{DC} = 0 + 3(-2\theta_D - \theta_C + 3R) = -3\theta_C + 9R$$

Two Joint Conditions and One Shear Condition (Fig. 156)

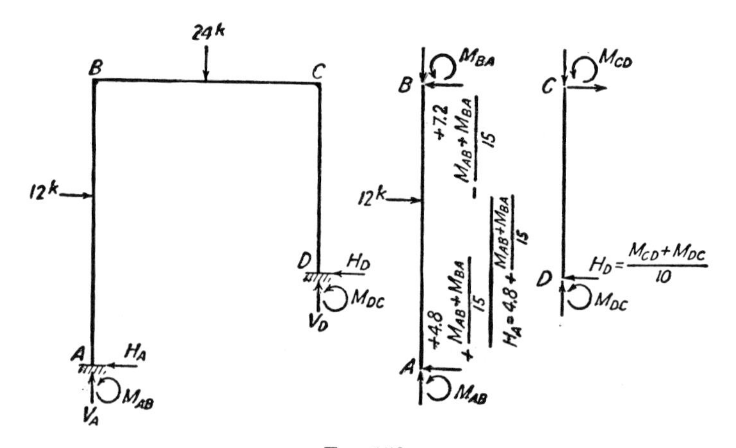

Fig. 156.

Joint B: $\quad M_{BA} + M_{BC} = 0 \qquad (a)$

Joint C: $\quad M_{CB} + M_{CD} = 0 \qquad (b)$

Shear: $\quad +12 - H_A - H_D = 0 \qquad (c)$

$$-14\theta_B - 5\theta_C + 4R = -10.08 \qquad (a')$$
$$-5\theta_B - 16\theta_C + 9R = +36 \qquad (b')$$
$$+4\theta_B + 9\theta_C - \tfrac{70}{3}R = -77.76 \qquad (c')$$

Solving,

$$\theta_B = +2.0281$$
$$\theta_C = -1.0391$$
$$R = +3.2794$$

Computations for End Moments

$$M_{AB} = +17.28 - (2)(+2.0281) + (4)(+3.2794) = +26.342 \text{ kip-ft}$$
$$M_{BA} = -25.92 - (4)(+2.0281) + (4)(+3.2794) = -20.914 \text{ kip-ft}$$
$$M_{BC} = +36 - (10)(+2.0281) - (5)(-1.0391) = +20.915 \text{ kip-ft}$$
$$M_{CB} = -36 - (10)(-1.0391) - (5)(+2.0281) = -35.750 \text{ kip-ft}$$
$$M_{CD} = -(6)(-1.0391) + (9)(+3.2794) = +35.750 \text{ kip-ft}$$
$$M_{DC} = -(3)(-1.0391) + (9)(+3.2794) = +32.632 \text{ kip-ft}$$

The free-body, shear, and moment diagrams of the individual members are shown in Fig. 157. The free-body diagram, moment diagram, and elastic curve of the whole frame are shown in Fig. 158.

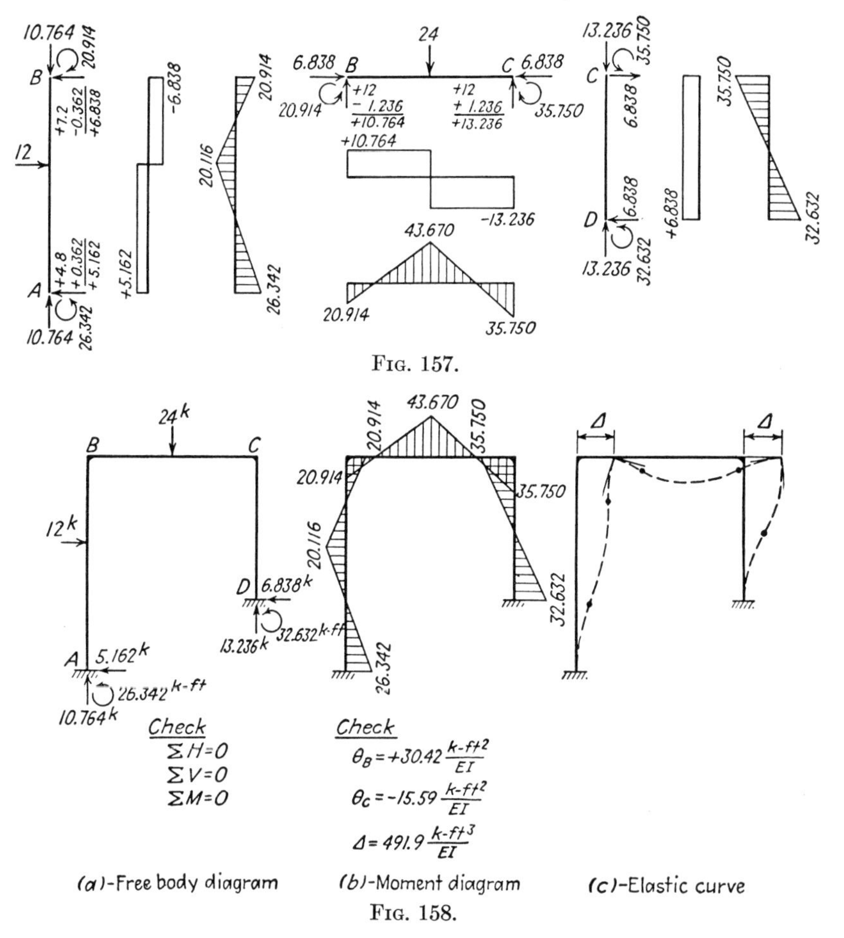

FIG. 157.

FIG. 158.

Example 63. Analyze the rigid frame shown in Fig. 159 by the slope-deflection method. Draw shear and moment diagrams. Sketch the deformed structure.

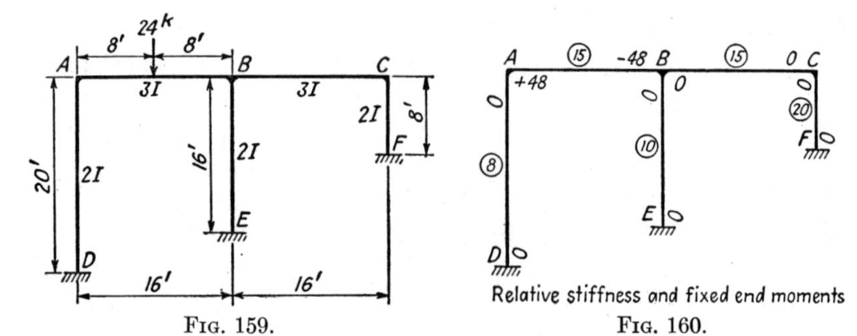

FIG. 159. FIG. 160.

Solution. *Relative Stiffness and Fixed-end Moments* (Fig. 160)

TABLE 63-1

AB, BC	$\frac{3I}{16} \times 80$	15
AD	$\left(\frac{2I}{20} = \frac{I}{10}\right) \times 80$	8
BE	$\left(\frac{2I}{16} = \frac{I}{8}\right) \times 80$	10
CF	$\left(\frac{2I}{8} = \frac{I}{4}\right) \times 80$	20

Span AB:

$$M_{FAB} = +\frac{(24)(16)}{8} = +48 \text{ kip-ft}$$

$$M_{FBA} = -48 \text{ kip-ft}$$

Relative Values of R (Fig. 161)

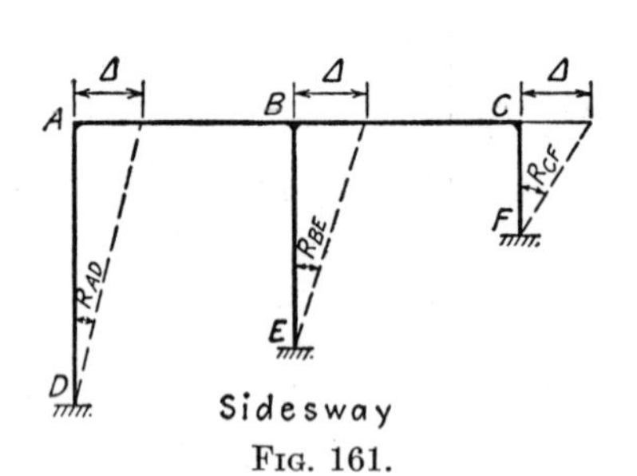

FIG. 161.

TABLE 63-2

R_{AD}	$\frac{\Delta}{20} \times 80$	$4R$
R_{BE}	$\frac{\Delta}{16} \times 80$	$5R$
R_{CF}	$\frac{\Delta}{8} \times 80$	$10R$

If $R_{AD} = 4R$, $R_{BE} = 5R$ and $R_{CF} = 10R$.

Slope-deflection Equations. The slope-deflection equations

$$M_{AB} = M_{FAB} + K_{AB}(-2\theta_A - \theta_B + R_{\text{rel}})$$
$$M_{BA} = M_{FBA} + K_{AB}(-2\theta_B - \theta_A + R_{\text{rel}})$$

will be used. θ_D, θ_E, and θ_F are known to be zero.

$$
\begin{aligned}
M_{AB} &= +48 + 15(-2\theta_A - \theta_B) = +48 - 30\theta_A - 15\theta_B \\
M_{BA} &= -48 + 15(-2\theta_B - \theta_A) = -48 - 30\theta_B - 15\theta_A \\
M_{BC} &= 0 + 15(-2\theta_B - \theta_C) = -30\theta_B - 15\theta_C \\
M_{CB} &= 0 + 15(-2\theta_C - \theta_B) = -30\theta_C - 15\theta_B \\
M_{AD} &= 0 + 8(-2\theta_A - \theta_D + 4R) = -16\theta_A + 32R \\
M_{DA} &= 0 + 8(-2\theta_D - \theta_A + 4R) - 8\theta_A + 32R \\
M_{BE} &= 0 + 10(-2\theta_B - \theta_E + 5R) = -20\theta_B + 50R \\
M_{EB} &= 0 + 10(-2\theta_E - \theta_B + 5R) = -10\theta_B + 50R \\
M_{CF} &= 0 + 20(-2\theta_C - \theta_F + 10R) = -40\theta_C + 200R \\
M_{FC} &= 0 + 20(-2\theta_F - \theta_C + 10R) = -20\theta_C + 200R
\end{aligned}
$$

Three Joint Conditions and One Shear Condition

Joint A: $\quad M_{AB} + M_{AD} = 0 \qquad (a)$

Joint B: $\quad M_{BA} + M_{BC} + M_{BE} = 0 \qquad (b)$

Joint C: $\quad M_{CB} + M_{CF} = 0 \qquad (c)$

Shear: $\quad -H_D - H_E - H_F = 0$ (Fig. 162) $\qquad (d)$

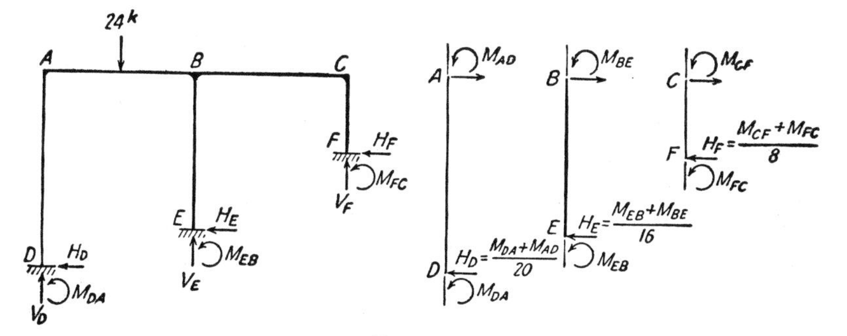

Fig. 162.

$$-46\theta_A - 15\theta_B \qquad\qquad + 32R = -48 \qquad (a')$$

$$-15\theta_A - 80\theta_B - 15\theta_C + 50R = +48 \qquad (b')$$

$$- 15\theta_B - 70\theta_C + 200R = 0 \qquad (c')$$

$$32\theta_A + 50\theta_B + 200\theta_C - \frac{4{,}756}{3}R = 0 \qquad (d')$$

Solving,

$$
\begin{aligned}
\theta_A &= +1.35820 \\
\theta_B &= -0.88703 \\
\theta_C &= +0.29470 \\
R &= +0.036617
\end{aligned}
$$

It is to be noted that extreme care must be exercised in solving the simultaneous equations. Care should be taken to maintain the significance of every number. It is better not to follow a definite pattern of solution in every case, because the easiest solution may vary from problem to problem. Much time can be saved, of course, by use of a computing machine.

Computations for End Moments

$$M_{AB} = +48 - (30)(+1.35820) - (15)(-0.88703) = +20.559 \text{ kip-ft}$$
$$M_{BA} = -48 - (30)(-0.88703) - (15)(+1.35820) = -41.762 \text{ kip-ft}$$
$$M_{BC} = -(30)(-0.88703) - (15)(+0.29470) - +22.190 \text{ kip-ft}$$
$$M_{CB} = -(30)(+0.29470) - (15)(-0.88703) = +4.464 \text{ kip-ft}$$
$$M_{AD} = -(16)(+1.35820) + (32)(+0.036617) = -20.559 \text{ kip-ft}$$
$$M_{DA} = -(8)(+1.35820) + (32)(+0.036617) = -9.694 \text{ kip-ft}$$
$$M_{BE} = -(20)(-0.88703) + (50)(+0.036617) = +19.572 \text{ kip-ft}$$
$$M_{EB} = -(10)(-0.88703) + (50)(+0.036617) = +10.701 \text{ kip-ft}$$
$$M_{CF} = -(40)(+0.29470) + (200)(+0.036617) = -4.465 \text{ kip-ft}$$
$$M_{FC} = -(20)(+0.29470) + (200)(+0.036617) = +1.429 \text{ kip-ft}$$

The free-body, shear, and moment diagrams of the individual members are shown in Fig. 163. The free-body diagram, moment diagram, and elastic curve of the whole frame are shown in Fig. 164.

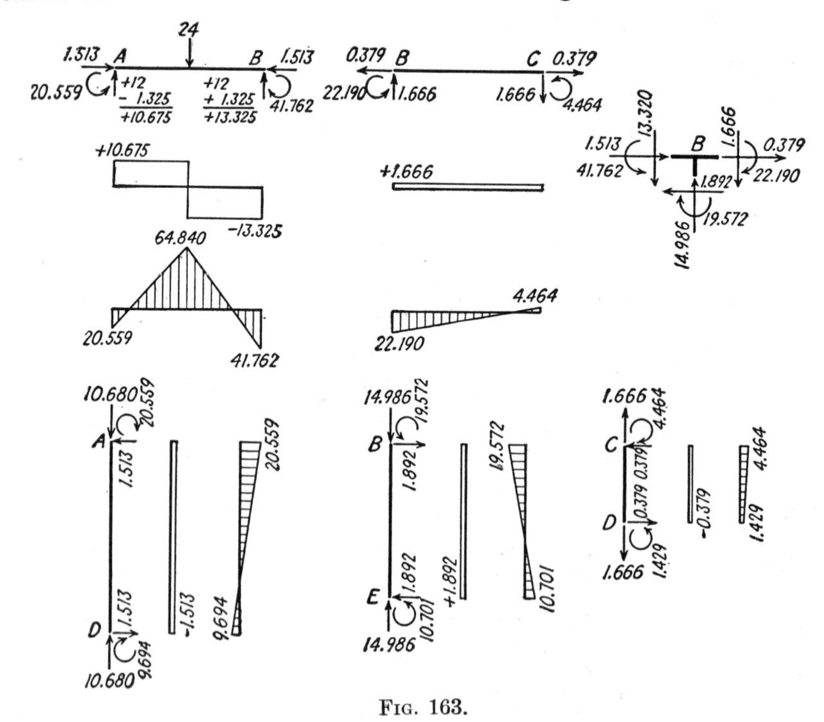

FIG. 163.

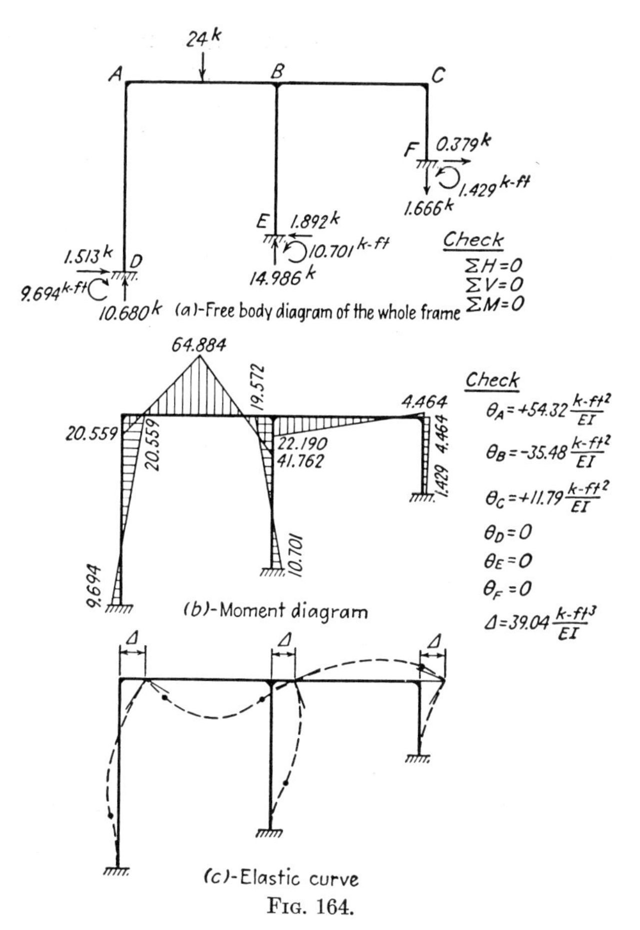

FIG. 164.

Example 64. Analyze the rigid frame shown in Fig. 165 by the slope-deflection method. Draw shear and moment diagrams. Sketch the deformed structure.

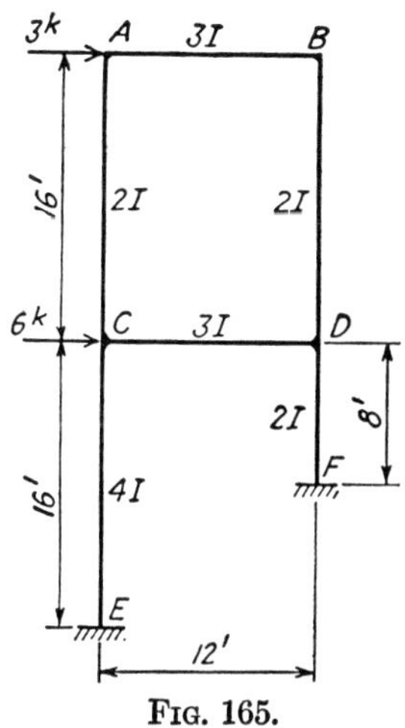

FIG. 165.

Solution. *Relative Stiffness and Fixed-end Moments* (Fig. 166)

TABLE 64-1

AB, CD	$\left(\frac{3I}{12} = \frac{I}{4}\right) \times 8$	2
AC, BD	$\left(\frac{2I}{16} = \frac{I}{8}\right) \times 8$	1
CE	$\left(\frac{4I}{16} = \frac{I}{4}\right) \times 8$	2
DF	$\left(\frac{2I}{8} = \frac{I}{4}\right) \times 8$	2

In this example, no load is applied on any member between the end joints; so fixed-end moments are zero for all members.

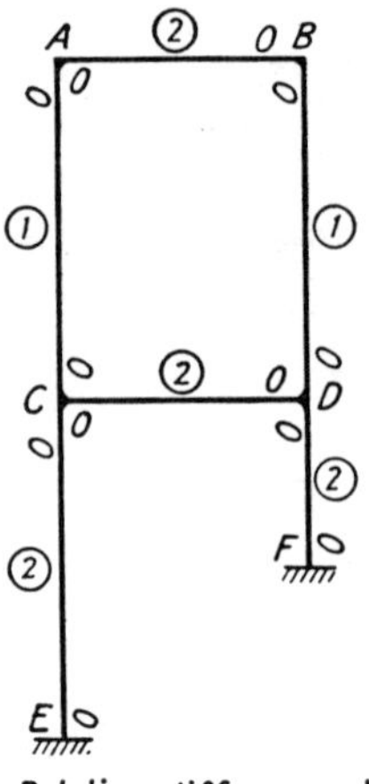

Relative stiffness and fixed end moments

FIG. 166.

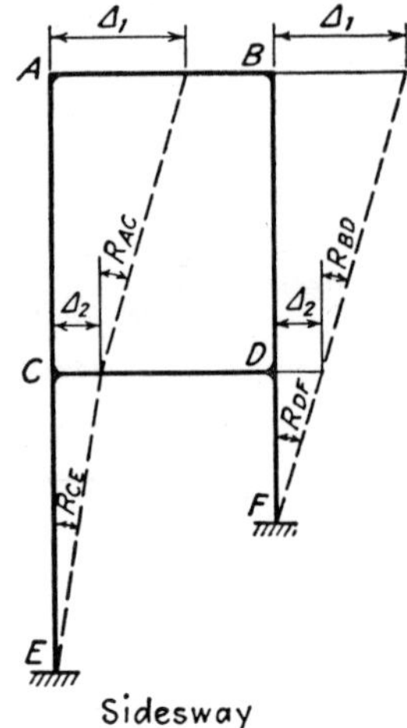

Sidesway

FIG. 167.

Relative Values of R (Fig. 167)

TABLE 64-2. UPPER STORY

R_{AC}	$\dfrac{\Delta_1 - \Delta_2}{16}$	R_1
R_{BD}	$\dfrac{\Delta_1 - \Delta_2}{16}$	R_1

If $R_{AC} = R_1$, $R_{BD} = R_1$.

TABLE 64-3. LOWER STORY

R_{CE}	$\dfrac{\Delta_2}{16}$	R_2
R_{DF}	$\dfrac{\Delta_2}{8}$	$2R_2$

If $R_{CE} = R_2$, $R_{DF} = 2R_2$.

Slope-deflection Equations. The slope-deflection equations

$$M_{AB} = M_{FAB} + K_{AB}(-2\theta_A - \theta_B + R_{\text{rel}})$$
$$M_{BA} = M_{FBA} + K_{AB}(-2\theta_B - \theta_A + R_{\text{rel}})$$

will be used. θ_E and θ_F are known to be zero.

$$\begin{aligned}
M_{AB} &= 0 + 2(-2\theta_A - \theta_B) = -4\theta_A - 2\theta_B \\
&= -(4)(2.5832) - (2)(1.8206) = -13.9740 \text{ kip-ft} \\
M_{BA} &= 0 + 2(-2\theta_B - \theta_A) = -4\theta_B - 2\theta_A \\
&= -(4)(1.8206) - (2)(2.5832) = -12.4488 \text{ kip-ft} \\
M_{CD} &= 0 + 2(-2\theta_C - \theta_D) = -4\theta_C - 2\theta_D \\
&= -(4)(3.0997) - (2)(6.1500) = -24.6988 \text{ kip-ft} \\
M_{DC} &= 0 + 2(-2\theta_D - \theta_C) = -4\theta_D - 2\theta_C \\
&= -(4)(6.1500) - (2)(3.0997) = -30.7994 \text{ kip-ft} \\
M_{AC} &= 0 + 1(-2\theta_A - \theta_C + R_1) = -2\theta_A - \theta_C + R_1 \\
&= -(2)(2.5832) - (3.0997) + (22.2401) = +13.9740 \text{ kip-ft}
\end{aligned}$$

$$\begin{aligned}
M_{CA} &= 0 + 1(-2\theta_C - \theta_A + R_1) = -2\theta_C - \theta_A + R_1 \\
&= -(2)(3.0997) - (2.5832) + (22.2401) = +13.4575 \text{ kip-ft} \\
M_{BD} &= 0 + 1(-2\theta_B - \theta_D + R_1) = -2\theta_B - \theta_D + R_1 \\
&= -(2)(1.8206) - (6.1500) + (22.2401) = +12.4489 \text{ kip-ft} \\
M_{DB} &= 0 + (1)(-2\theta_D - \theta_B + R_1) = -2\theta_D - \theta_B + R_1 \\
&= -(2)(6.1500) - (1.8206) + (22.2401) = +8.1195 \text{ kip-ft} \\
M_{CE} &= 0 + 2(-2\theta_C - \theta_E + R_2) = -4\theta_C + 2R_2 \\
&= -(4)(3.0997) + (2)(11.8199) = +11.2410 \text{ kip-ft} \\
M_{EC} &= 0 + 2(-2\theta_E - \theta_C + R_2) = -2\theta_C + 2R_2 \\
&= -(2)(3.0997) + (2)(11.8199) = +17.4404 \text{ kip-ft} \\
M_{DF} &= 0 + 2(-2\theta_D - \theta_F + 2R_2) = -4\theta_D + 4R_2 \\
&= -(4)(6.1500) + (4)(11.8199) = +22.6796 \text{ kip-ft} \\
M_{FD} &= 0 + 2(-2\theta_F - \theta_D + 2R_2) = -2\theta_D + 4R_2 \\
&= -(2)(6.1500) + (4)(11.8199) = +34.9796 \text{ kip-ft}
\end{aligned}$$

Four Joint Conditions and Two Shear Conditions

Joint A: $\quad M_{AB} + M_{AC} = 0 \quad$ (*a*)

Joint B: $\quad M_{BA} + M_{BD} = 0 \quad$ (*b*)

Joint C: $\quad M_{CA} + M_{CD} + M_{CE} = 0 \quad$ (*c*)

Joint D: $\quad M_{DB} + M_{DC} + M_{DF} = 0 \quad$ (*d*)

Shear condition—upper story: $\quad +3 - H_C - H_D = 0$ (Fig. 168) $\quad$ (*e*)

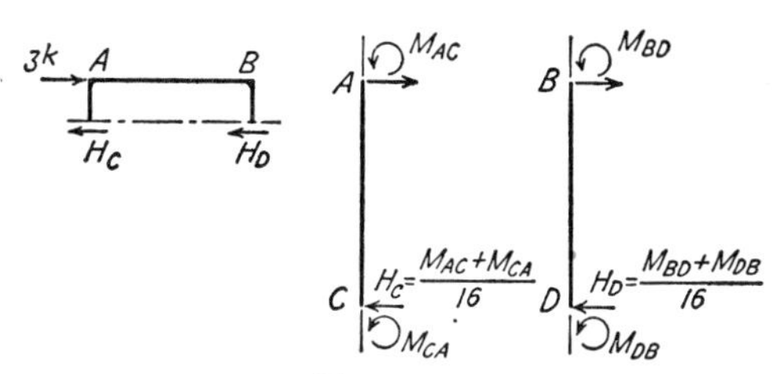

FIG. 168.

Shear condition—lower story: $\quad +9 - H_E - H_F = 0$ (Fig. 169) $\quad$ (*f*)

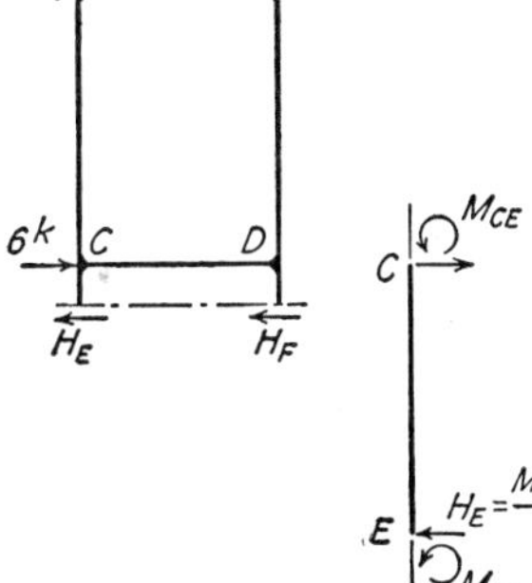

FIG. 169.

$$
\begin{array}{lr}
-6\theta_A - 2\theta_B - \theta_C + R_1 = 0 & (a') \\
-2\theta_A - 6\theta_B - \theta_D + R_1 = 0 & (b') \\
-\theta_A - 10\theta_C - 2\theta_D + R_1 + 2R_2 = 0 & (c') \\
-\theta_B - 2\theta_C - 10\theta_D + R_1 + 4R_2 = 0 & (d') \\
+\theta_A + \theta_B + \theta_C + \theta_D - \tfrac{4}{3}R_1 = -16 & (e') \\
+ 2\theta_C + 4\theta_D - \tfrac{20}{3}R_2 = -48 & (f')
\end{array}
$$

Solving,

$$
\begin{array}{ll}
\theta_A = +2.5832 & \theta_D = +6.1500 \\
\theta_B = +1.8206 & R_1 = +22.0401 \\
\theta_C = +3.0997 & R_2 = +11.8199
\end{array}
$$

The free-body, shear, and moment diagrams of the individual members are shown in Fig. 170. The free-body diagram, moment diagram, and elastic curve of the whole frame are shown in Fig. 171.

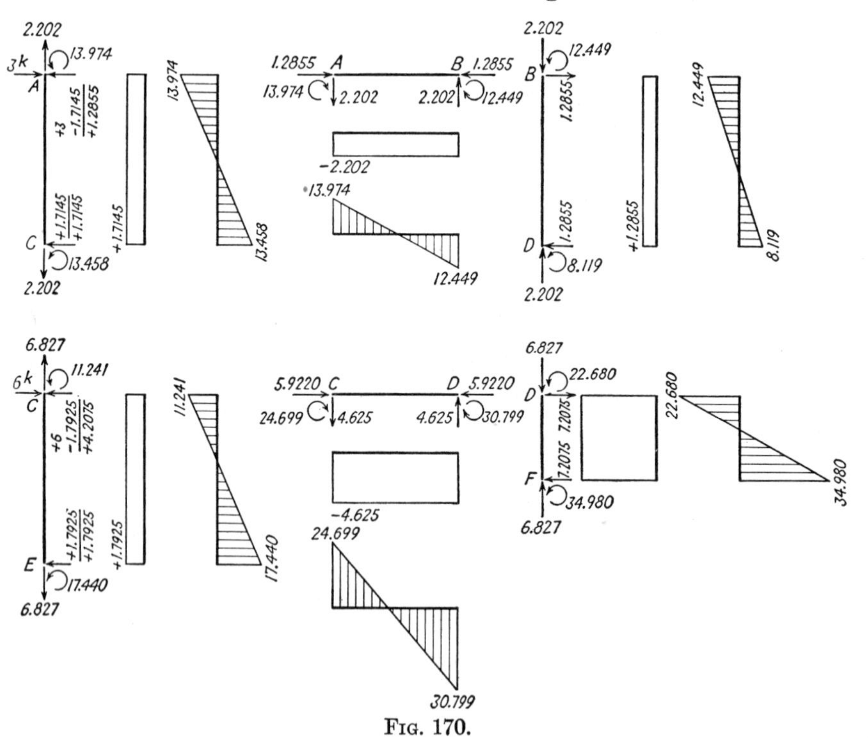

FIG. 170.

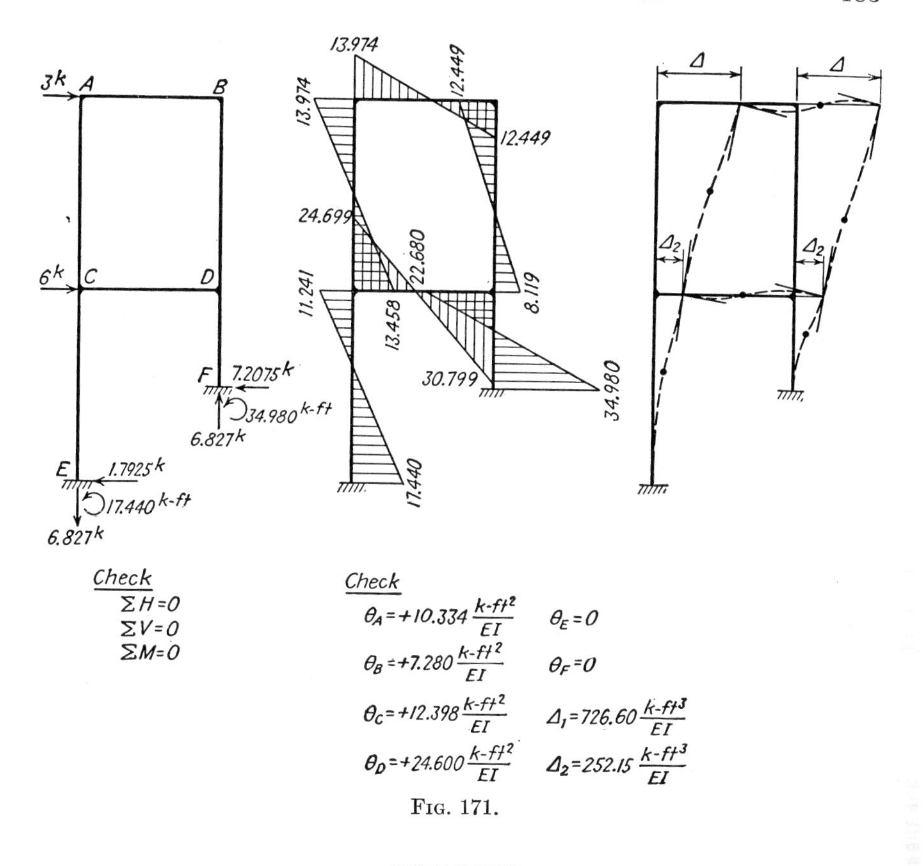

Check
ΣH=0
ΣV=0
ΣM=0

Check

$\theta_A = +10.334\,\frac{k\text{-}ft^2}{EI}$ $\theta_E = 0$

$\theta_B = +7.280\,\frac{k\text{-}ft^2}{EI}$ $\theta_F = 0$

$\theta_C = +12.398\,\frac{k\text{-}ft^2}{EI}$ $\Delta_1 = 726.60\,\frac{k\text{-}ft^3}{EI}$

$\theta_D = +24.600\,\frac{k\text{-}ft^2}{EI}$ $\Delta_2 = 252.15\,\frac{k\text{-}ft^3}{EI}$

FIG. 171.

EXERCISES

85 to 96. Analyze the rigid frame shown by the slope-deflection method. Draw shear and moment diagrams. Sketch the deformed structure.

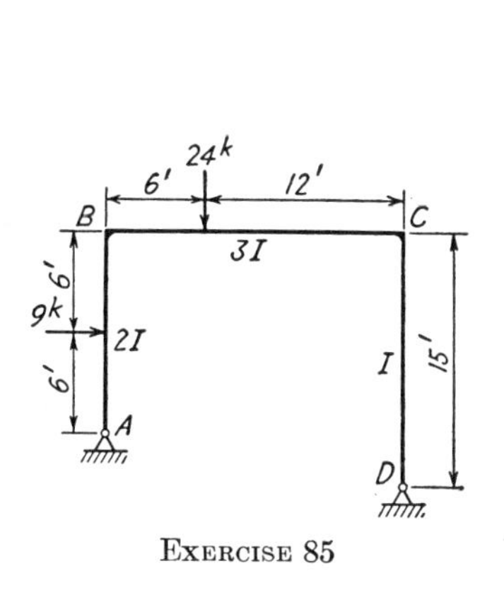

EXERCISE 85

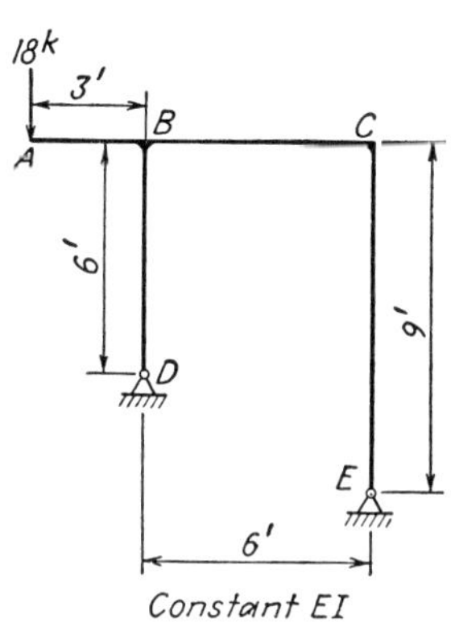

EXERCISE 86

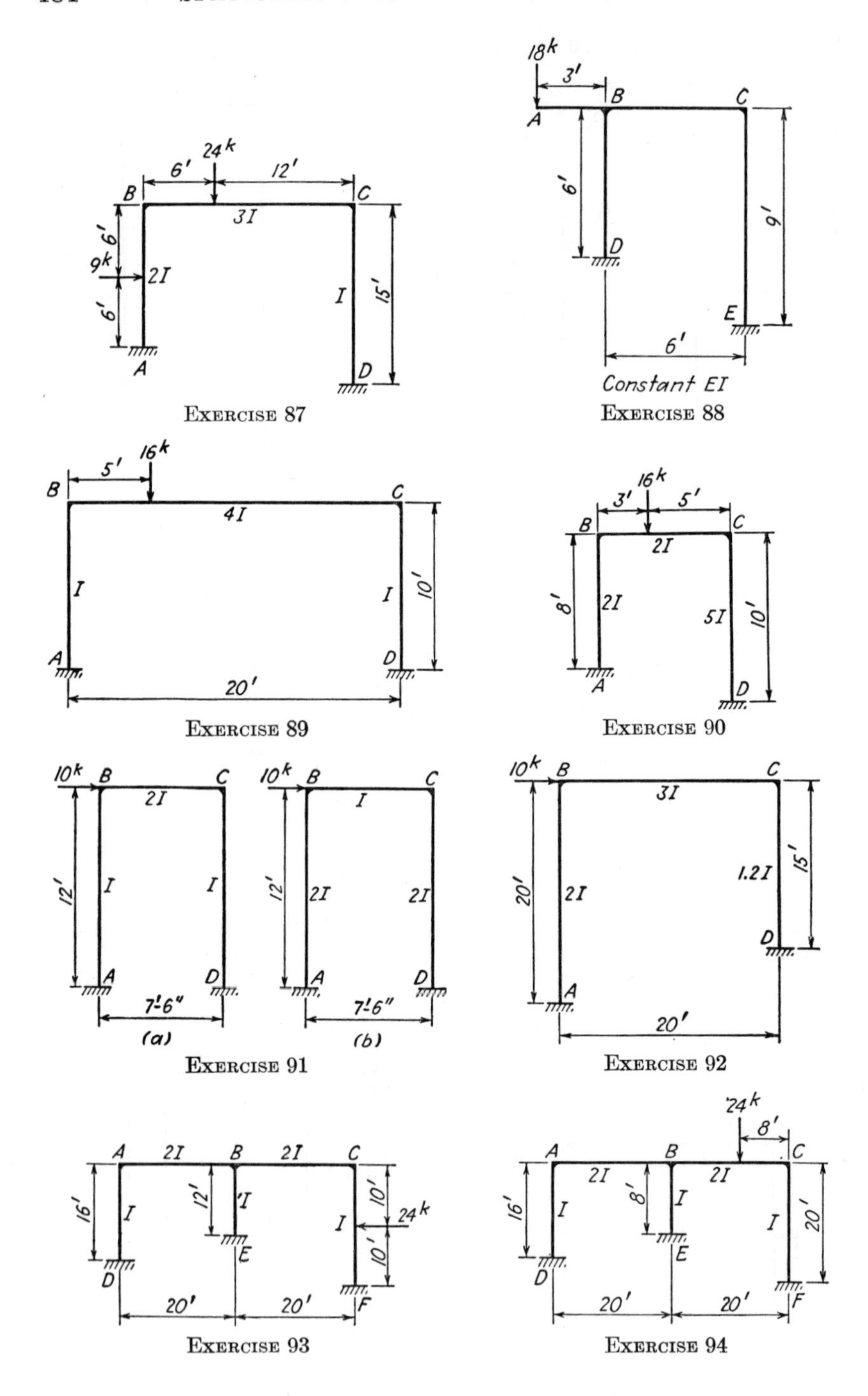

Exercise 87

Exercise 88

Exercise 89

Exercise 90

Exercise 91

Exercise 92

Exercise 93

Exercise 94

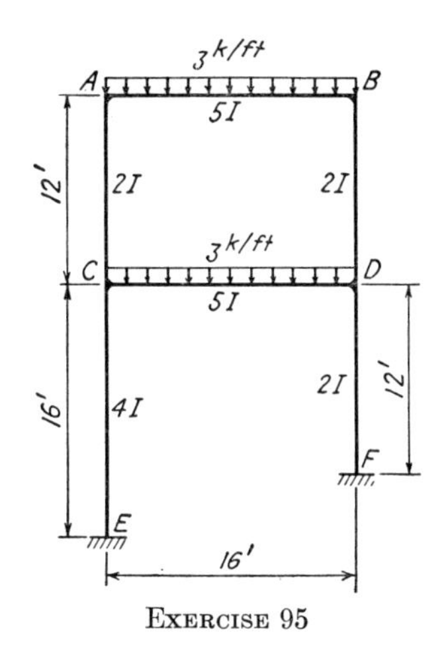

EXERCISE 95

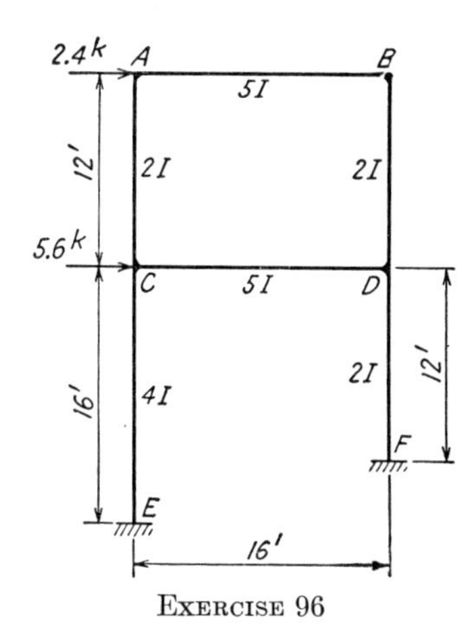

EXERCISE 96

42. Analysis of Statically Indeterminate Frames Due to Yielding of Supports. The induced reactions and the shears and moments in the members of a statically indeterminate frame due to yielding of supports can be found by the slope-deflection method. Take, for example, the rigid frame shown in Fig. 172. It is required to determine all end

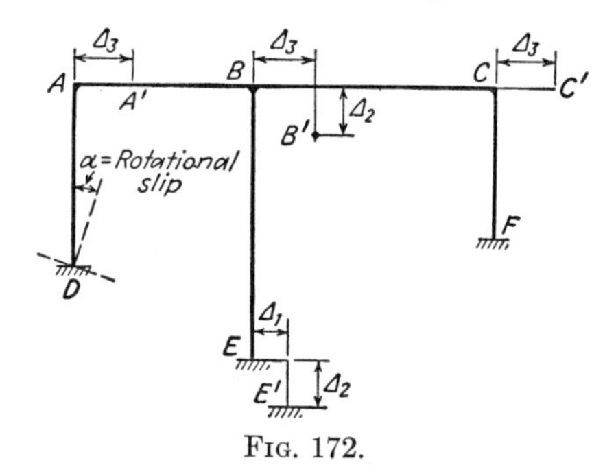

FIG. 172.

moments owing to a rotational slip of α rad of joint D and a horizontal movement Δ_1 and a vertical settlement Δ_2 at joint E. The slope deflection equations (66)

$$
\begin{aligned}
M_{AB} &= M_{FAB} + \frac{2EI}{L}(-2\theta_A - \theta_B + 3R_{AB}) \\
M_{BA} &= M_{FBA} + \frac{2EI}{L}(-2\theta_B - \theta_A + 3R_{AB})
\end{aligned}
\tag{66}
$$

are to be used, wherein all fixed-end moments are zero; values of EI/L must be expressed in kip-feet, and those of θ and R in radians. Before writing up the slope-deflection equations, it is known that $\theta_D = +\alpha$ rad,

$\theta_E = 0$, and $\theta_F = 0$. The unknowns are θ_A, θ_B, θ_C and Δ_3. The values of R are

$$R_{AB} = +\frac{\Delta_2}{L_{AB}} \qquad R_{BC} = -\frac{\Delta_2}{L_{BC}}$$

$$R_{AD} = +\frac{\Delta_3}{L_{AD}} \qquad R_{BE} = +\frac{\Delta_3 - \Delta_1}{L_{BE}} \qquad \text{and} \qquad R_{CF} = +\frac{\Delta_3}{L_{CF}}$$

The required conditions are furnished by the three joint conditions and the shear condition. Of course, generally the effect of one yielding component only is investigated at one time; the above discussion, however, and the following example, both containing more than one yielding component, serve to explain the procedure.

Example 65. By the slope-deflection method analyze the rigid frame shown in Fig. 173, due to a rotational slip of 0.002 rad clockwise and a vertical settlement of 0.45 in. at joint D. Draw shear and moment diagrams. Sketch the deformed structure.

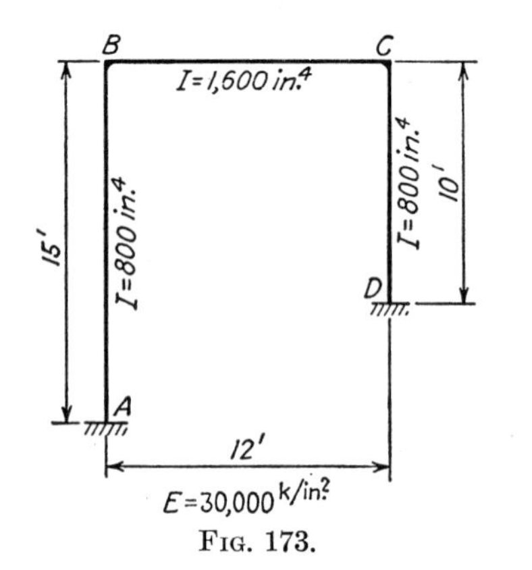

Fig. 173.

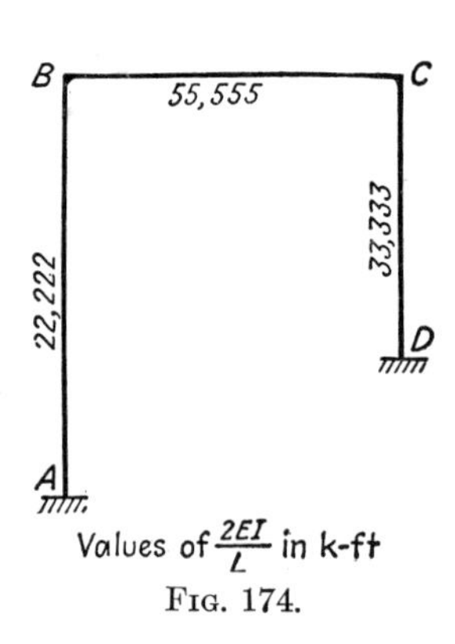

Fig. 174.

Solution. *Values of* $2EI/L$ (Fig. 174)

$$AB: \quad \frac{2EI}{L} = \frac{(2)(30{,}000)(800)\text{ kip-ft}}{(144)(15)} = 22{,}222\text{ kip-ft}$$

$$BC: \quad \frac{2EI}{L} = \frac{(2)(30{,}000)(1{,}600)\text{ kip-ft}}{(144)(12)} = 55{,}555\text{ kip-ft}$$

$$CD: \quad \frac{2EI}{L} = \frac{(2)(30{,}000)(800)\text{ kip-ft}}{(144)(10)} = 33{,}333\text{ kip-ft}$$

Values of R in radians (Fig. 175)

$$R_{AB} = +\Delta/180 \text{ rad}$$
$$R_{BC} = +0.45/144 \text{ rad}$$
$$R_{CD} = +\Delta/120 \text{ rad}$$

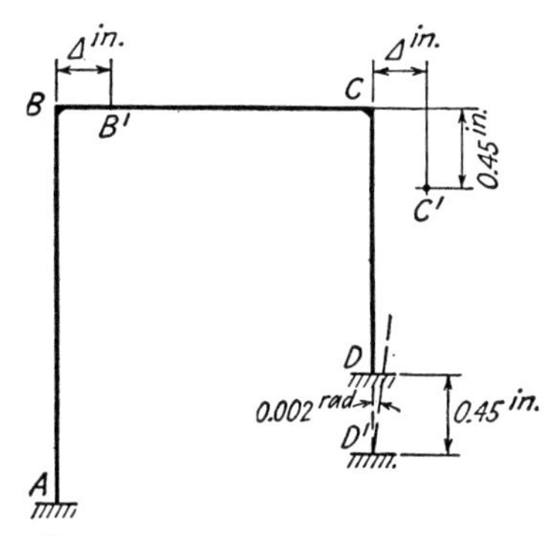

FIG. 175.

Slope-deflection Equations. It is known that $\theta_A = 0$; $\theta_D = +0.002$ rad.

$$M_{AB} = 0 + 22{,}222\left[-2\theta_A - \theta_B + 3\left(+\frac{\Delta}{180}\right)\right]$$
$$= -22{,}222\theta_B + 370.370\Delta$$
$$M_{BA} = 0 + 22{,}222\left[-2\theta_B - \theta_A + 3\left(+\frac{\Delta}{180}\right)\right]$$
$$= -44{,}444\theta_B + 370.370\Delta$$
$$M_{BC} = 0 + 55{,}555\left[-2\theta_B - \theta_C + 3\left(+\frac{0.45}{144}\right)\right]$$
$$= -111{,}111\theta_B - 55{,}555\theta_C + 520.828$$
$$M_{CB} = 0 + 55{,}555\left[-2\theta_C - \theta_B + 3\left(+\frac{0.45}{144}\right)\right]$$
$$= -111{,}111\theta_C - 55{,}555\theta_B + 520.828$$
$$M_{CD} = 0 + 33{,}333\left[-2\theta_C - \theta_D + 3\left(+\frac{\Delta}{120}\right)\right]$$
$$= -66{,}666\theta_C + 833.333\Delta - 66.667$$
$$M_{DC} = 0 + 33{,}333\left[-2\theta_D - \theta_C + 3\left(+\frac{\Delta}{120}\right)\right]$$
$$= -33{,}333\theta_C + 833.333\Delta - 133.333$$

Two Joint Conditions and One Shear Condition

Joint B: $M_{BA} + M_{BC} = 0$

Joint C: $M_{CB} + M_{CD} = 0$

Shear: $-H_A - H_D = 0$

$$-\left(\frac{M_{AB} + M_{BA}}{15}\right) - \left(\frac{M_{CD} + M_{DC}}{10}\right) = 0$$

Substituting,

$$-155{,}555\theta_B - 55{,}555\theta_C + 370.370\Delta = -520.828 \qquad (a)$$
$$-55{,}555\theta_B - 177{,}777\theta_C + 833.333\Delta = -454.161 \qquad (b)$$
$$-133{,}333\theta_B - 300{,}000\theta_C + 6{,}481.481\Delta = +600 \qquad (c)$$
$$+370.370\theta_B + 833.333\theta_C - 18.004\Delta = 1.667 \qquad (c')$$

Note that symmetry with respect to the diagonal can still be established by modifying (*c*) to the form of (*c*′).

Dividing (*a*) by 155,555,

$$-\theta_B - 0.357140\theta_C + 0.002381\Delta = -0.0033482 \qquad (d)$$

Dividing (*b*) by 55,555,

$$-\theta_B - 3.200000\theta_C + 0.015000\Delta = -0.0081749 \qquad (e)$$

Dividing (*c*) by 133,333,

$$-\theta_B - 2.250000\theta_C + 0.048611\Delta = +0.0045000 \qquad (f)$$

Subtracting (*e*) from (*d*),

$$+2.842860\theta_C - 0.012619\Delta = +0.0048267 \qquad (g)$$

Subtracting (*f*) from (*e*),

$$-0.950000\theta_C - 0.033611\Delta = -0.0126749 \qquad (h)$$

Dividing (*g*) by 2.84286,

$$+\theta_C - 0.0044388\Delta = +0.0016978 \qquad (i)$$

Dividing (*h*) by 0.95000,

$$-\theta_C - 0.0353800\Delta = -0.0133420 \qquad (j)$$

Adding (*i*) and (*j*)

$$-0.039819\Delta = -0.0116442$$
$$\Delta = +0.29243 \qquad (k)$$

Substituting (*k*) into (*i*),

$$\theta_C = +0.0029958 \qquad (l)$$

Substituting (*k*) and (*l*) into (*d*),

$$\theta_B = +0.0029746 \qquad (m)$$

The procedure used above in solving the system of simultaneous equations can be considered as a general method. Note that (*a*) to (*c*) are changed into (*d*) to (*f*), (*g*) and (*h*) into (*i*) and (*j*). When Δ is found in (*k*), its value is substituted in (*i*), not (*j*), to find θ_C because θ_C is predominant in (*i*). The values of Δ and θ_C thus found are substituted in (*d*), not (*e*) or (*f*), to find θ_B, for the same reason mentioned above.

Computations for End Moments

$$M_{AB} = -(22{,}222)(0.0029746) + (370.370)(0.29243)$$
$$= +42.205, \text{ or } +42.20 \text{ kip-ft}$$

$$\begin{aligned}
M_{BA} &= -(44{,}444)(0.0029746) + (370.370)(0.29243) \\
&= -23.897, \text{ or } -23.89 \text{ kip-ft} \\
M_{BC} &= -(111{,}111)(0.0029746) - (55{,}555)(0.0029958) + 520.828 \\
&= +23.884, \text{ or } +23.89 \text{ kip-ft} \\
M_{CB} &= -(111{,}111)(0.0029958) - (55{,}555)(0.0029746) + 520.828 \\
&= +22.706, \text{ or } +22.70 \text{ kip-ft} \\
M_{CD} &= -(66{,}666)(0.0029958) + (833.333)(0.29243) - 66.667 \\
&= -22.695, \text{ or } -22.70 \text{ kip-ft} \\
M_{DC} &= -(33{,}333)(0.0029958) + (833.33)(0.29243) - 133{,}333 \\
&= +10.499, \text{ or } +10.50 \text{ kip-ft}
\end{aligned}$$

The free-body, shear, and moment diagrams of the individual members are shown in Fig. 176. The free-body diagram, moment diagram, and elastic curve of the whole frame are shown in Fig. 177.

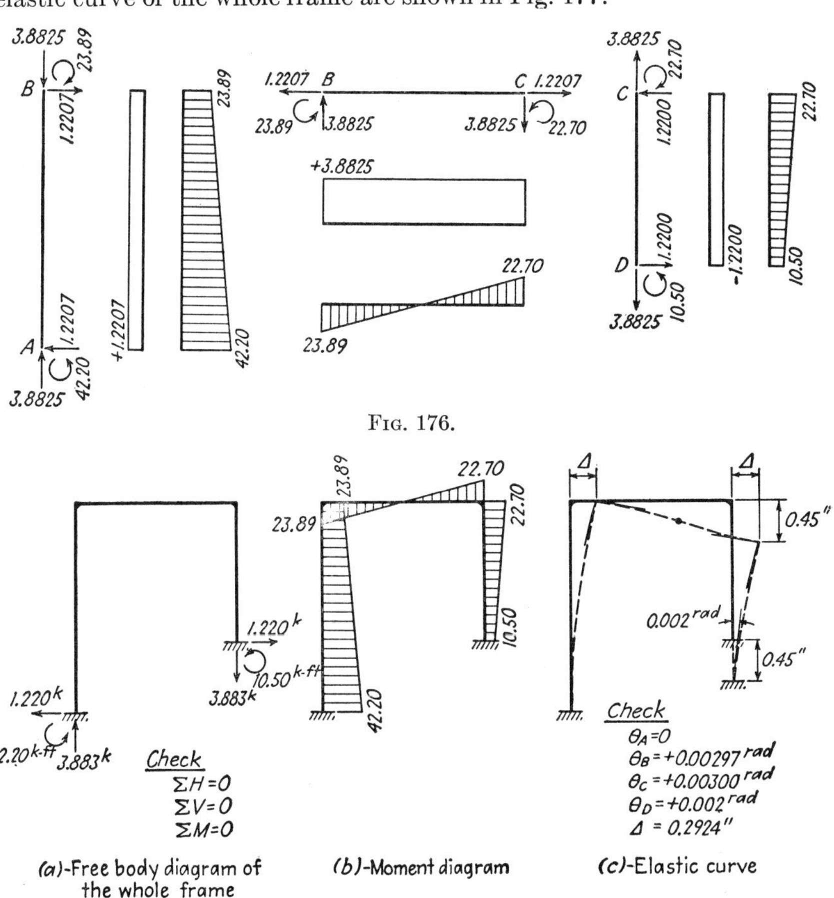

FIG. 176.

FIG. 177.

EXERCISES

97. Analyze the rigid frame shown owing to a vertical settlement of $\frac{1}{2}$ in. at the hinged support E. Draw shear and moment diagrams. Sketch the deformed structure.

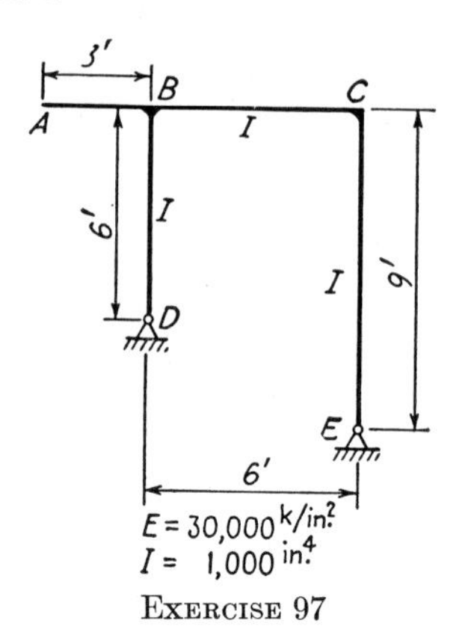

EXERCISE 97

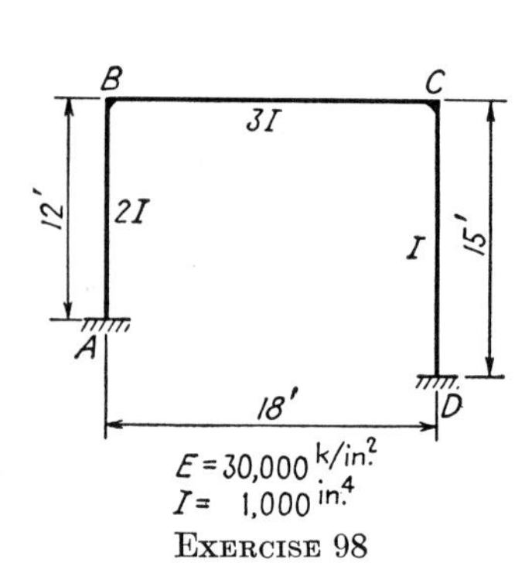

EXERCISE 98

98. Analyze the rigid frame shown owing to a rotational slip of 0.002 rad clockwise of joint A and a vertical settlement of $\frac{1}{2}$ in. at joint A. Draw shear and moment diagrams. Sketch the deformed structure.

43. Analysis of Gable Frames by the Slope-deflection Method. The general procedure of analyzing gable frames by the slope-deflection method is the same as heretofore discussed in this chapter. Because of the fact that the angles between members in the undeformed structure are not right angles, however, there are several items which need special consideration. These items can be illustrated by the following examples.

Example 66. Analyze the gable frame shown in Fig. 178 by the slope-deflection method. Draw shear and moment diagrams. Sketch the deformed structure.

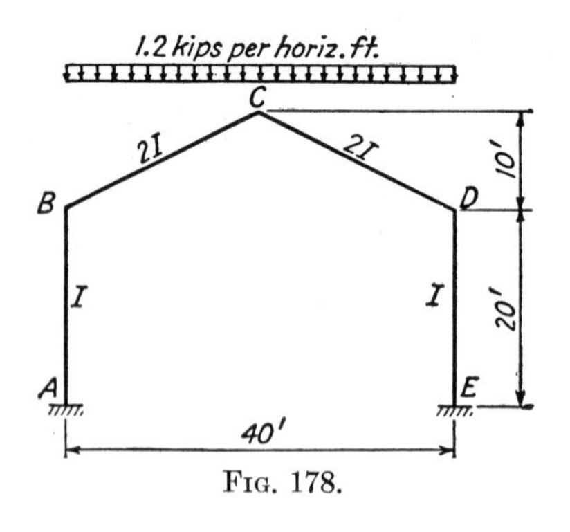

FIG. 178.

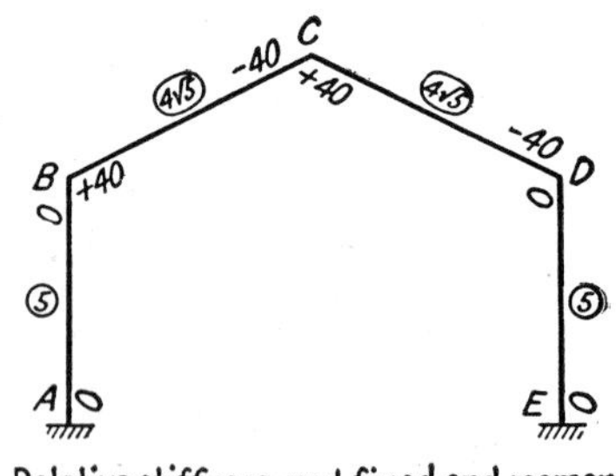

FIG. 179.

Solution. *Relative Stiffness* (Fig. 179)

TABLE 66-1

AB, DE..........	$\frac{I}{20} \times 100$	5
BC, CD..........	$\left(\frac{2I}{10\sqrt{5}} = \frac{\sqrt{5}I}{25}\right) \times 100$	$4\sqrt{5}$

Fixed-end Moments (Fig. 179). The magnitude of the fixed-end moment at one end of a span with length L, when subjected to a total

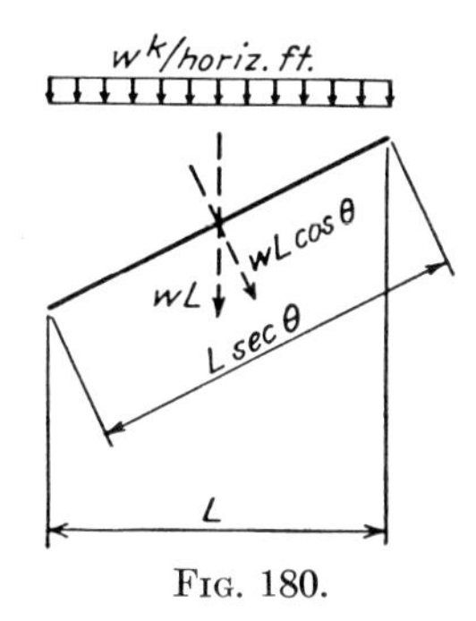

FIG. 180.

uniform load of W in a direction perpendicular to the axis of the member, is equal to $WL/12$. For the sloping member in Fig. 180,

$$\text{FEM} = \tfrac{1}{12}(wL \cos\theta)(L \sec\theta) = \tfrac{1}{12}wL^2$$

Thus the fixed-end moment for the sloping member is equal to that for a horizontal member with a span equal to the horizontal projection.

In this present example

$$M_{FBC} = +\frac{(1.2)(20)^2}{12} = +40 \text{ kip-ft}$$
$$M_{FCB} = -40 \text{ kip-ft}$$
$$M_{FCD} = +40 \text{ kip-ft}$$
$$M_{FDC} = -40 \text{ kip-ft}$$

Relative Values of R (Fig. 181). Before attempting to obtain the relative values of R, a *joint-displacement diagram* is drawn as shown in Fig. 181. One principle to be constantly kept in mind is that axial deformation in the members is neglected in the slope-deflection method and that, for the relative movement between the end joints of any member, only that in the direction perpendicular to the original axis of the member is permitted. In the present case, joints A and E are fixed.

Because of the symmetry of the properties of the frame and of the loading, joints B and D may move in the horizontal direction an amount Δ away from each other. Joint C, again because of symmetry, will have only vertical movement. A line $B'C''$ is drawn parallel and equal to BC. The perpendicular to $B'C''$ at C'' intersects a vertical line through C at C'.

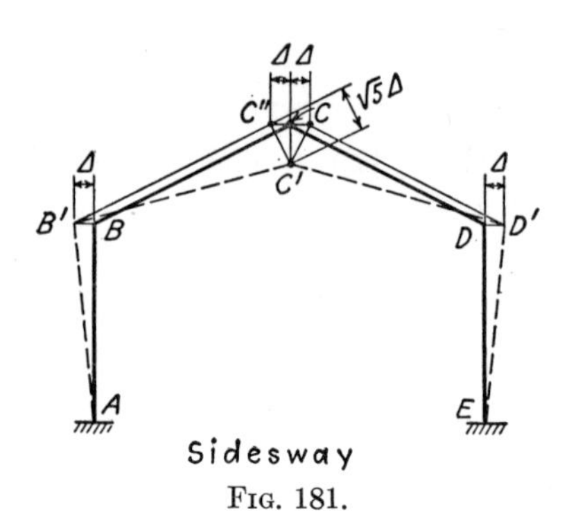

Fig. 181.

Thus the length of $B'C'$ is equal to that of BC. The joints B', C', and D' having been located, it is now possible to find the relative values of R. Note that, from triangle $CC'C''$, $C'C'' = \sqrt{5}\,\Delta$.

Table 66-2

R_{AB}	$-\frac{\Delta}{20} \times 20$	$-R$
R_{BC}	$\left(+\frac{C'C''}{10\sqrt{5}} = +\frac{\sqrt{5}\,\Delta}{10\sqrt{5}} = +\frac{\Delta}{10}\right) \times 20$	$+2R$
R_{CD}	$-\frac{\Delta}{10} \times 20$	$-2R$
R_{DE}	$+\frac{\Delta}{20} \times 20$	$+R$

If $R_{AB} = -R$, $R_{BC} = +2R$, $R_{CD} = -2R$, and $R_{DE} = +R$.

Slope-deflection Equations. It is known that (1) $\theta_A = \theta_E = 0$ (given), (2) $\theta_C = 0$ (because of symmetry), and (3) $\theta_D = -\theta_B$ (because of symmetry).

$$M_{AB} = 0 + 5(-2\theta_A - \theta_B - R) = -5\theta_B - 5R$$
$$M_{BA} = 0 + 5(-2\theta_B - \theta_A - R) = -10\theta_B - 5R$$
$$M_{BC} = +40 + 4\sqrt{5}\,(-2\theta_B - \theta_C + 2R) = +40 - 17.888\theta_B + 17.888R$$
$$M_{CB} = -40 + 4\sqrt{5}\,(-2\theta_C - \theta_B + 2R) = -40 - 8.944\theta_B + 17.888R$$

$$M_{CD} = +40 + 4\sqrt{5}\,(-2\theta_C - \theta_D - 2R) = +40 + 8.944\theta_B - 17.888R$$
$$M_{DC} = -40 + 4\sqrt{5}\,(-2\theta_D - \theta_C - 2R) = -40 + 17.888\theta_B - 17.888R$$
$$M_{DE} = 0 + 5(-2\theta_D - \theta_E + R) = +10\theta_B + 5R$$
$$M_{ED} = 0 + 5(-2\theta_E - \theta_D + R) = +5\theta_B + 5R$$

Joint and Shear Conditions. The joint condition at B requires

$$M_{BA} + M_{BC} = 0$$

The joint conditions at C and D will not yield independent equations because at the outset θ_D is made equal to $-\theta_B$ and θ_C to zero. The horizontal reactions at A and E already balance each other because $H_A + H_E = \dfrac{M_{AB} + M_{BA}}{20} + \dfrac{M_{DE} + M_{ED}}{20} = 0$. Besides the joint condition at B, therefore, one other condition is necessary for solving the

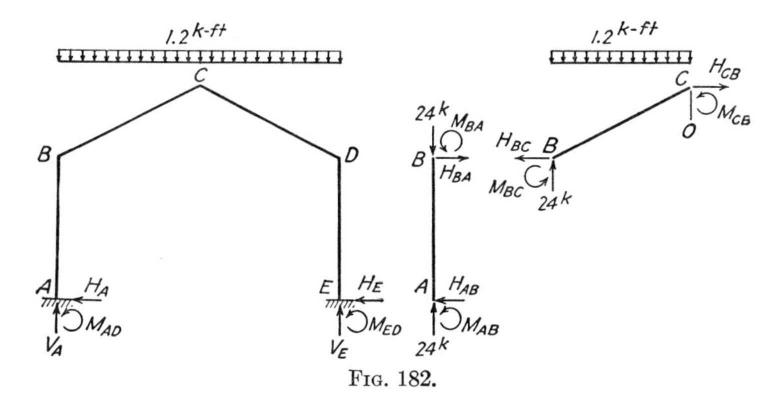

FIG. 182.

two unknowns θ_B and R. This condition is furnished by the requirement $H_{BA} = H_{BC}$ (Fig. 182), wherein

$$H_{BA} = \frac{M_{AB} + M_{BA}}{20} \quad \text{and} \quad H_{BC} = \frac{(M_{BC} + M_{CB}) - (24)(10)}{10}$$

Joint Condition at B: $\quad M_{BA} + M_{BC} = 0 \qquad (a)$

Shear Condition: $\quad H_{BA} = H_{BC} \qquad (b)$

$$-27.888\theta_B + 12.888R = -40 \qquad (a')$$
$$+38.664\theta_B - 81.552R = -480 \qquad (b')$$
$$+12.888\theta_B - 27.184R = -160 \qquad (b'')$$

Solving,

$$\theta_B = +5.3199$$
$$R = +8.4080$$

Computations for End Moments

$$M_{AB} = -(5)(5.3199) - (5)(8.4080) = -68.640 \text{ kip-ft}$$
$$M_{BA} = -(10)(5.3199) - (5)(8.4080) = -95.239 \text{ kip-ft}$$
$$M_{BC} = +40 - (17.888)(5.3199) + (17.888)(8.4080) = +95.240 \text{ kip-ft}$$
$$M_{CB} = -40 - (8.944)(5.3199) + (17.888)(8.4080) = +62.820 \text{ kip-ft}$$
$$M_{CD} = -62.820 \text{ kip-ft}$$
$$M_{DC} = +95.240 \text{ kip-ft}$$
$$M_{DE} = +95.239 \text{ kip-ft}$$
$$M_{ED} = +68.640 \text{ kip-ft}$$

The free-body, shear, and moment diagrams of the individual members are shown in Fig. 183. It is to be noted that the horizontal and vertical forces acting on BC (or CD) must first be replaced by forces perpendicular and parallel to the member BC (or CD) before shear and moment diagrams can be drawn in the usual way.

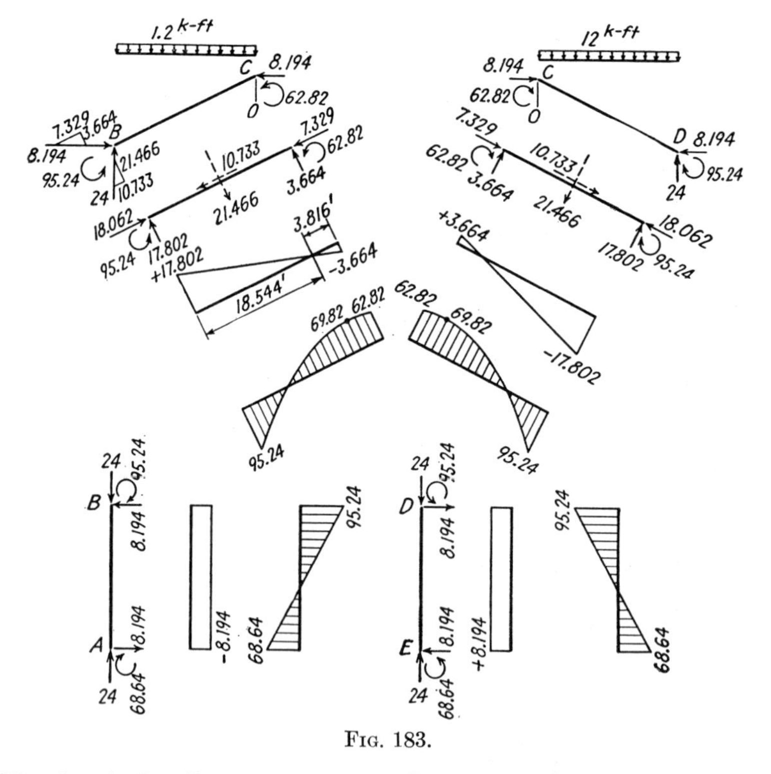

FIG. 183.

The free-body diagram, moment diagram, and elastic curve of the whole frame are shown in Fig. 184. The following checks on the conditions of geometry apply to Figs. 184*b* and *c*.

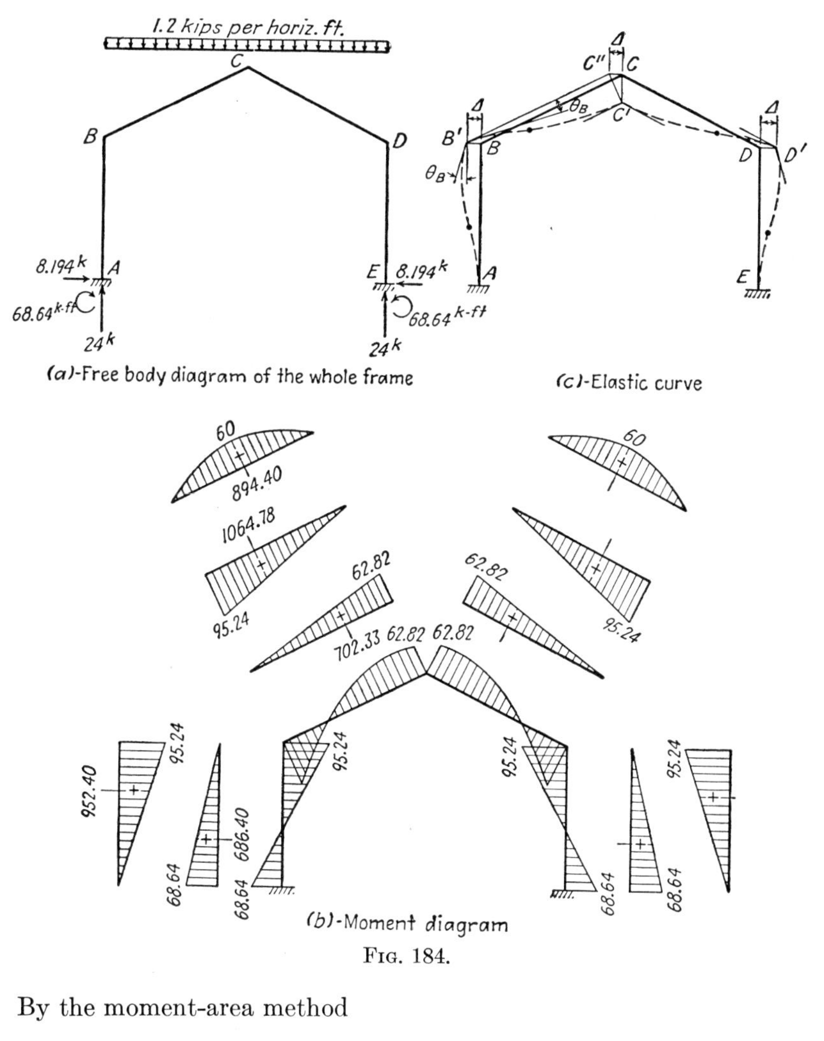

FIG. 184.

By the moment-area method

$$\theta_B = (952.40 - 686.40) = 266.0 \frac{\text{kip-ft}^2}{EI} \text{ clockwise}$$

$$BB' = (686.40 \times \tfrac{40}{3} - 952.40 \times \tfrac{20}{3}) = 2{,}802.7 \frac{\text{kip-ft}^3}{EI}$$

$$\theta_C = \theta_B - \left[\frac{894.40 + 702.33 - 1{,}064.78}{2EI}\right] = 0 \ (check)$$

$$C'C'' = 22.36\theta_B + \frac{\text{kip-ft}^3}{2EI}(22.36)\,(1{,}064.78 \times \tfrac{2}{3} - 894.40 \times \tfrac{1}{2} - 702.33 \times \tfrac{1}{3})$$

$$= \sqrt{5}\,(BB')\ (check)$$

Example 67. Analyze the gable frame shown in Fig. 185 by the slope-deflection method. Draw shear and moment diagrams. Sketch the deformed structure.

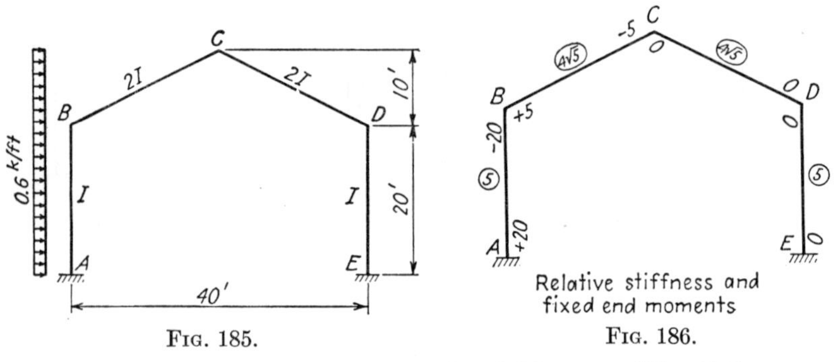

FIG. 185. FIG. 186.

Solution. *Relative Stiffness and Fixed-end Moments* (Fig. 186)

TABLE 67-1

AB, DE	$\frac{I}{20} \times 100$	5
BC, CD	$\left(\frac{2I}{10\sqrt{5}} = \frac{\sqrt{5}\,I}{25}\right) \times 100$	$4\sqrt{5}$

$$M_{FAB} = + \frac{(0.6)(20)^2}{12} = +20 \text{ kip-ft}$$

$$M_{FBA} = -20 \text{ kip-ft}$$

$$M_{FBC} = + \frac{(0.6)(10)^2}{12} = +5 \text{ kip-ft}$$

$$M_{FCB} = -5 \text{ kip-ft}$$

Relative Values of R (Fig. 187). The joint-displacement diagram showing the effect of sidesway is shown in Fig. 187. Joints A and E are fixed. Since the horizontal load acts only on the left side, joints B and D will move unequal horizontal distances Δ_1 and Δ_2 toward the right. Draw $B'C_1$ parallel and equal to BC. Draw $D'C_2$ parallel and equal to DC. The perpendicular lines to $B'C_1$ and $D'C_2$ at C_1 and C_2, respectively, intersect at C'. The joints B', C', and D' having been located, it is now possible to find the relative values of R. Note that, from triangle $C'C_1C_2$,

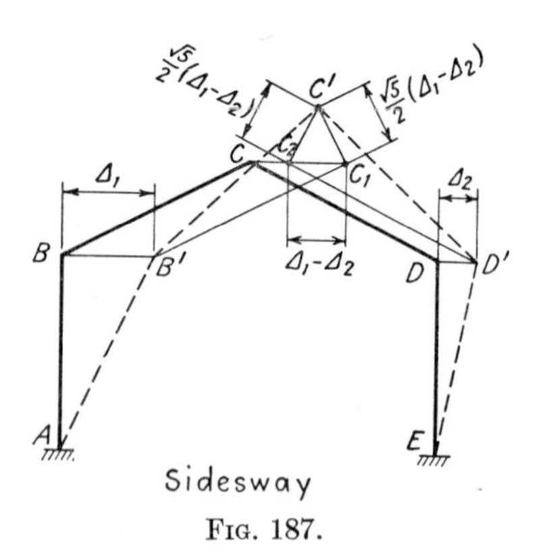

FIG. 187.

$$C'C_1 = C'C_2 = \frac{\sqrt{5}}{2}(\Delta_1 - \Delta_2)$$

TABLE 67-2

R_{AB}......	$+\frac{\Delta_1}{20} \times 20$	$+R_1$
R_{BC}......	$\left(-\frac{C'C_1}{10\sqrt{5}} = -\frac{\frac{\sqrt{5}}{2}(\Delta_1 - \Delta_2)}{10\sqrt{5}} = -\frac{\Delta_1 - \Delta_2}{20} \right) \times 20$	$-(R_1 - R_2)$
R_{CD}......	$+\frac{C'C_2}{10\sqrt{5}} = +\frac{\Delta_1 - \Delta_2}{20} \times 20$	$+(R_1 - R_2)$
R_{DE}......	$+\frac{\Delta_2}{20} \times 20$	$+R_2$

If $R_{AB} = +R_1$ and $R_{DE} = +R_2$, then $R_{BC} = -(R_1 - R_2)$ and $R_{CD} = +(R_1 - R_2)$.

Slope-deflection Equations. θ_A and θ_E are known to be zero.

$$
\begin{aligned}
M_{AB} &= +20 + 5(-2\theta_A - \theta_B + R_1) = +20 - 5\theta_B + 5R_1 \\
M_{BA} &= -20 + 5(-2\theta_B - \theta_A + R_1) = -20 - 10\theta_B + 5R_1 \\
M_{BC} &= +5 + 4\sqrt{5}\,(-2\theta_B - \theta_C - R_1 + R_2) \\
&= +5 - 17.888\theta_B - 8.944\theta_C - 8.944R_1 + 8.944R_2 \\
M_{CB} &= -5 + 4\sqrt{5}\,(-2\theta_C - \theta_B - R_1 + R_2) \\
&= -5 - 17.888\theta_C - 8.944\theta_B - 8.944R_1 + 8.944R_2 \\
M_{CD} &= 0 + 4\sqrt{5}\,(-2\theta_C - \theta_D + R_1 - R_2) \\
&= -17.888\theta_C - 8.944\theta_D + 8.944R_1 - 8.944R_2 \\
M_{DC} &= 0 + 4\sqrt{5}\,(-2\theta_D - \theta_C + R_1 - R_2) \\
&= -17.888\theta_D - 8.944\theta_C + 8.944R_1 - 8.944R_2 \\
M_{DE} &= 0 + 5(-2\theta_D - \theta_E + R_2) = -10\theta_D + 5R_2 \\
M_{ED} &= 0 + 5(-2\theta_E - \theta_D + R_2) = -5\theta_D + 5R_2
\end{aligned}
$$

Joint Conditions

Joint B: $M_{BA} + M_{BC} = 0$ (a)

Joint C: $M_{CB} + M_{CD} = 0$ (b)

Joint D: $M_{DC} + M_{DE} = 0$ (c)

$$-27.888\theta_B - 8.944\theta_C - 3.944R_1 + 8.944R_2 = +15 \qquad (a')$$

$$-8.944\theta_B - 35.776\theta_C - 8.944\theta_D = +5 \qquad (b')$$

$$8.944\theta_C - 27.888\theta_D + 8.944R_1 - 3.944R_2 = 0 \qquad (c')$$

Shear Conditions. Two shear conditions are required to correspond to the unknowns R_1 and R_2. If the gable frame is broken into two parts at joint C, the reactions of the left part on the right part must be equal and opposite to those of the right on the left. The moment requirement $M_{CB} = -M_{CD}$ is used to express the joint condition at C in the preceding paragraph. The other two requirements are:

1. That the horizontal reaction of the left part on the right part must be equal and opposite to that of the right on the left, or $H_{CB} = H_{CD}$ (Fig. 188).

2. That the vertical reaction of the left part on the right part must be equal and opposite to that of the right on the left, or $V_{CB} = V_{CD}$ (Fig. 188).

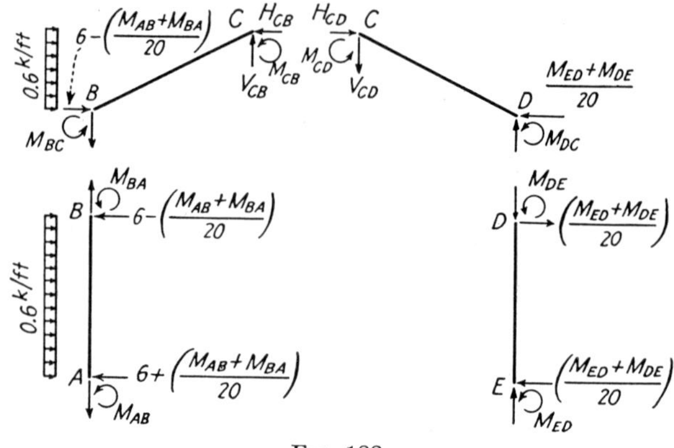

FIG. 188.

These two conditions can still be named *shear* conditions because the H's and V's noted above can be considered as vertical and horizontal shearing forces at section C. From free body BC (Fig. 188)

$$\Sigma H = 0, \qquad H_{CB} = 12 - \left(\frac{M_{AB} + M_{BA}}{20}\right)$$

$$\Sigma M_B = 0, \qquad V_{CB} = \frac{30 - H_{CB}(10) - (M_{BC} + M_{CB})}{20} = \frac{(M_{AB} + M_{BA})}{40} - \frac{(M_{BC} + M_{CB})}{20} - 4.5$$

From free body CD (Fig. 188)

$$\Sigma H = 0, \qquad H_{CD} = \left(\frac{M_{ED} + M_{DE}}{20}\right)$$

$$\Sigma M_D = 0, \qquad V_{CD} = \frac{H_{CD}(10) - (M_{CD} + M_{DC})}{20} = \frac{(M_{ED} + M_{DE})}{40} - \frac{(M_{CD} + M_{DC})}{20}$$

Shear condition $H_{CB} = H_{CD}$:

$$12 - \frac{(M_{AB} + M_{BA})}{20} = \frac{(M_{ED} + M_{DE})}{20}$$

$$M_{AB} + M_{BA} + M_{ED} + M_{DE} = +240$$

$$-15\theta_B - 15\theta_D + 10R_1 + 10R_2 = +240 \qquad (d')$$

Shear condition $V_{CB} = V_{CD}$:

$$\frac{(M_{AB} + M_{BA})}{40} - \frac{(M_{BC} + M_{CB})}{20} - 4.5 = \frac{(M_{ED} + M_{DE})}{40} - \frac{(M_{CD} + M_{DC})}{20}$$

$$(M_{AB} + M_{BA}) - (M_{ED} + M_{DE}) - 2(M_{BC} + M_{CB}) + 2(M_{CD} + M_{DC}) = +180$$

$$+38.664\theta_B - 38.664\theta_D + 81.552R_1 - 81.552R_2 = +180 \qquad (e')$$

Solving the five simultaneous equations (a') to (e'),

$$\theta_B = +2.3795$$
$$\theta_C = -1.8283$$
$$\theta_D = +4.3746$$
$$R_1 = +18.6421$$
$$R_2 = +15.4891$$

Computations for End Moments

$$M_{AB} = +20 - (5)(+2.3795) + (5)(+18.6421) = +101.312 \text{ kip-ft}$$
$$M_{BA} = -20 - (10)(+2.3795) + (5)(+18.6421) = +49.415 \text{ kip-ft}$$
$$M_{BC} = +5 - (17.888)(+2.3795) - (8.944)(-1.8283) - (8.944)(+18.6421) + (8.944)(+15.4891) = -49.413 \text{ kip-ft}$$
$$M_{CB} = -5 - (17.888)(-1.8283) - (8.944)(+2.3795) - (8.944)(+18.6421) + (8.944)(+15.4891) = -21.778 \text{ kip-ft}$$
$$M_{CD} = -(17.888)(-1.8283) - (8.944)(+4.3746) + (8.944)(+18.6421) - (8.944)(+15.4891) = +21.780 \text{ kip-ft}$$
$$M_{DC} = -(17.888)(+4.3746) - (8.944)(-1.8283) + (8.944)(+18.6421) - (8.944)(+15.4891) = -33.700 \text{ kip-ft}$$
$$M_{DE} = -(10)(+4.3746) + (5)(+15.4891) = +33.700 \text{ kip-ft}$$
$$M_{ED} = -(5)(+4.3746) + (5)(+15.4891) = +55.573 \text{ kip-ft}$$

The free-body, shear, and moment diagrams of the individual members are shown in Fig. 189. The free-body diagram, moment diagram, and elastic curve of the whole frame are shown in Fig. 190. The following checks on the conditions of geometry apply to Fig. 190*b* and *c*.

By the moment-area method

$$\theta_B = (1{,}013.10 - 494.10 - 400.00)\,\frac{\text{kip-ft}^2}{EI}$$

$$= 119.00\,\frac{\text{kip-ft}^2}{EI} \text{ clockwise}$$

$$\theta_C = [119.00 - \tfrac{1}{2}(111.80 + 552.40 - 243.50)]\,\frac{\text{kip-ft}^2}{EI}$$

$$= -91.35\,\frac{\text{kip-ft}^2}{EI}, \text{ or } 91.35\,\frac{\text{kip-ft}^2}{EI} \text{ counterclockwise}$$

$$\Delta_1 = (1{,}013.10 \times \tfrac{40}{3} - 494.10 \times \tfrac{20}{3} - 400 \times 10)\,\frac{\text{kip-ft}^3}{EI}$$

$$= 6{,}214.0\,\frac{\text{kip-ft}^3}{EI}$$

$$\theta_D = (555.70 - 337.00)\,\frac{\text{kip-ft}^2}{EI} = 218.70\,\frac{\text{kip-ft}^2}{EI}$$

$$\theta_C = [218.70 - \tfrac{1}{2}(376.77 + 243.50)]\,\frac{\text{kip-ft}^2}{EI}$$

$$= -91.44\,\frac{\text{kip-ft}^2}{EI}, \text{ or } 91.44\,\frac{\text{kip-ft}^2}{EI} \text{ counterclockwise } (check)$$

$$\Delta_2 = (555.70 \times \tfrac{40}{3} - 337.00 \times \tfrac{20}{3})\,\frac{\text{kip-ft}^3}{EI} = 5{,}162.7\,\frac{\text{kip-ft}^3}{EI}$$

$$\Delta_1 - \Delta_2 = 6{,}214.0 - 5{,}162.7 = 1{,}051.3\,\frac{\text{kip-ft}^3}{EI}$$

$$C_2C' = 22.36\theta_D - \left(243.50 \times \frac{22.36}{3} + 376.77 \times \tfrac{2}{3} \times 22.36\right)\frac{\text{kip-ft}^3}{2EI}$$

$$= \frac{\sqrt{5}}{2}(20 \times 218.70) - \frac{\sqrt{5}}{2}\left(\frac{1{,}623.3 + 5{,}023.6}{2}\right)$$

$$= \frac{\sqrt{5}}{2}(1{,}050.6) = \frac{\sqrt{5}}{2}(\Delta_1 - \Delta_2) \; (check)$$

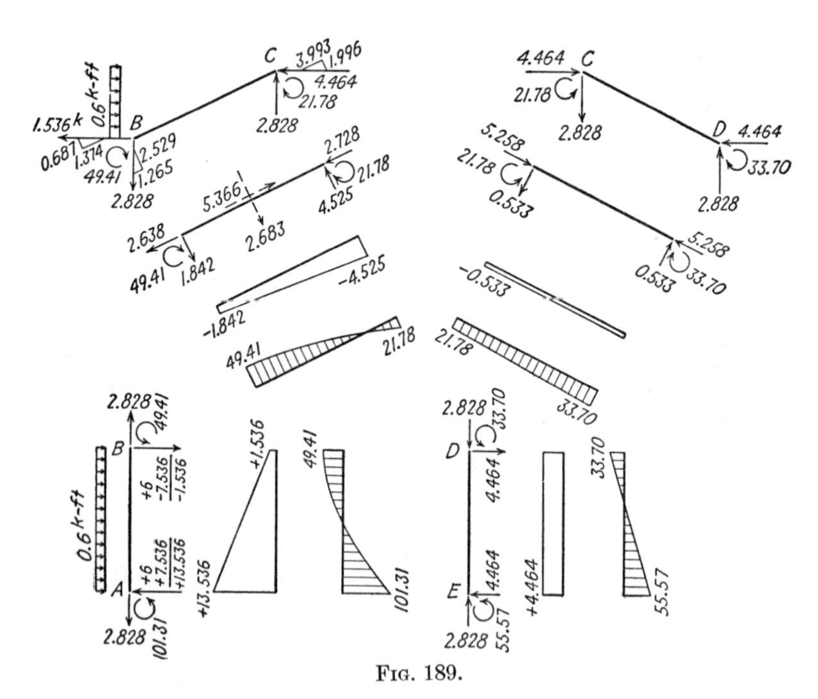

Fig. 189.

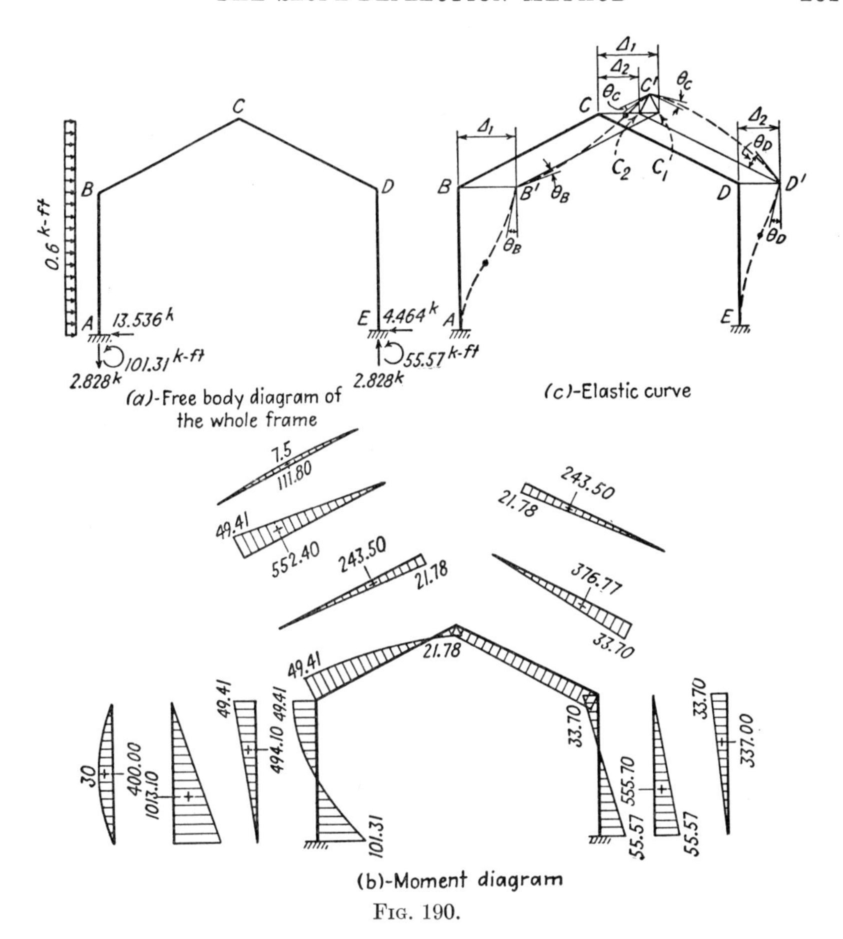

FIG. 190.

$$C_1C' = \left(111.80 \times 11.18 + 552.40 \times \tfrac{2}{3} \times 22.36 - 243.50 \times \frac{22.36}{3}\right) \frac{\text{kip-ft}^2}{2EI} - 22.36\theta_B$$

$$= \frac{\sqrt{5}}{2}\left(\frac{111.80 + 7{,}365.3 - 1{,}623.3}{2}\right) - \frac{\sqrt{5}}{2}(20 \times 119.00)$$

$$= \frac{\sqrt{5}}{2}(1{,}051.0) = \frac{\sqrt{5}}{2}(\Delta_1 - \Delta_2) \text{ (check)}$$

EXERCISES

99 to 102. Analyze the gable frame shown. Draw shear and moment diagrams. Sketch the deformed structure.

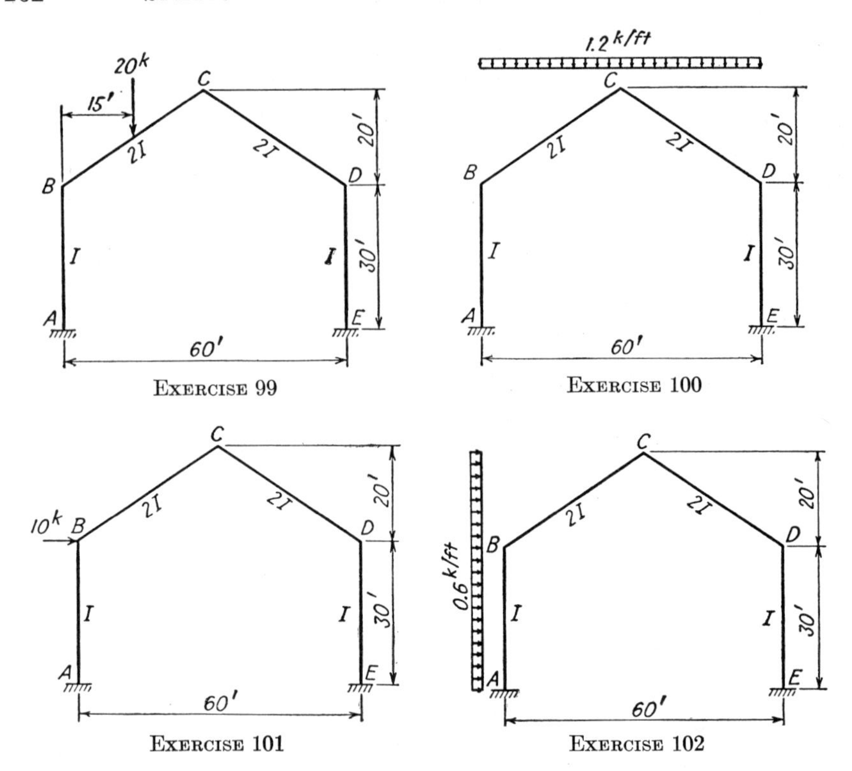

Exercise 99

Exercise 100

Exercise 101

Exercise 102

44. Slope-deflection Equations for Members with Variable Cross Sections. The slope-deflection equations

$$\begin{aligned} M_{AB} &= M_{FAB} + \frac{2EI}{L}(-2\theta_A - \theta_B + 3R) \\ M_{BA} &= M_{FBA} + \frac{2EI}{L}(-2\theta_B - \theta_A + 3R) \end{aligned} \qquad (66)$$

have been derived on the basis that the member AB has a constant cross section with moment of inertia equal to I. Now if the member AB has variable cross sections, the moment of inertia I will be variable; in other words, I has different values at different sections along the span. The slope-deflection equations for members with variable cross sections or variable I will be derived by the same general procedure as Eqs. (66).

The member AB shown in Fig. 191a is subjected to the applied loading and a settlement of Δ at support B. It is required to express the end moments M_{AB} and M_{BA} in terms of the applied loading, the values of θ_A, θ_B, and R, and the properties of the variable cross sections. As previously described, the member AB shown in Fig. 191a can be considered as the sum of those shown in Fig. 191b to e. Thus

$$\begin{aligned} M_{AB} &= M_{FAB} + M'_{FAB} + M'_{AB} \\ M_{BA} &= M_{FBA} + M'_{FBA} + M'_{BA} \end{aligned} \tag{71}$$

and

$$\begin{aligned} \theta_A &= -\theta_{A1} + \theta_{A2} \\ \theta_B &= +\theta_{B1} - \theta_{B2} \end{aligned} \tag{72}$$

M_{FAB} and M_{FBA} in Eqs. (71) can be found by the method of consistent deformation, with due regard to the fact that I is variable along the span. The method of column analogy is a shorter and more direct method, which will be treated later in Chap. IX.

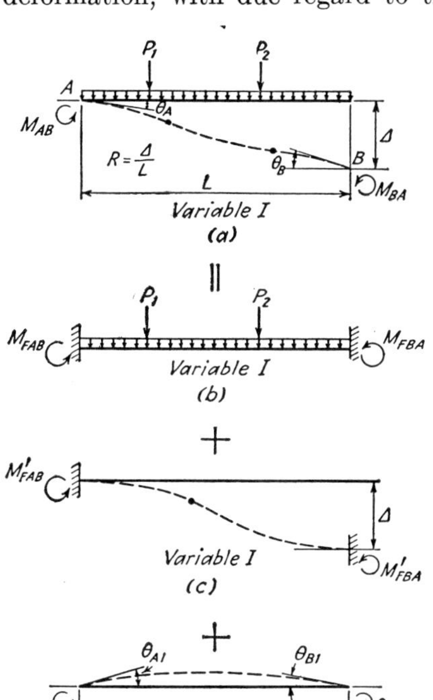

FIG. 191.

Before M'_{FAB}, M'_{FBA}, M'_{AB}, and M'_{BA} in Eqs. (71) can be expressed in terms of θ_A, θ_B, and R, it is necessary to solve the following two problems.

Problem I. Required to find ϕ_{AA} and ϕ_{BA} due to an end moment of 1 kip-ft counterclockwise applied at end A of span AB, with moment of

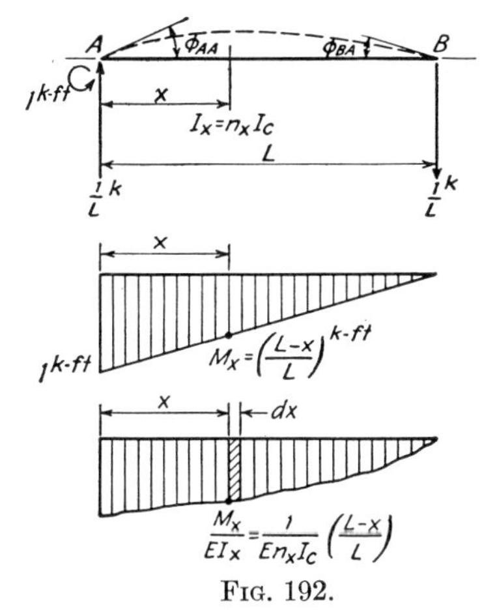

FIG. 192.

inertia $I_x = n_x I_c$, where I_c is a constant moment of inertia and n_x is a pure ratio which varies with x (Fig. 192).

Solution

$$\phi_{AA} = \frac{\text{moment of } M_x/EI_x \text{ diagram between } A \text{ and } B \text{ about } B}{L}$$

$$= \frac{1}{L}\int_0^L \left(\frac{M_x}{EI_x}\,dx\right)(L-x) = \frac{1}{L}\int_0^L \frac{(L-x)^2}{EI_cLn_x}\,dx$$

Let

$$C_1 = \frac{1}{L^3}\int_0^L \frac{(L-x)^2}{n_x}\,dx \tag{73}$$

Then

$$\phi_{AA} = C_1\frac{L}{EI_c} \tag{74}$$

$$\phi_{BA} = \frac{\text{moment of } M_x/EI_x \text{ diagram between } A \text{ and } B \text{ about } A}{L}$$

$$= \frac{1}{L}\int_0^L \left(\frac{M_x}{EI_x}\,dx\right)(x) = \frac{1}{L}\int_0^L \frac{(L-x)(x)}{EI_cLn_x}\,dx$$

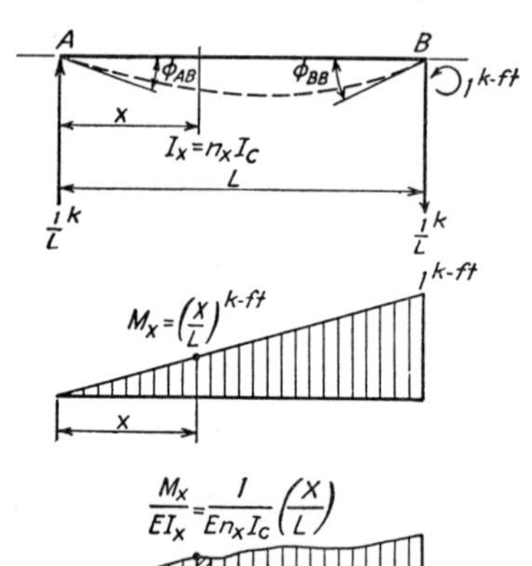

Fig. 193.

Let

$$C_2 = \frac{1}{L^3}\int_0^L \frac{(L-x)(x)}{n_x}\,dx \tag{75}$$

Then

$$\phi_{BA} = C_2\frac{L}{EI_c} \tag{76}$$

Problem II. Required to find ϕ_{AB} and ϕ_{BB} due to an end moment of 1 kip-ft counterclockwise applied at end B of span AB, with moment of inertia $I_x = n_xI_c$, where I_c is a constant moment of inertia and n_x is a pure ratio which varies with x (Fig. 193).

Solution

$$\phi_{AB} = \frac{\text{moment of } M_x/EI_x \text{ diagram between } A \text{ and } B \text{ about } B}{L}$$

$$= \frac{1}{L}\int_0^L \left(\frac{M_x}{EI_x}\,dx\right)(L-x) = \frac{1}{L}\int_0^L \frac{(L-x)(x)}{EI_cLn_x}\,dx$$

$$= C_2\frac{L}{EI_c} \tag{77}$$

$$\phi_{BB} = \frac{\text{moment of } M_x/EI_x \text{ diagram between } A \text{ and } B \text{ about } A}{L}$$

$$= \frac{1}{L}\int_0^L \left(\frac{M_x}{EI_x}\,dx\right)(x) = \frac{1}{L}\int_0^L \frac{(x)^2}{EI_cLn_x}\,dx$$

Let

$$C_3 = \frac{1}{L^3}\int_0^L \frac{(x)^2}{n_x}\,dx \tag{78}$$

Then

$$\phi_{BB} = C_3\frac{L}{EI_c} \tag{79}$$

Referring to Fig. 191d and using Eqs. (74) and (76),

$$\theta_{A1} = M'_{AB}\left(\frac{C_1L}{EI_c}\right) \qquad \theta_{B1} = M'_{AB}\left(\frac{C_2L}{EI_c}\right) \tag{80}$$

Referring to Fig. 191e and using Eqs. (77) and (79),

$$\theta_{A2} = M'_{BA}\left(\frac{C_2L}{EI_c}\right) \qquad \theta_{B2} = M'_{BA}\left(\frac{C_3L}{EI_c}\right) \tag{81}$$

Substituting Eqs. (80) and (81) into Eqs. (72),

$$\begin{aligned} \theta_A &= -M'_{AB}\left(\frac{C_1L}{EI_c}\right) + M'_{BA}\left(\frac{C_2L}{EI_c}\right) \\ \theta_B &= +M'_{AB}\left(\frac{C_2L}{EI_c}\right) - M'_{BA}\left(\frac{C_3L}{EI_c}\right) \end{aligned} \tag{82}$$

Solving Eqs. (82) for M'_{AB} and M'_{BA},

$$\begin{aligned} M'_{AB} &= +\frac{2EI_c}{L}\left[-\frac{C_3}{2(C_1C_3 - C_2^2)}\,\theta_A - \frac{C_2}{2(C_1C_3 - C_2^2)}\,\theta_B\right] \\ M'_{BA} &= +\frac{2EI_c}{L}\left[-\frac{C_1}{2(C_1C_3 - C_2^2)}\,\theta_B - \frac{C_2}{2(C_1C_3 - C_2^2)}\,\theta_A\right] \end{aligned} \tag{83}$$

Let

$$K_{AA} = +\frac{C_3}{2(C_1C_3 - C_2^2)} \qquad K_{AB} = +\frac{C_2}{2(C_1C_3 - C_2^2)}$$

$$K_{BB} = +\frac{C_1}{2(C_1C_3 - C_2^2)} \tag{84}$$

Then

$$\begin{aligned} M'_{AB} &= \frac{2EI_c}{L}\left[-K_{AA}\theta_A - K_{AB}\theta_B\right] \\ M'_{BA} &= \frac{2EI_c}{L}\left[-K_{BB}\theta_B - K_{AB}\theta_A\right] \end{aligned} \tag{85}$$

With M'_{AB} and M'_{BA} found in Eqs. (85), the next problem is to find M'_{FAB} and M'_{FBA} (Fig. 191c). The two conditions from which M'_{FAB} and M'_{FBA} can be found are (Fig. 194):

1. Deflection of B from tangent at $A = \Delta$.
2. Deflection of A from tangent at $B = \Delta$.

Using the first condition,

$$\Delta = \text{moment of } \frac{M_x}{EI_x} \text{ diagram between } A \text{ and } B \text{ about } B$$

$$= +\int_0^L \left(\frac{M'_{FAB}}{En_xI_c}\,\frac{L-x}{L}\,dx\right)(L-x) - \int_0^L \left(\frac{M'_{FBA}}{En_xI_c}\,\frac{x}{L}\,dx\right)(L-x)$$

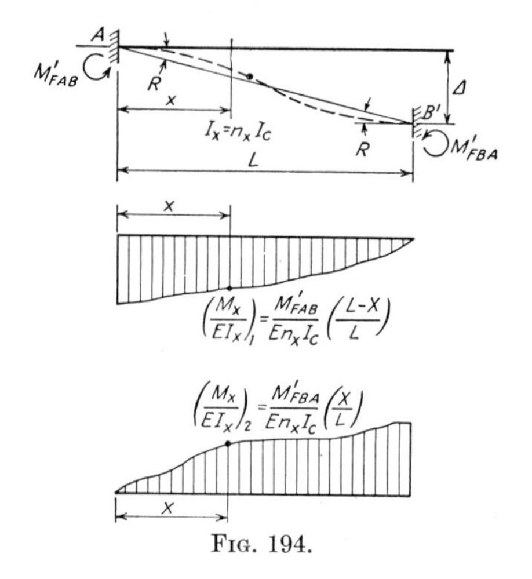

Fig. 194.

or

$$\Delta = +\frac{M'_{FAB}L^2}{EI_c}C_1 - \frac{M'_{FBA}L^2}{EI_c}C_2 \tag{86}$$

Using the second condition,

$$\Delta = \text{moment of } \frac{M_x}{EI_x} \text{ diagram between } A \text{ and } B \text{ about } A$$

$$\Delta = -\int_0^L \left(\frac{M'_{FAB}}{En_xI_c}\frac{L-x}{L}\,dx\right)(x) + \int_0^L \left(\frac{M'_{FBA}}{En_xI_c}\frac{x}{L}\,dx\right)(x)$$

or

$$\Delta = -\frac{M'_{FAB}L^2}{EI_c}C_2 + \frac{M'_{FBA}L^2}{EI_c}C_3 \tag{87}$$

Solving Eqs. (86) and (87) for M'_{FAB} and M'_{FBA} and recalling that $R = \Delta/L$,

$$\begin{aligned} M'_{FAB} &= +\frac{2EI_c}{L}\left[\frac{C_2 + C_3}{2(C_1C_3 - C_2^2)}R\right] = +\frac{2EI_c}{L}(K_{AA} + K_{AB})R \\ M'_{FBA} &= +\frac{2EI_c}{L}\left[\frac{C_1 + C_2}{2(C_1C_3 - C_2^2)}R\right] = +\frac{2EI_c}{L}(K_{BB} + K_{AB})R \end{aligned} \tag{88}$$

Substituting Eqs. (85) and (88) into Eqs. (71),

$$\begin{aligned} M_{AB} &= M_{FAB} + \frac{2EI_c}{L}[-K_{AA}\theta_A - K_{AB}\theta_B + (K_{AA} + K_{AB})R] \\ M_{BA} &= M_{FBA} + \frac{2EI_c}{L}[-K_{BB}\theta_B - K_{AB}\theta_A + (K_{BB} + K_{AB})R] \end{aligned} \tag{89}$$

where

$$K_{AA} = +\frac{C_3}{2(C_1C_3 - C_2^2)} \qquad K_{AB} = +\frac{C_2}{2(C_1C_3 - C_2^2)}$$

$$K_{BB} = +\frac{C_1}{2(C_1C_3 - C_2^2)}$$

and

$$C_1 = \frac{1}{L^3}\int_0^L \frac{(L-x)^2}{n_x}\,dx \qquad C_2 = \frac{1}{L^3}\int_0^L \frac{(x)(L-x)}{n_x}\,dx$$

$$C_3 = \frac{1}{L^3}\int_0^L \frac{(x)^2}{n_x}\,dx$$

Equations (89) are the slope-deflection equations for members with variable cross sections. It is to be noted that, for members with constant cross section, $n_x = 1$ for all values of x, $I_c = I$, $C_1 = \frac{1}{3}$, $C_2 = \frac{1}{6}$, $C_3 = \frac{1}{3}$, $K_{AA} = K_{BB} = 2$, $K_{AB} = 1$, and Eqs. (89) become Eqs. (66).

Example 68. Analyze the continuous beam shown in Fig. 195 by the slope-deflection method. Draw shear and moment diagrams, and sketch the elastic curve.

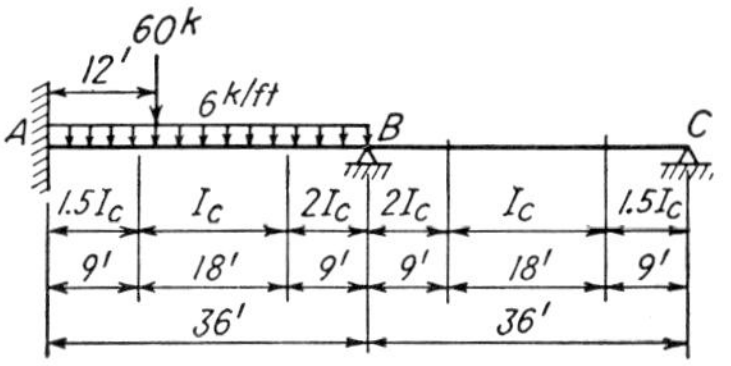

FIG. 195.

Solution. *For span AB:*

$$C_1 = \frac{1}{L^3}\int_0^L \frac{(L-x)^2}{n_x}\,dx$$

$$= \frac{1}{(36)^3}\left[\int_0^9 \frac{(36-x)^2}{1.5}\,dx + \int_9^{27}\frac{(36-x)^2}{1}\,dx + \int_{27}^{36}\frac{(36-x)^2}{2}\,dx\right]$$

$$= \frac{1}{(36)^3}\left\{\frac{1}{1.5}\left[1{,}296x - 36x^2 + \frac{x^3}{3}\right]_0^9 + \left[1{,}296x - 36x^2 + \frac{x^3}{3}\right]_9^{27} + \frac{1}{2}\left[1{,}296x - 36x^2 + \frac{x^3}{3}\right]_{27}^{36}\right\}$$

$$= \frac{(9)^3}{(36)^3}\left[\frac{1}{1.5}(16 - 4 + \tfrac{1}{3}) + (48 - 36 + 9 - 16 + 4 - \tfrac{1}{3}) + (\tfrac{1}{2})(64 - 64 + \tfrac{64}{3} - 48 + 36 - 9)\right]$$

$$= (\tfrac{1}{64})[\tfrac{74}{9} + \tfrac{26}{3} + \tfrac{1}{6}] = \frac{307}{1{,}152} = 0.266493$$

$$C_2 = \frac{1}{L^3}\int_0^L \frac{x(L-x)}{n_x}\,dx$$

$$= \frac{1}{(36)^3}\int_0^9 \frac{(x)(36-x)}{1.5}\,dx + \int_9^{27}\frac{(x)(36-x)}{1}\,dx + \int_{27}^{36}\frac{(x)(36-x)}{2}\,dx$$

$$= \frac{1}{(36)^3}\left\{\frac{1}{1.5}\left[18x^2 - \frac{x^3}{3}\right]_0^9 + \left[18x^2 - \frac{x^3}{3}\right]_9^{27} + \frac{1}{2}\left[18x^2 - \frac{x^3}{3}\right]_{27}^{36}\right\}$$

$$= \frac{(9)^3}{(36)^3}\left[\frac{1}{1.5}(2 - \tfrac{1}{3}) + (18 - 9 - 2 + \tfrac{1}{3}) + (\tfrac{1}{2})(32 - \tfrac{64}{3} - 18 + 9)\right]$$

$$= (\tfrac{1}{64})[\tfrac{10}{9} + \tfrac{22}{3} + \tfrac{5}{6}] = \frac{167}{1{,}152} = 0.144965$$

$$C_3 = \frac{1}{L^3}\int_0^L \frac{(x)^2}{n_x}\,dx$$

$$= \frac{1}{(36)^3}\left[\int_0^9 \frac{x^2}{1.5}\,dx + \int_9^{27} \frac{x^2}{1}\,dx + \int_{27}^{36} \frac{(x)^2}{2}\,dx\right]$$

$$= \frac{1}{(36)^3}\left\{\frac{1}{1.5}\left[\frac{x^3}{3}\right]_0^9 + \left[\frac{x^3}{3}\right]_9^{27} + \frac{1}{2}\left[\frac{x^3}{3}\right]_{27}^{36}\right\}$$

$$= \frac{(9)^3}{(36)^3}\left[\frac{1}{1.5}\,(\tfrac{1}{3}) + (9 - \tfrac{1}{3}) + (\tfrac{1}{2})(\tfrac{64}{3} - 9)\right]$$

$$= (\tfrac{1}{64})[\tfrac{2}{9} + \tfrac{26}{3} + \tfrac{37}{6}] = \frac{271}{1{,}152} = 0.235243$$

$$2(C_1C_3 - C_2^2) = 2[(0.266493)(0.235243) - (0.144965)^2] = 0.0833516$$

$$K_{AA} = +\frac{C_3}{2(C_1C_3 - C_2^2)} = +\frac{0.235243}{0.0833516} = +2.82230$$

$$K_{AB} = +\frac{C_2}{2(C_1C_3 - C_2^2)} = +\frac{0.144965}{0.0833516} = +1.73920$$

$$K_{BB} = +\frac{C_1}{2(C_1C_3 - C_2^2)} = +\frac{0.266493}{0.0833516} = +3.19722$$

For span BC:

$$C_1 = 0.235243 \qquad K_{BB} = +3.19722$$
$$C_2 = 0.144965 \qquad K_{BC} = +1.73920$$
$$C_3 = 0.266493 \qquad K_{CC} = +2.82230$$

Fixed-end Moments—Span AB. a. Due to Uniform Load (Fig. 196). By the method of consistent deformation

$$\theta_{A1} - \theta_{A2} - \theta_{A3} = 0$$
$$\theta_{B1} - \theta_{B2} - \theta_{B3} = 0$$

By the conjugate-beam method

$$EI_c\theta_{A1} = R'_A$$

$$= \frac{1}{36}\left[\frac{1}{1.5}\int_0^9 (108x - 3x^2)(36 - x)\,dx\right.$$
$$\left. + \int_9^{27} (108x - 3x^2)(36 - x)\,dx + \frac{1}{2}\int_{27}^{36} (108x - 3x^2)(36 - x)\,dx\right]$$
$$= 10{,}350.28$$

$$EI_c\theta_{B1} = R'_B$$

$$= \frac{1}{36}\left[\frac{1}{1.5}\int_0^9 (108x - 3x^2)(x)\,dx\right.$$
$$\left. + \int_9^{27} (108x - 3x^2)(x)\,dx + \frac{1}{2}\int_{27}^{36} (108x - 3x^2)(x)\,dx\right]$$
$$= 9{,}940.22$$

By definition of C_1, C_2, and C_3

$$EI_c\theta_{A2} = M_AC_1L = (0.266493)(36)M_A$$
$$EI_c\theta_{B2} = M_AC_2L = (0.144965)(36)M_A$$
$$EI_c\theta_{A3} = M_BC_2L = (0.144965)(36)M_B$$
$$EI_c\theta_{B3} = M_BC_3L = (0.235243)(36)M_B$$

The equations for consistent deformation now become

$$10{,}350.28 - (0.266493)(36)M_A - (0.144965)(36)M_B = 0$$
$$9{,}940.22 - (0.144965)(36)M_A - (0.235243)(36)M_B = 0$$

Solving,

$$M_A = 662.42 \text{ kip-ft}$$
$$M_B = 765.55 \text{ kip-ft}$$

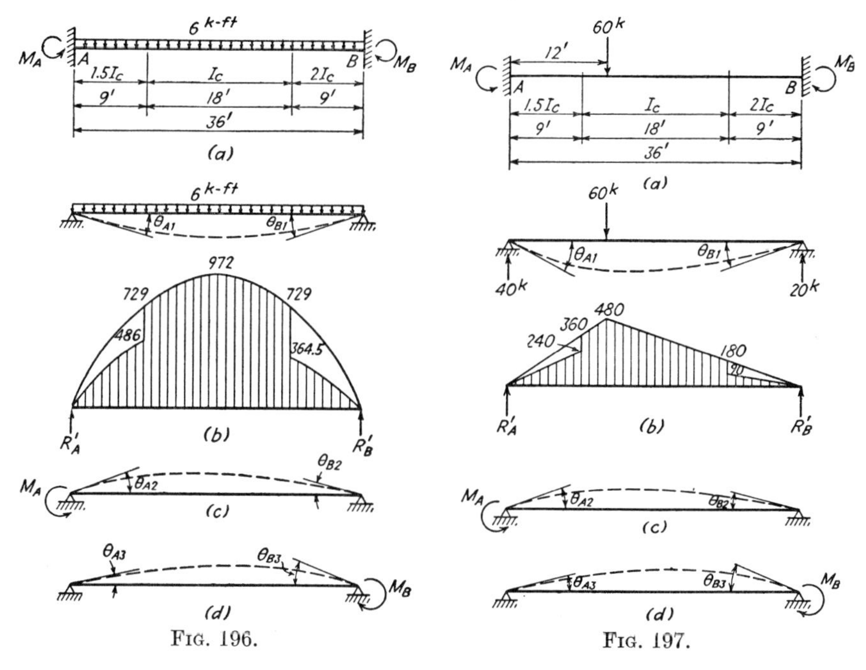

FIG. 196. FIG. 197.

b. Due to the Concentrated Load (Fig. 197). By the method of consistent deformation

$$\theta_{A1} - \theta_{A2} - \theta_{A3} = 0$$
$$\theta_{B1} - \theta_{B2} - \theta_{B3} = 0$$

By the conjugate-beam method

$$EI_c\theta_{A1} = 4{,}282.5$$
$$EI_c\theta_{B1} = 3{,}412.5$$

By definition of C_1, C_2, and C_3

$$EI_c\theta_{A2} = (0.266493)(36)M_A \qquad EI_c\theta_{B2} = (0.144965)(36)M_A$$
$$EI_c\theta_{A3} = (0.144965)(36)M_B \qquad EI_c\theta_{B3} = (0.235243)(36)M_B$$

The equations for consistent deformation now become

$$4{,}282.5 - (0.266493)(36)M_A - (0.144965)(36)M_B = 0$$
$$3{,}412.5 - (0.144965)(36)M_A - (0.235243)(36)M_B = 0$$

Solving,

$$M_A = 341.75 \text{ kip-ft}$$
$$M_B = 192.35 \text{ kip-ft}$$

Slope-deflection Equations. Since joints A, B, and C do not move, R for both members AB and BC is zero. $2EI_c/L$ is the same for AB and BC; so it is considered to be unity. θ_A is known to be zero.

$$M_{FAB} = +662.42 + 341.75 = +1{,}004.17 \text{ kip-ft}$$
$$M_{FBA} = -765.55 - 192.35 = -957.90 \text{ kip-ft}$$
$$M_{FBC} = 0$$
$$M_{FCB} = 0$$

Using the slope-deflection equations (89),

$$M_{AB} = +1{,}004.17 + 1(-2.82230\theta_A - 1.73920\theta_B)$$
$$= +1{,}004.17 - 1.73920\theta_B$$
$$M_{BA} = -957.90 + 1(-3.19722\theta_B - 1.73920\theta_A)$$
$$= -957.90 - 3.19722\theta_B$$
$$M_{BC} = 0 + 1(-3.19722\theta_B - 1.73920\theta_C)$$
$$= -3.19722\theta_B - 1.73920\theta_C$$
$$M_{CB} = 0 + 1(-2.82230\theta_C - 1.73920\theta_B)$$
$$= -2.82230\theta_C - 1.73920\theta_B$$

Joint Conditions

Joint B: $M_{BA} + M_{BC} = 0$ $\qquad -6.394444\theta_B - 1.73920\theta_C = +957.90$ (*a*)

Joint C: $M_{CB} = 0$ $\qquad -1.73920\theta_B - 2.82230\theta_C = 0$ (*b*)

Solving,

$$\theta_B = -179.996$$
$$\theta_C = +110.901$$

Computations for End Moments

$$M_{AB} = +1{,}004.17 - (1.73920)(-179.996) = +1{,}317.17 \text{ kip-ft}$$
$$M_{BA} = -957.90 - (3.19722)(-179.996) = -382.51 \text{ kip-ft}$$
$$M_{BC} = (-3.19722)(-179.996) - (1.73920)(+110.901)$$
$$= +382.51 \text{ kip-ft}$$
$$M_{CB} = (-2.82230)(+110.901) - (1.73920)(-179.996) = 0$$

The reactions, shear and moment diagrams, and the elastic curve are shown in Fig. 198.

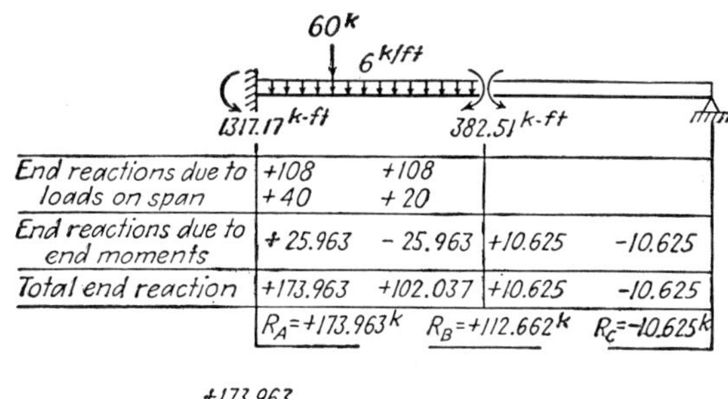

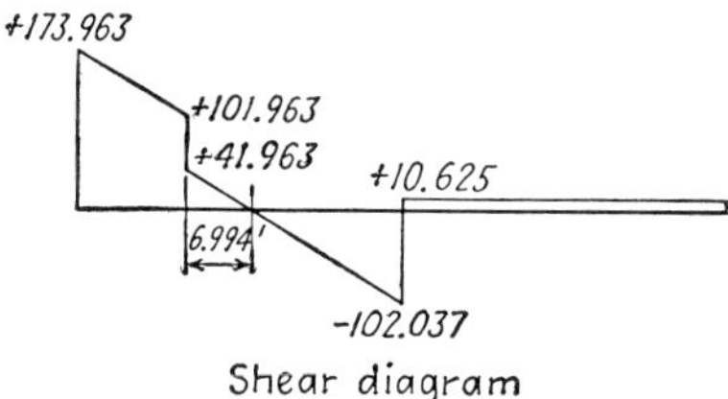

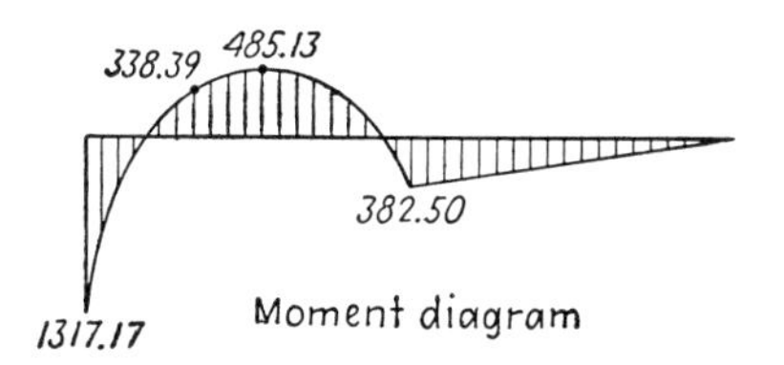

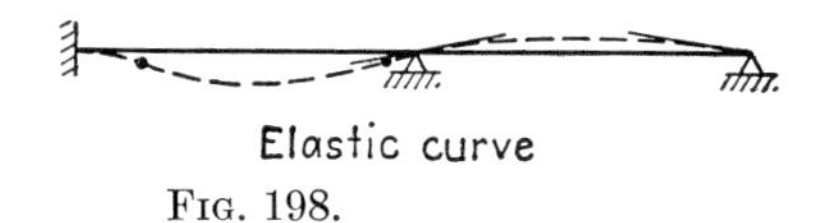

FIG. 198.

Check on Continuity. A check on continuity may be had by finding the rotations at the supports.

TABLE 68-1. VALUES OF $EI_c\theta$ (+ FOR CLOCKWISE, − FOR COUNTERCLOCKWISE)

	A	B	B	C
Due to uniform load...	+10,350.28	−9,940.22		
Due to concentrated load..............	+ 4,282.50	−3,412.50		
Due to end moments..	−12,636.60 − 1,996.22	+6,873.97 +3,239.38	−3,239.38	+1,966.22
	− 0.04 (*Check:* θ_A = 0)	−3,239.37 (*Check*)	−3,239.38	+1,966.22

For example, $C_1L(1{,}317.17) = 12{,}636.60$

Example 69. By the slope-deflection method analyze the continuous beam shown in Fig. 199 owing to a vertical settlement of 0.450 in. at support B.

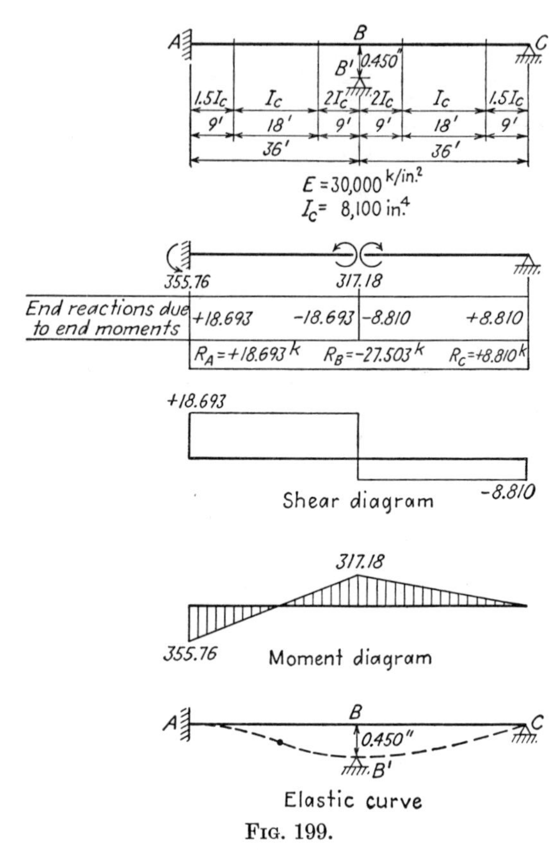

Fig. 199.

Solution. The slope-deflection equations

$$M_{AB} = M_{FAB} + \frac{2EI_c}{L}[-K_{AA}\theta_A - K_{AB}\theta_B + (K_{AA} + K_{AB})R]$$
$$M_{BA} = M_{FBA} + \frac{2EI_c}{L}[-K_{BB}\theta_B - K_{AB}\theta_A + (K_{BB} + K_{AB})R] \qquad (89)$$

will be used.

For span AB: $M_{FAB} = M_{FBA} = 0$

$$\frac{2EI_c}{L} = \frac{(2)(30{,}000)(8{,}100)}{(144)(36)} = 93{,}750 \text{ kip-ft}$$

From Example 68

$$K_{AA} = 2.82230 \qquad K_{AB} = 1.73920 \qquad K_{BB} = 3.19722$$

$$R = + \frac{0.450}{(36)(12)} = +0.00104167 \text{ rad}$$

For span BC: $M_{FBC} = M_{FCB} = 0$

$$\frac{2EI_c}{L} = \frac{(2)(30{,}000)(8{,}100)}{(144)(36)} = 93{,}750 \text{ kip-ft}$$

$$K_{BB} = 3.19722 \qquad K_{BC} = 1.73920 \qquad K_{CC} = 2.82230$$

$$R = - \frac{0.450}{(36)(12)} = -0.00104167 \text{ rad}$$

Slope-deflection Equations

$$\begin{aligned}
M_{AB} &= 0 + 93{,}750[-2.822230\theta_A - 1.739200\theta_B + 4.56150(+0.00104167)] \\
&= +445.46 - 163{,}050\theta_B \\
M_{BA} &= 0 + 93{,}750[-3.197220\theta_B - 1.739200\theta_A + 4.93642(+0.00104167)] \\
&= +482.07 - 299{,}739\theta_B \\
M_{BC} &= 0 + 93{,}750[-3.197220\theta_B - 1.739200\theta_C + 4.93642(-0.00104167)] \\
&= -482.07 - 299{,}739\theta_B - 163{,}050\theta_C \\
M_{CB} &= 0 + 93{,}750[-2.822300\theta_C - 1.739200\theta_B + 4.56150(-0.00104167)] \\
&= -445.46 - 264{,}591\theta_C - 163{,}050\theta_B
\end{aligned}$$

Joint Conditions

Joint B: $M_{BA} + M_{BC} = 0 \qquad -599{,}478\theta_B - 163{,}050\theta_C = 0 \qquad (a)$

Joint C: $M_{CB} = 0 \qquad -163{,}050\theta_B - 264{,}591\theta_C = +445.46 \qquad (b)$

Solving,

$$\theta_B = +0.00055011$$
$$\theta_C = -0.00202257$$

Computations for End Moments

$$\begin{aligned}
M_{AB} &= +445.46 - (163{,}050)(+0.00055011) = +355.76 \text{ kip-ft} \\
M_{BA} &= +482.07 - (299{,}739)(-0.00055011) = +317.18 \text{ kip-ft} \\
M_{BC} &= -482.07 - (299{,}739)(+0.00055011) - (163{,}050)(-0.00202257) \\
&= -317.18 \text{ kip-ft} \\
M_{CB} &= -445.46 - (264{,}591)(-0.00202257) - (163{,}050)(+0.00055011) \\
&= 0
\end{aligned}$$

The reactions, shear and moment diagrams, and the elastic curve are shown in Fig. 199.

Check on Continuity. A check on continuity may be had by finding the rotations at the supports.

TABLE 69-1. VALUES OF θ IN RADIANS

	A	*B*	*B*	*C*
Due to settlement......	+0.00104167	+0.00104167	−0.00104167	−0.00104167
Due to end moments...	−0.00202256 +0.00098091	+0.00110022 −0.00159177	+0.00159177	−0.00098091
	+0.00000002 (*Check* $\theta_A = 0$)	+0.00055012	+0.00055010 (*Check*)	−0.00202258

For example, $355.76 \dfrac{C_1 L}{EI_c} = \dfrac{(355.76)(36)(144)}{(30{,}000)(8{,}100)} (0.266493) = 0.00202256$

EXERCISES

103 and 104. Analyze the continuous beam shown. Draw shear and moment diagrams. Sketch the elastic curve.

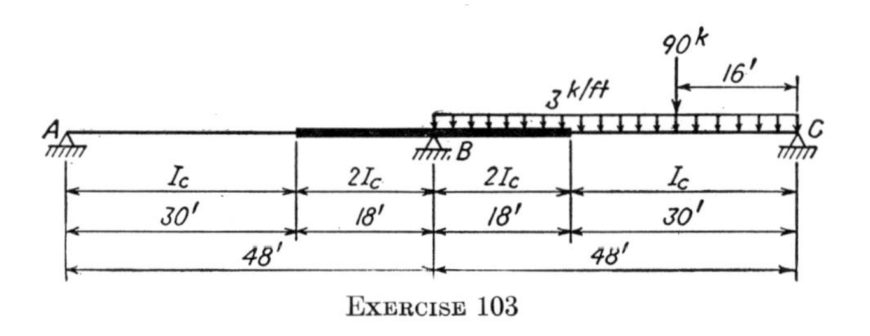

EXERCISE 103

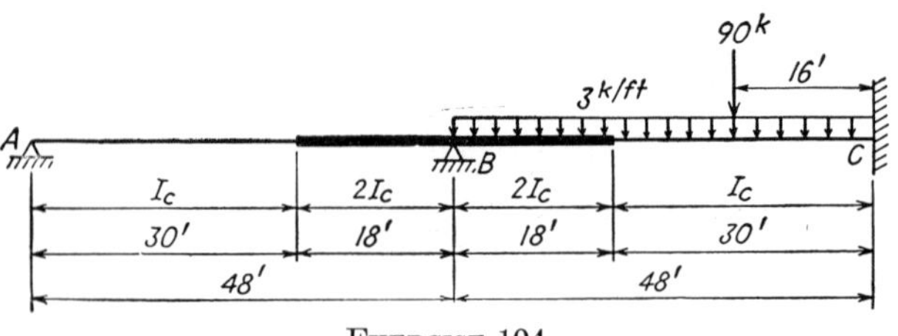

EXERCISE 104

105 and 106. Analyze the continuous beam shown owing to a settlement of $\frac{1}{2}$ in. at the support B. Draw shear and moment diagrams. Sketch the elastic curve.

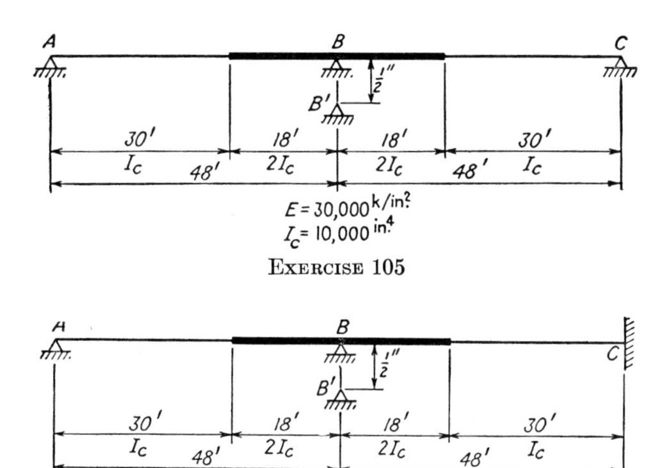

EXERCISE 105

$E = 30{,}000$ k/in.²
$I_c = 10{,}000$ in.⁴

EXERCISE 106

CHAPTER VIII

THE MOMENT-DISTRIBUTION METHOD

45. General Description of the Moment-distribution Method. The moment-distribution method can be used to analyze all types of statically indeterminate beams or rigid frames. Essentially it consists in solving the simultaneous equations in the slope-deflection method by successive approximations. In order to develop the method, it will be helpful to consider the following problem: If a clockwise moment of M_A kip-ft is applied at the simple support of a straight member of constant cross section simply supported at one end and fixed at the other, find the rotation θ_A at the simple support and the moment M_B at the fixed end (Fig. 200). The method of consistent deformation will be used. The condition of geometry required is, in this case,

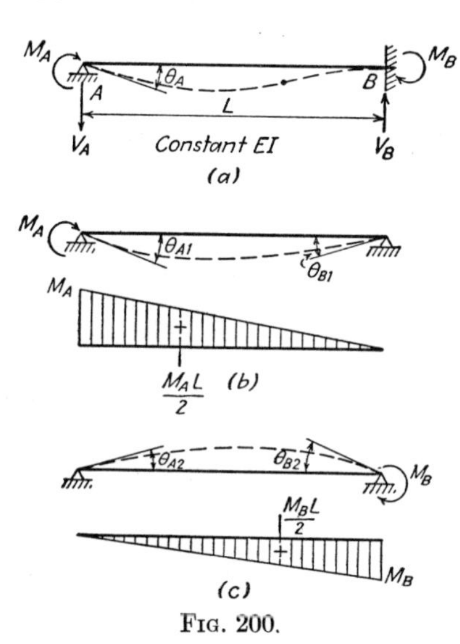

FIG. 200.

$$\theta_B = 0, \qquad \text{or} \qquad \theta_{B1} = \theta_{B2} \tag{90}$$

By the conjugate-beam method

$$\theta_{B1} = \frac{M_A L}{6EI} \qquad \theta_{B2} = \frac{M_B L}{3EI} \tag{91}$$

Substituting Eq. (91) into Eq. (90),

$$\frac{M_A L}{6EI} = \frac{M_B L}{3EI} \qquad M_B = \tfrac{1}{2} M_A \tag{92}$$

Also,

$$\theta_A = \theta_{A1} - \theta_{A2} = \frac{M_A L}{3EI} - \frac{M_B L}{6EI} = \frac{M_A L}{3EI} - \frac{(\frac{1}{2}M_A)L}{6EI} = \frac{L}{4EI} M_A \tag{93}$$

Solving for M_A in Eq. (93),

$$M_A = \frac{4EI}{L} \theta_A \tag{94}$$

Thus, for a span AB which is simply supported at A and fixed at B, a clockwise rotation of θ_A can be effected by applying a clockwise moment of $M_A = (4EI/L)\theta_A$ at A, which in turn induces a clockwise moment of $M_B = \frac{1}{2}M_A$ on the member at B. The expression $4EI/L$ is usually called the *stiffness factor*, which is defined as the moment required to be applied at A to cause a rotation of 1 rad at A of a span AB simply supported at A and fixed at B; the number $+\frac{1}{2}$ is the *carry-over factor*, which is the ratio of the moment induced at B to the moment applied at A. Note that the same sign convention is used in the moment-distribution method as in the slope-deflection method.

Now take the continuous beam ABC shown in Fig. 201*a*. If the joints A, B, and C are to be restrained against rotation, the moments as

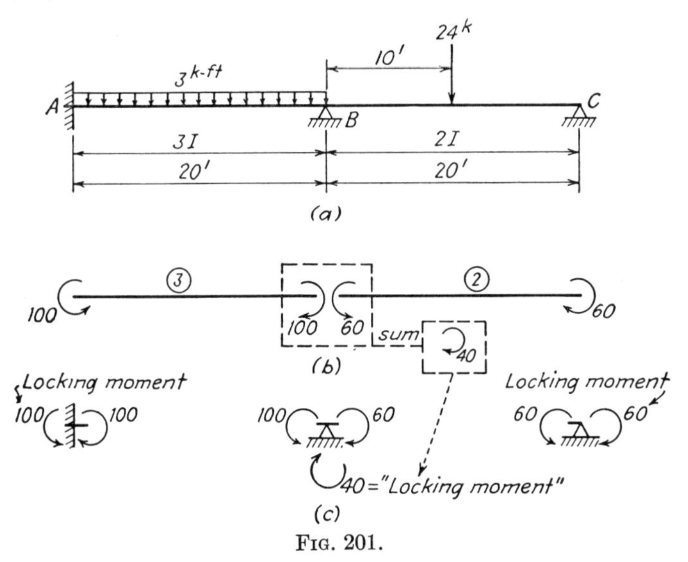

Fig. 201.

shown in Fig. 201*b* must be applied, which are in fact the fixed-end moments on spans AB and BC. The restraining moments required to hold the joints against rotation are (1) 100 kip-ft counterclockwise at A, (2) $100 - 60 = 40$ kip-ft clockwise at B, and (3) 60 kip-ft clockwise at C. These restraining moments are sometimes called the "locking" moments to "lock" the joints against rotation. Note that the joint B shown as a free body in Fig. 201*c* is in equilibrium under the action of the fixed-end moments, which are opposite in direction to those acting on the members, and the locking moment. The procedure may be described as follows: First lock all three joints. Then release joint B only. Joint B, now under the action of 40 kip-ft counterclockwise, will rotate a certain amount in the counterclockwise direction, which will in turn induce counterclockwise moments at B to act on BA and BC in amounts proportional to the *stiffness factors* of each, with a sum of 40 kip-ft. The relative stiffness factors of BA and BC are 3 and 2. Thus $3/(3 + 2) = 0.600$ times 40 kip-ft, or 24 kip-ft counterclockwise, will act on BA, and $2/(2 + 3) = 0.400$ times 40 kip-ft, or 16 kip-ft counterclockwise, will act on BC. The numbers 0.600 and 0.400 are usually called the *distribution factors.* Now lock joint B in its new position, and release joint C, which will rotate a certain amount in the counterclockwise direction. This rotation must be such as to induce a counterclockwise moment of 60 kip-ft to act on CB at C. Joint A is a fixed support; so it need not be released at all. Thus the first cycle of the moment distribution has been completed (see the adjoining moment-distribution table). To summarize, all joints are first locked by locking moments +100, −100, +60, −60 acting on all members, then joints B and C are released in succession, and the "balancing" moments are 0, +24, +16, and +60 (joint A is a fixed support).

MOMENT-DISTRIBUTION TABLE

Joint		*A*	*B*		*C*
Member		*AB*	*BA*	*BC*	*CB*
Distribution factors		——	0.600	0.400	1.000
1st cycle	FEM Balance	+100 ——	−100 + 24	+60 +16	−60 +60
2d cycle	Carry-over Balance	+ 12 ——	—— − 18	+30 −12	+ 8 − 8
3d cycle	Carry-over Balance	− 9 ——	—— + 2.4	− 4 + 1.6	− 6 + 6
4th cycle	Carry-over Balance	+ 1.2 ——	—— − 1.8	+ 3 − 1.2	+ 0.8 − 0.8
5th cycle	Carry-over Balance	− 0.9 ——	—— + 0.24	− 0.4 + 0.16	− 0.6 + 0.6
		and so on, to any desired degree of accuracy			
Total end moments (5 cycles)		+103.3	−93.16	+93.16	0

When a balancing moment of +24 kip-ft is placed at B of span AB, one-half of this amount, or +12 kip-ft, is induced at A. In the same manner, one-half of the +16 kip-ft, or +8 kip-ft, acts at C, and one-half of the +60 kip-ft, or +30 kip-ft, acts at B. These moments +12, 0, +30, +8 are called the "carry-over" moments, which are *kept out* during the first balancing and now are considered as new locking moments to lock the joints in positions after the first rotations. Then the joints B and C are released for the second time, and the second rotations at B and C induce balancing moments as shown in the second cycle and carry-over moments as shown in the first line of the third cycle. The same process can be repeated for as many cycles as desired to bring the balancing or carry-over moments to very small magnitudes. Thus any degree of accuracy can be obtained, and the work required decreases as the required accuracy decreases. The final, or total, end moments are obtained by adding all numbers in the respective columns.

Thus the moment-distribution method consists in successively locking and releasing the joints; the first locking moments are the fixed-end moments due to the applied loading, after the first balancing; the successive locking moments are the carry-over moments which are induced to act at the other ends of the respective spans by the balancing moments.

The reader is advised to read this article *again* and *again* while working through the rest of this chapter.

46. Application of the Moment-distribution Method to the Analysis of Statically Indeterminate Beams. The application of the moment-distribution method as described in general in the preceding article to the analysis of statically indeterminate beams will be illustrated by the following examples.

Example 70. Analyze the continuous beam shown in Fig. 202*a* by the moment-distribution method. Draw shear and moment diagrams. Sketch the elastic curve.

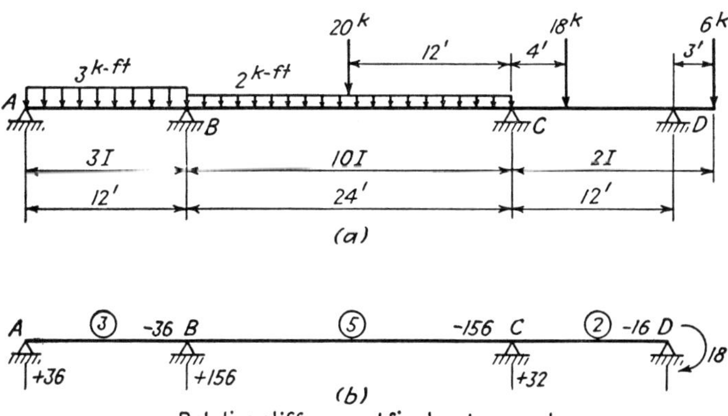

Fig. 202.

Solution. *Relative Stiffness and Fixed-end Moments* (Fig. 202*b*)

TABLE 70-1. RELATIVE STIFFNESS

AB............	$\left(\frac{3I}{12} = \frac{I}{4}\right) \times 12$	3
BC............	$\left(\frac{10I}{24} = \frac{5I}{12}\right) \times 12$	5
CD............	$\left(\frac{2I}{12} = \frac{I}{6}\right) \times 12$	2

$$M_{FAB} = +\frac{(3)(12)^2}{12} = +36 \text{ kip-ft}$$
$$M_{FBA} = -36 \text{ kip-ft}$$
$$M_{FBC} = +\frac{(2)(24)^2}{12} + \frac{(20)(24)}{8} = +156 \text{ kip-ft}$$
$$M_{FCB} = -156 \text{ kip-ft}$$
$$M_{FCD} = +\frac{(18)(4)(8)^2}{(12)^2} = +32 \text{ kip-ft}$$
$$M_{FDC} = -\frac{(18)(8)(4)^2}{(12)^2} = -16 \text{ kip-ft}$$

In the moment-distribution table (see Table 70-2) the distribution factors are first computed. The DF (distribution factor) at joint A (or D) is 1.000 on member AB (or DC) because there is only one member entering the joint. The DF's at joint B are $3/(3 + 5) = 0.375$ on member BA and $5/(3 + 5) = 0.625$ on member BC. The DF's at joint C are $5/(5 + 2) = 0.714$ on member CB and $2/(5 + 2) = 0.286$ on member CD. In cycle 1, the locking moments or the unbalanced moments are the FEM (fixed-end moments), which are then entered in the table. However, note that at joint D, there is a clockwise moment of 18 kip-ft acting, *clockwise* moment acting on the joint being taken as *positive*. Joints A, B, C, and D are then released and balanced. The "unbalance" at A is +36.00; so the balancing moment is −36.00. The unbalance at B is $-36.00 + 156.00 = +120.00$; so the balancing moments are $-(0.375)(+120.00) = -45.00$ and $-(0.625)(+120.00) = -75.00$ on BA and BC, respectively. The same explanation applies to joints C and D. Note that all moments are kept to two decimal places in this solution. The carry-overs as shown in the first line of the second cycle are $+\frac{1}{2}$ times the balancing moments placed at the far ends of the respective members in the preceding line. These carry-overs are the new unbalances, which are balanced in the same manner as before. The process is repeated, keeping all figures to two decimal places. It is observed that in this example the total end moments at all places are not

TABLE 70-2. MOMENT DISTRIBUTION

Joint		*A*	*B*		*C*		*D*	
Member		*AB*	*BA*	*BC*	*CB*	*CD*	*DC*	
K		3	3	5	5	2	2	
Cycle	DF	1.000	0.375	0.625	0.714	0.286	1.000	
1	FEM	+36.00	−36.00	+156.00	−156.00	+32.00	−16.00	+18.00
	Bal.	−36.00	−45.00	−75.00	+88.54	+35.46	−2.00	
2	CO	−22.50	−18.00	+44.27	−37.50	−1.00	+17.73	
	Bal.	+22.50	−9.85	−16.42	+27.49	+11.01	−17.73	
3	CO	−4.92	+11.25	+13.74	−8.21	−8.86	+5.50	
	Bal.	+4.92	−9.37	−15.62	+12.19	+4.88	−5.50	
4	CO	−4.68	+2.46	+6.10	−7.81	−2.75	+2.44	
	Bal.	+4.68	−3.21	−5.35	+7.54	+3.02	−2.44	
5	CO	−1.60	+2.34	+3.77	−2.60	−1.22	+1.51	
	Bal.	+1.60	−2.29	−3.82	+2.78	+1.12	−1.51	
6	CO	−1.14	+0.80	+1.39	−1.91	−0.76	+0.56	
	Bal.	+1.14	−0.82	−1.37	+1.91	+0.76	−0.56	
7	CO	−0.41	+0.57	+0.96	−0.68	−0.28	+0.38	
	Bal.	+0.41	−0.57	−0.96	+0.68	+0.28	−0.38	
Total		0	−107.69	+107.69	−73.66	+73.66	−18.00	+18.00
Check:								
Change		−36.00	−71.69	−48.31	+82.34	+41.66	−2.00	
$-\frac{1}{2}$ (change)		+35.84	+18.00	−41.17	+24.15	+1.00	−20.83	
Sum		−0.16	−53.69	−89.48	+106.49	+42.66	+22.83	
$\theta_{rel} = \text{sum}/(-K)$		+0.053	+17.90	+17.90	−21.30	−21.33	+11.42	
			(*Check*)		(*Check*)			

For explanation of check, see Art. 47.

materially affected by the seventh cycle of moment distribution. This suggests that no further moment distribution is needed. By adding the moments in the respective columns the total end moments are obtained. These total end moments check fairly well with those obtained by the slope-deflection method in Example 54. The reactions, shear and moment diagrams, and the elastic curve have been worked out previously in Example 54; so they will not be repeated here.

In carrying out the work in the moment-distribution table, it is advisable to first put down the *signs* in any one line all at the same time and then the numerical values. This helps the computer to concentrate on one operation at a time. Note also that the sum of the balancing moments placed at the ends of members meeting in one joint at one time must be equal to the total unbalance at that joint. This will ensure that the sum of the total end moments acting on all members meeting at any one joint is zero.

Example 71. Analyze the continuous beam shown in Fig. 203*a* by the moment-distribution method. Draw shear and moment diagrams. Sketch the elastic curve.

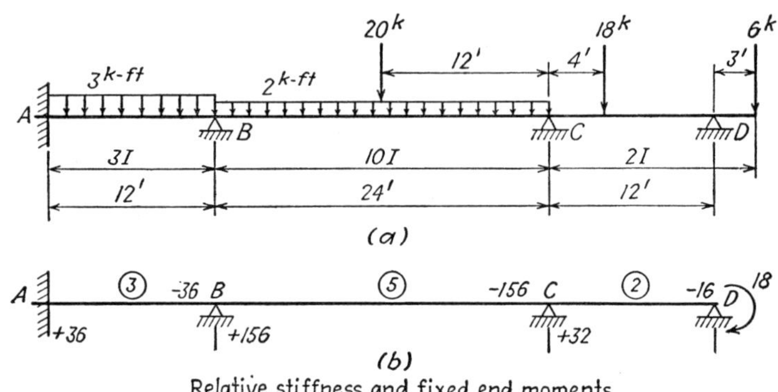

FIG. 203.

Solution. *Relative Stiffness and Fixed-end Moments* (Fig. 203*b*)

TABLE 71-1. RELATIVE STIFFNESS

AB	$\left(\frac{3I}{12} = \frac{I}{4}\right) \times 12$	3
BC	$\left(\frac{10I}{24} = \frac{5I}{12}\right) \times 12$	5
CD	$\left(\frac{2I}{12} = \frac{I}{6}\right) \times 12$	2

$$M_{FAB} = +\frac{(3)(12)^2}{12} = +36 \text{ kip-ft}$$

$$M_{FBA} = -36 \text{ kip-ft}$$

$$M_{FBC} = +\frac{(2)(24)^2}{12} + \frac{(20)(24)}{8} = +156 \text{ kip-ft}$$

$$M_{FCB} = -156 \text{ kip-ft}$$

$$M_{FCD} = +\frac{(18)(4)(8)^2}{(12)^2} = +32 \text{ kip-ft}$$

$$M_{FDC} = -\frac{(18)(8)(4)^2}{(12)^2} = -16 \text{ kip-ft}$$

TABLE 71-2. MOMENT DISTRIBUTION

Joint		*A*	*B*		*C*		*D*	
Member		*AB*	*BA*	*BC*	*CB*	*CD*	*DC*	
K		3	3	5	5	2	2	
Cycle	DF	——	0.375	0.625	0.714	0.286	1.000	
1	FEM	+36.00	− 36.00	+156.00	−156.00	+32.00	−16.00	+18.00
	Bal.	——	− 45.00	− 75.00	+ 88.54	+35.46	− 2.00	
2	CO	−22.50	——	+ 44.27	− 37.50	− 1.00	+17.73	
	Bal.	——	− 16.60	− 27.67	+ 27.49	+11.01	−17.73	
3	CO	− 8.30	——	+ 13.74	− 13.84	− 8.86	+ 5.50	
	Bal.	——	− 5.15	− 8.59	+ 16.21	+ 6.49	− 5.50	
4	CO	− 2.58	——	+ 8.10	− 4.30	− 2.75	+ 3.24	
	Bal.	——	− 3.04	− 5.06	+ 5.04	+ 2.01	− 3.24	
5	CO	− 1.52	——	+ 2.52	− 2.53	− 1.62	+ 1.00	
	Bal.	——	− 0.94	− 1.58	+ 2.96	+ 1.19	− 1.00	
6	CO	− 0.47	——	+ 1.48	− 0.79	− 0.50	+ 0.60	
	Bal.	——	− 0.56	− 0.92	+ 0.92	+ 0.37	− 0.60	
7	CO	− 0.28	——	+ 0.46	− 0.46	− 0.30	+ 0.18	
	Bal.	——	− 0.17	− 0.29	+ 0.54	+ 0.22	− 0.18	
8	CO	− 0.08	——	+ 0.27	− 0.14	− 0.09	+ 0.11	
	Bal.	——	− 0.10	− 0.17	+ 0.16	+ 0.07	− 0.11	
9	CO	− 0.05	——	+ 0.08	− 0.08	− 0.06	+ 0.04	
	Bal.	——	− 0.03	− 0.05	+ 0.10	+ 0.04	− 0.04	
10	CO	− 0.02	——	+ 0.05	− 0.02	− 0.02	+ 0.02	
	Bal.	——	− 0.02	− 0.03	+ 0.03	+ 0.01	− 0.02	
11	CO	− 0.01	——	+ 0.02	− 0.02	− 0.01	——	
	Bal.	——	− 0.01	− 0.01	+ 0.02	+ 0.01	——	
12	CO	——	——	+ 0.01	——	——	——	
	Bal.	——	——	− 0.01	——	——	——	
Total		+ 0.19	−107.62	+107.62	− 73.67	+73.67	−18.00	+18.00
Check:								
Change		−35.81	− 71.62	− 48.38	+ 82.33	+41.67	− 2.00	
$-\frac{1}{2}$ (change)		+35.81	+ 17.90	− 41.16	+ 24.19	+ 1.00	−20.84	
Sum		0	− 53.72	− 89.54	+106.52	+42.67	−22.84	
θ_{rel} = sum/(−*K*)		0	+ 17.91	+ 17.91	− 21.30	−21.33	+11.42	
		(*Check*)	(*Check*)		(*Check*)			

For explanation of check, see Art. 47.

In the moment-distribution table (see Table 71-2) the distribution factors are first computed. Since joint A is a fixed support, it can resist any moment assigned to it and therefore need not be released; in such a case there should be no distribution factor applicable to joint A. The distribution factors at joints B, C, and D are determined as previously discussed. The fixed-end moments and the moment of +18 kip-ft at the outside of joint D are placed at the beginning of the first cycle. Joint A needs no balancing; so the balancing moments in all cycles are zero, for which a *dash* is used in the table. The moment distribution is carried out as usual. By keeping all figures to two decimal places, the table comes to an "automatic" stop at the end of the twelfth cycle. The total end moments as thus obtained by adding the moments in the respective columns check fairly well with those in the slope-deflection solution. The reactions, shear and moment diagrams, and the elastic curve have been worked out before in Example 55; so they will not be repeated here.

47. Check on Moment Distribution. The moment-distribution table begins with relative stiffnesses and fixed-end moments and ends with final end moments. A check can be made to ensure that the correct end moments have been obtained on the basis of the relative stiffnesses and fixed-end moments at the head of the table. Of course, the first obvious check is to see whether or not the moments are balanced at each interior joint and the moment is zero at each exterior simple support; this check is on the conditions of *statics*. There remains the check on the conditions of *geometry*. This can be made by finding the relative values of the rotation at each joint. In the slope-deflection equations

$$\begin{aligned} M_{AB} &= M_{FAB} + K_{\text{rel}}(-2\theta_A - \theta_B) \\ M_{BA} &= M_{FBA} + K_{\text{rel}}(-2\theta_B - \theta_A) \end{aligned} \qquad (62)$$

the end moments are expressed in terms of the fixed-end moments and the end rotations. Conversely, the end rotations can be expressed in terms of the fixed-end moments and the final end moments. Solving Eqs. (62) for θ_A and θ_B,

$$\begin{aligned} (\theta_A)_{\text{rel}} &= \frac{(M_{AB} - M_{FAB}) - \frac{1}{2}(M_{BA} - M_{FBA})}{-1.5K_{\text{rel}}} \\ (\theta_B)_{\text{rel}} &= \frac{(M_{BA} - M_{FBA}) - \frac{1}{2}(M_{AB} - M_{FAB})}{-1.5K_{\text{rel}}} \end{aligned} \qquad (95)$$

Since only the relative values of θ_A and θ_B are expressed in Eqs. (95), the pure number 1.5 in the denominator can be dropped off. Thus

$$\begin{aligned} (\theta_A)_{\text{rel}} &= \frac{(M_{AB} - M_{FAB}) - \frac{1}{2}(M_{BA} - M_{FBA})}{(-K_{\text{rel}})} \\ (\theta_B)_{\text{rel}} &= \frac{(M_{BA} - M_{FBA}) - \frac{1}{2}(M_{AB} - M_{FAB})}{(-K_{\text{rel}})} \end{aligned} \qquad (96)$$

In Eqs. (96), $(M_{AB} - M_{FAB})$ and $(M_{BA} - M_{FBA})$ will be called the change in the moment from the fixed-end moment to the final end moment. Thus the relative value of the rotation θ at any one end of a member is equal to the change in the moment at the near end minus one-half of the change in the moment at the far end and then divided by $(-K_{rel})$. Or

$$\theta_{\text{near end}} = \frac{(\text{change})_{\text{near end}} + (-\frac{1}{2})(\text{change})_{\text{far end}}}{(-K_{rel})} \tag{97}$$

By the use of Eq. (97) the relative values of the rotation at the ends of each member can be computed; the check on the conditions of geometry will be that the rotations at the ends of all members meeting in one joint should be equal and that the rotation at the fixed support should be zero.

At the end of Table 70-2 for moment distribution the check made by applying Eq. (97) is shown. The relative values of the rotations at joints A, B, C, and D are $+0.053$, $+17.90$, -21.30, and $+11.42$, respectively. The same is done for the moment distribution in Table 71-2, wherein the relative values of the rotations at joints A, B, C, and D are 0, $+17.91$, -21.30 and $+11.42$, respectively.

It is to be noted again that the above check has nothing to do with the correctness of the values of the relative stiffness and the fixed-end moments used at the beginning of the moment-distribution table; *i.e.*, if the latter are wrong, the answers would be correspondingly in error even though they met the test of the check.

EXERCISES

107 to 110. Solve Exercises 72 to 75 by the moment-distribution method.

48. Stiffness Factor at the Near End of a Member When the Far End Is Hinged. The stiffness factor has been defined as the moment required to rotate the tangent to the elastic curve at the near end of a member through one radian when the far end is *fixed;* for a member with constant cross section, this stiffness factor is $4EI/L$ (Fig. 204*a*). Now, if the far end is hinged instead of being fixed, the moment required to rotate the tangent at the near end through 1 rad will be $3EI/L$ instead of $4EI/L$ (Fig. 204*b*). This can easily be derived by the conjugate-beam method. From Fig. 204*b*

$$EI\theta_A = R'_A = \frac{M_A L}{3} \qquad \text{or} \qquad M_A = \frac{3EI}{L}\,\theta_A = \frac{3}{4}\left(\frac{4EI}{L}\right)\theta_A \tag{98}$$

Thus the stiffness factor at the near end when the far end is hinged is $3EI/L$, or three-fourths of that when the far end is fixed.

In Fig. 205 are shown four members AE, BE, CE, and DE, meeting at a rigid joint E. If A, B, and D are fixed and C is hinged, any unbalanced moment at joint E will cause a certain amount of rotation at joint E, or the unbalance will distribute itself into four parts to act on the ends of EA, EB, EC, and ED in the ratio of K_{AE}, K_{BE}, $\frac{3}{4}(K_{CE})$, and K_{DE}. One-half of the balancing moments placed at E on members EA,

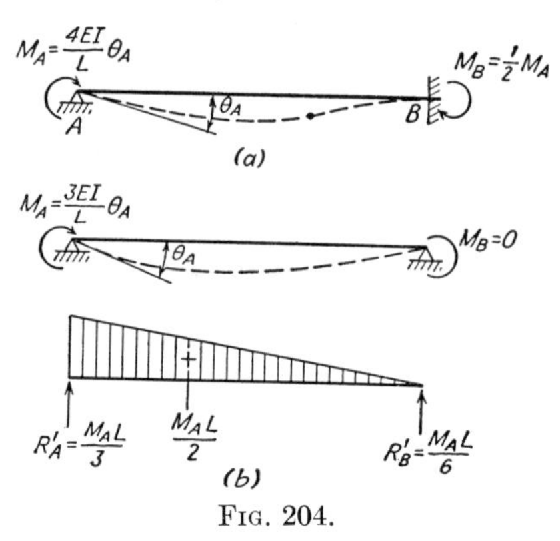

Fig. 204.

EB, and ED will then be carried to A, B, and D; but no carry-over to the hinge is necessary, because by modifying the stiffness factor of EC to three-fourths of its usual value, the fact that the moment at C should always be zero has already been taken care of. If, however, the "unbalance" at E is distributed in the ratio of K_{AE}, K_{BE}, K_{CE}, and K_{DE}, one-half of the balancing moment placed at E on EC must be carried over to C

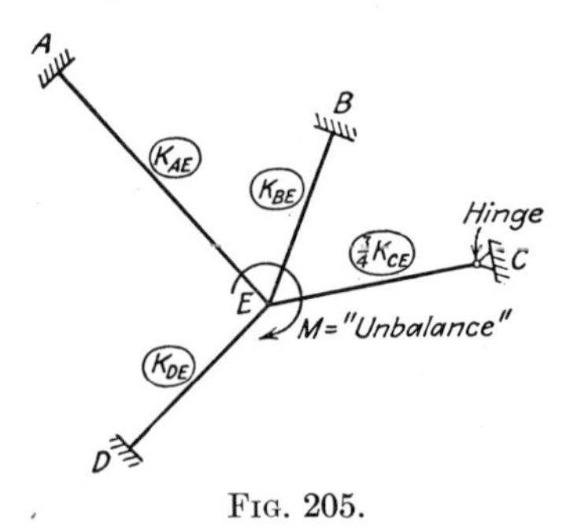

Fig. 205.

and joint C must be balanced in every cycle of moment distribution. This latter procedure has been followed in dealing with the exterior supports A and D in Example 70 and the exterior support D in Example 71. The alternate work on moment distribution by modifying the stiffness of the members with exterior simple or hinged supports is shown in the accompanying tables for Examples 70 and 71.

MOMENT DISTRIBUTION BY MODIFIED STIFFNESS METHOD, FOR EXAMPLE 70

Joint		*A*	*B*		*C*		*D*	
Member		*AB*	*BA*	*BC*	*CB*	*CD*	*DC*	
K		3	3	5	5	2	2	
Modified *K*		2.25	2.25	5	5	1.5	1.5	
Cycle	DF	——	0.310	0.690	0.769	0.231	——	
1	FEM	+36.00	− 36.00	+156.00	−156.00	+32.00	−16.00	+18.00
	Bal.	−36.00	− 37.20	− 82.80	+ 95.36	+28.64	− 2.00	
2	CO	↑	− 18.00	+ 47.68	− 41.40	− 1.00	↑	
	Bal.		− 9.20	− 20.48	+ 32.61	+ 9.79		
3	CO		——	+ 16.30	− 10.24	——		
	Bal.		− 5.05	− 11.25	+ 7.87	+ 2.37		
4	CO		——	+ 3.94	− 5.62	——		
	Bal.		− 1.22	− 2.72	+ 4.32	+ 1.30		
5	CO		——	+ 2.16	− 1.36	——		
	Bal.		− 0.67	− 1.49	+ 1.05	+ 0.31		
6	CO		——	+ 0.52	− 0.74	——		
	Bal.		− 0.16	− 0.36	+ 0.57	+ 0.17		
7	CO		——	+ 0.28	− 0.18	——		
	Bal.		− 0.09	− 0.19	+ 0.14	+ 0.04		
8	CO		——	+ 0.07	− 0.10	——		
	Bal.		− 0.02	− 0.05	+ 0.08	+ 0.02		
9	CO		——	+ 0.04	− 0.02	——		
	Bal.		− 0.01	− 0.03	+ 0.02	——		
10	CO		——	+ 0.01	− 0.02	——		
	Bal.	↓	——	− 0.01	+ 0.02	——	↓	
Total		0	−107.62	+107.62	− 73.64	+73.64	−18.00	+18.00
Check:								
Change		−36.00	− 71.62	− 48.38	+ 82.36	+41.64	− 2.00	
$-\frac{1}{2}$ (change)		+35.81	+ 18.00	− 41.18	+ 24.19	+ 1.00	−20.82	
Sum		− 0.19	− 53.62	− 89.56	+106.55	+42.64	−22.82	
θ_{rel} = sum/($-K$)		+ 0.063	+ 17.87	+ 17.91	− 21.31	−21.32	+11.41	
			(*Check*)		(*Check*)			

MOMENT DISTRIBUTION BY MODIFIED STIFFNESS METHOD, FOR EXAMPLE 71

Joint		A	B		C		D	
Member		*AB*	*BA*	*BC*	*CB*	*CD*	*DC*	
K		3	3	5	5	2	2	
Modified K		3	3	5	5	1.5	1.5	
Cycle	DF	———	0.375	0.625	0.769	0.231	———	
1	FEM	+36.00	− 36.00	+156.00	−156.00	+32.00	−16.00	+18.00
	Bal.	———	− 45.00	− 75.00	+ 95.36	+28.64	− 2.00	
2	CO	−22.50	———	+ 47.68	− 37.50	− 1.00		
	Bal.	———	− 17.88	− 29.80	+ 29.61	+ 8.89		
3	CO	− 8.94	———	+ 14.80	− 14.90	———		
	Bal.	———	− 5.55	− 9.25	+ 11.46	+ 3.44		
4	CO	− 2.78	———	+ 5.73	− 4.62	———		
	Bal.	———	− 2.15	− 3.58	+ 3.55	+ 1.07		
5	CO	− 1.08	———	+ 1.78	− 1.79	———		
	Bal.	———	− 0.67	− 1.11	+ 1.38	+ 0.41		
6	CO	− 0.34	———	+ 0.69	− 0.56	———		
	Bal.	———	− 0.26	− 0.43	+ 0.43	+ 0.13		
7	CO	− 0.13	———	+ 0.22	− 0.22	———		
	Bal.	———	− 0.08	− 0.14	+ 0.17	+ 0.05		
8	CO	− 0.04	———	+ 0.08	− 0.07	———		
	Bal.	———	− 0.03	− 0.05	+ 0.05	+ 0.02		
9	CO	− 0.02	———	+ 0.02	− 0.02	———		
	Bal.	———	− 0.01	− 0.01	+ 0.02	———		
10	CO	———	———	+ 0.01	———	———		
	Bal.	———	———	− 0.01	———	———		
Total		+ 0.17	−107.63	+107.63	− 73.65	+73.65	−18.00	+18.00
Check:								
Change		−35.83	− 71.63	− 48.37	+ 82.35	+41.65	− 2.00	
$-\frac{1}{2}$ (change)		+35.82	+ 17.92	− 41.18	+ 24.18	+ 1.00	−20.82	
Sum		0	− 53.71	− 89.55	+106.53	+42.65	−22.82	
θ_{rel} = sum/($-K$)		0	+ 17.90	+ 17.91	− 21.31	−21.32	+11.41	
		(*Check*)	(*Check*)		(*Check*)			

In the table for Example 70 the modified values of the relative stiffnesses are $3 \times \frac{3}{4} = 2.25$ and $2 \times \frac{3}{4} = 1.5$, respectively, for spans *AB* and *CD*. The simple supports *A* and *D* need to be released only once in the first cycle; thus no distribution factor is shown. Note the vertical arrow drawn under joints *A* and *D* after the first cycle. By drawing such lines the computer is reminded not to carry any more moment to the joints *A* and *D*. The distribution factors at joint *B* are $2.25/(2.25 + 5) = 0.310$ and $5/(2.25 + 5) = 0.690$, while those at joint *C* are $5/(5 + 1.5) = 0.769$ and $1.5/(5 + 1.5) = 0.231$. The FEM's are as before. After the first cycle no further work need be done for joints *A* and *D*. In the check, $\theta_{rel} = \text{sum}/(-K)$, the unmodified values of the relative stiffness should be used. The discrepancy of θ_{rel} values at joint *B*, for instance, is due to the fact that there are only three significant figures in the distribution factors.

In the table for Example 71 the relative stiffness of span *CD* only is modified to become $2 \times \frac{3}{4} = 1.5$. Note, then, that joint *D* is balanced only once in the first cycle, because for the span *CD* there are no carry-overs to joint *D* owing to the balancing moments at joint *C*. In general, some amount of arithmetical work can be saved in moment distribution by modifying the relative stiffness of members with one exterior simple or hinged support; thereby the simple or hinged supports need to be balanced only *once*, and no carry-overs will ever be brought back to them. This procedure is recommended.

EXERCISES

111. Work the moment distribution in Exercise 107 by modifying the stiffness of spans *AB* and *CD*.

112. Work the moment distribution in Exercise 108 by modifying the stiffness of span *AB*.

49. Analysis of Statically Indeterminate Beams Due to Yielding of Supports. The moment-distribution method can be used to analyze statically indeterminate beams due to the yielding of supports. The physical concept involved is that the joints are first locked against rota-

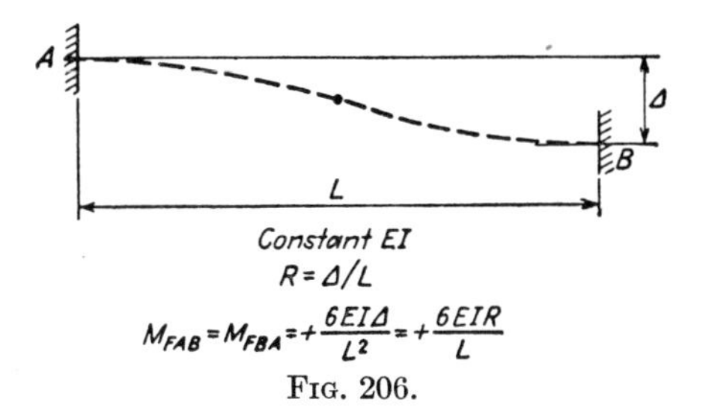

Fig. 206.

tion and then displaced to conform with the amount of yielding; the locking moments acting on the ends of each member will be the fixed-end moments as derived in Eq. (65), which are repeated here and shown in Fig. 206. Then the joints are released or balanced; the carry-overs become the next unbalances, the joints are again balanced, and so on. In other words, the fixed-end moments due to the movement of one end relative to the other in a direction perpendicular to the original direction of the member are treated in exactly the same manner as those due to the applied loadings.

Example 72. By the moment-distribution method analyze the continuous beam shown in Fig. 207 due to a vertical settlement of $\frac{1}{2}$ in. at support B.

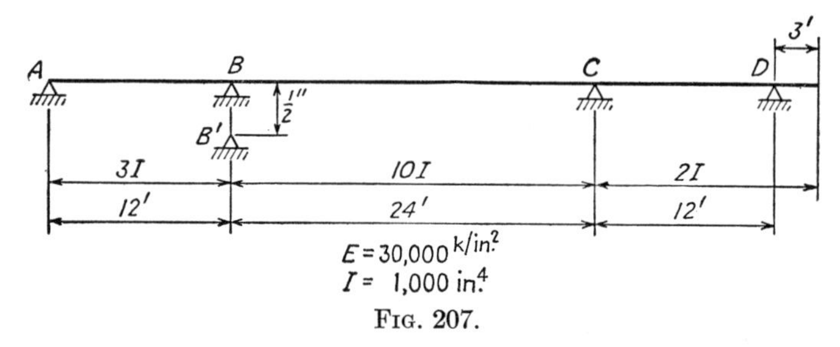

FIG. 207.

Solution. *Relative Stiffness and Fixed-end Moments*

TABLE 72-1. RELATIVE STIFFNESS

AB	$\left(\frac{3I}{12} = \frac{I}{4}\right) \times 12$	3
BC	$\left(\frac{10I}{24} = \frac{5I}{12}\right) \times 12$	5
CD	$\left(\frac{2I}{12} = \frac{I}{6}\right) \times 12$	2

$$M_{FAB} = M_{FBA} = +\frac{6EIR}{L}$$

$$= +\frac{(6)(30{,}000)(3{,}000)}{(144)(12)}\left(\frac{0.500}{144}\right) \text{ kip-ft}$$

$$= 1{,}085.07 \text{ kip-ft}$$

$$M_{FBC} = M_{FCB} = +\frac{6EIR}{L}$$

$$= +\frac{(6)(30{,}000)(10{,}000)}{(24)(144)}\left(\frac{-0.500}{288}\right)$$

$$= -904.22 \text{ kip-ft}$$

$$M_{FCD} = M_{FDC} = 0$$

Moment Distribution (see Table 72-2)

TABLE 72-2. MOMENT DISTRIBUTION

Joint		A	B		C		D
Member		AB	BA	BC	CB	CD	DC
K		3	3	5	5	2	2
Cycle	DF	1.000	0.375	0.625	0.714	0.286	1.000
1	FEM	+1,085.1	+1,085.1	−904.2	−904.2	———	———
	Bal.	−1,085.1	− 67.8	−113.1	+645.6	+258.6	———
2	CO	− 33.9	− 542.6	+322.8	− 56.6	———	+129.3
	Bal.	+ 33.9	+ 82.4	+137.4	+ 40.4	+ 16.2	−129.3
3	CO	+ 41.2	+ 17.0	+ 20.2	+ 68.7	− 64.6	+ 8.1
	Bal.	− 41.2	− 14.0	− 23.2	− 2.9	− 1.2	− 8.1
4	CO	− 7.0	− 20.6	− 1.4	− 11.6	− 4.0	− 0.6
	Bal.	+ 7.0	+ 8.2	+ 13.8	+ 11.1	+ 4.5	+ 0.6
5	CO	+ 4.1	+ 3.5	+ 5.6	+ 6.9	+ 0.3	+ 2.2
	Bal.	− 4.1	− 3.4	− 5.7	− 5.1	− 2.1	− 2.2
6	CO	− 1.7	− 2.0	− 2.6	− 2.8	− 1.1	− 1.0
	Bal.	+ 1.7	+ 1.7	+ 2.9	+ 2.8	+ 1.1	+ 1.0
7	CO	+ 0.8	+ 0.8	+ 1.4	+ 1.4	+ 0.5	+ 0.5
	Bal.	− 0.8	− 0.8	− 1.4	− 1.4	− 0.5	− 0.5
Total		0	+ 547.5	−547.5	−207.7	+207.7	0
Check:							
Change		−1,085.1	− 537.6	+356.7	+696.5	+207.7	0
$-\frac{1}{2}$ (change)		+ 268.8	+ 542.6	−348.2	−178.4	0	−103.8
Sum		− 816.3	+ 5.0	+ 8.5	+518.1	+207.7	−103.8
θ_{rel} = sum/(−K)		+ 272.1	− 1.7	− 1.7	−103.6	−103.8	+ 51.9
			(*Check*)		(*Check*)		

Moment Distribution, Alternate Solution (see Table 72-3)

TABLE 72-3. MOMENT DISTRIBUTION, ALTERNATE SOLUTION

Joint		A	B		C		D
Member		AB	BA	BC	CB	CD	DC
K		3	3	5	5	2	2
Modified K		2.25	2.25	5	5	1.5	1.5
Cycle	DF	——	0.310	0.690	0.769	0.231	——
1	FEM	+1,085.1	+1,085.1	−904.2	−904.2	——	——
	Bal.	−1,085.1	− 56.1	−124.8	+695.3	+208.9	——
2	CO	↑	− 542.6	+347.6	− 62.4	——	↑
	Bal.		+ 60.5	+134.5	+ 48.0	+ 14.4	
3	CO		——	+ 24.0	+ 67.2	——	
	Bal.		− 7.4	− 16.6	− 51.7	− 15.5	
4	CO		——	− 25.8	− 8.3	——	
	Bal.		+ 8.0	+ 17.8	+ 6.4	+ 1.9	
5	CO		——	+ 3.2	+ 8.9	——	
	Bal.		− 1.0	− 2.2	− 6.8	− 2.1	
6	CO		——	− 3.4	− 1.1	——	
	Bal.		+ 1.1	+ 2.3	+ 0.8	+ 0.3	
7	CO		——	+ 0.4	+ 1.2	——	
	Bal.		− 0.1	− 0.3	− 0.9	− 0.3	
8	CO		——	− 0.4	− 0.2	——	
	Bal.		+ 0.1	+ 0.3	+ 0.2	——	
9	CO		——	+ 0.1	+ 0.2	——	
	Bal.	↓	——	− 0.1	− 0.2	——	↓
Total		0	+ 547.6	−547.6	−207.6	+207.6	0
Check:							
Change		−1,085.1	− 537.5	+356.6	+696.6	+207.6	0
$-\frac{1}{2}$ (change)		+ 268.8	+ 542.6	−348.3	−178.3	0	−103.8
Sum		− 816.3	+ 5.1	+ 8.3	+518.3	+207.6	−103.8
θ_{rel} = sum/($-K$)		+ 272.1	− 1.7	− 1.7	−103.7	−103.8	+ 51.9
			(*Check*)		(*Check*)		

For reactions, shear and moment diagrams, and the elastic curve, see Example 56.

Example 73. By the moment-distribution method analyze the continuous beam shown in Fig. 208 due to a vertical settlement of ½ in. at support B.

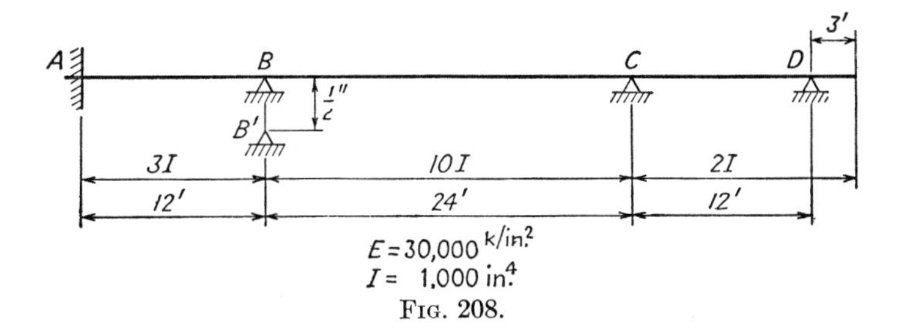

FIG. 208.

Solution. *Relative Stiffness and Fixed-end Moments*

TABLE 73-1. RELATIVE STIFFNESS

AB	$\left(\frac{3I}{12} = \frac{I}{4}\right) \times 12$	3
BC	$\left(\frac{10I}{24} = \frac{5I}{12}\right) \times 12$	5
CD	$\left(\frac{2I}{12} = \frac{I}{6}\right) \times 12$	2

$$M_{FAB} = M_{FBA} = +\frac{6EIR}{L}$$

$$= +\frac{(6)(30{,}000)(3{,}000)}{(12)(144)}\left(\frac{0.500}{144}\right) \text{ kip-ft}$$

$$= +\ 1{,}085.07 \text{ kip-ft}$$

$$M_{FBC} = M_{FCB} = +\frac{6EIR}{L}$$

$$= +\frac{(6)(30{,}000)(10{,}000)}{(24)(144)}\left(\frac{-0.500}{288}\right) \text{ kip-ft}$$

$$= -904.22 \text{ kip-ft}$$

$$M_{FCD} = M_{FDC} = 0$$

Moment Distribution (see Table 73-2)

TABLE 73-2. MOMENT DISTRIBUTION

Joint		A	B		C		D
Member		AB	BA	BC	CB	CD	DC
K		3	3	5	5	2	2
Cycle	DF	—	0.375	0.625	0.714	0.286	1.000
1	FEM	+1,085.1	+1,085.1	−904.2	−904.2	—	—
	Bal.	—	− 67.8	−113.1	+645.6	+258.6	—
2	CO	− 33.9	—	+322.8	− 56.6	—	+129.3
	Bal.	—	− 121.0	−201.8	+ 40.4	+ 16.2	−129.3
3	CO	− 60.5	—	+ 20.2	−100.9	− 64.6	+ 8.1
	Bal.	—	− 7.6	− 12.6	+118.2	+ 47.3	− 8.1
4	CO	− 3.8	—	+ 59.1	− 6.3	− 4.0	+ 23.6
	Bal.	—	− 22.2	− 36.9	+ 7.4	+ 2.9	− 23.6
5	CO	− 11.1	—	+ 3.7	− 18.4	− 11.8	+ 1.4
	Bal.	—	− 1.4	− 2.3	+ 21.6	+ 8.6	− 1.4
6	CO	− 0.7	—	+ 10.8	− 1.2	− 0.7	+ 4.3
	Bal.	—	− 4.0	− 6.8	+ 1.4	+ 0.5	− 4.3
7	CO	− 2.0	—	+ 0.7	− 3.4	− 2.2	+ 0.2
	Bal.	—	− 0.3	− 0.4	+ 4.0	+ 1.6	− 0.2
8	CO	− 0.2	—	+ 2.0	− 0.2	− 0.1	+ 0.8
	Bal.	—	− 0.8	− 1.2	+ 0.2	+ 0.1	− 0.8
9	CO	− 0.4	—	+ 0.1	− 0.6	− 0.4	—
	Bal.	—	—	− 0.1	+ 0.7	+ 0.3	—
10	CO	—	—	+ 0.4	—	—	+ 0.2
	Bal.	—	− 0.2	− 0.2	—	—	− 0.2
11	CO	− 0.1	—	—	− 0.1	− 0.1	—
	Bal.	—	—	—	+ 0.1	+ 0.1	—
Total		+ 972.4	+ 859.8	−859.8	−252.3	+252.3	0
Check:							
Change		− 112.7	− 225.3	+ 44.4	+651.9	+252.3	0
$-\frac{1}{2}$ (change)		+ 112.6	+ 56.4	−326.0	− 22.2	0	−126.2
Sum		0	− 168.9	−281.6	+629.7	+252.3	−126.2
θ_{rel} = sum/$(-K)$		0	+ 56.3	+ 56.3	−125.9	−126.2	+ 63.1
		(*Check*)	(*Check*)		(*Check*)		

Moment Distribution, Alternate Solution (see Table 73-3)

TABLE 73-3. MOMENT DISTRIBUTION, ALTERNATE SOLUTION

Joint		*A*	*B*		*C*		*D*
Member		*AB*	*BA*	*BC*	*CB*	*CD*	*DC*
K		3	3	5	5	2	2
Modified *K*		3	3	5	5	1.5	1.5
Cycle	DF	——	0.375	0.625	0.769	0.231	——
1	FEM	+1,085.1	+1,085.1	−904.2	−904.2	——	——
	Bal.	——	− 67.8	−113.1	+695.3	+208.9	——
2	CO	− 33.9	——	+347.6	− 56.6	——	
	Bal.	——	− 130.4	−217.2	+ 43.5	+ 13.1	
3	CO	− 65.2	——	+ 21.8	−108.6	——	
	Bal.	——	− 8.2	− 13.6	+ 83.5	+ 25.1	
4	CO	− 4.1	——	+ 41.8	− 6.8	——	
	Bal.	——	− 15.7	− 26.1	+ 5.2	+ 1.6	
5	CO	− 7.8	——	+ 2.6	− 13.0	——	
	Bal.	——	− 1.0	− 1.6	+ 10.0	+ 3.0	
6	CO	− 0.5	——	+ 5.0	− 0.8	——	
	Bal.	——	− 1.9	− 3.1	+ 0.6	+ 0.2	
7	CO	− 1.0	——	+ 0.3	− 1.6	——	
	Bal.	——	− 0.1	− 0.2	+ 1.2	+ 0.4	
8	CO	——	——	+ 0.6	− 0.1	——	
	Bal.	——	− 0.2	− 0.4	+ 0.1	——	
9	CO	− 0.1	——	——	− 0.2	——	
	Bal.	——	——	——	+ 0.2	——	
10	CO	——	——	+ 0.1	——	——	
	Bal.	——	——	− 0.1	——	——	
Total		+ 972.5	+ 859.8	−859.8	−252.3	+252.3	0
Check:							
Change		− 112.6	− 225.3	+ 44.4	+651.9	+252.3	0
$-\frac{1}{2}$ (change)		+ 112.6	+ 56.3	−326.0	− 22.2	0	−126.2
Sum		0	− 169.0	−281.6	−629.7	+252.3	−126.2
θ_{rel} = sum/(−*K*)		0	+ 56.3	+ 56.3	−125.9	−126.2	+ 63.1
		(*Check*)	(*Check*)		(*Check*)		

For reactions, shear and moment diagrams, and the elastic curve, see Example 57.

EXERCISES

113 and 114. Solve Exercises 76 and 77 by the moment-distribution method.

50. Application of the Moment-distribution Method to the Analysis of Statically Indeterminate Frames. *Case* I—*without Joint Movements.* The application of the moment-distribution method to the analysis of statically indeterminate frames wherein no joint movements or "side-sway" is involved is very similar to that of beams as discussed in the previous articles, except that in the case of frames there are frequently more than two members meeting in one joint. In such cases the unbalance at any joint is distributed to the ends of the several members meeting at the joint in the ratio of their relative stiffnesses. There are several ways in which the work for the moment distribution may be arranged, but the tabular form in which all members meeting at the same joint are grouped together is used here and is suggested as the most convenient form.

Example 74. Analyze the rigid frame shown in Fig. 209*a* by the moment-distribution method. Draw shear and moment diagrams. Sketch the deformed structure.

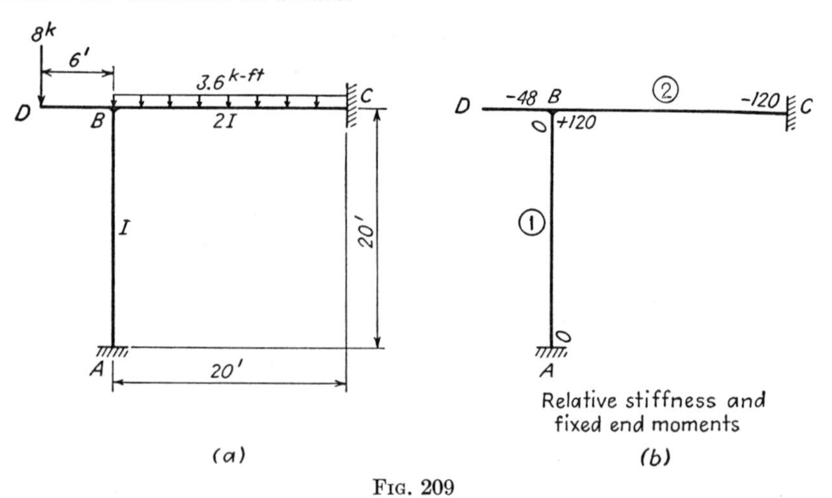

FIG. 209

Solution. $R = 0$ for all members.

Relative Stiffness and Fixed-end Moments (Fig. 209*b*)

TABLE 74-1. RELATIVE STIFFNESS

AB..........	$\frac{I}{20} \times 20$	1
BC..........	$\frac{2I}{20} \times 20$	2

$$M_{FBC} = + \frac{(3.6)(20)^2}{12} = +120 \text{ kip-ft}$$
$$M_{FCB} = -120 \text{ kip-ft}$$

Moment Distribution (see Table 74-2)

TABLE 74-2. MOMENT DISTRIBUTION

Joint..........		*A*	*B*			*C*
Member..........		*AB*	*BA*	*BC*	*BD*	*CB*
K..........		1	1	2	——	2
Cycle	DF	——	0.333	0.667	——	——
1	FEM Bal.	—— ——	—— −24.0	+120.0 − 48.0	−48.0 ↑	−120.0 ——
2	CO Bal.	−12.0 ——	—— ——	—— ——	↓	− 24.0 ——
Total..........		−12.0	−24.0	+ 72.0	−48.0	−144.0
Check: Change.......... $-\frac{1}{2}$ (change).......... Sum.......... θ_{rel} = sum/(−*K*)..........		 −12.0 +12.0 0 0 (*Check*)	 −24.0 + 6.0 −18.0 +18.0 (*Check*)	 − 48.0 + 12.0 − 36.0 + 18.0		 − 24.0 + 24.0 0 0 (*Check*)

The overhang BD, being a cantilever, has no rigidity and is not treated as a member. However, a clockwise moment of 48 kip-ft acts on BD at B; or a counterclockwise moment of 48 kip-ft acts on the joint B. This moment is entered in the moment-distribution table (Table 74-2) under BD as -48 kip-ft. The moment distribution comes to a natural stop at the end of the second cycle.

For reactions, shear and moment diagrams, and the elastic curve, see Example 58.

Example 75. Analyze the rigid frame shown in Fig. 210*a* by the moment-distribution method. Draw shear and moment diagrams. Sketch the deformed structure.

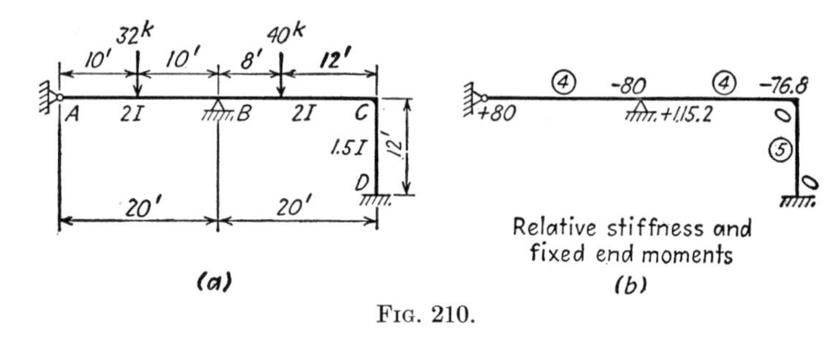

FIG. 210.

Solution. *Relative Stiffness and Fixed-end Moments* (Fig. 210*b*)

TABLE 75-1. RELATIVE STIFFNESS

AB	$\left(\frac{2I}{20} = \frac{I}{10}\right) \times 40$	4
BC	$\left(\frac{2I}{20} = \frac{I}{10}\right) \times 40$	4
CD	$\left(\frac{1.5I}{12} = \frac{I}{8}\right) \times 40$	5

$$M_{FAB} = + \frac{(32)(20)}{8} = +80 \text{ kip-ft}$$

$$M_{FBA} = -80 \text{ kip-ft}$$

$$M_{FBC} = + \frac{(40)(8)(12)^2}{(20)^2} = +115.2 \text{ kip-ft}$$

$$M_{FCB} = - \frac{(40)(12)(8)^2}{(20)^2} = -76.8 \text{ kip-ft}$$

$$M_{FCD} = M_{FDC} = 0$$

Moment Distribution (see Table 75-2)

TABLE 75-2. MOMENT DISTRIBUTION

Joint		A	B		C		D
Member		AB	BA	BC	CB	CD	DC
K		4	4	4	4	5	5
Cycle	DF	1.000	0.500	0.500	0.4444	0.5556	——
1	FEM	+80.00	− 80.00	+115.20	−76.80	——	——
	Bal.	−80.00	− 17.60	− 17.60	+34.13	+42.67	——
2	CO	− 8.80	− 40.00	+ 17.06	− 8.80	——	+21.34
	Bal.	+ 8.80	+ 11.47	+ 11.47	+ 3.91	+ 4.89	——
3	CO	+ 5.74	+ 4.40	+ 1.96	+ 5.74	——	+ 2.44
	Bal.	− 5.74	− 3.18	− 3.18	− 2.55	− 3.19	——
4	CO	− 1.59	− 2.87	− 1.28	− 1.59	——	− 1.60
	Bal.	+ 1.59	+ 2.08	+ 2.07	+ 0.71	+ 0.88	——
5	CO	+ 1.04	+ 0.80	+ 0.36	+ 1.04	——	+ 0.44
	Bal.	− 1.04	− 0.58	− 0.58	− 0.46	− 0.58	——
6	CO	− 0.29	− 0.52	− 0.23	− 0.29	——	− 0.29
	Bal.	+ 0.29	+ 0.37	+ 0.38	+ 0.13	+ 0.16	——
7	CO	+ 0.18	+ 0.14	+ 0.06	+ 0.19	——	+ 0.08
	Bal.	− 0.18	− 0.10	− 0.10	− 0.08	− 0.11	——
8	CO	− 0.05	− 0.09	− 0.04	− 0.05	——	− 0.06
	Bal.	+ 0.05	+ 0.07	+ 0.06	+ 0.02	+ 0.03	——
9	CO	+ 0.04	+ 0.02	+ 0.01	+ 0.03	——	+ 0.02
	Bal.	− 0.04	− 0.02	− 0.01	− 0.01	− 0.02	——
10	CO	− 0.01	− 0.02	——	——	——	− 0.01
	Bal.	+ 0.01	+ 0.01	+ 0.01	——	——	——
Total		0	−125.62	+125.62	−44.73	+44.73	+22.36
Check:							
Change		−80.00	− 45.62	+ 10.42	+32.07	+44.73	+22.36
$-\frac{1}{2}$ (change)		+22.81	+ 40.00	− 16.04	− 5.21	−11.18	−22.36
Sum		−57.19	− 5.62	− 5.62	+26.86	+33.55	0
θ_{rel} = sum/(−K)		+14.30	+ 1.40	+ 1.40	− 6.72	− 6.71	0
			(*Check*)		(*Check*)		(*Check*)

Moment Distribution, Alternate Solution (see Table 75-3)

TABLE 75-3. MOMENT DISTRIBUTION, ALTERNATE SOLUTION

Joint		A	B		C		D
Member		AB	BA	BC	CB	CD	DC
K		4	4	4	4	5	5
Modified K		3	3	4	4	5	5
Cycle	DF	——	0.4286	0.5714	0.4444	0.5556	——
1	FEM	+80.00	− 80.00	+115.20	−76.80	——	——
	Bal.	−80.00	− 15.09	− 20.11	+34.13	+42.69	——
2	CO	↑	− 40.00	+ 17.06	−10.06	——	+21.34
	Bal.		+ 9.83	+ 13.11	+ 4.47	+ 5.59	——
3	CO		——	+ 2.24	+ 6.56	——	+ 2.80
	Bal.		− 0.96	− 1.28	− 2.92	− 3.64	——
4	CO		——	− 1.46	− 0.64	——	− 1.82
	Bal.		+ 0.63	+ 0.83	+ 0.28	+ 0.36	——
5	CO		——	+ 0.14	+ 0.42	——	+ 0.18
	Bal.		− 0.06	− 0.08	− 0.19	− 0.23	——
6	CO		——	− 0.10	− 0.04	——	− 0.12
	Bal.		+ 0.04	+ 0.06	+ 0.02	+ 0.02	——
7	CO		——	+ 0.01	+ 0.03	——	+ 0.01
	Bal.		——	− 0.01	− 0.01	− 0.02	——
8	CO		——	——	——	——	− 0.01
	Bal.	↓	——	——	——	——	——
Total		0	−125.61	+125.61	−44.75	+44.75	+22.38
Check:							
Change		−80.00	− 45.61	+ 10.41	+32.05	+44.75	+22.38
$-\frac{1}{2}$ (change)		+22.80	+ 40.00	− 16.02	− 5.20	−11.19	−22.38
Sum		−57.20	− 5.61	− 5.61	+26.85	+33.56	0
$\theta_{rel} = \text{sum}/(-K)$		+14.30	+ 1.40	+ 1.40	− 6.71	− 6.71	0
			(*Check*)		(*Check*)		(*Check*)

For reactions, shear and moment diagrams, and the elastic curve, see Example 59.

Example 76. Analyze the rigid frame shown in Fig. 211*a* by the moment-distribution method. Draw shear and moment diagrams. Sketch the deformed structure.

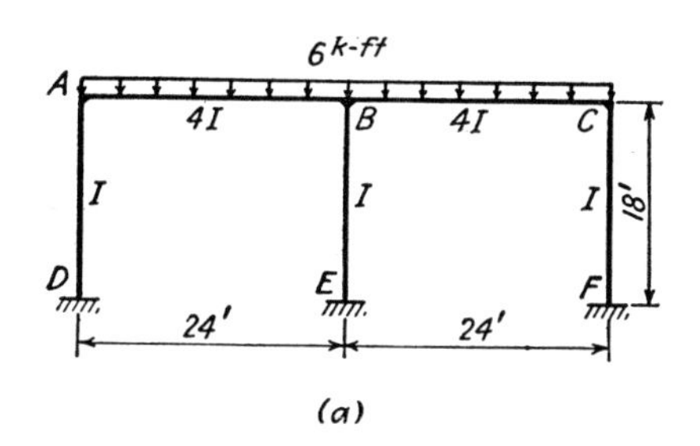

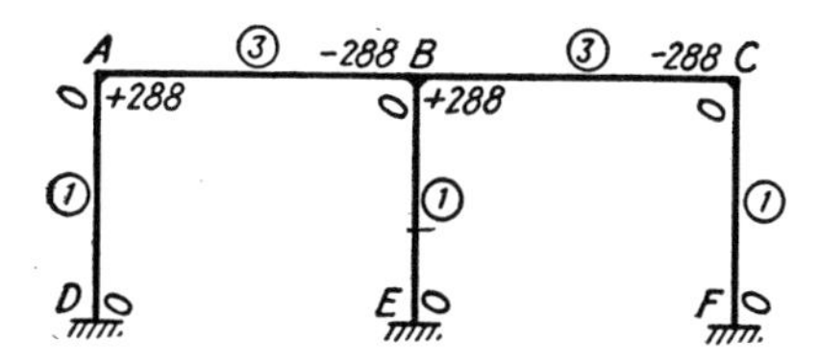

Relative stiffness and fixed end moments
(b).

FIG. 211.

Solution. $R = 0$ for all members.

Relative Stiffness and Fixed-end Moments (Fig. 211*b*)

TABLE 76-1. RELATIVE STIFFNESS

AB, *BC*................	$\left(\frac{4I}{24} = \frac{I}{6}\right) \times 18$	3
AD, *BE*, *CF*............	$\frac{I}{18} \times 18$	1

$$M_{FAB} = M_{FBC} = +\frac{(6)(24)^2}{12} = +288 \text{ kip-ft}$$

$$M_{FBA} = M_{FCB} = -288 \text{ kip-ft}$$

Moment Distribution (see Table 76-2).

TABLE 76-2. MOMENT DISTRIBUTION

Joint		*A*		*B*			*C*		*D*	*E*	*F*
Member		*AB*	*AD*	*BA*	*BC*	*BE*	*CB*	*CF*	*DA*	*EB*	*FC*
K		3	1	3	3	1	3	1	1	1	1
Cycle	DF	0.750	0.250	0.429	0.429	0.142	0.750	0.250	—	—	—
1	FEM	+288.0	—	−288.0	+288.0	—	−288.0	—	—	—	—
	Bal.	−216.0	−72.0	—	—	—	+216.0	+72.0	—	—	—
2	CO	—	—	−108.0	+108.0	—	—	—	−36.0	—	+36.0
	Bal.	—	—	—	—	—	—	—	—	—	—
Total		+ 72.0	−72.0	−396.0	+396.0	—	− 72.0	+72.0	−36.0	—	+36.0
Check:											
Change		−216.0	−72.0	−108.0	+108.0	—	+216.0	+72.0	−36.0	—	+36.0
$-\frac{1}{2}$ (change)		+ 54.0	+18.0	+108.0	−108.0	—	− 54.0	−18.0	+36.0	—	−36.0
Sum		−162.0	−54.0	0	0	0	+162.0	+54.0	0	0	0
θ_{rel} = sum/($-K$)		+ 54.0	+54.0	0	0	0	− 54.0	−54.0	0	0	0
		(*Check*)		(*Check*)			(*Check*)		(*Check*)	(*Check*)	(*Check*)

For reactions, shear and moment diagrams, and the elastic curve, see Example 60.

EXERCISES

115 to 121. Solve Exercises 78 to 84 by the moment-distribution method.

51. Application of the Moment-distribution Method to the Analysis of Statically Indeterminate Frames. *Case* II—*with Joint Movements.* The application of the moment-distribution method to the analysis of statically indeterminate frames in which sidesway or joint movements are involved consists in the following:

1. The joints are first held against sidesway. The fixed-end moments as caused by the applied loading are distributed, and a first set of balanced end moments are obtained.
2. The unloaded frame is assumed to have a certain amount of sidesway which will cause a set of fixed-end moments. These fixed-end moments are then distributed, and a second set of balanced end moments are obtained.
3. The resulting set of end moments can be obtained by adding the first set and the product of a ratio and the second set, the ratio being determined by use of the shear condition, as will be explained.

Take, for example, the rigid frame shown in Fig. 212a. It is required to analyze this statically indeterminate frame by the moment-distribu-

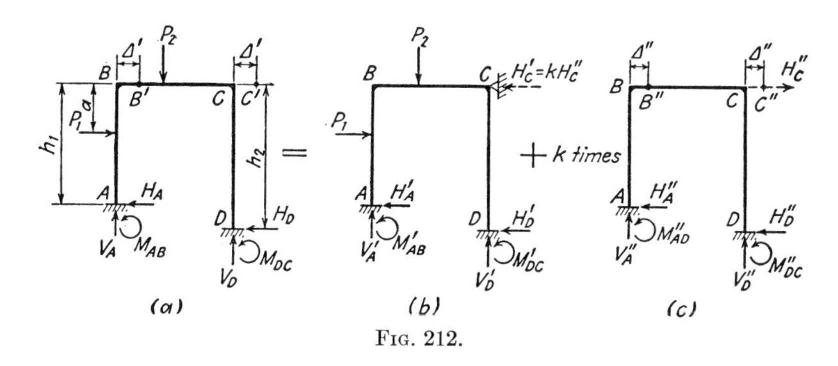

FIG. 212.

tion method. The given frame shown in Fig. 212a is equivalent to the sum of Fig. 212b and Fig. 212c. In Fig. 212b the joints B and C are held against sidesway by the fictitious support at C, the horizontal reaction of which is denoted as H'_C. If the fictitious support at C is removed, the force H'_C would act at joint C. In Fig. 212c, Δ'' is the sidesway caused by any arbitrary force H''_C. If H'_C is equal to kH''_C, where k is the unknown ratio, the actual amount of the sidesway, Δ', must be equal to $k\Delta''$. Let M'_{AB}, M'_{BA}, M'_{BC}, M'_{CB}, M'_{CD}, and M'_{DC} be the balanced moments obtained by distributing the fixed-end moments due to the applied loading while only permitting joints B and C to rotate but holding their locations (Fig. 212*b*). Let M''_{AB}, M''_{BA}, M''_{BC}, M''_{CB}, M''_{CD}, and M''_{DC} be the balanced moments obtained by distributing the fixed-end moments due to any assumed amount Δ'' of the horizontal movement of joint B or C. The shear condition required of the frame shown in Fig. 212*a* is

$$H_A + H_D = P_1$$

Since

$$H_A = \frac{M_{AB} + M_{BA}}{h_1} + \frac{P_1 a}{h_1} \quad \text{and} \quad H_D = \frac{M_{CD} + M_{DC}}{h_2}$$

the shear condition becomes

$$\frac{M_{AB} + M_{BA}}{h_1} + \frac{P_1 a}{h_1} + \frac{M_{CD} + M_{DC}}{h_2} = P_1 \tag{99}$$

Also, by superposition,

$$\begin{aligned} M_{AB} &= M'_{AB} + k(M''_{AB}) & M_{BA} &= M'_{BA} + k(M''_{BA}) \\ M_{BC} &= M'_{BC} + k(M''_{BC}) & M_{CB} &= M'_{CB} + k(M''_{CB}) \\ M_{CD} &= M'_{CD} + k(M''_{CD}) & M_{DC} &= M'_{DC} + k(M''_{DC}) \end{aligned} \tag{100}$$

By substituting Eqs. (100) into Eq. (99),

$$\frac{(M'_{AB} + M'_{BA}) + k(M''_{AB} + M''_{BA})}{h_1} + \frac{P_1 a}{h_1} + \frac{(M'_{CD} + M'_{DC}) + k(M''_{CD} + M''_{DC})}{h_2} = P_1 \tag{101}$$

The unknown ratio k can then be found by solving Eq. (101). Once k is known, all end moments acting on the frame of Fig. 212*a* can be found from Eqs. (100).

Where two or more unknown movements of sidesway are involved, the resulting set of end moments can be expressed as the sum of (1) the balanced end moments by distributing the fixed-end moments due to the applied loading, and (2) the products of an unknown ratio and the balanced end moments found by distributing the fixed-end moments due to a certain amount of the first movement in sidesway, and (3) the products of a *second* unknown ratio and the balanced end moments due to a certain amount of the *second* movement in sidesway, and so on. The unknown ratios are determined from the shear conditions.

The procedure discussed above will be illustrated by the following examples.

Example 77. Analyze the rigid frame shown in Fig. 213*a* by the moment-distribution method. Draw shear and moment diagrams. Sketch the deformed structure.

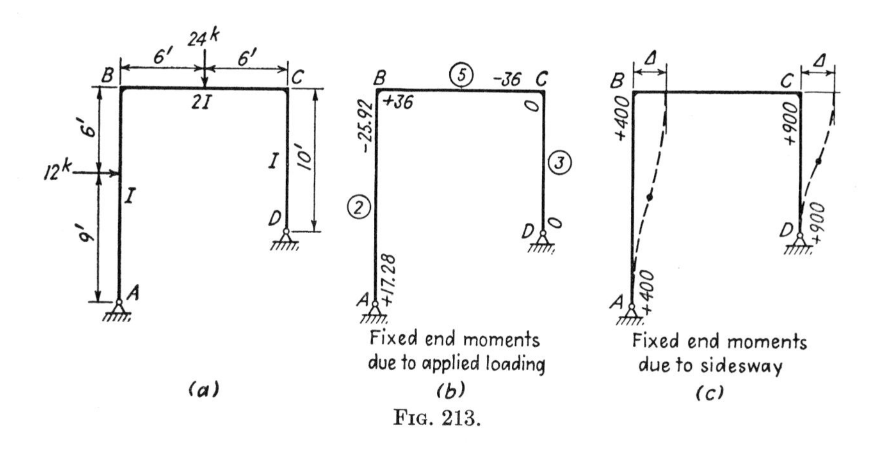

FIG. 213.

Solution. *Relative Stiffness*

TABLE 77-1. RELATIVE STIFFNESS

AB............	$\frac{I}{15} \times 30$	2
BC............	$\left(\frac{2I}{12} = \frac{I}{6}\right) \times 30$	5
CD............	$\frac{I}{10} \times 30$	3

Distribution of Fixed-end Moments Due to the Applied Loading (see Fig. 213*b*)

$$M_{FAB} = + \frac{(12)(9)(6)^2}{(15)^2} = +17.28 \text{ kip-ft}$$

$$M_{FBA} = - \frac{(12)(6)(9)^2}{(15)^2} = -25.92 \text{ kip-ft}$$

$$M_{FBC} = + \frac{(24)(12)}{8} = +36 \text{ kip-ft}$$

$$M_{FCB} = -36 \text{ kip-ft}$$

$$M_{FCD} = M_{FDC} = 0$$

For moment distribution of these fixed-end moments, see Table 77-2.

TABLE 77-2. DISTRIBUTION OF FEM DUE TO THE APPLIED LOADING

Joint		*A*	*B*		*C*		*D*
Member		*AB*	*BA*	*BC*	*CB*	*CD*	*DC*
K		2	2	5	5	3	3
Modified K		1.5	1.5	5	5	2.25	2.25
Cycle	DF	——	0.231	0.769	0.690	0.310	——
1	FEM	+17.28	−25.92	+36.00	−36.00	——	——
	Bal.	−17.28	− 2.33	− 7.75	+24.84	+11.16	——
2	CO	↑	− 8.64	+12.42	− 3.88	——	↑
	Bal.		− 0.87	− 2.91	+ 2.68	+ 1.20	
3	CO		——	+ 1.34	− 1.46	——	
	Bal.		− 0.31	− 1.03	+ 1.01	+ 0.45	
4	CO		——	+ 0.50	− 0.52	——	
	Bal.		− 0.12	− 0.38	+ 0.36	+ 0.16	
5	CO		——	+ 0.18	− 0.19	——	
	Bal.		− 0.04	− 0.14	+ 0.13	+ 0.06	
6	CO		——	+ 0.06	− 0.07	——	
	Bal.		− 0.01	− 0.05	+ 0.05	+ 0.02	
7	CO		——	+ 0.02	− 0.02	——	
	Bal.		——	− 0.02	+ 0.01	+ 0.01	
8	CO		——	——	− 0.01	——	
	Bal.	↓	——	——	+ 0.01	——	↓
Total		0	−38.24	+38.24	−13.06	+13.06	0
Check:							
Change		−17.28	−12.32	+ 2.24	+22.94	+13.06	0
$-\frac{1}{2}$ (change)		+ 6.16	+ 8.64	−11.47	− 1.12	0	−6.53
Sum		−11.22	− 3.68	− 9.23	+21.82	+13.06	−6.53
θ_{rel} = sum/$(-K)$		+ 5.56	+ 1.84	+ 1.85	− 4.36	− 4.35	+2.18
			(*Check*)		(*Check*)		

Distribution of Fixed-end Moments Due to Sidesway (see Fig. 213c)

TABLE 77-3. FEM DUE TO SIDESWAY

	Relative magnitudes		
$M_{FAB} = M_{FBA}$	$+\ 6EI\Delta/(15)^2$	$+\frac{1}{225} \times 90{,}000$	$+400$
$M_{FCD} = M_{FDC}$	$+6EI\Delta/(10)^2$	$+\frac{1}{100} \times 90{,}000$	$+900$

Note that only the relative magnitudes of the fixed-end moments due to a certain amount of sidesway Δ are required.

For moment distribution of these fixed-end moments, see Table 77-4.

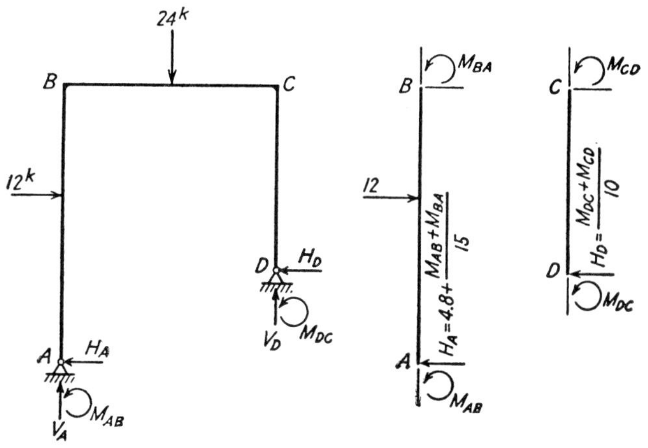

FIG. 214.

Determination of Ratio. The shear condition (Fig. 214) is

$$H_A + H_D = 12 \qquad \text{or} \qquad \left(4.8 + \frac{M_{AB} + M_{BA}}{15}\right) + \left(\frac{M_{CD} + M_{DC}}{10}\right) = 12$$

Simplifying,

$$2(M_{AB} + M_{BA}) + 3(M_{CD} + M_{DC}) = 216$$

Substituting

$$\begin{aligned} M_{AB} &= 0 + k(0) = 0 \\ M_{BA} &= -38.24 + k(+188.1) = -38.24 + 188.1k \\ M_{CD} &= +13.06 + k(+316.6) = +13.06 + 316.6k \\ M_{DC} &= 0 + k(0) = 0 \end{aligned}$$

into the above equation and solving for the ratio k,

$$2(-38.24 + 188.1k) + 3(+13.06 + 316.6k) = 216$$
$$k = +0.1910$$

TABLE 77-4. DISTRIBUTION OF FEM DUE TO SIDESWAY

Joint		A	B		C		D
Member		*AB*	*BA*	*BC*	*CB*	*CD*	*DC*
K		2	2	5	5	3	3
Modified K		1.5	1.5	5	5	2.25	2.25
Cycle	DF	——	0.231	0.769	0.690	0.310	——
1	FEM Bal.	+400.0 −400.0	+400.0 − 92.4	—— −307.6	—— −621.0	+900.0 −279.0	+900.0 −900.0
2	CO Bal.	↑	−200.0 +117.9	−310.5 +392.6	−153.8 +416.6	−450.0 +187.2	↑
3	CO Bal.		—— − 48.1	+208.3 −160.2	+196.3 −135.4	—— − 60.9	
4	CO Bal.		—— + 15.6	− 67.7 + 52.1	− 80.1 + 55.3	—— + 24.8	
5	CO Bal.		—— − 6.4	+ 27.6 − 21.2	+ 26.0 − 17.9	—— − 8.1	
6	CO Bal.		—— + 2.1	− 9.0 + 6.9	− 10.6 + 7.3	—— + 3.3	
7	CO Bal.		—— − 0.8	+ 3.6 − 2.8	+ 3.4 − 2.3	—— − 1.1	
8	CO Bal.		—— − 0.3	− 1.2 + 0.9	− 1.4 + 1.0	—— + 0.4	
9	CO Bal.		—— − 0.1	+ 0.5 − 0.4	+ 0.4 − 0.3	—— − 0.1	
10	CO Bal.	↓	—— ——	− 0.2 + 0.2	− 0.2 + 0.1	—— + 0.1	↓
Total		0	+188.1	−188.1	−316.6	+316.6	0
Check:							
Change		−400.0	−211.9	−188.1	−316.6	−583.4	−900.0
$-\frac{1}{2}$ (change)		+106.0	+200.0	+158.3	+ 94.0	+450.0	+291.7
Sum		−294.0	− 11.9	− 29.8	−222.6	−133.4	−618.3
θ_{rel} = sum/($-K$)		+147.0	+ 6.0	+ 6.0	+ 44.5	+ 44.5	+206.1
			(*Check*)		(*Check*)		

Combination of the Two Sets of Balanced Moments (see Table 77-5)

TABLE 77-5. COMBINATION OF THE TWO SETS OF BALANCED MOMENTS

Joint	*A*	*B*		*C*		*D*
Member	*AB*	*BA*	*BC*	*CB*	*CD*	*DC*
FEM from Table 77-2	+17.28	−25.92	+36.00	−36.00	——	——
0.1910 times FEM from Table 77-4	+76.40	+76.40	——	——	+171.90	+171.90
Total FEM	+93.68	+50.48	+36.00	−36.00	+171.90	+171.90
Balanced moments from Table 77-2	——	−38.24	+38.24	−13.06	+ 13.06	——
0.1910 times balanced moments (Table 77-4)	——	+35.93	−35.93	−60.47	+ 60.47	——
Total balanced moments	——	− 2.31	+ 2.31	−73.53	+ 73.53	——
Check:						
Change	−93.68	−52.79	−33.69	−37.53	− 98.37	−171.90
$-\frac{1}{2}$ (change)	+26.40	+46.84	+18.76	+16.84	+ 85.95	+ 49.18
Sum	−67.28	− 5.95	−14.93	−20.69	− 12.42	−122.72
θ_{rel} = sum/(−*K*)	+33.64	+ 2.98	+ 2.99	+ 4.14	+ 4.14	+ 40.91
		(*Check*)		(*Check*)		

For reactions, shear and moment diagrams, and the elastic curve, see Example 61.

Example 78. Analyze the rigid frame shown in Fig. 215*a* by the moment-distribution method. Draw shear and moment diagrams. Sketch the deformed structure.

Solution. *Relative Stiffness*

TABLE 78-1. RELATIVE STIFFNESS

AB	$\frac{I}{15} \times 30$	2
BC	$\left(\frac{2I}{12} = \frac{I}{6}\right) \times 30$	5
CD	$\frac{I}{10} \times 30$	3

Distribution of FEM Due to the Applied Loading (see Fig. 215*b*)

$$M_{FAB} = + \frac{(12)(9)(6)^2}{(15)^2} = +17.28 \text{ kip-ft}$$

$$M_{FBA} = -\frac{(12)(6)(9)^2}{(15)^2} = -25.92 \text{ kip-ft}$$

$$M_{FBC} = +\frac{(24)(12)}{8} = +36 \text{ kip-ft}$$

$$M_{FCB} = -36 \text{ kip-ft}$$

$$M_{FCD} = M_{FDC} = 0$$

For moment distribution of these fixed-end moments, see Table 78-2.

TABLE 78-2. DISTRIBUTION OF FEM DUE TO THE APPLIED LOADING

Joint		*A*	*B*		*C*		*D*
Member		*AB*	*BA*	*BC*	*CB*	*CD*	*DC*
K		2	2	5	5	3	3
Cycle	DF	——	0.286	0.714	0.625	0.375	——
1	FEM	+17.28	−25.92	+36.00	−36.00	——	——
	Bal.	——	− 2.88	− 7.20	+22.50	+13.50	——
2	CO	− 1.44	——	+11.25	− 3.60	——	+6.75
	Bal.	——	− 3.22	− 8.03	+ 2.25	+ 1.35	——
3	CO	− 1.61	——	+ 1.12	− 4.02	——	+0.68
	Bal.	——	− 0.32	− 0.80	+ 2.51	+ 1.51	——
4	CO	− 0.16	——	+ 1.25	− 0.40	——	+0.75
	Bal.	——	− 0.36	− 0.89	+ 0.25	+ 0.15	——
5	CO	− 0.18	——	+ 0.12	− 0.44	——	+0.08
	Bal.	——	− 0.03	− 0.09	+ 0.28	+ 0.16	——
6	CO	− 0.02	——	+ 0.14	− 0.04	——	+0.08
	Bal.	——	− 0.04	− 0.10	+ 0.02	+ 0.02	——
7	CO	− 0.02	——	+ 0.02	− 0.05	——	+0.01
	Bal.	——	− 0.01	− 0.01	+ 0.03	+ 0.02	——
8	CO	——	——	+ 0.02	——	——	+0.01
	Bal.	——	− 0.01	− 0.01	——	——	——
Total		+13.85	−32.79	+32.79	−16.71	+16.71	+8.36
Check:							
Change		− 3.43	− 6.87	− 3.21	+19.29	+16.71	+8.36
$-\frac{1}{2}$ (change)		+ 3.43	+ 1.72	− 9.64	+ 1.60	− 4.18	−8.36
Sum		0	− 5.15	−12.85	+20.89	+12.53	0
θ_{rel} = sum/($-K$)		0	+ 2.58	+ 2.57	− 4.18	− 4.18	0
		(*Check*)	(*Check*)		(*Check*)		(*Check*)

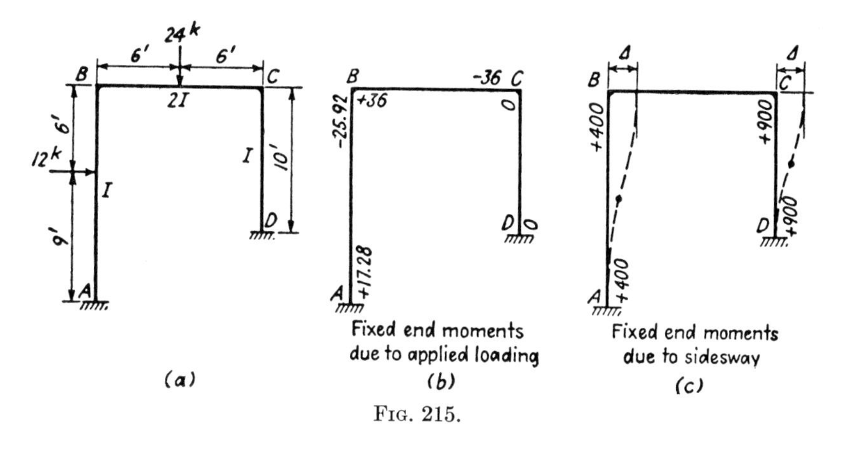

FIG. 215.

Distribution of FEM Due to Sidesway (see Fig. 215c)

TABLE 78-3. FEM DUE TO SIDESWAY

$M_{FAB} = M_{FBA}$..........	$+\dfrac{6EI\Delta}{(15)^2}$	$+\frac{1}{225} \times 90{,}000$	+400
$M_{FBC} = M_{FCB}$..........	$+\dfrac{6EI\Delta}{(10)^2}$	$+\frac{1}{100} \times 90{,}000$	+900

Note that only the relative magnitudes of the fixed-end moments due to a certain amount of sidesway Δ are required.

For moment distribution of these fixed-end moments, see Table 78-4.

Determination of Ratio. The shear condition (Fig. 216) is

$$H_A + H_D = 12 \qquad \text{or} \qquad \left(4.8 + \frac{M_{AB} + M_{BA}}{15}\right) + \left(\frac{M_{CD} + M_{DC}}{10}\right) = 12$$

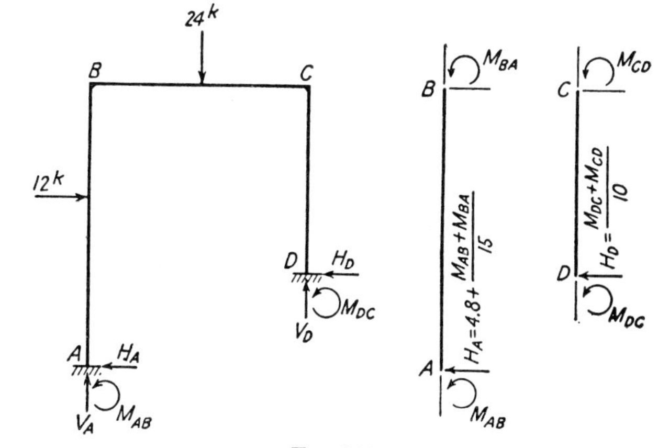

FIG. 216.

TABLE 78-4. DISTRIBUTION OF FEM DUE TO SIDESWAY

Joint		A	B		C		D
Member		AB	BA	BC	CB	CD	DC
K		2	2	5	5	3	3
Cycle	DF	——	0.286	0.714	0.625	0.375	——
1	FEM	+400.0	+400.0	——	——	+900.0	+900.0
	Bal.	——	−114.4	−285.6	−562.5	−337.5	——
2	CO	− 57.2	——	−281.2	−142.8	——	−168.8
	Bal.	——	+ 80.4	+200.8	+ 89.2	+ 53.6	——
3	CO	+ 40.2	——	+ 44.6	+100.4	——	+ 26.8
	Bal.	——	− 12.8	− 31.8	− 62.8	− 37.6	——
4	CO	− 6.4	——	− 31.4	− 15.9	——	− 18.8
	Bal.	——	+ 9.0	+ 22.4	+ 9.9	+ 6.0	——
5	CO	+ 4.5	——	+ 5.0	+ 11.2	——	+ 3.0
	Bal.	——	− 1.4	− 3.6	− 7.0	− 4.2	——
6	CO	− 0.7	——	− 3.5	− 1.8	——	− 2.1
	Bal.	——	+ 1.0	+ 2.5	+ 1.1	+ 0.7	——
7	CO	− 0.5	——	+ 0.6	+ 1.2	——	+ 0.4
	Bal.	——	− 0.2	− 0.4	− 0.8	− 0.4	——
8	CO	− 0.1	——	− 0.4	− 0.2	——	− 0.2
	Bal.	——	+ 0.1	+ 0.3	+ 0.1	+ 0.1	——
9	CO	——	——	——	+ 0.2	——	——
	Bal.	——	——	——	− 0.1	− 0.1	——
Total		+380.8	+361.7	−361.7	−580.6	+580.6	+740.3
Check:							
Change		− 19.2	− 38.3	−361.7	−580.6	−319.4	−159.7
$-\frac{1}{2}$ (change)		+ 19.2	+ 9.6	+290.3	+180.8	+ 79.8	+159.7
Sum		0	− 28.7	− 71.4	−399.8	−239.6	0
θ_{rel} = sum/($-K$)		0	+ 14.4	+ 14.3	+ 80.0	+ 79.9	0
		(*Check*)	(*Check*)		(*Check*)		(*Check*)

Simplifying,

$$2(M_{AB} + M_{BA}) + 3(M_{CD} + M_{DC}) = 216$$

Substituting

$$M_{AB} = +13.85 + k(+380.8)$$
$$M_{BA} = -32.79 + k(+361.7)$$
$$M_{CD} = +16.71 + k(+580.6)$$
$$M_{DC} = +8.36 + k(+740.3)$$

into the above equation and solving for the ratio k,

$$(2)(-18.94 + 742.5k) + (3)(25.07 + 1{,}320.9k) = 216$$
$$k = +0.03280$$

Combination of the Two Sets of Balanced Moments (see Table 78-5)

TABLE 78-5. COMBINATION OF THE TWO SETS OF BALANCED MOMENTS

Joint	*A*	*B*		*C*		*D*
Member	*AB*	*BA*	*BC*	*CB*	*CD*	*DC*
FEM from Table 78-2	+17.28	−25.92	+36.00	−36.00	——	——
0.03280 times FEM from Table 78-4	+13.12	+13.12	——	——	+29.52	+29.52
Total FEM	+30.40	−12.80	+36.00	−36.00	+29.52	+29.52
Balanced moments from Table 78-2	+13.85	−32.79	+32.79	−16.71	+16.71	+ 8.36
0.03280 times balanced moments (Table 78-4)	+12.49	+11.86	−11.86	−19.04	+19.04	+24.28
Total balanced moments	+26.34	−20.93	+20.93	−35.74	+35.74	+32.64
Check:						
Change	− 4.06	− 8.13	−15.07	+ 0.25	+ 6.23	+ 3.12
$-\frac{1}{2}$ (change)	+ 4.06	+ 2.03	− 0.12	+ 7.54	− 1.56	− 3.12
Sum	0	− 6.10	−15.19	+ 7.79	+ 4.67	0
θ_{rel} = sum/(−K)	0	+ 3.05	+ 3.04	− 1.56	− 1.56	0
	(*Check*)	(*Check*)		(*Check*)		(*Check*)

For reactions, shear and moment diagrams, and the elastic curve, see Example 62.

Example 79. Analyze the rigid frame shown in Fig. 217*a* by the moment-distribution method. Draw shear and moment diagrams. Sketch the deformed structure.

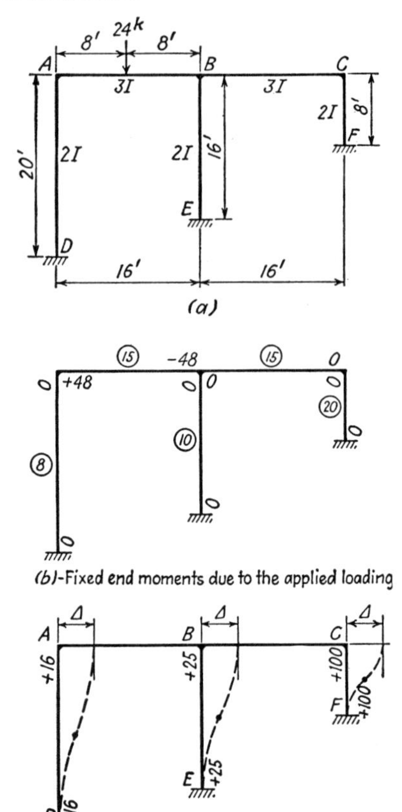

FIG. 217.

Solution. *Relative Stiffness*

TABLE 79-1. RELATIVE STIFFNESS

AB, *BC*	$\frac{3I}{16} \times 80$	15
AD	$\left(\frac{2I}{20} = \frac{I}{10}\right) \times 80$	8
BE	$\left(\frac{2I}{16} = \frac{I}{8}\right) \times 80$	10
CF	$\left(\frac{2I}{8} = \frac{I}{4}\right) \times 80$	20

TABLE 79-2. DISTRIBUTION OF FEM DUE TO THE APPLIED LOADING

Joint		*A*		*B*			*C*		*D*	*E*	*F*
Member		*AB*	*AD*	*BA*	*BC*	*BE*	*CB*	*CF*	*DA*	*EB*	*FC*
K		15	8	15	15	10	15	20	8	10	20
Cycle	DF	0.652	0.348	0.375	0.375	0.250	0.429	0.571	—	—	—
1	FEM	+48.00	—	−48.00	—	—	—	—	—	—	—
	Bal.	−31.30	−16.70	+18.00	+18.00	+12.00	—	—	—	—	—
2	CO	+ 9.00	—	−15.65	—	—	+ 9.00	—	− 8.35	+6.00	—
	Bal.	− 5.87	− 3.13	+ 5.87	+ 5.87	+ 3.91	− 3.86	−5.14	—	—	—
3	CO	+ 2.94	—	− 2.94	− 1.93	—	+ 2.94	—	− 1.56	+1.96	−2.57
	Bal.	− 1.92	− 1.02	+ 1.82	+ 1.83	+ 1.22	− 1.26	−1.68	—	—	—
4	CO	+ 0.91	—	− 0.96	− 0.63	—	+ 0.92	—	− 0.51	+0.61	−0.84
	Bal.	− 0.59	− 0.32	+ 0.60	+ 0.59	+ 0.40	− 0.39	−0.53	—	—	—
5	CO	+ 0.30	—	− 0.30	− 0.20	—	+ 0.30	—	− 0.16	+0.20	−0.26
	Bal.	− 0.20	− 0.10	+ 0.19	+ 0.19	+ 0.12	− 0.13	−0.17	—	—	—
6	CO	+ 0.10	—	− 0.10	− 0.06	—	+ 0.10	—	− 0.05	+0.06	−0.08
	Bal.	− 0.06	− 0.04	+ 0.06	+ 0.06	+ 0.04	− 0.04	−0.06	—	—	—
7	CO	+ 0.03	—	− 0.03	− 0.02	—	+ 0.03	—	− 0.02	+0.02	−0.03
	Bal.	− 0.02	− 0.01	+ 0.02	+ 0.02	+ 0.01	− 0.01	−0.02	—	—	—
8	CO	+ 0.01	—	− 0.01	—	—	+ 0.01	—	—	—	−0.01
	Bal.	− 0.01	—	+ 0.01	—	—	—	−0.01	—	—	—
Total		+21.32	−21.32	−41.42	+23.72	+17.70	+ 7.61	−7.61	−10.65	+8.85	−3.79
Check:											
Change		−26.68	−21.32	+ 6.58	+23.72	+17.70	+ 7.61	−7.61	−10.65	+8.85	−3.79
$-\frac{1}{2}$ (change)		− 3.29	+ 5.32	+13.34	− 3.80	− 4.42	−11.86	+1.90	+10.66	−8.85	+3.80
Sum		−29.97	−16.00	+19.92	+19.92	+13.28	− 4.25	−5.71	0	0	0
θ_{rel} = sum/$(-K)$		+ 2.00	+ 2.00	− 1.33	− 1.33	− 1.33	+ 0.28	+0.28	0	0	0
		(*Check*)		(*Check*)			(*Check*)		(*Check*)	(*Check*)	(*Check*)

Distribution of FEM Due to the Applied Loading (see Fig. 217*b*)

$$M_{FAB} = + \frac{(24)(16)}{8} = +48 \text{ kip-ft}$$

$$M_{FBA} = -48 \text{ kip-ft}$$

For moment distribution of these fixed-end moments, see Table 79-2.

Distribution of FEM Due to Sidesway (see Fig. 217*c*)

TABLE 79-3. FEM DUE TO SIDESWAY

$M_{FAD} = M_{FDA}$	$+\frac{6E(2I)\Delta}{(20)^2}$	$+\frac{1}{400} \times 6{,}400$	+16
$M_{FBE} = M_{FEB}$	$+\frac{6E(2I)\Delta}{(16)^2}$	$+\frac{1}{256} \times 6{,}400$	+25
$M_{FCF} = M_{FFC}$	$+\frac{6E(2I)\Delta}{(8)^2}$	$+\frac{1}{64} \times 6{,}400$	+100

Note that only the relative magnitudes of the fixed-end moments due to a certain amount of sidesway Δ are required.

For moment distribution of these fixed-end moments, see Table 79-4.

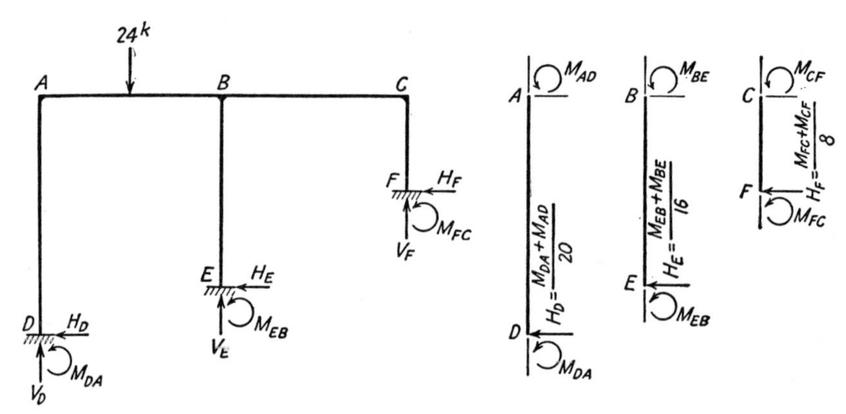

FIG. 218.

Determination of Ratio. The shear condition (Fig. 218) is

$$H_D + H_E + H_F = 0 \qquad \text{or} \qquad \left(\frac{M_{DA} + M_{AD}}{20}\right) + \left(\frac{M_{EB} + M_{BE}}{16}\right) + \left(\frac{M_{FC} + M_{CF}}{8}\right) = 0$$

Simplifying,

$$4(M_{DA} + M_{AD}) + 5(M_{EB} + M_{BE}) + 10(M_{CF} + M_{FC}) = 0$$

TABLE 79-4. DISTRIBUTION OF FEM DUE TO SIDESWAY

Joint		*A*		*B*			*C*		*D*	*E*	*F*
Member		*AB*	*AD*	*BA*	*BC*	*BE*	*CB*	*CF*	*DA*	*EB*	*FC*
K		15	8	15	15	10	15	20	8	10	20
Cycle	DF	0.652	0.348	0.375	0.375	0.250	0.429	0.571	—	—	—
1	FEM	—	+16.00	—	—	+25.00	—	+100.00	+16.00	+25.00	+100.00
	Bal.	−10.43	− 5.57	− 9.37	− 9.38	− 6.25	−42.90	− 57.10	—	—	—
2	CO	− 4.68	—	− 5.22	−21.45	—	− 4.69	—	− 2.78	− 3.12	− 28.55
	Bal.	+ 3.05	+ 1.63	+10.00	+10.00	+ 6.67	+ 2.01	+ 2.68	—	—	—
3	CO	+ 5.00	—	+ 1.52	+ 1.00	—	+ 5.00	—	+ 0.82	+ 3.34	+ 1.34
	Bal.	− 3.26	− 1.74	− 0.94	− 0.95	− 0.63	− 2.15	− 2.85	—	—	—
4	CO	− 0.47	—	− 1.63	− 1.08	—	− 0.48	—	− 0.87	− 0.32	− 1.42
	Bal.	+ 0.31	+ 0.16	+ 1.01	+ 1.02	+ 0.68	+ 0.21	+ 0.27	—	—	—
5	CO	+ 0.50	—	+ 0.16	+ 0.10	—	+ 0.51	—	+ 0.08	+ 0.34	+ 0.14
	Bal.	− 0.33	− 0.17	− 0.10	− 0.10	− 0.06	− 0.22	− 0.29	—	—	—
6	CO	− 0.05	—	− 0.16	− 0.11	—	− 0.05	—	− 0.08	− 0.03	− 0.14
	Bal.	+ 0.03	+ 0.02	+ 0.10	+ 0.10	+ 0.07	+ 0.02	+ 0.03	—	—	—
7	CO	+ 0.05	—	+ 0.02	+ 0.01	—	+ 0.05	—	+ 0.01	+ 0.04	+ 0.02
	Bal.	− 0.03	− 0.02	− 0.01	− 0.01	− 0.01	− 0.02	− 0.03	—	—	—
8	CO	—	—	− 0.02	− 0.01	—	—	—	− 0.01	—	− 0.02
	Bal.	—	—	+ 0.01	+ 0.01	+ 0.01	—	—	—	—	—
Total		−10.31	+10.31	− 4.63	−20.85	+25.48	−42.71	+ 42.71	+13.17	+25.25	+ 71.37
Check:											
Change		−10.31	− 5.69	− 4.63	−20.85	+ 0.48	−42.71	− 57.29	− 2.83	+ 0.25	− 28.63
$-\frac{1}{2}$ (change)		+ 2.32	+ 1.42	+ 5.15	+21.36	− 0.12	+10.42	+ 14.32	+ 2.84	− 0.24	+ 28.64
Sum		− 7.99	− 4.27	+ 0.52	+ 0.51	+ 0.36	−32.29	− 42.97	0	0	0
θ_{rel} = sum/(−K)		+ 0.53	+ 0.53	− 0.035	− 0.034	− 0.036	+ 2.15	+ 2.15	0	0	0
		(*Check*)			(*Check*)		(*Check*)		(*Check*)	(*Check*)	(*Check*)

TABLE 79-5. COMBINATION OF THE TWO SETS OF BALANCED MOMENTS

Joint	A		B			C		D	E	F
Member	*AB*	*AD*	*BA*	*BC*	*BE*	*CB*	*CF*	*DA*	*EB*	*FC*
FEM from Table 79-2	+48.00	——	−48.00	——	——	——	——	——	——	——
0.07332 times FEM from Table 79-4	——	+ 1.17	——	——	+ 1.83	——	+ 7.33	+ 1.17	+ 1.83	+7.33
Total FEM	+48.00	+ 1.17	−48.00	——	+ 1.83	——	+ 7.33	+ 1.17	+ 1.83	+7.33
Balanced moments from Table 79-2	+21.32	−21.32	−41.42	+23.72	+17.70	+ 7.61	− 7.61	−10.65	+ 8.85	−3.79
0.07332 times balanced moments (Table 79-4)	− 0.76	+ 0.76	− 0.34	− 1.53	+ 1.87	− 3.13	+ 3.13	+ 0.96	+ 1.85	+5.23
Total balanced moments	+20.56	−20.56	−41.76	+22.19	+19.57	+ 4.48	− 4.48	− 9.69	+10.70	+1.44
Check:										
Change	−27.44	−21.73	+ 6.24	+22.19	+17.74	+ 4.48	−11.81	−10.86	+ 8.87	−5.89
$-\frac{1}{2}$ (change)	− 3.12	+ 5.43	+13.72	− 2.24	− 4.44	−11.10	+ 2.94	+10.86	− 8.87	+5.90
Sum	−30.56	−16.30	+19.96	+19.95	+13.30	− 6.62	− 8.87	0	0	0
θ_{rel} = sum/(−K)	+ 2.04	+ 2.04	− 1.33	− 1.33	− 1.33	+ 0.44	+ 0.44	0	0	0
	(*Check*)			(*Check*		(*Check*)		(*Check*)	(*Check*)	(*Check*)

Substituting

$$
\begin{aligned}
M_{DA} &= -10.65 + k(13.17)\\
M_{AD} &= -21.32 + k(10.31)\\
M_{EB} &= +8.85 + k(25.25)\\
M_{BE} &= +17.70 + k(25.48)\\
M_{FC} &= -3.79 + k(71.37)\\
M_{CF} &= -7.61 + k(42.71)
\end{aligned}
$$

into the above equation and solving for the ratio k,

$$4(-31.97 + 23.48k) + 5(+26.55 + 50.73k) + 10(-11.40 + 114.08k) = 0$$

$$k = +0.07332$$

Combination of the Two Sets of Balanced Moments (see Table 79-5).

For reactions, shear and moment diagrams, and the elastic curve, see Example 63.

Example 80. Analyze the rigid frame shown in Fig. 219*a* by the moment-distribution method. Draw shear and moment diagrams. Sketch the deformed structure.

Solution. *Relative Stiffness*

TABLE 80-1. RELATIVE STIFFNESS

AB, CD	$\left(\frac{3I}{12} = \frac{I}{4}\right) \times 8$	2
AC, BD	$\left(\frac{2I}{16} = \frac{I}{8}\right) \times 8$	1
CE	$\left(\frac{4I}{16} = \frac{I}{4}\right) \times 8$	2
DF	$\left(\frac{2I}{8} = \frac{I}{4}\right) \times 8$	2

Distribution of FEM Due to Sidesway in Direction AB (see Fig. 219*b*)

TABLE 80-2. FEM DUE TO SIDESWAY IN DIRECTION AB

$M_{FAC} = M_{FCA}$	$+\frac{6E(2I)\Delta_1}{(16)^2}$	$+\frac{3}{64} \times \frac{6{,}400}{3}$	+100
$M_{FBD} = M_{FDB}$	$+\frac{6E(2I)\Delta_1}{(16)^2}$	$+\frac{3}{64} \times \frac{6{,}400}{3}$	+100

Note that only the relative magnitudes of the fixed-end moments due to a certain amount of sidesway Δ_1 are required.

For moment distribution of these fixed-end moments, see Table 80-3.

TABLE 80-3. DISTRIBUTION OF FEM DUE TO SIDESWAY IN DIRECTION AB

Joint		A		B		C			D			E	F
Member		AB	AC	BA	BD	CA	CD	CE	DB	DC	DF	EC	FD
K		2	1	2	1	1	2	2	1	2	2	2	2
Cycle	DF	0.667	0.333	0.667	0.333	0.200	0.400	0.400	0.200	0.400	0.400	—	—
1	FEM	—	+100.0	—	+100.0	+100.0	—	—	+100.0	—	—	—	—
	Bal.	−66.7	− 33.3	−66.7	− 33.3	− 20.0	−40.0	−40.0	− 20.0	−40.0	−40.0	—	—
2	CO	−33.3	− 10.0	−33.3	− 10.0	− 16.6	−20.0	—	− 16.6	−20.0	—	−20.0	−20.0
	Bal.	+28.9	+ 14.4	+28.9	+ 14.4	+ 7.3	+14.6	+14.7	+ 7.3	+14.6	+14.7	—	—
3	CO	+14.4	+ 3.6	+14.4	+ 3.6	+ 7.2	+ 7.3	—	+ 7.2	+ 7.3	—	+ 7.3	+ 7.3
	Bal.	−12.0	− 6.0	−12.0	− 6.0	− 2.9	− 5.8	− 5.8	− 2.9	− 5.8	− 5.8	—	—
4	CO	− 6.0	− 1.4	− 6.0	− 1.4	− 3.0	− 2.9	—	− 3.0	− 2.9	—	− 2.9	− 2.9
	Bal.	+ 4.9	+ 2.5	+ 4.9	+ 2.5	+ 1.2	+ 2.4	+ 2.3	+ 1.2	+ 2.4	+ 2.3	—	—
5	CO	+ 2.4	+ 0.6	+ 2.4	+ 0.6	+ 1.2	+ 1.2	—	+ 1.2	+ 1.2	—	+ 1.2	+ 1.2
	Bal.	− 2.0	− 1.0	− 2.0	− 1.0	− 0.5	− 0.9	− 1.0	− 0.5	− 0.9	− 1.0	—	—
6	CO	− 1.0	− 0.2	− 1.0	− 0.2	− 0.5	− 0.5	—	− 0.5	− 0.5	—	− 0.5	− 0.5
	Bal.	+ 0.8	+ 0.4	+ 0.8	+ 0.4	+ 0.2	+ 0.4	+ 0.4	+ 0.2	+ 0.4	+ 0.4	—	—
7	CO	+ 0.4	+ 0.1	+ 0.4	+ 0.1	+ 0.2	+ 0.2	—	+ 0.2	+ 0.2	—	+ 0.2	+ 0.2
	Bal.	− 0.3	− 0.2	− 0.3	− 0.2	− 0.1	− 0.2	− 0.1	− 0.1	− 0.2	− 0.1	—	—
8	CO	− 0.2	—	− 0.2	—	− 0.1	− 0.1	—	− 0.1	− 0.1	—	—	—
	Bal.	+ 0.1	+ 0.1	+ 0.1	+ 0.1	—	+ 0.1	+ 0.1	—	+ 0.1	+ 0.1	—	—
Total		−69.6	+ 69.6	+69.6	+ 69.6	+ 73.6	+44.2	−29.4	+ 73.6	−44.2	−29.4	−14.7	−14.7
Check: Change		−69.6	− 30.4	−69.6	− 30.4	− 26.4	−44.2	−29.4	− 26.4	−44.2	−29.4	−14.7	−14.7
$-\frac{1}{2}$ (change)		+34.8	+ 13.2	+34.8	+ 13.2	+ 15.2	+22.1	+ 7.4	+ 15.2	+22.1	+ 7.4	+14.7	+14.7
Sum		−34.8	− 17.2	−34.8	− 17.2	− 11.2	−22.1	−22.0	− 11.2	−22.1	−22.0	0	0
$\theta_{rel} = \text{sum}/(-K)$		+17.4	+ 17.2	+17.4	+ 17.2	+ 11.2	+11.0	+11.2	+ 11.2	+11.0	+11.0	0	0
		(*Check*)		(*Check*)			(*Check*)			(*Check*)		(*Check*)	(*Check*)

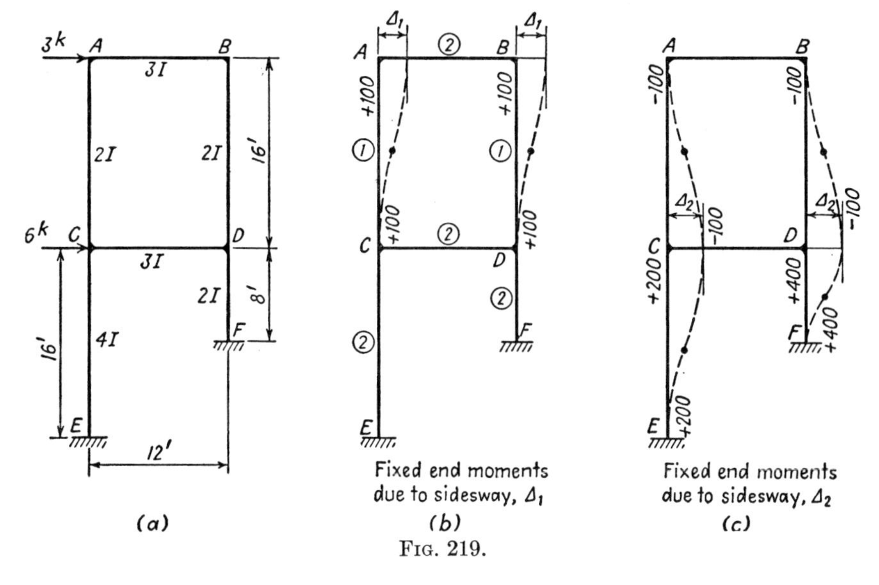

FIG. 219.

Distribution of FEM Due to Sidesway in Direction CD (see Fig. 219c)

TABLE 80-4. FEM DUE TO SIDESWAY IN DIRECTION CD

$M_{FAC} = M_{FCA}$........	$-\dfrac{6E(2I)\Delta_2}{(16)^2}$	$-\dfrac{3}{64} \times \dfrac{6{,}400}{3}$	-100
$M_{FBD} = M_{FDB}$........	$-\dfrac{6E(2I)\Delta_2}{(16)^2}$	$-\dfrac{3}{64} \times \dfrac{6{,}400}{3}$	-100
$M_{FCE} = M_{FEC}$........	$+\dfrac{6E(4I)\Delta_2}{(16)^2}$	$+\dfrac{3}{32} \times \dfrac{6{,}400}{3}$	$+200$
$M_{FDF} = M_{FFD}$........	$+\dfrac{6E(2I)\Delta_2}{(8)^2}$	$+\dfrac{3}{16} \times \dfrac{6{,}400}{3}$	$+400$

For moment distribution of these fixed-end moments, see Table 80-5.

Determination of Ratio. The shear conditions in the upper and lower stories are, respectively (Fig. 220),

$$H_C + H_D = 3 \qquad \text{or} \qquad \frac{M_{AC} + M_{CA}}{16} + \frac{M_{BD} + M_{DB}}{16} = 3$$

$$H_E + H_F = 9 \qquad \text{or} \qquad \frac{M_{CE} + M_{EC}}{16} + \frac{M_{DF} + M_{FD}}{8} = 9$$

Simplifying,

$$M_{AC} + M_{CA} + M_{BD} + M_{DB} = 48$$
$$M_{CE} + M_{EC} + 2(M_{DF} + M_{FD}) = 144$$

TABLE 80-5. DISTRIBUTION OF FEM DUE TO SIDESWAY IN DIRECTION CD

Joint		A		B		C			D			E	F
Member		AB	AC	BA	BD	CA	CD	CE	DB	DC	DF	EC	FD
K		2	1	2	1	1	2	2	1	2	2	2	2
Cycle	DF	0.667	0.333	0.667	0.333	0.200	0.400	0.400	0.200	0.400	0.400	—	—
1	FEM	—	−100.0	—	−100.0	−100.0	—	+200.0	−100.0	—	+400.0	+200.0	+400.0
	Bal.	+66.7	+ 33.3	+66.7	+ 33.3	− 20.0	−40.0	− 40.0	− 60.0	−120.0	−120.0	—	—
2	CO	+33.3	− 10.0	+33.3	− 30.0	+ 16.6	−60.0	—	+ 16.6	− 20.0	—	− 20.0	− 60.0
	Bal.	−15.5	− 7.8	− 2.2	− 1.1	+ 8.7	+17.3	+ 17.4	+ 0.7	+ 1.3	+ 1.4	—	—
3	CO	− 1.1	+ 4.4	− 7.8	+ 0.4	− 3.9	+ 0.6	—	− 0.6	+ 8.6	—	+ 8.7	+ 0.7
	Bal.	− 2.2	− 1.1	+ 4.9	+ 2.5	+ 0.7	+ 1.3	+ 1.3	− 1.6	− 3.2	− 3.2	—	—
4	CO	+ 2.4	+ 0.4	− 1.1	− 0.8	− 0.6	− 1.6	—	+ 1.2	+ 0.6	—	+ 0.6	− 1.6
	Bal.	− 1.9	− 0.9	+ 1.3	+ 0.6	+ 0.4	+ 0.9	+ 0.9	− 0.4	− 0.7	− 0.7	—	—
5	CO	+ 0.6	+ 0.2	− 1.0	− 0.2	− 0.4	− 0.4	—	+ 0.3	+ 0.4	—	+ 0.4	− 0.4
	Bal.	− 0.5	− 0.3	+ 0.8	+ 0.4	+ 0.2	+ 0.3	+ 0.3	− 0.1	− 0.3	− 0.3	—	—
6	CO	+ 0.4	+ 0.1	− 0.2	—	− 0.2	− 0.2	—	+ 0.2	+ 0.2	—	+ 0.2	− 0.2
	Bal.	− 0.3	− 0.2	+ 0.1	+ 0.1	+ 0.1	+ 0.2	+ 0.1	− 0.1	− 0.2	− 0.1	—	—
7	CO	—	—	− 0.2	—	− 0.1	− 0.1	—	—	+ 0.1	—	—	—
	Bal.	—	—	+ 0.1	+ 0.1	—	+ 0.1	+ 0.1	—	− 0.1	—	—	—
Total		+81.9	− 81.9	+94.7	− 94.7	− 98.5	−81.6	+180.1	−143.8	−133.3	+277.1	+189.9	+338.5
Check:													
Change		+81.9	+ 18.1	+94.7	+ 5.3	+ 1.5	−81.6	− 19.9	− 43.8	−133.3	−122.9	− 10.1	− 61.5
$-\frac{1}{2}$ (change)		−47.4	− 0.8	−41.0	− 21.9	− 9.0	+66.6	+ 5.0	− 2.6	+ 40.8	+ 30.8	+ 10.0	+ 61.5
Sum		+34.5	+ 17.3	+53.7	+ 27.2	− 7.5	−15.0	− 14.9	− 46.4	− 92.5	− 92.1	0	0
θ_{rel} = sum/($-K$)		−17.2	− 17.3	−26.8	− 27.2	+ 7.5	+ 7.5	+ 7.5	+ 46.4	+ 46.2	+ 46.0	0	0
		(*Check*)		(*Check*)			(*Check*)			(*Check*)		(*Check*)	(*Check*)

TABLE 80-6. COMBINATION OF THE TWO SETS OF BALANCED MOMENTS

Joint	A		B		C			D			E	F
Member	AB	AC	BA	BD	CA	CD	CE	DB	DC	DF	EC	FD
0.3403 times FEM in Table 80-3	—	+34.0	—	+34.0	+34.0	—	—	+34.0	—	—	—	—
0.1181 times FEM in Table 80-5	—	−11.8	—	−11.8	−11.8	—	+23.6	−11.8	—	+47.2	+23.6	+47.2
FEM	—	+22.2	—	+22.2	+22.2	—	+23.6	+22.2	—	+47.2	+23.6	+47.2
0.3403 times balanced moments in Table 80-3	−23.7	+23.7	−23.7	+23.7	+25.0	−15.0	−10.0	+25.0	−15.0	−10.0	− 5.0	− 5.0
0.1181 times balanced moments in Table 80-5	+ 9.7	− 9.7	+11.2	−11.2	−11.6	− 9.6	+21.2	−17.0	−15.7	+32.7	+22.4	+40.0
Balanced moments	−14.0	+14.0	−12.5	+12.5	+13.4	−24.6	+11.2	+ 8.0	−30.7	+22.7	+17.4	+35.0
Check:												
Change	−14.0	− 8.2	−12.5	− 9.7	− 8.8	−24.6	−12.4	−14.2	−30.7	−24.5	− 6.2	−12.2
$-\frac{1}{2}$ (change)	+ 6.2	+ 4.4	+ 7.0	+ 7.1	+ 4.1	+15.4	+ 3.1	+ 4.8	+12.3	+ 6.1	+ 6.2	+12.2
Sum	− 7.8	− 3.8	− 5.5	− 2.6	− 4.7	− 9.2	− 9.3	− 9.4	−18.4	−18.4	0	0
θ_{rel} = sum/$(-K)$	+ 3.9	+ 3.8	+ 2.7	+ 2.6	+ 4.7	+ 4.6	+ 4.6	+ 9.4	+ 9.2	+ 9.2	0	0
	(*Check*)		(*Check*)			(*Check*)			(*Check*)		(*Check*)	(*Check*)

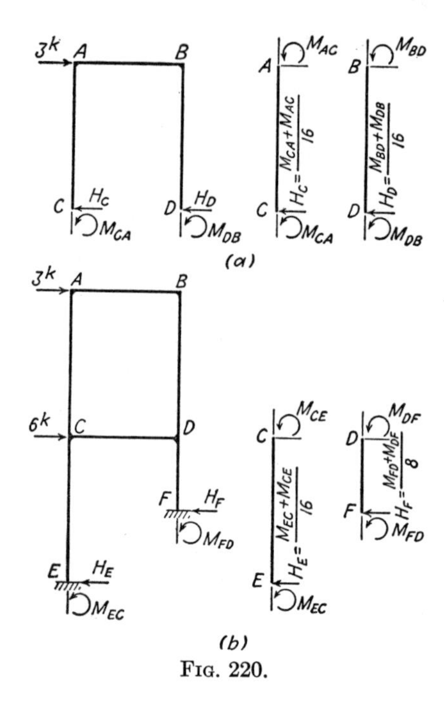

FIG. 220.

Substituting

$$
\begin{aligned}
M_{AC} &= +69.6k_1 - 81.9k_2 & M_{CE} &= -29.4k_1 + 180.1k_2 \\
M_{CA} &= +73.6k_1 - 98.5k_2 & M_{EC} &= -14.7k_1 + 189.9k_2 \\
M_{BD} &= +69.6k_1 - 94.7k_2 & M_{DF} &= -29.4k_1 + 277.1k_2 \\
M_{DB} &= +73.6k_1 - 143.8k_2 & M_{FD} &= -14.7k_1 + 338.5k_2
\end{aligned}
$$

into the above equations and solving for the ratios k_1 and k_2,

$$
\begin{aligned}
286.4k_1 - 418.9k_2 &= 48 \\
-132.3k_1 + 1{,}601.2k_2 &= 144 \\
k_1 &= +0.3403 \\
k_2 &= +0.1181
\end{aligned}
$$

Combination of the Two Sets of Balanced Moments (see Table 80-6).

For reactions, shear and moment diagrams, and the elastic curve, see Example 64.

EXERCISES

122 to 133. Solve Exercises 85 to 96 by the moment-distribution method.

52. Analysis of Statically Indeterminate Frames Due to Yielding of Supports. The moment-distribution method can readily be used to analyze statically indeterminate frames due to the yielding of supports.

The procedure is to distribute and balance the fixed-end moments due to each yielding. If the yielding is a rotational slip of a certain joint, the fixed-end moments are the moments required to maintain this rotational slip. If the yielding is a linear displacement, the fixed-end moments are the moments required to hold the tangents at the joints fixed in direction. Thus the end moments required to maintain a rotation of

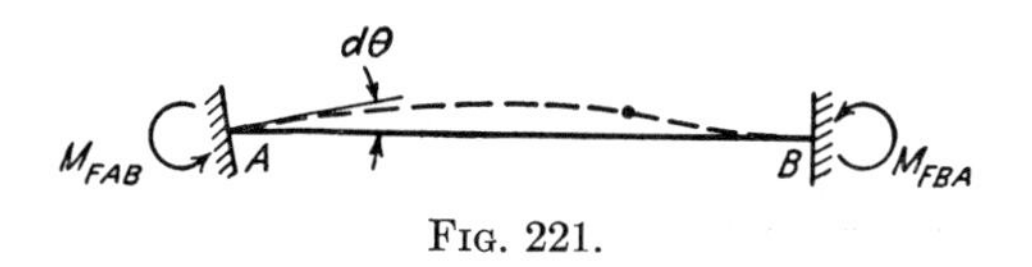

Fig. 221.

$d\theta$ rad in the counterclockwise direction at end A of member AB are (Fig. 221),

$$M_{FAB} = +\frac{4EI}{L}\,d\theta \qquad \text{and} \qquad M_{FBA} = +\frac{2EI}{L}\,d\theta \tag{102}$$

The fixed-end moments required to hold the tangents at ends A and B of member AB when B settles an amount of Δ relative to A are (Fig. 222)

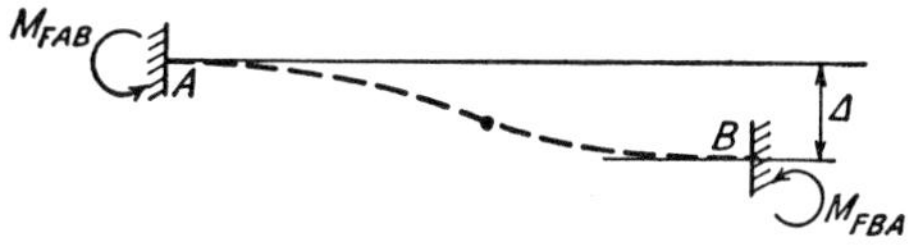

Fig. 222.

$$M_{FAB} = +\frac{6EI\Delta}{L^2} \qquad M_{FBA} = +\frac{6EI\Delta}{L^2} \tag{103}$$

Equations (102) and (103) can be easily verified by the use of the slope-deflection equations

$$\begin{aligned} M_{AB} &= M_{FAB} + \frac{2EI}{L}\left(-2\theta_A - \theta_B + \frac{3\Delta}{L}\right) \\ M_{BA} &= M_{FBA} + \frac{2EI}{L}\left(-2\theta_B - \theta_A + \frac{3\Delta}{L}\right) \end{aligned} \tag{66}$$

Example 81. By the moment-distribution method analyze the rigid frame shown in Fig. 223a owing to a rotational slip of 0.002 rad clockwise of joint D and a vertical settlement of 0.45 in. at joint D.

$$E = 30{,}000 \text{ kips/in.}^2 \qquad I = 800 \text{ in.}^4$$

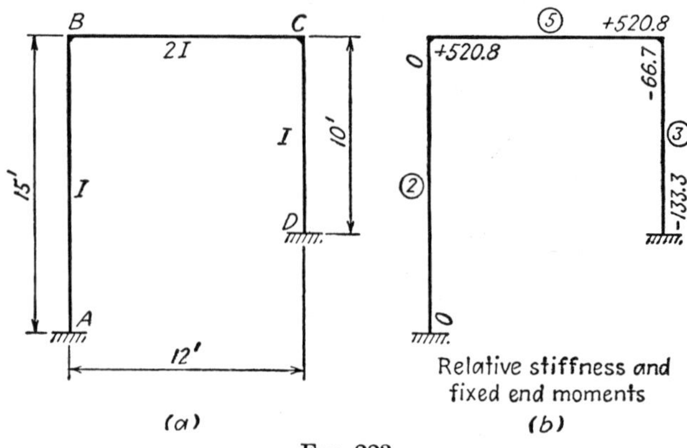

FIG. 223.

Solution. *Relative Stiffness* (Fig. 223*b*)

TABLE 81-1. RELATIVE STIFFNESS

AB...........	$\frac{I}{15} \times 30$	2
BC...........	$\left(\frac{2I}{12} = \frac{I}{6}\right) \times 30$	5
CD...........	$\frac{I}{10} \times 30$	3

FEM Due to a Clockwise Rotation of 0.002 *Rad at D* (Fig. 223*b*)

$$M_{FDC} = +\frac{4EI\,d\theta}{L} = +\frac{(4)(30{,}000)(800)(-0.002)}{(144)(10)} = -133.3 \text{ kip-ft}$$

$$M_{FCD} = +\frac{2EI\,d\theta}{L} = +\frac{(2)(30{,}000)(800)(-0.002)}{(144)(10)} = -66.7 \text{ kip-ft}$$

FEM Due to a Vertical Settlement of 0.45 *In. at D* (Fig. 223*b*)

$$M_{FBC} = +\frac{6EI\Delta}{L^2} = +\frac{(6)(30{,}000)(1{,}600)(0.45)}{(1{,}728)(12)^2} = +520.8 \text{ kip-ft}$$

$$M_{FCB} = +520.8 \text{ kip-ft}$$

The distribution of the fixed-end moments computed above is shown in Table 81-2. Note that the relative value of the rotation at D is found to be zero in the check. This does not mean that the tangent at D remains vertical; it means that since the tangent at D had the initial rotational slip of 0.002 rad in the clockwise direction, no further rotation has occurred.

$U_0U_1 \quad \Sigma M_{L_1} = 0$

$$U_0U_1\left(\frac{20}{3}\right) = 10 \times 16$$

$$U_0U_1 = \frac{48}{2} = 24$$

$$\frac{24}{\sqrt{676}} L_2U_3 = 24$$

$$U_2U_3 = \frac{24}{24}\sqrt{676} = 26$$

10K

U_0 8' U_1 8' U_2 8' U_3

(5)

L_2

10'

L_1

L_0

U_2U_3

$$\Sigma M_{L_2} = 0$$

$$U_2U_3 \frac{10}{3} = 10 \times 8$$

$$U_2U_3 = 24$$

TABLE 81-2. DISTRIBUTION OF FEM DUE TO YIELDING OF SUPPORTS

Joint		A	B		C		D
Member		AB	BA	BC	CB	CD	DC
K		2	2	5	5	3	3
Cycle	DF	——	0.286	0.714	0.625	0.375	——
1	FEM	——	——	+520.8	+520.8	− 66.7	−133.3
	Bal.	——	−148.9	−371.9	−283.8	−170.3	——
2	CO	−74.4	——	−141.9	−186.0	——	− 85.2
	Bal.	——	+ 40.6	+101.3	+116.2	+ 69.8	——
3	CO	+20.3	——	+ 58.1	+ 50.6	——	+ 34.9
	Bal.	——	− 16.6	− 41.5	− 31.6	− 19.0	——
4	CO	− 8.3	——	− 15.8	− 20.8	——	− 9.5
	Bal.	——	+ 4.5	+ 11.3	+ 13.0	+ 7.8	——
5	CO	+ 2.2	——	+ 6.5	+ 5.6	——	+ 3.9
	Bal.	——	− 1.9	− 4.6	− 3.5	− 2.1	——
6	CO	− 1.0	——	− 1.8	− 2.3	——	− 1.0
	Bal.	——	+ 0.5	+ 1.3	+ 1.4	+ 0.9	——
7	CO	+ 0.2	——	+ 0.7	+ 0.6	——	+ 0.4
	Bal.	——	− 0.2	− 0.5	− 0.4	− 0.2	——
8	CO	− 0.1	——	− 0.2	− 0.2	——	− 0.1
	Bal.	——	+ 0.1	+ 0.1	+ 0.1	+ 0.1	——
Total		−61.1	−121.9	+121.9	+179.7	−179.7	−189.9
Check:							
Change		−61.1	+121.9	−398.9	−341.1	−113.0	− 56.6
$-\frac{1}{2}$ (change)		+61.0	+ 30.6	+170.6	+199.4	+ 28.3	+ 56.5
Sum		0	− 91.3	−228.3	−141.7	− 84.7	0
θ_{rel} = sum/(−K)		0	+ 45.6	+ 45.7	+ 28.3	+ 28.2	0
		(*Check*)	(*Check*)		(*Check*)		(*Check*)

Distribution of FEM Due to Sidesway (see Table 78-4 in Example 78)

TABLE 81-3. DISTRIBUTION OF FEM DUE TO SIDESWAY

Joint	*A*	*B*		*C*		*D*
Member	*AB*	*BA*	*BC*	*CB*	*CD*	*DC*
FEM	+400.0	+400.0	——	——	+900.0	+900.0
Balanced moments	+380.8	+361.7	−361.7	−580.6	+580.6	+740.3

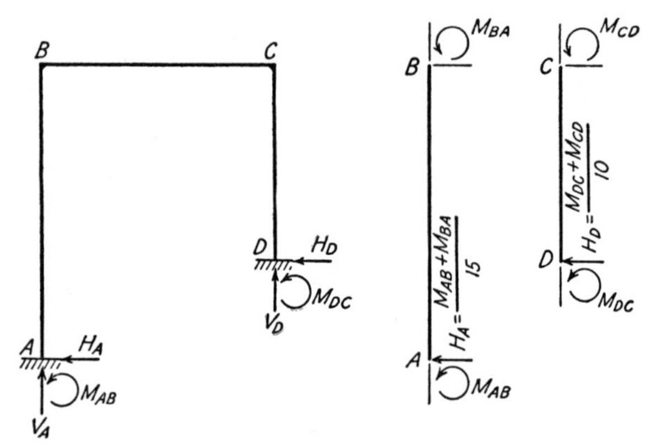

FIG. 224.

Determination of Ratio. The shear condition (Fig. 224) is

$$H_A + H_D = 0 \qquad \text{or} \qquad \left(\frac{M_{AB} + M_{BA}}{15}\right) + \left(\frac{M_{DC} + M_{CD}}{10}\right) = 0$$

Simplifying,

$$2(M_{AB} + M_{BA}) + 3(M_{CD} + M_{DC}) = 0$$

Substituting

$$\begin{aligned} M_{AB} &= -61.1 + 380.8k \\ M_{BA} &= -121.9 + 361.7k \\ M_{CD} &= -179.7 + 580.6k \\ M_{DC} &= -189.9 + 740.3k \end{aligned}$$

into the above equation and solving for the ratio k,

$$2(-183.0 + 742.5k) + 3(-369.6 + 1{,}320.9k) = 0$$
$$k = +0.2707$$

Combination of the Two Sets of Balanced Moments (see Table 81-4)

TABLE 81-4. COMBINATION OF THE TWO SETS OF BALANCED MOMENTS

Joint	A	B		C		D
Member	AB	BA	BC	CB	CD	DC
FEM from Table 81-2	——	——	+520.8	+520.8	− 66.7	−133.3
+0.2707 times FEM in Table 81-3	+108.3	+108.3	——	——	+243.6	+243.6
Total FEM	+108.3	+108.3	+520.8	+520.8	+176.9	+110.3
Balanced moments from Table 81-2	− 61.1	−121.9	+121.9	+179.7	−179.7	−189.9
+0.2707 times balanced moments in Table 81-3	+103.1	+ 97.9	− 97.9	−157.2	+157.2	+200.4
Total balanced moments	+ 42.0	− 24.0	+ 24.0	+ 22.5	− 22.5	+ 10.5
Check:						
Change	− 66.3	−132.3	−496.8	−498.3	−199.4	− 99.8
$-\frac{1}{2}$ (change)	+ 66.2	+ 33.2	+249.2	+248.4	+ 49.9	+ 99.7
Sum	0	− 99.1	−247.6	−249.9	−149.5	0
$\theta_{rel} = \text{sum}/(-K)$	0	+ 49.6	+ 49.5	+ 50.0	+ 49.8	0
	(*Check*)	(*Check*)		(*Check*)		(*Check*)

For reactions, shear and moment diagrams, and the elastic curve, see Example 65.

EXERCISES

134 and 135. Solve Exercises 97 and 98 by the moment-distribution method.

53. Analysis of Gable Frames by the Moment-distribution Method. The moment-distribution method can be used to analyze gable frames in the same manner as described in the preceding articles. The comments in Art. 43 regarding fixed-end moments and relative amounts of sidesway still apply in the present method as in the slope-deflection method.

Example 82. Analyze the gable frame shown in Fig. 225*a* by the moment-distribution method. Draw shear and moment diagrams. Sketch the deformed structure.

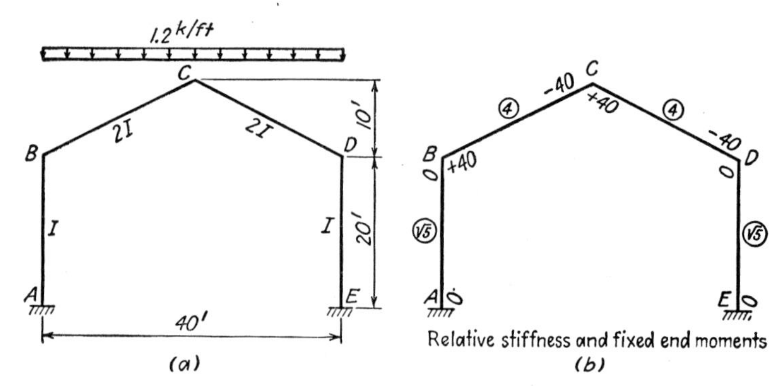

FIG. 225.

Solution. *Relative Stiffness* (see Fig. 225*b*)

TABLE 82-1. RELATIVE STIFFNESS

AB, DE	$\frac{I}{20} \times 20\sqrt{5}$	$\sqrt{5}$
BC, CD	$\frac{2I}{10\sqrt{5}} \times 20\sqrt{5}$	4

Fixed-end Moments (see Example 66)

$$M_{FBC} = M_{FCD} = +\frac{(1.2)(20)^2}{12} = +40 \text{ kip-ft}$$
$$M_{FCB} = M_{FDC} = -40 \text{ kip-ft}$$

For moment distribution of these fixed-end moments, see Table 82-2.

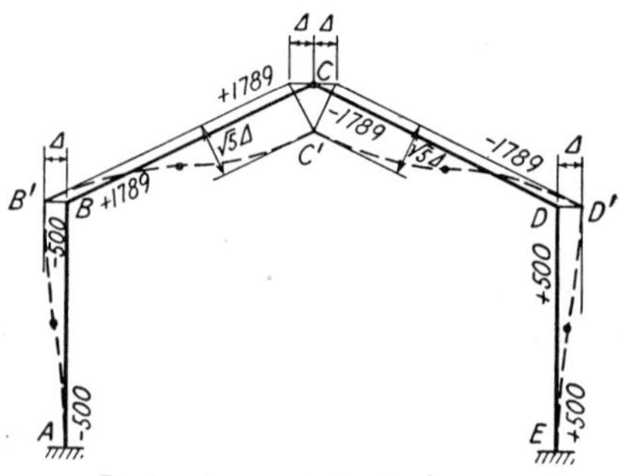

FIG. 226.

TABLE 82-2. DISTRIBUTION OF FEM DUE TO THE APPLIED LOADING

Joint		A	B		C		D		E
Member		AB	BA	BC	CB	CD	DC	DE	ED
K		$\sqrt{5}$	$\sqrt{5}$	4	4	4	4	$\sqrt{5}$	$\sqrt{5}$
Cycle	DF	——	0.3585	0.6415	0.5000	0.5000	0.6415	0.3585	——
1	FEM	——	——	+40.00	−40.00	+40.00	−40.00	——	——
	Bal.	——	−14.34	−25.66	——	——	+25.66	+14.34	——
2	CO	−7.17	——	——	−12.83	+12.83	——	——	+7.17
	Bal.	——	——	——	——	——	——	——	——
Total		−7.17	−14.34	+14.34	−52.83	+52.83	−14.34	+14.34	+7.17
Check:									
Change		−7.17	−14.34	−25.66	−12.83	+12.83	+25.66	+14.34	+7.17
$-\frac{1}{2}$(change)		+7.17	+ 3.58	+ 6.42	+12.83	−12.83	− 6.42	− 3.58	−7.17
Sum		0	−10.76	−19.24	0	0	+19.24	+10.76	0
θ_{rel} = sum/(−K)		0	+ 4.81	+ 4.81	0	0	− 4.81	− 4.81	0
		(*Check*)	(*Check*)		(*Checks*)		(*Check*)		(*Check*)

Distribution of FEM Due to Sidesway (*Fig.* 226) (see Example 66)

TABLE 82-3. FEM DUE TO SIDESWAY

$M_{FAB} = M_{FBA}$	$-\dfrac{6E(I)(\Delta)}{(20)^2} \times \dfrac{2{,}000}{6}$	-5	$-\ 500$
$M_{FBC} = M_{FCB}$	$+\dfrac{6E(2I)(\sqrt{5}\,\Delta)}{(10\sqrt{5})^2} \times \dfrac{2{,}000}{6}$	$+8\sqrt{5}$	$+1{,}789$
$M_{FCD} = M_{FDC}$	$-\dfrac{6E(2I)(\sqrt{5}\,\Delta)}{(10\sqrt{5})^2} \times \dfrac{2{,}000}{6}$	$-8\sqrt{5}$	$-1{,}789$
$M_{FDE} = M_{FED}$	$+\dfrac{6E(I)(\Delta)}{(20)^2} \times \dfrac{2{,}000}{6}$	$+5$	$+\ 500$

For moment distribution of these fixed-end moments, see Table 82-4.
Determination of Ratio. The shear condition is (Fig. 227)

$$H_{BA} + H_{BC} = 0 \qquad \text{or} \qquad \frac{M_{AB} + M_{BA}}{20} + \frac{240 - (M_{BC} + M_{CB})}{10} = 0$$

Simplifying, $(M_{AB} + M_{BA}) - 2(M_{BC} + M_{CB}) = -480$

TABLE 82-4. DISTRIBUTION OF FEM DUE TO SIDESWAY

Joint		A	B		C		D		E
Member		*AB*	*BA*	*BC*	*CB*	*CD*	*DC*	*DE*	*ED*
K		$\sqrt{5}$	$\sqrt{5}$	4	4	4	4	$\sqrt{5}$	$\sqrt{5}$
Cycle	DF	———	0.3585	0.6415	0.5000	0.5000	0.6415	0.3585	———
1	FEM	−500	−500	+1,789	+1,789	−1,789	−1,789	+500	+500
	Bal.	———	−462	− 827	———	———	+ 827	+462	———
2	CO	−231	———	———	− 414	+ 414	———	———	+231
	Bal.	———	———	———	———	———	———	———	———
Total		−731	−962	+ 962	+1,375	−1,375	− 962	+962	+731
Check:									
Change		−231	−462	− 827	− 414	+ 414	+ 827	+462	+231
$-\frac{1}{2}$ (change)		+231	+116	+ 207	+ 414	− 414	− 207	−116	−231
Sum		0	−346	− 620	0	0	+ 620	+346	0
θ_{rel} = sum/(−*K*)		0	+155	+ 155	0	0	− 155	−155	0
		(*Check*)	(*Check*)		(*Check*)		(*Check*)		(*Check*)

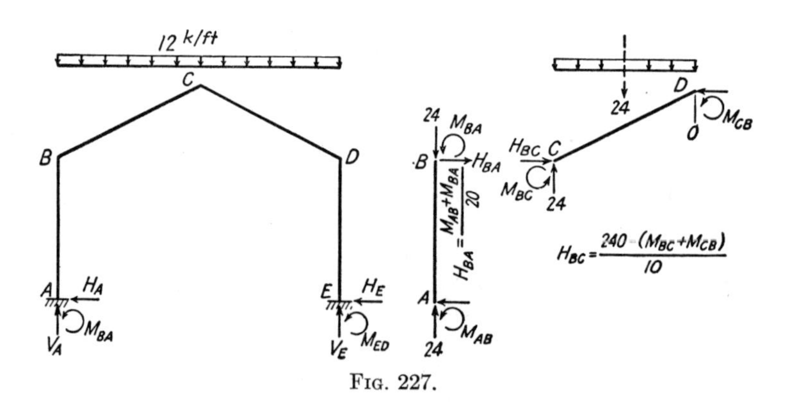

FIG. 227.

Substituting

$$M_{AB} = -7.17 - 731k$$
$$M_{BA} = -14.34 - 962k$$
$$M_{BC} = +14.34 + 962k$$
$$M_{CB} = -52.83 + 1{,}375k$$

TABLE 82-5. COMBINATION OF THE TWO SETS OF BALANCED MOMENTS

Joint	A	B		C		D		E
Member	*AB*	*BA*	*BC*	*CB*	*CD*	*DC*	*DE*	*ED*
FEM from Table 82-2	——	——	+ 40.00	− 40.00	+ 40.00	− 40.00	——	——
+0.0841 times FEM in Table 82-4	−42.05	−42.05	+150.45	+150.45	−150.45	−150.45	+42.05	+42.05
Total FEM	−42.05	−42.05	+190.45	+110.45	−110.45	−190.45	+42.05	+42.05
Balanced moments from Table 82-2	− 7.17	−14.34	+ 14.34	− 52.83	+ 52.83	− 14.34	+14.34	+ 7.17
+0.0841 times balanced moments in Table 82-4	−61.48	−80.90	+ 80.90	+115.64	−115.64	− 80.90	+80.90	+61.48
Total balanced moments	−68.65	−95.24	+ 95.24	+ 62.81	− 62.81	− 95.24	+95.24	+68.65
Check:								
Change	−26.60	−53.19	− 95.21	− 47.64	+ 47.64	+ 95.21	+53.19	+26.60
$-\frac{1}{2}$ (change)	+26.60	+13.30	+ 23.82	+ 47.60	− 47.60	− 23.82	−13.30	−26.60
Sum	0	−39.89	− 71.39	0	0	+ 71.39	+39.89	0
θ_{rel} = sum/(−K)	0	+17.84	+ 17.85	0	0	− 17.85	−17.84	0
	(*Check*)	(*Check*)		(*Check*)		(*Check*)		(*Check*)

into the above equation and solving for the ratio k,

$$(-21.51 - 1{,}693k) - 2(-38.49 + 2{,}337k) = -480$$
$$k = +0.0841$$

Combination of the Two Sets of Balanced Moments (see Table 82-5).

For reactions, shear and moment diagrams, and the elastic curve, see Example 66.

Example 83. Analyze the gable frame shown in Fig. 228*a* by the moment-distribution method. Draw shear and moment diagrams. Sketch the deformed structure.

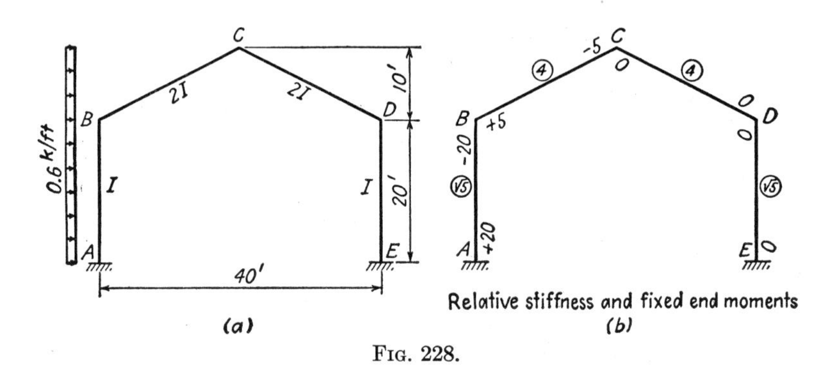

FIG. 228.

Solution. *Relative Stiffness* (Fig. 228*b*)

TABLE 83-1. RELATIVE STIFFNESS

AB, DE..........	$\frac{I}{20} \times 20\sqrt{5}$	$\sqrt{5}$
BC, CD.....	$\frac{2I}{10\sqrt{5}} \times 20\sqrt{5}$	4

Fixed-end Moments (Fig. 228*b*)

$$M_{FAB} = +\frac{(0.6)(20)^2}{12} = +20 \text{ kip-ft}$$
$$M_{FBA} = -20 \text{ kip-ft}$$
$$M_{FBC} = +\frac{(0.6)(10)^2}{12} = +5 \text{ kip-ft}$$
$$M_{FCB} = -5 \text{ kip-ft}$$

Distribution of FEM Due to the Applied Loading (see Table 83-2)

TABLE 83-2. DISTRIBUTION OF FEM DUE TO THE APPLIED LOADING

Joint		*A*	*B*		*C*		*D*		*E*
Member		*AB*	*BA*	*BC*	*CB*	*CD*	*DC*	*DE*	*ED*
K		$\sqrt{5}$	$\sqrt{5}$	4	4	4	4	$\sqrt{5}$	$\sqrt{5}$
Cycle	DF	——	0.3585	0.6415	0.5000	0.5000	0.6415	0.3585	——
1	FEM	+20.00	−20.00	+5.00	−5.00	——	——	——	——
	Bal.	——	+5.38	+9.62	+2.50	+2.50	——	——	——
2	CO	+2.69	——	+1.25	+4.81	——	+1.25	——	——
	Bal.	——	−0.45	−0.80	−2.41	−2.40	−0.80	−0.45	——
3	CO	−0.22	——	+1.20	−0.40	−0.40	−1.20	——	−0.22
	Bal.	——	+0.43	+0.77	+0.40	+0.40	+0.77	+0.43	——
4	CO	+0.22	——	+0.20	+0.38	+0.38	+0.20	——	+0.22
	Bal.	——	−0.07	−0.13	−0.38	−0.38	−0.13	−0.07	——
5	CO	−0.04	——	−0.19	−0.06	−0.06	−0.19	——	−0.04
	Bal.	——	+0.07	+0.12	+0.06	+0.06	+0.12	+0.07	——
6	CO	+0.04	——	+0.03	+0.06	+0.06	+0.03	——	+0.04
	Bal.	——	−0.01	−0.02	−0.06	−0.06	−0.02	−0.01	——
7	CO	——	——	−0.03	−0.01	−0.01	−0.03	——	——
	Bal.	——	+0.01	+0.02	+0.01	+0.01	+0.02	+0.01	——
8	CO	——	——	——	+0.01	+0.01	——	——	——
	Bal.	——	——	——	−0.01	−0.01	——	——	——
Total		+22.69	−14.64	+14.64	−0.10	+0.10	+0.02	−0.02	0
Check:									
Change		+2.69	+5.36	+9.64	+4.90	+0.10	+0.02	−0.02	0
$-\frac{1}{2}$ (change)		−2.68	−1.34	−2.45	−4.82	−0.01	−0.05	0	+0.01
Sum		0	+4.02	+7.19	+0.08	+0.09	−0.03	−0.02	0
$\theta_{rel} = \text{sum}/(-K)$		0	−1.80	−1.80	−0.02	−0.02	+0.01	+0.01	0
		(*Check*)	(*Check*)		(*Check*)		(*Check*)		(*Check*)

Distribution of FEM Due to Sidesway at B (Fig. 229)

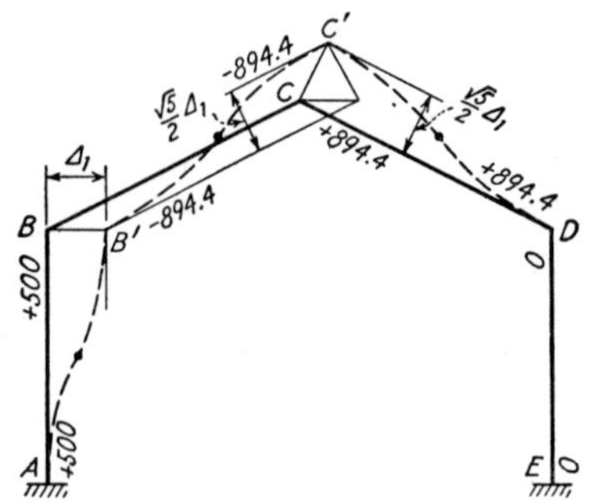

Fixed end moments due to sidesway at *B*

FIG. 229.

TABLE 83-3. FEM DUE TO SIDESWAY AT *B*

$M_{FAB} = M_{FBA}$	$+\frac{6E(I)\Delta_1}{(20)^2} \times \frac{2{,}000}{6}$	$+5$	$+500.0$
$M_{FBC} = M_{FCB}$	$-\frac{6E(2I)(\sqrt{5}\,\Delta_1/2)}{(10\sqrt{5})^2} \times \frac{2{,}000}{6}$	$-4\sqrt{5}$	-894.4
$M_{FCD} = M_{FDC}$	$+\frac{6E(2I)(\sqrt{5}\,\Delta_1/2)}{(10\sqrt{5})^2} \times \frac{2{,}000}{6}$	$+4\sqrt{5}$	$+894.4$
$M_{FDE} = M_{FED}$	0	0	0

For moment distribution of these fixed-end moments, see Table 83-4.

Distribution of FEM Due to Sidesway at D (Fig. 230)

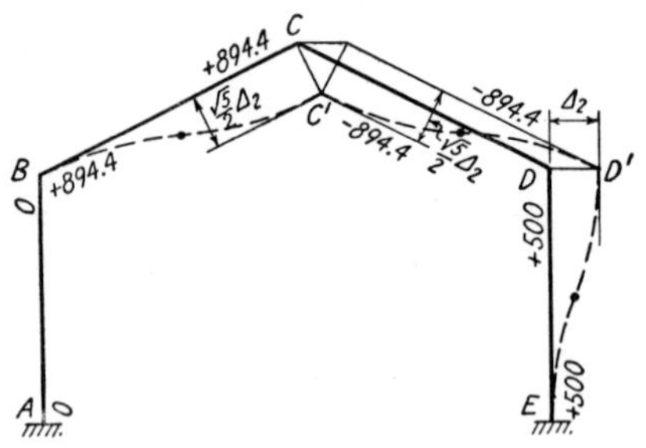

Fixed end moments due to sidesway at *D*

FIG. 230.

TABLE 83-4. DISTRIBUTION OF FEM DUE TO SIDESWAY AT B

Joint		A	B		C		D		E
Member		AB	BA	BC	CB	CD	DC	DE	ED
K		$\sqrt{5}$	$\sqrt{5}$	4	4	4	4	$\sqrt{5}$	$\sqrt{5}$
Cycle	DF	—	0.3585	0.6415	0.5000	0.5000	0.6415	0.3585	—
1	FEM	+500.0	+500.0	−894.4	−894.4	+894.4	+894.4	—	—
	Bal.	—	+141.4	+253.0	—	—	−573.8	−320.6	—
2	CO	+ 70.7	—	—	+126.5	−286.9	—	—	−160.3
	Bal.	—	—	—	+ 80.2	+ 80.2	—	—	—
3	CO	—	—	+ 40.1	—	—	+ 40.1	—	—
	Bal.	—	− 14.4	− 25.7	—	—	− 25.7	− 14.4	—
4	CO	− 7.2	—	—	− 12.8	− 12.8	—	—	− 7.2
	Bal.	—	—	—	+ 12.8	+ 12.8	—	—	—
5	CO	—	—	+ 6.4	—	—	+ 6.4	—	—
	Bal.	—	− 2.3	− 4.1	—	—	− 4.1	− 2.3	—
6	CO	− 1.2	—	—	− 2.0	− 2.0	—	—	− 1.2
	Bal.	—	—	—	+ 2.0	+ 2.0	—	—	—
7	CO	—	—	+ 1.0	—	—	+ 1.0	—	—
	Bal.	—	− 0.4	− 0.6	—	—	− 0.6	− 0.4	—
8	CO	− 0.2	—	—	− 0.3	− 0.3	—	—	− 0.2
	Bal.	—	—	—	+ 0.3	+ 0.3	—	—	—
9	CO	—	—	+ 0.2	—	—	+ 0.2	—	—
	Bal.	—	− 0.1	− 0.1	—	—	− 0.1	− 0.1	—
Total		+562.1	+624.2	−624.2	−687.7	+687.7	+337.8	−337.8	−168.9
Check:									
Change		+ 62.1	+124.2	+270.2	+206.7	−206.7	−556.6	−337.8	−168.9
$-\frac{1}{2}$ (change)		− 62.1	− 31.0	−103.4	−135.1	+278.3	+103.4	+ 84.4	+168.9
Sum		0	+ 93.2	+166.8	+ 71.6	+ 71.6	−453.2	−253.4	0
θ_{rel} = sum/ $(-K)$		0	− 41.7	− 41.7	− 17.9	− 17.9	+113.3	+113.3	0
		(*Check*)	(*Check*)		(*Check*)		(*Check*)		(*Check*)

TABLE 83-5. FEM DUE TO SIDESWAY AT D

$M_{FAB} = M_{FBA}$........	0	0	0
$M_{FBC} = M_{FCB}$........	$+\dfrac{6E(2I)(\sqrt{5}\,\Delta_2/2)}{(10\sqrt{5})^2} \times \dfrac{2{,}000}{6}$	$+4\sqrt{5}$	+894.4
$M_{FCD} = M_{FDC}$........	$-\dfrac{6E(2I)(\sqrt{5}\,\Delta_2/2)}{(10\sqrt{5})^2} \times \dfrac{2{,}000}{6}$	$-4\sqrt{5}$	−894.4
$M_{FDE} = M_{FED}$........	$+\dfrac{6E(I)(\Delta_2)}{(20)^2} \times \dfrac{2{,}000}{6}$	+5	+500.0

For moment distribution of these fixed-end moments see Table 83-6.

TABLE 83-6. DISTRIBUTION OF FEM DUE TO SIDESWAY AT D

Joint..........		A	B		C		D		E
Member........		AB	BA	BC	CB	CD	DC	DE	ED
K............		$\sqrt{5}$	$\sqrt{5}$	4	4	4	4	$\sqrt{5}$	$\sqrt{5}$
Cycle	DF	——	0.3585	0.6415	0.5000	0.5000	0.6415	0.3585	——
1	FEM	——	——	+894.4	+894.4	−894.4	−894.4	+500.0	+500.0
					(Refer to Table 83-4)				
Total.........		−168.9	−337.8	+337.8	+687.7	−687.7	−624.2	+624.2	+562.1

Determination of the Ratios k_1 and k_2

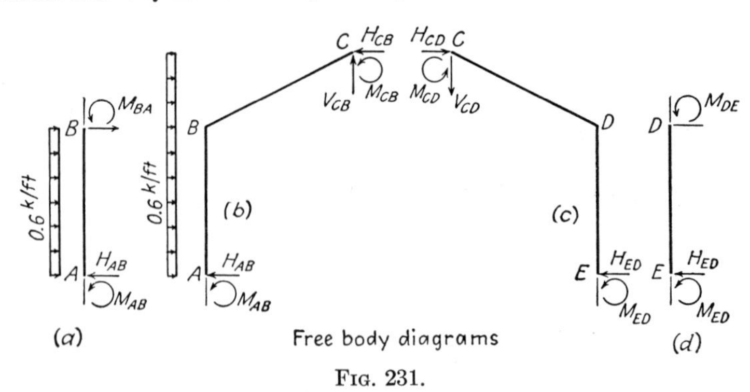

FIG. 231.

Fig. 231*a*: $\Sigma M_B = 0 \qquad H_{AB} = 6 + \dfrac{M_{AB} + M_{BA}}{20}$

Fig. 231*b*: $\Sigma F_x = 0 \qquad H_{CB} = 18 - H_{AB} = 12 - \dfrac{M_{AB} + M_{BA}}{20}$

Fig. 231*d*: $\Sigma M_D = 0 \qquad H_{ED} = \dfrac{M_{DE} + M_{ED}}{20}$

Fig. 231*c*: $\Sigma F_x = 0 \qquad H_{CD} = H_{ED} = \dfrac{M_{DE} + M_{ED}}{20}$

Equating H_{CB} and H_{CD},

$$12 - \frac{M_{AB} + M_{BA}}{20} = \frac{M_{DE} + M_{ED}}{20}$$

or

$$M_{AB} + M_{BA} + M_{DE} + M_{ED} = 240 \qquad (a)$$

Fig. 231*b*: $\Sigma M_A = 0$

$$V_{CB} = \frac{270 - H_{CB}(30) - (M_{AB} + M_{CB})}{20} = \frac{M_{AB} + 3M_{BA} - 2M_{CB} - 180}{40}$$

Fig. 231*c*: $\Sigma M_E = 0$

$$V_{CD} = \frac{30H_{CD} - (M_{CD} + M_{ED})}{20} = \frac{M_{ED} + 3M_{DE} - 2M_{CD}}{40}$$

Equating V_{CB} and V_{CD},

$$M_{AB} + 3M_{BA} - 2M_{CB} - 180 = M_{ED} + 3M_{DE} - 2M_{CD}$$

or

$$(M_{AB} - M_{ED}) + 3(M_{BA} - M_{DE}) - 2(M_{CB} - M_{CD}) = 180 \qquad (b)$$

Substituting

$$\begin{aligned}
M_{AB} &= +22.69 + 562.1k_1 - 168.9k_2 \\
M_{BA} &= -14.64 + 624.2k_1 - 337.8k_2 \\
M_{CB} &= -0.10 - 687.7k_1 + 687.7k_2 \\
M_{CD} &= +0.10 + 687.7k_1 - 687.7k_2 \\
M_{DE} &= -0.02 - 337.8k_1 + 624.2k_2 \\
M_{ED} &= 0 - 168.9k_1 + 562.1k_2
\end{aligned}$$

into the two conditions (*a*) and (*b*) above,

$$\begin{aligned}
679.6k_1 + 679.6k_2 &= 231.97 \\
6{,}367.8k_1 - 6{,}367.8k_2 &= 200.77
\end{aligned}$$

Solving for k_1 and k_2,

$$\begin{aligned}
k_1 &= +0.18643 \\
k_2 &= +0.15490
\end{aligned}$$

Combination of the Three Sets of Balanced Moments (see Table 83-7)

TABLE 83-7. COMBINATION OF THE THREE SETS OF BALANCED MOMENTS

Joint	A	B		C		D		E
Member	AB	BA	BC	CB	CD	DC	DE	ED
FEM (Table 83-2)	+ 20.0	− 20.00	+ 5.00	− 5.00	——	——	——	——
+0.18643 times FEM (Table 83-4)	+ 93.22	+ 93.22	−166.74	−166.74	+166.74	+166.74	——	——
+0.15490 times FEM (Table 83-6)	——	——	+138.54	+138.54	−138.54	−138.54	+77.45	+77.45
Total FEM	+113.22	+ 73.22	− 23.20	− 33.20	+ 28.20	+ 28.20	+77.45	+77.45
Balanced moments (Table 83-2)	+ 22.69	− 14.64	+ 14.64	− 0.10	+ 0.10	+ 0.02	− 0.02	0
+0.18643 times balanced moments (Table 83-4)	+104.79	+116.37	−116.37	−128.21	+128.21	+ 62.98	−62.98	−31.49
+0.15490 times balanced moments (Table 83-6)	− 26.16	− 52.32	+ 52.32	+106.52	−106.52	− 96.69	+96.69	+87.07
Total balanced moments	+101.32	+ 49.41	− 49.41	− 21.79	+ 21.79	− 33.69	+33.69	+55.58
Check:								
Change	− 11.90	− 23.81	− 26.21	+ 11.41	− 6.41	− 61.89	−43.76	−21.87
$-\frac{1}{2}$ (change)	+ 11.90	+ 5.95	− 5.70	+ 13.10	+ 30.94	+ 3.20	+10.94	+21.88
Sum	0	− 17.86	− 31.91	+ 24.51	+ 24.53	− 58.69	−32.82	0
θ_{rel} = sum/(−K)	0	+ 7.99	+ 7.98	− 6.13	− 6.13	+ 14.67	+14.68	0
	(*Check*)	(*Check*)		(*Check*)		(*Check*)		(*Check*)

For reactions, shear and moment diagrams, and the elastic curve, see Example 67.

EXERCISES

136 to 139. Solve Exercises 99 to 102 by the moment-distribution method.

54. Stiffness and Carry-over Factors for Members with Variable Cross Sections. The moment-distribution method can be applied to the analysis of rigid frames in which some or all members have variable cross sections. The stiffness and carry-over factors for members with variable cross sections will obviously be different from those for members with constant cross section, which are $4EI/L$ and $\frac{1}{2}$, respectively. In Fig. 232 is shown the member AB with moment of inertia $I_x = n_x I_c$, where I_c is a constant moment of inertia and n_x is a pure ratio which varies with x. The end moments M'_{AB} and M'_{BA} required to maintain a slope of θ_A at A and zero at B are $M'_{AB} = S_A\theta_A$ and $M'_{BA} = C_{AB}M'_{AB}$ in which S_A is the stiffness factor at A and C_{AB} is the carry-over factor from A to B. Similarly, in Fig. 233, the end moments M''_{BA} and M''_{AB} required to maintain a slope of θ_B at B and zero at A are $M''_{BA} = S_B\theta_B$ and $M''_{AB} = C_{BA}M''_{BA}$, in which S_B is the stiffness factor at B and C_{BA} is the carry-over factor from B to A.

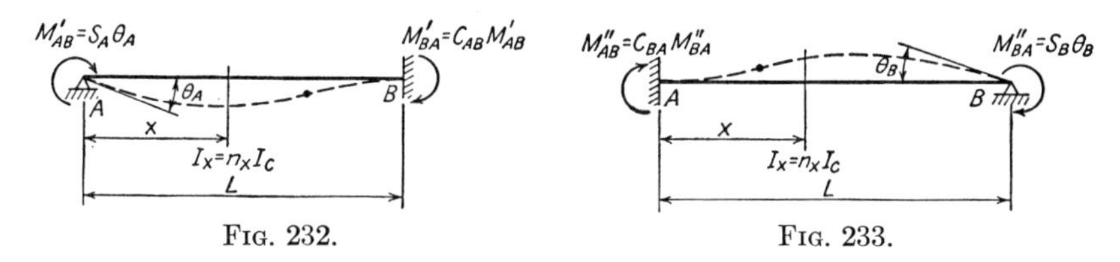

FIG. 232. FIG. 233.

The slope-deflection equations for members with variable cross sections, as previously derived in Art. 44, are

$$
\begin{aligned}
M_{AB} &= M_{FAB} + \frac{2EI_c}{L}[-K_{AA}\theta_A - K_{AB}\theta_B + (K_{AA} + K_{AB})R] \\
M_{BA} &= M_{FBA} + \frac{2EI_c}{L}[-K_{BB}\theta_B - K_{AB}\theta_A + (K_{BB} + K_{AB})R]
\end{aligned}
\tag{89}
$$

where

$$
K_{AA} = +\frac{C_3}{2(C_1C_3 - C_2^2)} \qquad C_1 = \frac{1}{L^3}\int_0^L \frac{(L-x)^2}{n_x}\,dx
$$

$$
K_{AB} = +\frac{C_2}{2(C_1C_3 - C_2^2)} \qquad C_2 = \frac{1}{L^3}\int_0^L \frac{(x)(L-x)}{n_x}\,dx
$$

$$
K_{BB} = +\frac{C_1}{2(C_1C_3 - C_2^2)} \qquad C_3 = \frac{1}{L^3}\int_0^L \frac{(x)^2}{n_x}\,dx
$$

The values of M'_{AB} and M'_{BA}, as defined in the preceding paragraph, can be obtained by substituting $M_{FAB} = M_{FBA} = 0$, $\theta_B = 0$, and $R = 0$ into the right-hand sides of Eqs. (89). Thus

$$
M'_{AB} = S_A\theta_A = \frac{2EI_c}{L}[-K_{AA}\theta_A] \tag{104}
$$

and

$$
M'_{BA} = C_{AB}M'_{AB} = \frac{2EI_c}{L}[-K_{AB}\theta_A] \tag{105}
$$

From Eq. (104), without regard to signs,

$$
S_A = K_{AA}\left(\frac{2EI_c}{L}\right) \tag{106}
$$

Substituting Eq. (104) into Eq. (105), with due regard to signs,

$$C_{AB} = +\frac{K_{AB}}{K_{AA}} \quad \text{or} \quad \frac{C_2}{C_3} \tag{107}$$

Similarly it can be shown that

$$S_B = K_{BB}\left(\frac{2EI_c}{L}\right) \tag{108}$$

and

$$C_{BA} = +\frac{K_{AB}}{K_{BB}} \quad \text{or} \quad \frac{C_2}{C_1} \tag{109}$$

Thus in Eqs. (106) to (109) are given the expressions for stiffness and carry-over factors for a member with variable cross sections. It is to be noted that, for unsymmetrical members, the stiffness factors at A and at B are different, and so are the carry-over factors from A to B and from B to A. These stiffness and carry-over factors can be more easily found by the method of column analogy, which will be treated in Chap. IX.

In Fig. 234 is shown the member AB with variable cross sections, simply supported at A and B. The end moment M'_{AB} required to main-

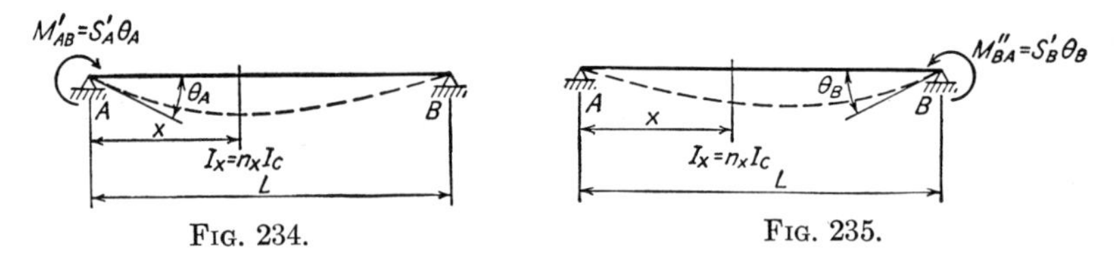

FIG. 234. FIG. 235.

tain a slope of θ_A at A is $M'_{AB} = S'_A\theta_A$ in which S'_A is the modified stiffness factor at A. By substituting $M_{FAB} = M_{FBA} = 0$ and $R = 0$ into the slope-deflection equations (89),

$$M'_{AB} = S'_A\theta_A = \frac{2EI_c}{L}[-K_{AA}\theta_A - K_{AB}\theta_B] \tag{110}$$

$$M'_{BA} = 0 = \frac{2EI_c}{L}[-K_{BB}\theta_B - K_{AB}\theta_A] \tag{111}$$

From Eq. (111),

$$\theta_B = -\frac{K_{AB}}{K_{BB}}\theta_A = -C_{BA}\theta_A \tag{112}$$

Substituting Eq. (112) into (110) and solving for S'_A,

$$S'_A = \frac{2EI_c}{L}(-K_{AA}\theta_A)(1 - C_{AB}C_{BA}) = S_A(1 - C_{AB}C_{BA}) \tag{113}$$

Similarly, referring to Fig. 235,

$$S'_B = S_B(1 - C_{AB}C_{BA}) \tag{114}$$

In Eqs. (113) and (114) are given the expressions for the modified stiffness factors at the near end when the far end is simply supported or hinged for a member with variable cross sections.

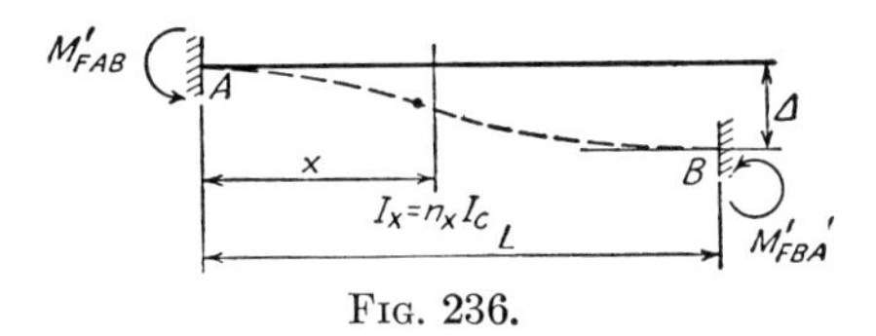

FIG. 236.

The fixed-end moments M'_{FAB} and M'_{FBA} due to a settlement of Δ ($R = \Delta/L$) at B relative to A can be found by substituting

$$M_{FAB} = M_{FBA} = 0$$

and $\theta_A = \theta_B = 0$ into the right-hand sides of Eqs. (89). Thus, referring to Fig. 236,

$$\begin{aligned} M'_{FAB} &= \frac{2EI_c}{L}(K_{AA} + K_{AB})R \qquad \text{or} \qquad S_A(1 + C_{AB})R \\ M'_{FBA} &= \frac{2EI_c}{L}(K_{BB} + K_{AB})R \qquad \text{or} \qquad S_B(1 + C_{BA})R \end{aligned} \tag{115}$$

In the check process for moment distribution the formula

$$(\theta_{\text{rel}})_{\text{near}} = \frac{(\text{change at near end}) - \frac{1}{2}(\text{change at far end})}{-(\text{relative stiffness})}$$

has been used in the case of members with constant cross section. The corresponding formula for members with variable cross sections can be found by solving Eqs. (89) for θ_A and θ_B, which are

$$\begin{aligned} \theta_A &= \frac{(M_{AB} - M_{FAB}) - C_{BA}(M_{BA} - M_{FBA})}{-S_A(1 - C_{AB}C_{BA})} \\ \theta_B &= \frac{(M_{BA} - M_{FBA}) - C_{AB}(M_{AB} - M_{FAB})}{-S_B(1 - C_{AB}C_{BA})} \end{aligned} \tag{116}$$

Example 84. Analyze the continuous beam shown in Fig. 237 by the moment-distribution method. Draw shear and moment diagrams. Sketch the elastic curve.

Solution. $R = 0$ for members AB and BC. From Example 68

$$M_{FAB} = +1{,}004.17 \text{ kip-ft}$$
$$M_{FBA} = -957.90 \text{ kip-ft}$$
$$K_{AA} = +2.82230 \qquad K_{BB} = +3.19722$$
$$K_{AB} = +1.73920 \qquad K_{BC} = +1.73920$$
$$K_{BB} = +3.19722 \qquad K_{CC} = +2.82230$$

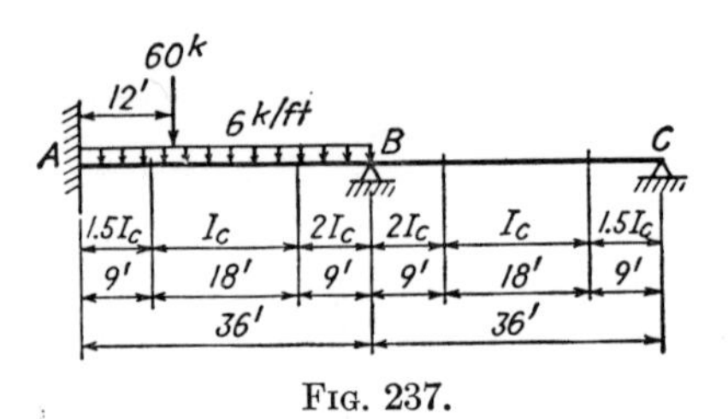

FIG. 237.

Stiffness and Carry-over Factors

Span AB: $S_A = \dfrac{2EI_c}{L} K_{AA} = 2.82230 \left(\dfrac{2EI_c}{36}\right) = \left(\dfrac{EI_c}{18}\right)(2.82230)$

$$S_B = \frac{2EI_c}{L} K_{BB} = 3.19722 \left(\frac{2EI_c}{36}\right) = \left(\frac{EI_c}{18}\right)(3.19722)$$

$$C_{AB} = \frac{K_{AB}}{K_{AA}} = \frac{1.73920}{2.82230} = 0.61623$$

$$C_{BA} = \frac{K_{AB}}{K_{BB}} = \frac{1.73920}{3.19722} = 0.54397$$

Span BC: $S_B = \dfrac{2EI_c}{L} K_{BB} = 3.19722 \left(\dfrac{2EI_c}{36}\right) = \left(\dfrac{EI_c}{18}\right)(3.19722)$

$$S_C = \frac{2EI_c}{L} K_{CC} = 2.82230 \left(\frac{2EI_c}{36}\right) = \left(\frac{EI_c}{18}\right)(2.82230)$$

$$C_{BC} = \frac{K_{BC}}{K_{BB}} = \frac{1.73920}{3.19722} = 0.54397$$

$$C_{CB} = \frac{K_{BC}}{K_{CC}} = \frac{1.73920}{2.82230} = 0.61623$$

Modified stiffness at B of span BC

$$= S_B[1 - C_{BC}C_{CB}]$$
$$= \left(\frac{EI_c}{18}\right)(3.19722)[1 - (0.54397)(0.61623)]$$
$$= \left(\frac{EI_c}{18}\right)(2.12548)$$

Moment Distribution (*Regular Method*) (see Table 84-1)

TABLE 84-1. MOMENT DISTRIBUTION (REGULAR METHOD)

Joint		A	B		C
Member		AB	BA	BC	CB
Stiffness		$\left(\frac{EI_c}{18}\right)$ (2.82230)	$\left(\frac{EI_c}{18}\right)$ (3.19722)	$\left(\frac{EI_c}{18}\right)$ (3.19722)	$\left(\frac{EI_c}{18}\right)$ (2.82230)
DF		——	0.500	0.500	1.000
Cycle	COF	0.616	0.544	0.544	0.616
1	FEM Bal.	+1004 ——	−958 +479	—— +479	—— ——
2	CO Bal.	+ 261 ——	—— ——	—— ——	+261 −261
3	CO Bal.	—— ——	—— + 81	−161 + 80	—— ——
4	CO Bal.	+ 44 ——	—— ——	—— ——	+ 44 − 44
5	CO Bal.	—— ——	—— + 14	− 27 + 13	—— ——
6	CO Bal.	+ 7 ——	—— ——	—— ——	+ 7 − 7
7	CO Bal.	—— ——	—— + 2	− 4 + 2	—— ——
8	CO Bal.	+ 1 ——	—— ——	—— ——	+ 1 − 1
Total		+1317	−382	+382	0
Check:					
Change		+ 313	+576	+382	——
−(COF)(change)		− 313	−193	——	−208
Sum		0	+383	+382	−208
θ*		0	$-\left(\frac{18}{EI_c}\right)$ (181)	$-\left(\frac{18}{EI_c}\right)$ (181)	$+\left(\frac{18}{EI_c}\right)$ (111)
		(*Check*)	(*Check*)		

$$* \; \theta_A = \frac{\text{sum}}{-S_A(1 - C_{AB}C_{BA})}, \quad \theta_B = \frac{\text{sum}}{-S_B(1 - C_{AB}C_{BA})}.$$

Moment Distribution (*Modified Stiffness Method*) (see Table 84-2)

TABLE 84-2. MOMENT DISTRIBUTION (MODIFIED STIFFNESS METHOD)

Joint		A	B		C
Member		AB	BA	BC	CB
Stiffness		$\left(\frac{EI_c}{18}\right)$ (2.82230)	$\left(\frac{EI_c}{18}\right)$ (3.19722)	$\left(\frac{EI_c}{18}\right)$ (3.19722)	$\left(\frac{EI_c}{18}\right)$ (2.82230)
Modified stiffness		$\left(\frac{EI_c}{18}\right)$ (2.82230)	$\left(\frac{EI_c}{18}\right)$ (3.19722)	$\left(\frac{EI_c}{18}\right)$ (2.12548)	——
DF		——	0.6007	0.3993	——
Cycle	COF	0.616	0.544	0.544	0.616
1	FEM Bal.	+1004 ——	−958 +575	—— +383	—— ——
2	CO Bal.	+ 313 ——	—— ——	—— ——	↕
Total		+1,317	−383	+383	0
Check		Same as in regular method			

For reactions, shear and moment diagrams, and the elastic curve, see Example 68.

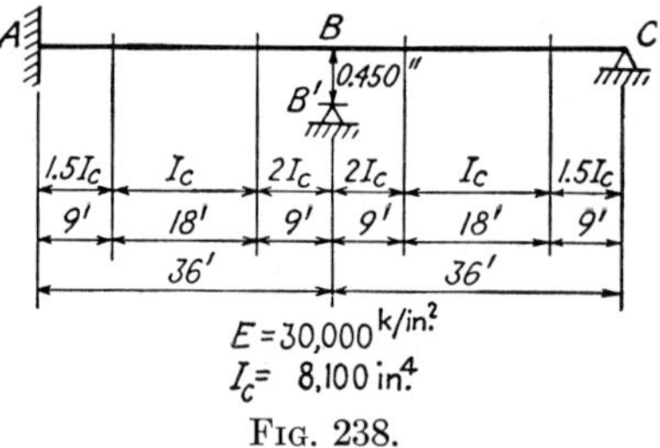

FIG. 238.

Example 85. By the moment-distribution method analyze the continuous beam shown in Fig. 238 due to a settlement of 0.450 in. at the support B. Draw shear and moment diagrams. Sketch the elastic curve.

Solution. For values of K_{AA}, K_{AB}, and K_{BB} in span AB and K_{BB}, K_{BC}, and K_{CC} in span BC, see Example 84.

$$M_{FAB} = +(K_{AA} + K_{AB})\left(\frac{2EI_cR}{L}\right)$$

$$= +(2.82230 + 1.73920)\left(\frac{2 \times 30{,}000 \times 8{,}100}{144 \times 36}\right)\left(\frac{0.450}{36 \times 12}\right)$$

$$= +445.5 \text{ kip-ft}$$

TABLE 85-1. MOMENT DISTRIBUTION (REGULAR METHOD)

Joint		A	B		C
Member		AB	BA	BC	CB
Stiffness		$\left(\frac{EI_c}{18}\right)$ (2.82230)	$\left(\frac{EI_c}{18}\right)$ (3.19722)	$\left(\frac{EI_c}{18}\right)$ (3.19722)	$\left(\frac{EI_c}{18}\right)$ (2.82230)
DF		———	0.500	0.500	1.000
Cycle	COF	0.616	0.544	0.544	0.616
1	FEM Bal.	+445.5 ———	+482.1 ———	−482.1 ———	−445.5 +455.5
2	CO Bal.	——— ———	——— −137.2	+274.4 −137.2	——— ———
3	CO Bal.	− 74.6 ———	——— ———	——— ———	− 74.6 + 74.6
4	CO Bal.	——— ———	——— − 23.0	+ 46.0 − 23.0	——— ———
5	CO Bal.	− 12.5 ———	——— ———	——— ———	− 12.5 + 12.5
6	CO Bal.	——— ———	——— − 3.8	+ 7.7 − 3.9	——— ———
7	CO Bal.	− 2.1 ———	——— ———	——— ———	− 2.1 + 2.1
8	CO Bal.	——— ———	——— − 0.7	+ 1.3 − 0.6	——— ———
9	CO Bal.	− 0.4 ———	——— ———	——— ———	− 0.4 + 0.4
10	CO Bal.	——— ———	——— − 0.1	+ 0.2 − 0.1	——— ———
Total		+355.9	+317.3	−317.3	0
Check:					
Change		− 89.6	−164.8	+164.8	+445.5
−(COF)(change)		+ 89.6	+ 55.2	−274.4	− 89.6
Sum		0	−109.6	−109.6	+355.9
θ*		0	$+\left(\frac{18}{EI_c}\right)$ (51.6)	$+\left(\frac{18}{EI_c}\right)$ (51.6)	$-\left(\frac{18}{EI_c}\right)$ (189.7)
		(*Check*)	(*Check*)		

$$* \ \theta_A = \frac{\text{sum}}{-S_A(1 - C_{AB}C_{BA})}, \quad \theta_B = \frac{\text{sum}}{-S_B(1 - C_{AB}C_{BA})}.$$

$$M_{FBA} = +(K_{BB} + K_{AB})\left(\frac{2EI_cR}{L}\right)$$

$$= +(3.19722 + 1.73920)\left(\frac{2 \times 30{,}000 \times 8{,}100}{144 \times 36}\right)\left(\frac{0.450}{36 \times 12}\right)$$

$$= +482.1 \text{ kip-ft}$$

$$M_{FBC} = -482.1 \text{ kip-ft}$$

$$M_{FCB} = -445.5 \text{ kip-ft}$$

Moment Distribution (*Regular Method*) (see Table 85-1).
Moment Distribution (*Modified Stiffness Method*) (see Table 85-2)

TABLE 85-2. MOMENT DISTRIBUTION (MODIFIED STIFFNESS METHOD)

Joint		*A*	*B*		*C*
Member		*AB*	*BA*	*BC*	*CB*
Stiffness		$\left(\frac{EI_c}{18}\right)$ (2.82230)	$\left(\frac{EI_c}{18}\right)$ (3.19722)	$\left(\frac{EI_c}{18}\right)$ (3.19722)	$\left(\frac{EI_c}{18}\right)$ (2.82230)
Modified stiffness		$\left(\frac{EI_c}{18}\right)$ (2.82230)	$\left(\frac{EI_c}{18}\right)$ (3.19722)	$\left(\frac{EI_c}{18}\right)$ (2.12548)	——
DF		——	0.6007	0.3993	——
Cycle	COF	0.616	0.544	0.544	0.616
1	FEM Bal.	+455.5 ——	+482.1 ——	−482.1 ——	−445.5 +445.5
2	CO Bal.	—— ——	—— −164.8	+274.4 −109.6	↑
3	CO Bal.	− 89.6 ——	—— ——	—— ——	↓
Total		+355.9	+317.3	−317.3	0
Check		Same as in regular method			

For reactions, shear and moment diagrams, and the elastic curve, see Example 69.

EXERCISES

140 to 143. Solve Exercises 103 to 106 by the moment-distribution method.

CHAPTER IX

THE METHOD OF COLUMN ANALOGY

55. General Introduction. The method of column analogy is most useful in the analysis of beams and curved members with two fixed supports and of rigid frames, either with two fixed supports or "closed" with one "cell" inside, as will be explained later. The analysis of these structures, which are statically indeterminate to the third degree, by the method of consistent deformation involves the solution of three simultaneous equations. In the method of column analogy, however, the moment at any point in the given structure can be computed directly. Probably the most extensive use of the method is in the computation of stiffness and carry-over factors for members with variable cross sections and in the analysis of closed rigid frames and fixed arches. In the subsequent articles of this chapter, the method will be gradually developed and illustrated from the relatively simple cases to the more complicated.

56. The Method of Column Analogy Applied to Prismatic Members. The use of the method of column analogy to determine the fixed-end moments of a prismatic member due to the applied loading will now be developed.

Let it be required to find the fixed-end moments M_A and M_B acting on the ends of beam AB, which is fixed at both ends and subjected to the applied loading as shown in Fig. 239*a*. The moment diagram of the given fixed-ended beam, as discussed previously, is the sum of the moment diagram due to the applied loading, if acting on a simple beam AB (Fig. 239*b*), and that due to the end moments (Fig. 239*c*). The conditions from which M_A and M_B can be determined, when the method of consistent deformation is used, are:

1. Change of slope between A and $B = 0$; or
   Sum of area of moment diagrams between A and $B = 0$ (note that EI is constant); or
   Area of moment diagram of Fig. 239*b* = area of moment diagram of Fig. 239*c*
2. Deflection of B from tangent at $A = 0$; or
   Sum of moment of moment diagrams between A and B about $B = 0$; or

$$\begin{bmatrix}\text{Moment of moment diagram} \\ \text{of Fig. 239}b\text{ about }B\end{bmatrix} = \begin{bmatrix}\text{moment of moment diagram} \\ \text{of Fig. 239}c\text{ about }B\end{bmatrix}$$

Now if an imaginary short column with a cross section as shown in Fig. 239*e* is so visualized that the loading acting on the top of the column is the moment diagram shown as Fig. 239*b* and the pressure acting on the bottom is the moment diagram shown as Fig. 239*c*, it is obvious that this

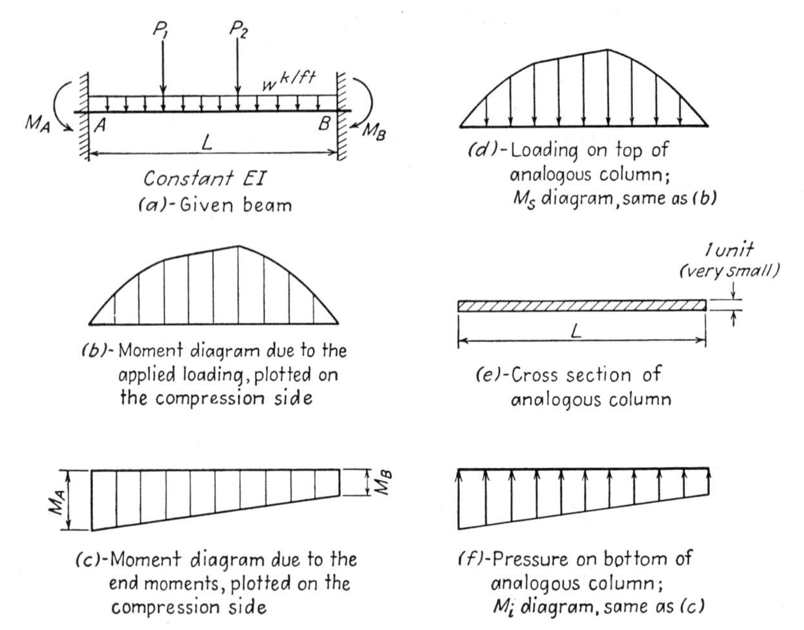

FIG. 239.

column is in equilibrium by reason of the two conditions stated above, which are (1) total load on the top is equal to the total pressure at the bottom and (2) moment of load about B is equal to the moment of pressure about B. Thus if the loading diagram (Fig. 239*d*) is known, the pressure diagram (Fig. 239*f*) can be determined.

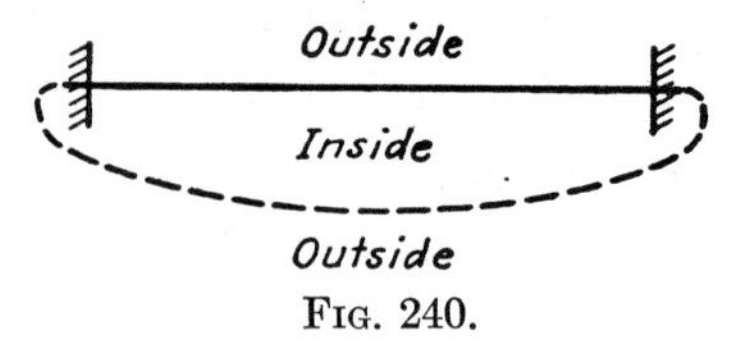

FIG. 240.

It is necessary to establish a sign convention to be followed in subsequent work. This sign convention includes (Fig. 240):

1. Loading on top of column is downward if M_s (statical moment, or moment due to the applied loading on simple beam AB as determined by the laws of statics) is positive, which means that it causes compression on the outside.

2. Upward pressure on bottom of column, M_i (indeterminate moment, or moment to be determined to satisfy the conditions of geometry), is positive.

3. Moment at any point of the given fixed-ended beam is equal to $M = M_s - M_i$, which is positive if it causes compression on the outside.

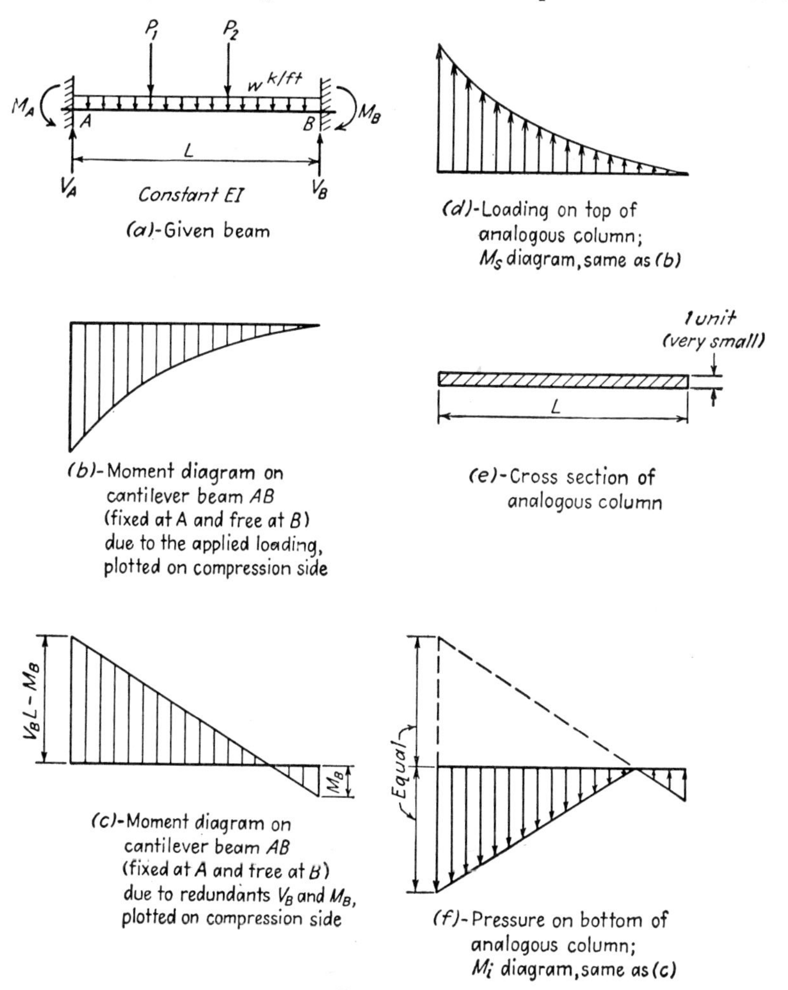

Fig. 241.

If, in applying the method of consistent deformation, the cantilever beam AB fixed at A and free at B is chosen as the basic determinate beam, the moment diagram of the given beam will be the sum of the moment diagrams shown in Fig. 241*b* and *c*. However, the same two conditions of geometry, *i.e.*, (1) change of slope between A and $B = 0$ and (2) deflection of B from tangent at $A = 0$, can be applied to the moment diagrams of Fig. 241*b* and *c* as previously; consequently the method of column analogy as described in Fig. 241*d* to *f* can still be applied.

Thus in finding the fixed-end moments acting on the ends of a prismatic member due to the applied loading by the method of column analogy, it is necessary only to determine the pressure, or M_i, at the two ends when the analogous column is loaded with the M_s diagram; M_A or M_B can then be found by the relation $M = M_s - M_i$. It is to be noted that the sign convention must be *strictly* followed.

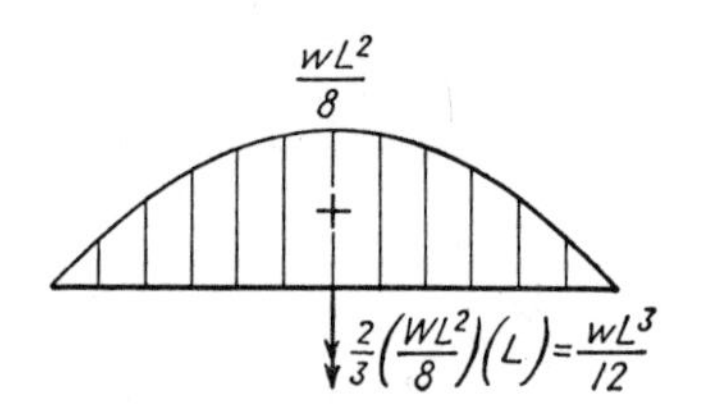

(a)-Loading on top of analogous column; M_s diagram

Example 86. Determine the fixed-end moments for the beam shown in Fig. 242 by the method of column analogy.

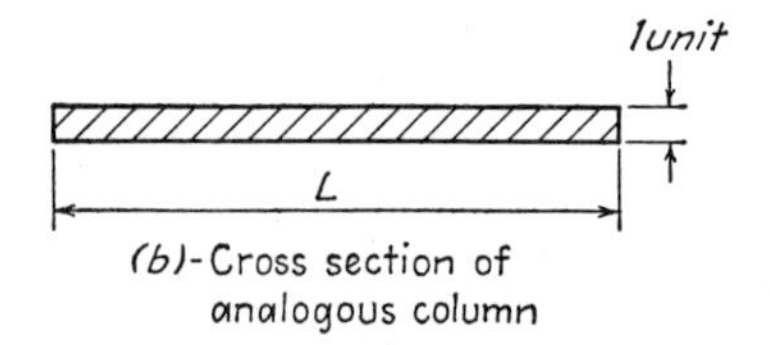

(b)-Cross section of analogous column

Solution (refer to Fig. 243). The simple beam moment diagram due to the uniform load is applied as a *downward* loading on the top of the analogous column because this moment causes *compression* on the *outside* (or

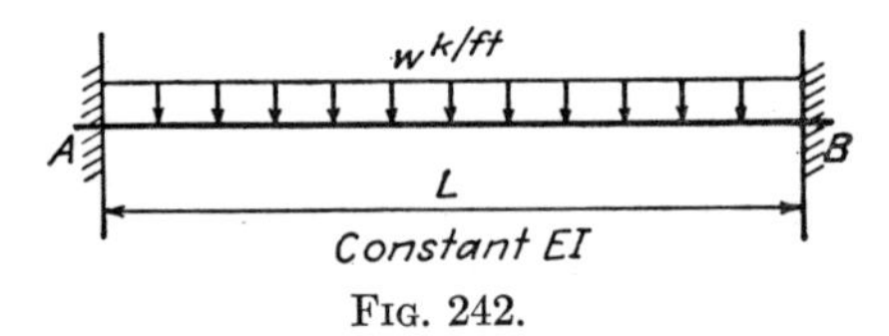

FIG. 242.

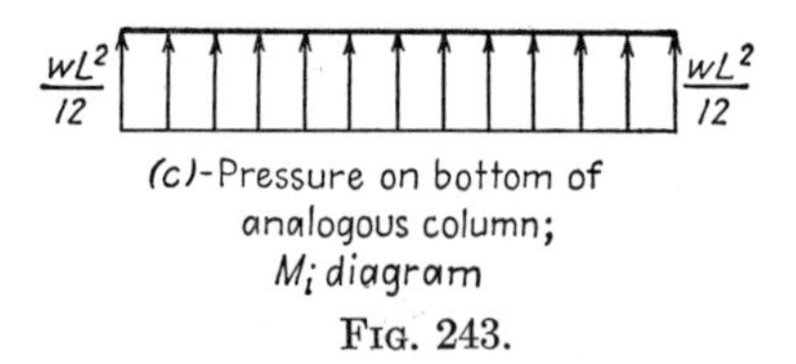

(c)-Pressure on bottom of analogous column; M_i diagram

FIG. 243.

top) of the beam. The pressure along the base of the column is constant in this case and is equal to the total load divided by the area of the column.

$$\text{Pressure} = \frac{wL^3}{12} \div L = \frac{wL^2}{12}$$

Thus, at A or B,

$$M_s = 0$$

$$M_i = +\frac{wL^2}{12}$$

$$M_A = M_B = M_s - M_i = 0 - \frac{wL^2}{12} = -\frac{wL^2}{12}$$

The minus sign for M_A (or M_B) indicates that M_A (or M_B) is in such a direction as to cause compression on the inside (or bottom) of the beam at A (or B).

Alternate Solution (refer to Fig. 244). The moment diagram due to the uniform load acting on a cantilever beam fixed at A and free at B is applied as an *upward* loading on the top of the analogous column

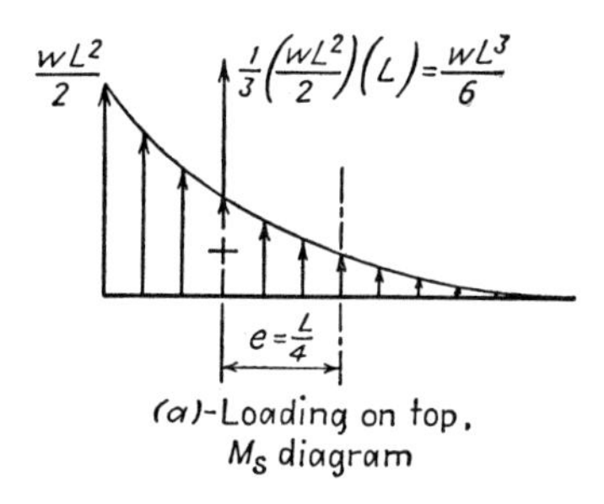

(a)-Loading on top, M_s diagram

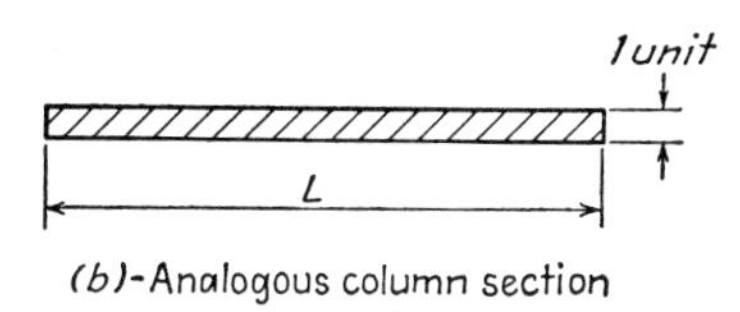

(b)-Analogous column section

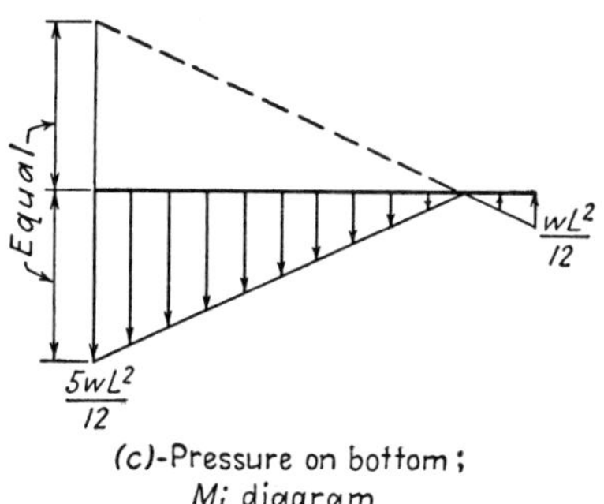

(c)-Pressure on bottom; M_i diagram

FIG. 244.

because this moment causes compression on the inside (or bottom) of the beam. The pressures at A and B can be found by the formula

$$p = \left[\frac{P}{A}\right] \pm \left[\frac{Mc}{I}\right]$$

$$M_s \text{ at } A = -\frac{wL^2}{2}$$

$$M_i \text{ at } A = -\frac{(wL^3/6)}{(1)(L)} - \frac{(wL^3/6)(L/4)}{(1)(L)^2/6}$$

$$= -\frac{wL^2}{6} - \frac{wL^2}{4} = -\frac{5wL^2}{12}$$

$$M_A = (M_s \text{ at } A) - (M_i \text{ at } A)$$
$$= \left(-\frac{wL^2}{2}\right) - \left(-\frac{5wL^2}{12}\right) = -\frac{wL^2}{12}$$
$$M_s \text{ at } B = 0$$
$$M_i \text{ at } B = -\frac{(wL^3/6)}{(1)(L)} + \frac{(wL^3/6)(L/4)}{(1)(L)^2/6}$$
$$= -\frac{wL^2}{6} + \frac{wL^2}{4} = +\frac{wL^2}{12}$$
$$M_B = (M_s \text{ at } B) - (M_i \text{ at } B)$$
$$= (0) - \left(+\frac{wL^2}{12}\right) = -\frac{wL^2}{12}$$

Note that in determining the pressure, or M_i, at A or B the upward load when acting at the centroid of the column section causes negative pressure, while the clockwise overturning moment causes negative pressure at A and positive pressure at B.

Example 87. Determine the fixed-end moments for the beam shown in Fig. 245 by the method of column analogy.

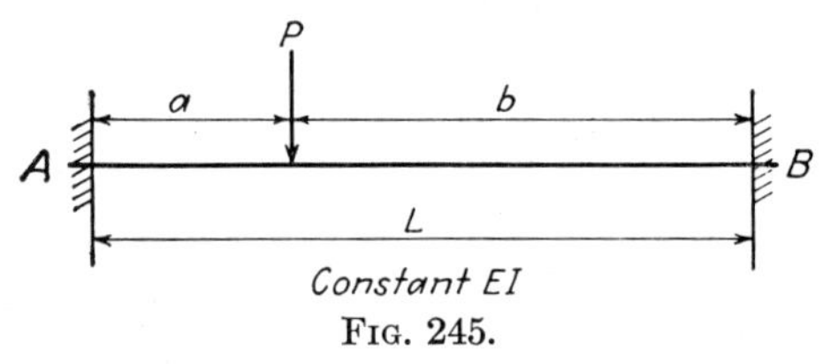

FIG. 245.

Solution (refer to Fig. 246)

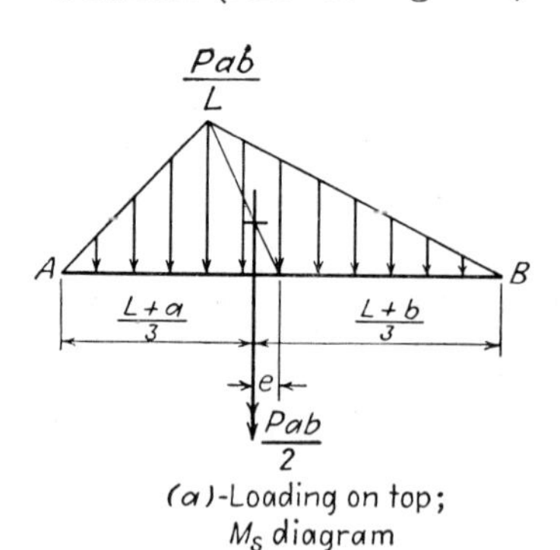

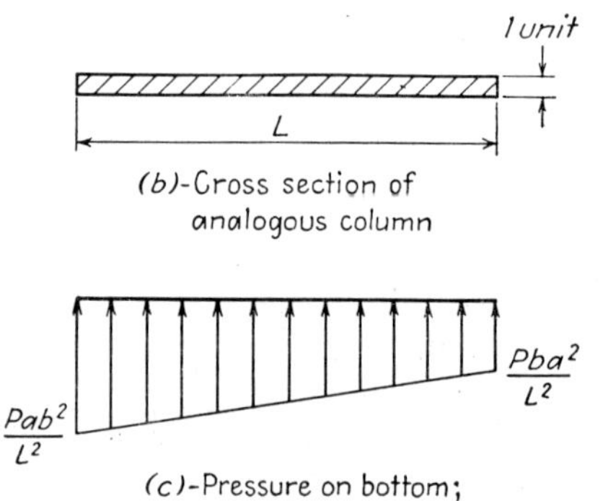

FIG. 246.

Total downward load on the column
= area of simple beam-moment diagram

$$= \frac{1}{2}\left(\frac{Pab}{L}\right)(L) = \frac{Pab}{2}$$

$$\text{Eccentricity } (e) = \frac{1}{3}\left(\frac{L}{2} - a\right)$$

$$\frac{P}{A} = \frac{(Pab/2)}{(L)} = \frac{Pab}{2L}$$

$$\frac{Mc}{I} = \frac{(Pab/2)[\frac{1}{3}(L/2 - a)]}{(1)(L)^2/6} = \frac{Pab(L/2 - a)}{L^2}$$

$$M_s \text{ at } A = 0$$

$$M_i \text{ at } A = \left[\frac{P}{A}\right] + \left[\frac{Mc}{I}\right] = \frac{Pab}{2L} + \frac{Pab(L/2 - a)}{L^2}$$

$$= \frac{Pab}{2L^2}(L + L - 2a) = +\frac{Pab^2}{L^2}$$

$$M_A = (M_s \text{ at } A) - (M_i \text{ at } A)$$

$$= 0 - \left(+\frac{Pab^2}{L^2}\right) = -\frac{Pab^2}{L^2}$$

$$M_s \text{ at } B = 0$$

$$M_i \text{ at } B = \left[\frac{P}{A}\right] - \left[\frac{Mc}{I}\right] = \frac{Pab}{2L} - \frac{Pab(L/2 - a)}{L^2}$$

$$= \frac{Pab}{2L^2}(L - L + 2a) = +\frac{Pba^2}{L^2}$$

$$M_B = (M_s \text{ at } B) - (M_i \text{ at } B)$$

$$= 0 - \left(+\frac{Pba^2}{L^2}\right) = -\frac{Pba^2}{L^2}$$

The minus sign for M_A (or M_B) indicates that M_A (or M_B) is in such a direction as to cause compression on the inside (or bottom) of the beam at A (or B).

Alternate Solution (refer to Fig. 247)

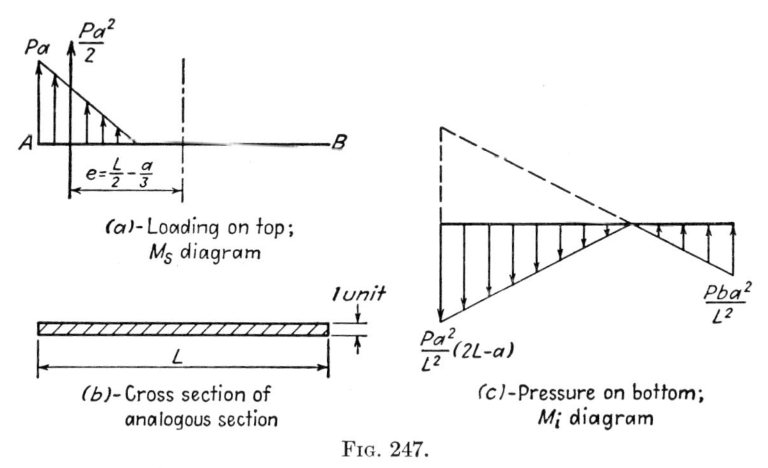

FIG. 247.

Total upward load on the column

$$= \text{area of cantilever beam-moment diagram}$$

$$= \frac{Pa^2}{2}$$

$$\text{Eccentricity } (e) = \frac{L}{2} - \frac{a}{3}$$

$$\frac{P}{A} = -\frac{Pa^2/2}{L} = -\frac{Pa^2}{2L}$$

$$\frac{Mc}{I} = \frac{(Pa^2/2)(L/2 - a/3)}{(1)(L)^2/6} = \frac{Pa^2}{2L^2}(3L - 2a)$$

$$M_s \text{ at } A = -Pa$$

$$M_i \text{ at } A = \left[\frac{P}{A}\right] - \left[\frac{Mc}{I}\right] = -\frac{Pa^2}{2L} - \frac{Pa^2}{2L^2}(3L - 2a)$$

$$= -\frac{Pa^2}{2L^2}(L + 3L - 2a) = -\frac{Pa^2}{L^2}(2L - a)$$

$$M_A = [M_s \text{ at } A] - [M_i \text{ at } A]$$

$$= [-Pa] - \left[-\frac{Pa^2}{L^2}(2L - a)\right]$$

$$= -\frac{Pa}{L^2}(L^2 - 2La + a^2)$$

$$= -\frac{Pab^2}{L^2}$$

$$M_s \text{ at } B = 0$$

$$M_i \text{ at } B = \left[\frac{P}{A}\right] + \left[\frac{Mc}{I}\right] = -\frac{Pa^2}{2L} + \frac{Pa^2}{2L^2}(3L - 2a)$$

$$= +\frac{Pa^2}{2L^2}(-L + 3L - 2a) = +\frac{Pba^2}{L^2}$$

$$M_B = (M_s \text{ at } B) - (M_i \text{ at } B)$$

$$= (0) - \left(+\frac{Pba^2}{L^2}\right) = -\frac{Pba^2}{L^2}$$

EXERCISES

144 to 148. Determine the fixed-end moments for the beam shown by the method of column analogy. Check the solution by using a different basic determinate structure.

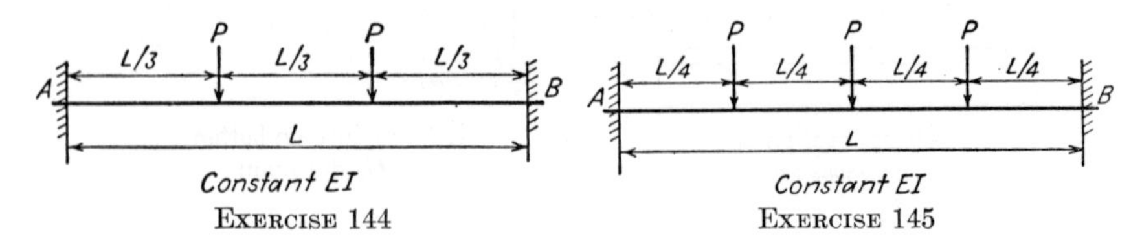

EXERCISE 144 EXERCISE 145

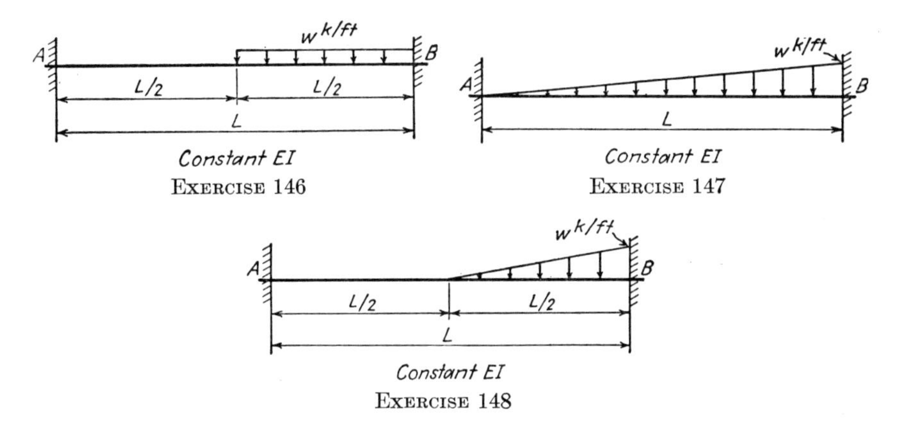

EXERCISE 146

EXERCISE 147

EXERCISE 148

57. Stiffness and Carry-over Factors of Prismatic Members. The stiffness factor K at A of span AB has been defined as the moment to be applied at A in order to rotate the tangent at A by one radian when the end B is fixed. The carry-over factor (COF) from A to B has been defined as the ratio of the fixed-end moment at B to the moment applied at A under the above conditions. In Fig. 248 are shown the span AB

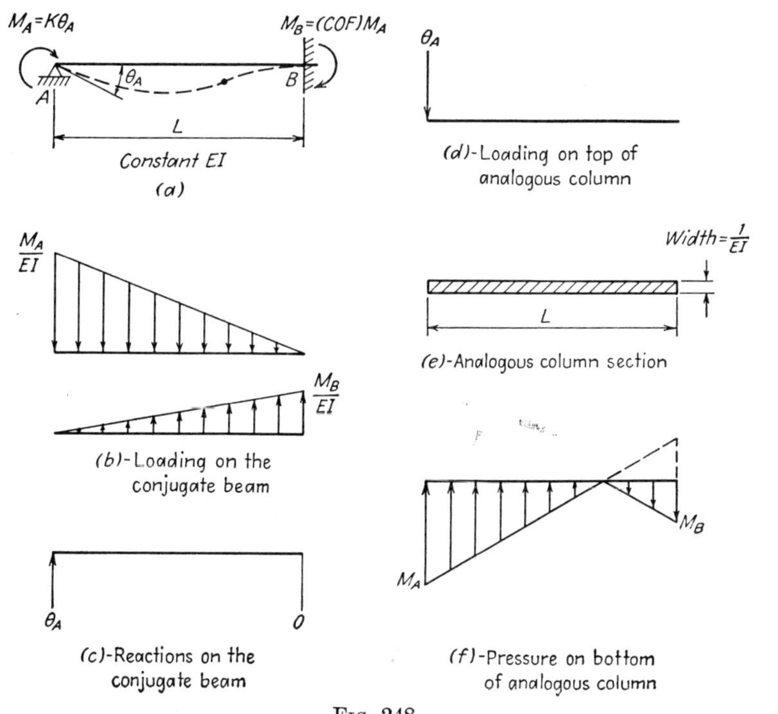

FIG. 248.

with constant EI and the loading and reaction diagrams when the conjugate-beam method is applied. Now, if the reactions to the conjugate beam are considered as the loads (actually there is only one load, θ_A) on an analogous column and the M_A/EI plus M_B/EI diagrams are considered as pressures on the bottom of the column, the column is obviously in equilibrium. In order that the pressures at A and B may be M_A and M_B directly instead of M_A/EI and M_B/EI, it is found convenient to call the width of the analogous-column section $1/EI$ instead of unity. Thus M_A and M_B can be found *directly* as the pressures at A and B, whereas two simultaneous equations are involved when the method of consistent deformation is used. The solution of the stiffness factor K and the carry-over factor COF of prismatic members by the method of column analogy is summarized below (see Fig. 248*d* to *f*).

$$\text{Area of analogous-column section} = \left(\frac{1}{EI}\right)(L) = \frac{L}{EI}$$

$$\text{Downward load on column} = \theta_A \text{ at } A$$

$$\text{Eccentricity } (e) = \frac{L}{2}$$

$$M_A = \left[\frac{P}{A}\right] + \left[\frac{Mc}{I}\right] = \frac{\theta_A}{(L/EI)} + \frac{(\theta_A)(L/2)}{(1/EI)(L)^2/6} = \frac{4EI}{L}(\theta_A) = K\theta_A$$

Thus

$$\text{Stiffness factor } (K) = \frac{4EI}{L}$$

$$M_B = \left[\frac{P}{A}\right] - \left[\frac{Mc}{I}\right] = \frac{\theta_A}{(L/EI)} - \frac{(\theta_A)(L/2)}{(1/EI)(L)^2/6} = -\frac{2EI}{L}\theta_A = -\tfrac{1}{2}M_A$$

The minus sign for M_B means that M_B and M_A have opposite signs according to the sign convention used in the method of column analogy; or M_A and M_B should have the same sign according to the sign convention used in the slope-deflection or moment-distribution methods. Thus

$$\text{Carry-over factor (COF)} = \frac{M_B}{M_A} = +\tfrac{1}{2}$$

58. The Method of Column Analogy Applied to Straight Members with Variable Cross Sections. The method of column analogy can be used to determine the fixed-end moments on a straight member with variable cross sections, when it is subjected to applied loads in the transverse direction.

Let it be required to find the fixed-end moments M_A and M_B acting on the ends of beam AB with variable cross sections, due to the applied loading as shown in Fig. 249*a*. When the method of consistent deforma-

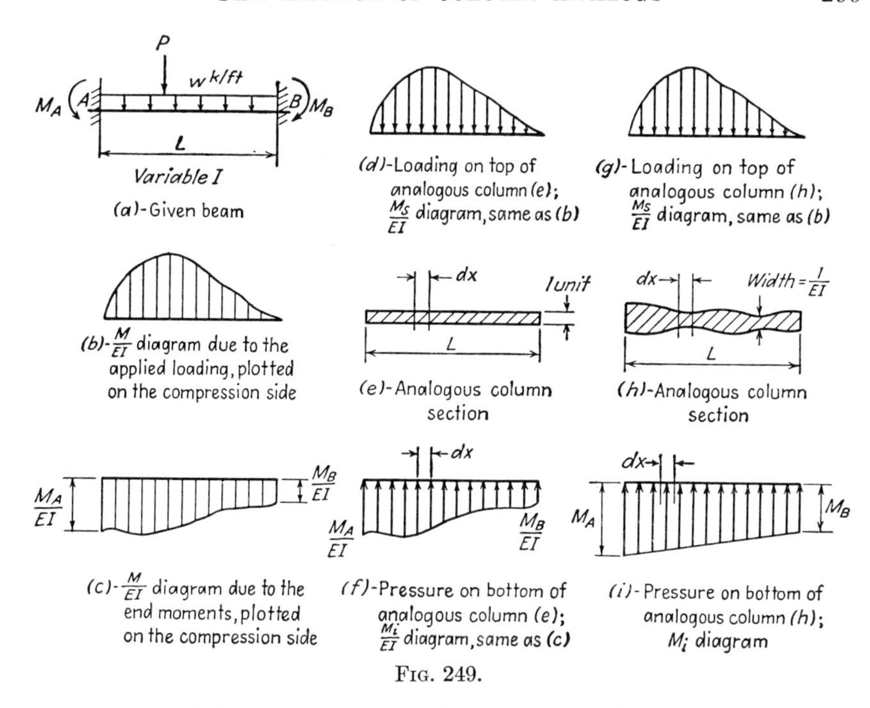

FIG. 249.

tion is used and if a simple beam AB is chosen as the basic determinate beam, the conditions from which M_A and M_B can be determined are

1. Change of slope between A and $B = 0$; or
   Area of Fig. 249b = area of Fig. 249c
2. Deflection of B from tangent at $A = 0$; or

$$\begin{bmatrix}\text{Moment of area of Fig.}\\ \text{249}b\text{ about }B\end{bmatrix} = \begin{bmatrix}\text{moment of area of Fig.}\\ \text{249}c\text{ about }B\end{bmatrix}$$

Now, if an imaginary short column with a cross section as shown in Fig. 249e is subjected to the loading on top and the pressure on bottom as shown in Fig. 249d and f, respectively, it is obvious that this column is in equilibrium by reason of the two conditions stated above. It will be proved that the "loading-section-pressure" diagrams of Fig. 249d to f can be replaced by those of Fig. 249g to i. The loading diagrams of Fig. 249d and g being the same, it remains necessary only to show that the upward reactions over any division dx at the base of the respective columns are also the same, which is obviously true because

$$\begin{bmatrix}\left(\dfrac{M_i}{EI}\right) dx\ (1\text{ unit})\\ \text{in Fig. 249}e\text{ and }f\end{bmatrix} = \begin{bmatrix}(M_i)\ dx\left(\dfrac{1}{EI}\right)\\ \text{in Fig. 249}h\text{ and }i\end{bmatrix}$$

The loading-section-pressure system of Fig. 249*g* to *i* will be used to find the fixed-end moments M_A and M_B.

The same sign convention as stated in Art. 56 should again be *strictly* followed.

The discussion in Art. 56 regarding the use of the moment diagram due to action of the applied loads on a cantilever beam fixed at *A* and free at *B* as the M_s diagram applied equally well in the present problem of a fixed-ended beam with variable cross sections.

Example 88. Determine the fixed-end moments for the beam shown in Fig. 250*a* by the method of column analogy.

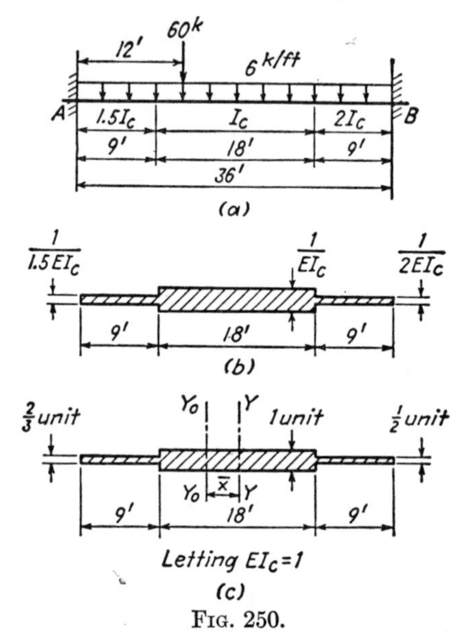

FIG. 250.

First Solution. *Properties of the Analogous-column Section.* The length of the analogous-column section is taken the same as that of the span of the given beam and the width as equal to $1/EI$. Since values of EI change along the span, the width of the analogous-column section changes correspondingly, as shown in Fig. 250*b*. It will be convenient to let $EI_c = 1$, so that the column section shown in Fig. 250*c* can be used.

Area of column section $= (9)(\frac{2}{3}) + (18)(1) + (9)(\frac{1}{2}) = 28.5$

$$\bar{x} = \frac{(6)(-13.5) + (18)(0) + (4.5)(+13.5)}{28.5} = \frac{-81 + 60.75}{28.5} = -0.71 \text{ ft}$$

$$I_{y_0y_0} = \frac{(\frac{2}{3})(9)^3}{12} + (6)(12.79)^2 + \frac{(1)(18)^3}{12} + (18)(0.71)^2 + \frac{(\frac{1}{2})(9)^3}{12} + (4.5)(14.21)^2$$

$$= 2{,}456.2$$

The properties of the analogous-column section are summarized in Fig. 252*b*.

Loads on Top of Analogous Column. The simple beam AB is chosen as the determinate beam. The M_s diagrams are shown in Fig. 251*b*.

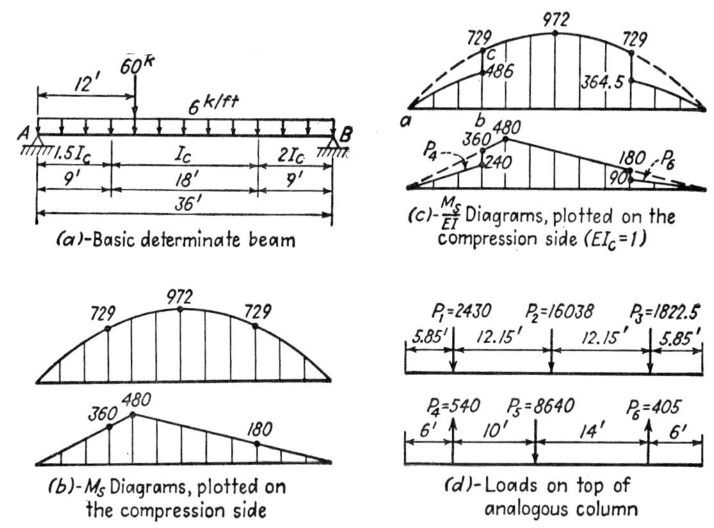

FIG. 251.

The M_s/EI diagrams are shown in Fig. 251*c*; note again that EI_c is taken equal to unity. The loads on the top of the analogous column should be equal to the area of the M_s/EI diagrams shown in Fig. 251*c*.

Let P_1, P_2, and P_3 be the areas of the M_s/EI diagram due to the uniform load over the 9 ft, 18 ft, and 9 ft sections, respectively.

$$\text{Area of } (abc) = \int_0^9 (108x - 3x^2)\,dx = 3{,}645$$

$$P_1 = \tfrac{2}{3}[\text{area } (abc)] = (\tfrac{2}{3})(3{,}645) = 2{,}430$$

$$P_3 = \tfrac{1}{2}[\text{area } (abc)] = (\tfrac{1}{2})(3{,}645) = 1{,}822.5$$

Moment of area (abc) about a vertical axis through A

$$= \int_0^9 (108x - 3x^2)(x)\,dx = 21{,}323.25$$

Distance between centroid of area (abc) and the left support

$$= \frac{21{,}323.25}{3{,}645} = 5.85 \text{ ft}$$

$$P_2 = (\tfrac{2}{3})(972)(36) - 2[\text{area of } (abc)]$$
$$= 23{,}328 - (2)(3{,}645) = 16{,}038$$

Let the M_s/EI diagram due to the concentrated load be the area (P_5) of the triangle with base = 36 and altitude = 480 minus the two small triangles indicated by P_4 and P_6 in Fig. 251*c*.

$P_5 = (\frac{1}{2})(36)(480) = 8{,}640$

Location of P_5 from the left support $= \dfrac{36 + 12}{3} = 16$ ft

$P_4 = (\frac{1}{2})(9)(120) = 540$
$P_6 = (\frac{1}{2})(9)(90) = 405$

Referring to Fig. 251*d*,

Total load on the column
$= P_1 + P_2 + P_3 - P_4 + P_5 - P_6$
$= 2{,}430 + 16{,}038 + 1{,}822.5 - 540 + 8{,}640 - 405$
$= 27{,}985.5$

Total moment of the loads about centroid G
$= (2{,}430)(11.44) - (16{,}038)(0.71) - (1{,}822.5)(12.86)$
$\quad - (540)(11.29) + (8{,}640)(1.29) + (405)(12.71)$
$= 27{,}799.2 - 11{,}387.0 - 23{,}437.4 - 6{,}096.6 + 11{,}145.6 + 5{,}147.6$
$= 3{,}171.4$ counterclockwise

Referring to Fig. 252,

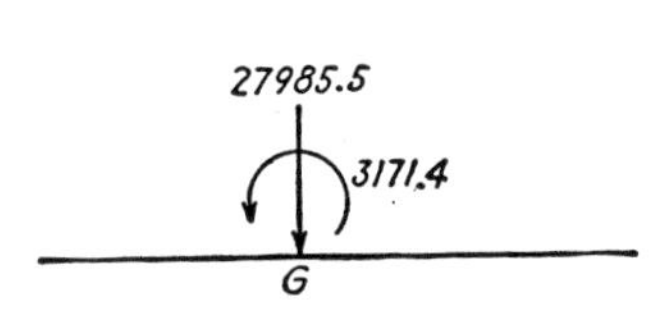

(*a*)-Loads on top of analogous column

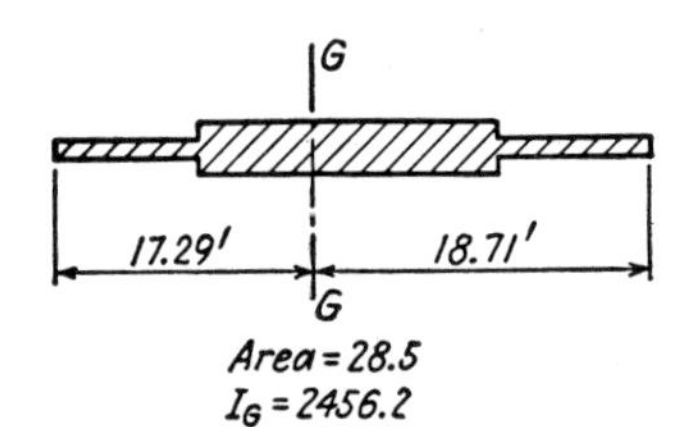

(*b*)-Properties of the column section

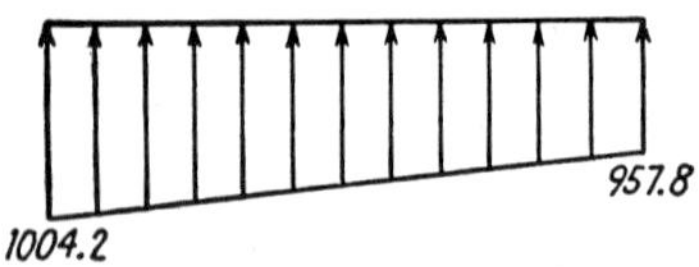

(*c*)-Pressure on bottom of analogous column

FIG. 252.

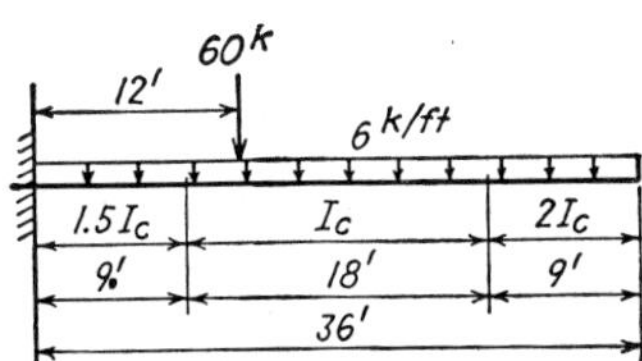

(*a*)-Basic determinate beam

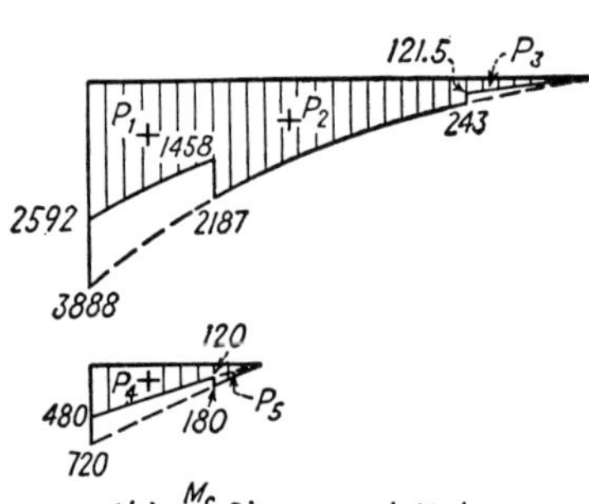

(*b*)-$\frac{M_S}{EI}$ Diagrams, plotted on the compression side ($EI_C = 1$)

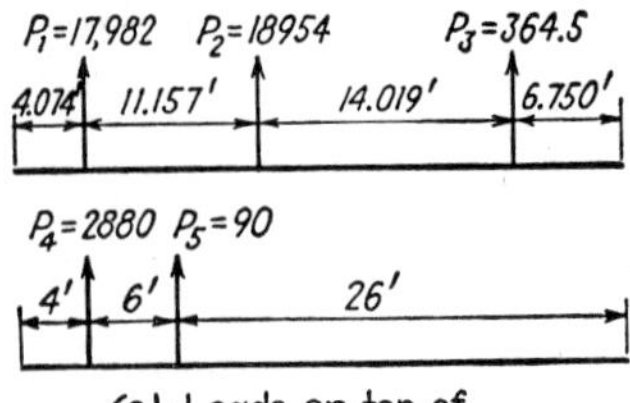

(*c*)-Loads on top of analogous column

FIG. 253.

M_s at $A = 0$

$$M_i \text{ at } A = \left[\frac{P}{A}\right] + \left[\frac{Mc}{I}\right] = +\frac{27{,}985.5}{28.5} + \frac{(3{,}171.4)(17.29)}{2{,}456.2}$$
$$= +981.9 + 22.3 = +1{,}004.2$$

$M_A = (M_s \text{ at } A) - (M_i \text{ at } A) = (0) - (+1{,}004.2) = -1{,}004.2$ kip-ft

M_s at $B = 0$

$$M_i \text{ at } B = \left[\frac{P}{A}\right] - \left[\frac{Mc}{I}\right] = +\frac{27{,}985.5}{28.5} - \frac{(3{,}171.4)(18.71)}{2{,}456.2}$$
$$= +981.9 - 24.1 = +957.8$$

$M_B = (M_s \text{ at } B) - (M_i \text{ at } B) = 0 - (+957.8) = -957.8$ kip-ft

The minus sign for M_A or M_B shows that M_A or M_B should be in such a direction as to cause compression on the inside or lower side at A or B.

Second Solution. An alternate solution in which the cantilever beam AB, fixed at A and free at B, is chosen as the basic determinate beam will now be shown.

Loads on Top of Analogous Column (Fig. 253)

$$P_1 = \frac{2}{3}\int_{27}^{36} (3x^2)\, dx = 17{,}982$$

$$\text{Moment of } P_1 \text{ about the free end} = \frac{2}{3}\int_{27}^{36} (3x^2)(x)\, dx = 574{,}087.5$$

$$\text{Distance from } P_1 \text{ to the free end} = \frac{574{,}087.5}{17{,}982} = 31.926 \text{ ft}$$

$$P_2 = \int_{9}^{27} (3x^2)\, dx = 18{,}954$$

$$\text{Moment of } P_2 \text{ about the free end} = \int_{9}^{27} (3x^2)(x)\, dx = 393{,}660$$

$$\text{Distance from } P_2 \text{ to the free end} = \frac{393{,}660}{18{,}954} = 20.769 \text{ ft}$$

$P_3 = (\frac{1}{3})(121.5)(9) = 364.5$

Distance from P_3 to the free end $= (\frac{3}{4})(9) = 6.750$ ft

$P_4 = (\frac{1}{2})(12)(480) = 2{,}880$

$P_5 = (\frac{1}{2})(60)(3) = 90$

Referring to Fig. 253*c*,

$$\text{Total load on the column} = P_1 + P_2 + P_3 + P_4 + P_5$$
$$= 17{,}982 + 18{,}954 + 364.5 + 2{,}880 + 90$$
$$= 40{,}270.5$$

Total moment of the loads about centroid G

$$= (17{,}982)(13.216) + (18{,}954)(2.059) - (364.5)(11.960)$$
$$+ (2{,}880)(13.29) + (90)(7.29)$$
$$= 237{,}650 + 39{,}026 - 4{,}359 + 38{,}275 + 656$$
$$= 311{,}248 \text{ clockwise}$$

Referring to Fig. 254,

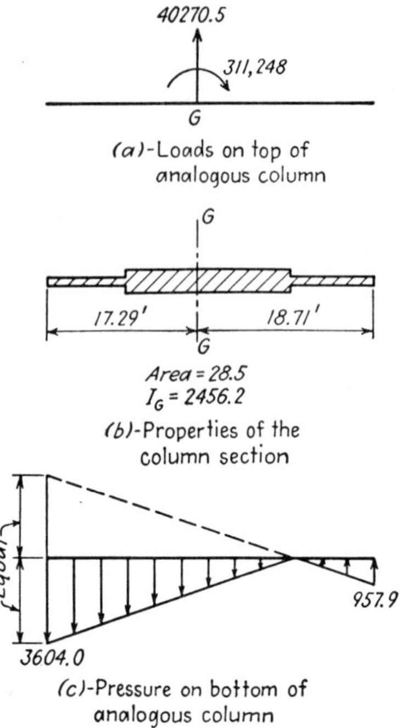

Fig. 254.

$$M_s \text{ at } A = -4{,}608$$

$$M_i \text{ at } A = \left[\frac{P}{A}\right] - \left[\frac{Mc}{I}\right] = -\frac{40{,}270.5}{28.5} - \frac{(311{,}248)(17.29)}{2{,}456.2}$$

$$= -1{,}413.0 - 2{,}191.0 = -3{,}604.0$$

$$M_A = (M_s \text{ at } A) - (M_i \text{ at } A) = (-4{,}608) - (-3{,}604.0)$$

$$= -1{,}004.0 \text{ kip-ft}$$

$$M_s \text{ at } B = 0$$

$$M_i \text{ at } B = \left[\frac{P}{A}\right] + \left[\frac{Mc}{I}\right] = -\frac{40{,}270.5}{28.5} + \frac{(311{,}248)(18.71)}{(2{,}456.2)}$$

$$= -1{,}413.0 + 2{,}370.9 = +957.9$$

$$M_B = (M_s \text{ at } B) - (M_i \text{ at } B) = 0 - (+957.9) = -957.9 \text{ kip-ft}$$

It is to be noted that there are other ways of choosing the basic determinate beam, such as a cantilever beam fixed at B and free at A or a fixed-ended beam with two internal hinges inserted where the sections change abruptly.

EXERCISES

149 and 150. Determine the fixed-end moments for the beam shown by the method of column analogy. Check the solution by using a different basic determinate structure.

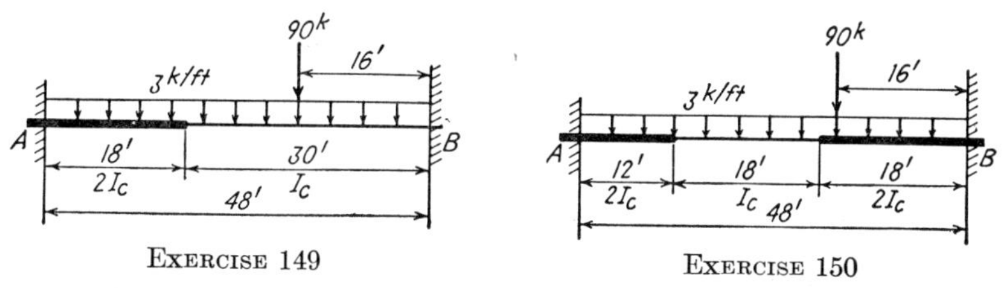

EXERCISE 149 EXERCISE 150

59. Stiffness and Carry-over Factors of Straight Members with Variable Cross Sections. The stiffness and carry-over factors of straight members with variable cross sections can be determined in the same manner as in the case of prismatic members, except that the width of the analogous-column section is variable along the length instead of being equal to the constant $1/EI$.

Example 89. Determine the stiffness factors at A and B and the carry-over factors from A to B and from B to A for the straight member with variable cross sections as shown in Fig. 255.

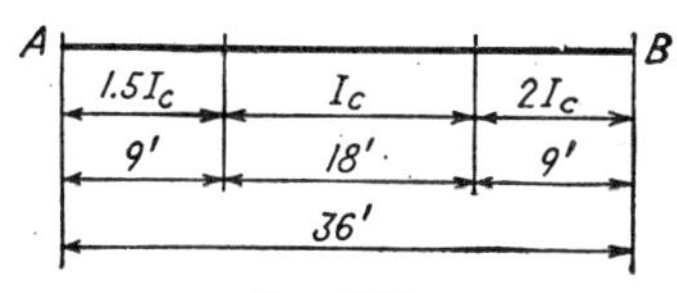

FIG. 255.

Solution. *Stiffness Factor at A and Carry-over Factor from A to B.* A load of 1 rad is applied at the left edge of the analogous column, the properties of which have been determined in Example 88. The load through the centroid and the moment due to eccentricity are shown in Fig. 256.

$$M_A = \text{pressure at } A$$
$$= \left[\frac{P}{A}\right] + \left[\frac{Mc}{I}\right]$$
$$= EI_c\left(\frac{1}{28.5}\right) + EI_c\,\frac{(17.29)(17.29)}{2{,}456.2}$$
$$= \frac{EI_c}{L}\left(\frac{36}{28.5}\right) + \frac{EI_c}{L}\,\frac{(17.29)(17.29)(36)}{2{,}456.2}$$
$$= \frac{EI_c}{L}(1.2632 + 4.3815) = 5.6447\left(\frac{EI_c}{L}\right)$$

$$M_B = \text{pressure at } B$$
$$= \left[\frac{P}{A}\right] - \left[\frac{Mc}{I}\right]$$
$$= EI_c\left(\frac{1}{28.5}\right) - EI_c\,\frac{(17.29)(18.71)}{2{,}456.2}$$

$$= \frac{EI_c}{L}\left(\frac{36}{28.5}\right) - \frac{EI_c}{L}\frac{(17.29)(18.71)(36)}{2{,}456.2}$$

$$= \frac{EI_c}{L}(1.2632 - 4.7414) = -3.4782\left(\frac{EI_c}{L}\right)$$

Thus

$$\text{Stiffness factor at } A = 5.6447\left(\frac{EI_c}{L}\right)$$

$$\text{Carry-over factor from } A \text{ to } B = \frac{3.4782}{5.6447} = 0.6162$$

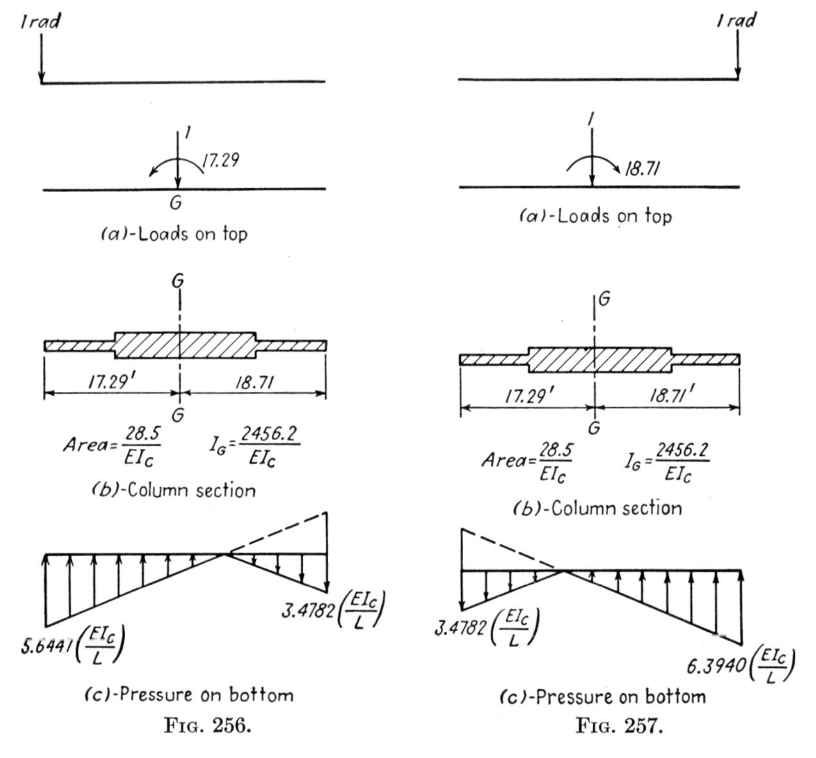

FIG. 256.

FIG. 257.

Stiffness Factor at B and Carry-over Factor from B to A. A load of 1 rad is applied at the right edge of the analogous column. The load through the centroid and the moment due to eccentricity are shown in Fig. 257.

$$M_B = \text{pressure at } B$$

$$= \left[\frac{P}{A}\right] + \left[\frac{Mc}{I}\right]$$

$$= EI_c\left(\frac{1}{28.5}\right) + EI_c\frac{(18.71)(18.71)}{2{,}456.2}$$

$$= \frac{EI_c}{L}\left(\frac{36}{28.5}\right) + \frac{EI_c}{L}\frac{(18.71)(18.71)(36)}{2{,}456.2}$$

$$= \frac{EI_c}{L}(1.2632 + 5.1308) = 6.3940\left(\frac{EI_c}{L}\right)$$

$$M_A = \text{pressure at } A$$

$$= \left[\frac{P}{A}\right] - \left[\frac{Mc}{I}\right]$$

$$= EI_c\left(\frac{1}{28.5}\right) - EI_c\frac{(18.71)(17.29)}{2{,}456.2}$$

$$= \frac{EI_c}{L}\left(\frac{36}{28.5}\right) - \frac{EI_c}{L}\frac{(18.71)(17.29)(36)}{2{,}456.2}$$

$$= \frac{EI_c}{L}(1.2632 - 4.7414) = -3.4782\left(\frac{EI_c}{L}\right)$$

Thus

$$\text{Stiffness factor at } B = 6.3940\left(\frac{EI_c}{L}\right)$$

$$\text{Carry-over factor from } B \text{ to } A = \frac{3.4782}{6.3940} = 0.5440$$

EXERCISES

151 and 152. Determine the stiffness factors at A and at B and the carry-over factors from A to B and from B to A for the straight member with variable cross sections shown in the figure.

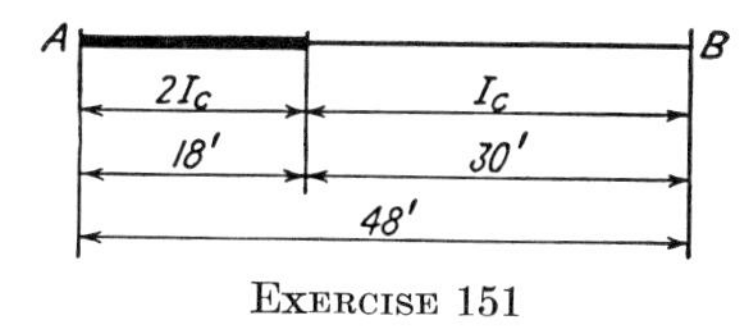

EXERCISE 151

A
B
2Ic
Ic
2Ic
12'
18'
18'
48'

EXERCISE 152

60. The Method of Column Analogy Applied to Quadrangular Frames with One Axis of Symmetry. The method of column analogy can be used to analyze quadrangular frames with one axis of symmetry, when such frames have two fixed supports and are subjected to the applied loadings. A typical quadrangular frame fulfilling the requirements mentioned above is shown in Fig. 258. It is to be noted that the axis of symmetry refers only to the properties of the frame, and not to the applied loads. The analysis of a frame shown in Fig. 258 by the method of consistent deformation requires the solution of three simultaneous equations.

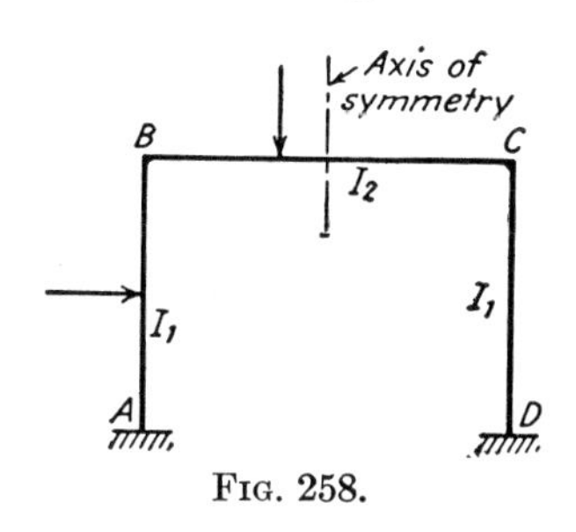

FIG. 258.

In the method of column analogy, however, the moment at any point of the frame can be determined by a *direct* procedure.

A simple geometric problem will be discussed first. If the curved line $APCB$ is "opened up" at the point P by a small angle $d\phi$ as shown in Fig. 259, it is apparent that the area PCB will rotate about point P

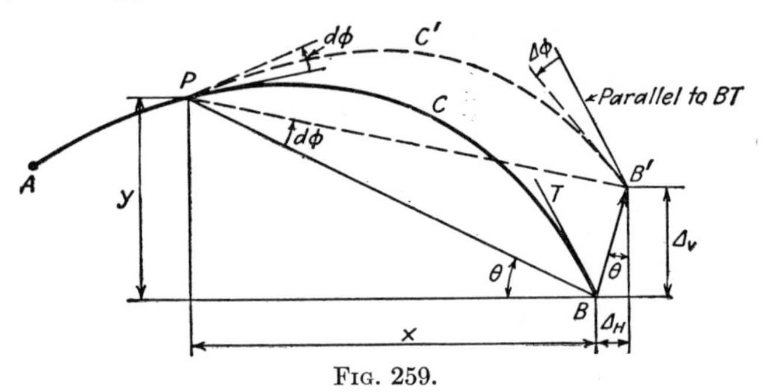

FIG. 259.

through an angle $d\phi$ in the counterclockwise direction, thus taking the new position $PC'B'$. If the angle $d\phi$ is very small, BB' can be assumed perpendicular to PB. Let Δ_ϕ, Δ_H, and Δ_V be the rotation of the tangent and the horizontal and vertical deflections at B. It can be seen that

$$\Delta_\phi = d\phi \text{ counterclockwise} \tag{117}$$

$$\begin{aligned} \Delta_H &= (BB')(\sin\theta) \\ &= (PB\, d\phi)(\sin\theta) \\ &= (PB \sin\theta)\, d\phi \\ &= (+y)\, d\phi = y\, d\phi \text{ to the right} \end{aligned} \tag{118}$$

$$\begin{aligned} \Delta_V &= (BB')(\cos\theta) \\ &= (PB\, d\phi)(\cos\theta) \\ &= (PB \cos\theta)\, d\phi \\ &= (-x)\, d\phi = -x\, d\phi \text{ upward} \end{aligned} \tag{119}$$

Note that x and y are the coordinates of P referred to B as origin. Thus, in Fig. 259, x is negative, and y is positive; so $(-x)$, which is then a positive quantity, should be substituted for $(PB \cos \theta)$ in Eq. (119).

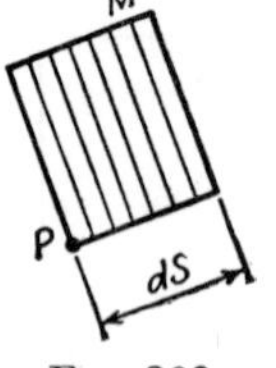

FIG. 260.

Now consider a curved member $APCB$ (see also Fig. 259) which is rigid (cannot be deformed) except for an infinitesimal segment, ds, at P on which there is a moment M acting (Fig. 260). M is considered positive if it causes compression on the outside (or the convex side). The curved member, being compressed on the outside at P over the elastic segment ds, will be "opened up." The rotation of the tangent at P, $d\phi$, is equal to $M\, ds/EI$, according to Eq. (24) derived in Art. 13. By substituting $d\phi = M\,ds/EI$ into Eqs. (117), (118), and (119), the rotation of the tangent and the horizontal and vertical deflections at B, due to the action of M over ds at P, are found to be

$$\begin{aligned} \Delta_\phi \text{ at } B &= +\frac{M\,ds}{EI} \\ \Delta_H \text{ at } B &= +\frac{My\,ds}{EI} \\ \Delta_V \text{ at } B &= -\frac{Mx\,ds}{EI} \end{aligned} \tag{120}$$

in which x and y are the coordinates of P referred to B as origin. Equations (120), which are derived for the general case of a curved member, can, of course, be applied to a quadrangular member such as $ABCD$ in Fig. 261*a*.

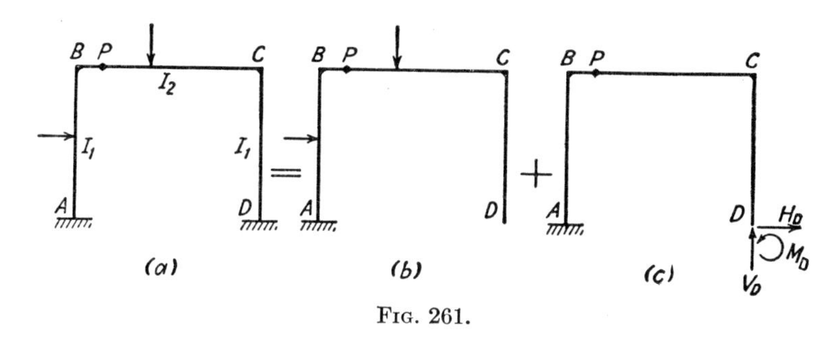

FIG. 261.

The analysis of the quadrangular frame shown in Fig. 261*a* by the method of column analogy will now be taken up. As discussed previously in the method of consistent deformation, the given frame of Fig. 261*a* may be considered equivalent to the sum of the two cantilever frames shown in Fig. 261*b* and *c*. The moment at any point, P, in the frame of Fig. 261*a* is equal to the sum of the moments at P in the frames of Fig. 261*b* and *c*. Let the moment at P in the frame of Fig. 261*b* be M_s, or the statical moment in the determinate cantilever structure. Note again that all moments are considered positive if they cause compression on the outside. Thus, for the frame of Fig. 261*a* and at the point P,

$$M = M_s + (M_D + H_D y - V_D x) \tag{121}$$

in which x and y are the coordinates of P referred to D as origin. Note that, in this present case, x in Eq. (121) is a negative quantity for all positions of P on the frame. The rotation and horizontal and vertical deflections at D due to M over an infinitesimal segment ds at P are

$$\begin{aligned}
\Delta_\phi &= +\frac{M\,ds}{EI} = +\frac{(M_s + M_D + H_D y - V_D x)\,ds}{EI} \\
\Delta_H &= +\frac{My\,ds}{EI} = +\frac{(M_s + M_D + H_D y - V_D x)y\,ds}{EI} \\
\Delta_V &= -\frac{Mx\,ds}{EI} = -\frac{(M_s + M_D + H_D y - V_D x)x\,ds}{EI}
\end{aligned} \tag{122}$$

By the conditions required in the geometry of deformation, the total Δ_ϕ, Δ_H, and Δ_V at D due to the action of M over all the segments throughout the members AB, BC, and CD should be equal to zero. Thus

$$\begin{aligned}
\text{Total } \Delta_\phi \text{ at } D &= +\sum \frac{M\,ds}{EI} = +\sum \frac{M_s\,ds}{EI} + M_D \sum \frac{ds}{EI} \\
&\quad + H_D \sum \frac{y\,ds}{EI} - V_D \sum \frac{x\,ds}{EI} = 0 \\
\text{Total } \Delta_H \text{ at } D &= +\sum \frac{My\,ds}{EI} = +\sum \frac{M_s y\,ds}{EI} + M_D \sum \frac{y\,ds}{EI} \\
&\quad + H_D \sum \frac{y^2\,ds}{EI} - V_D \sum \frac{xy\,ds}{EI} = 0 \\
\text{Total } \Delta_V \text{ at } D &= -\sum \frac{Mx\,ds}{EI} = -\sum \frac{M_s x\,ds}{EI} - M_D \sum \frac{x\,ds}{EI} \\
&\quad - H_D \sum \frac{xy\,ds}{EI} + V_D \sum \frac{x^2\,ds}{EI} = 0
\end{aligned} \tag{123}$$

in which x and y are the coordinates referred to D as origin. Equations (123) are in effect the three simultaneous equations which would be obtained if the method of consistent deformation were used.

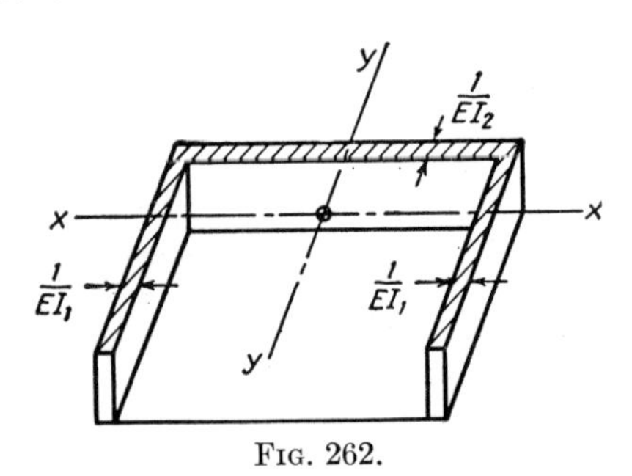

Fig. 262.

The elastic center, point O, will be defined as the centroid of the analogous-column section, the cross section of which is the same as the shape of the given structure with a thickness at any point equal to $1/EI$ (see Fig. 262). Referring to the axes x-x and y-y (y-y is the axis of symmetry), let A, I_{xy}, I_x, and I_y be the area and the moments of inertia of the analogous-column section. Then, by definition,

$$A = \sum \frac{ds}{EI}$$
$$A\bar{x} = \sum \frac{x\,ds}{EI} = 0$$
$$A\bar{y} = \sum \frac{y\,ds}{EI} = 0$$
$$I_{xy} = \sum \frac{xy\,ds}{EI} = 0^* \qquad (124)$$
$$I_x = \sum \frac{y^2\,ds}{EI}$$
$$I_y = \sum \frac{x^2\,ds}{EI}$$

Now if to the given structure of Fig. 263*a* a rigid arm *OD* joining the elastic center *O* and the fixed support *D* is attached, the equivalent

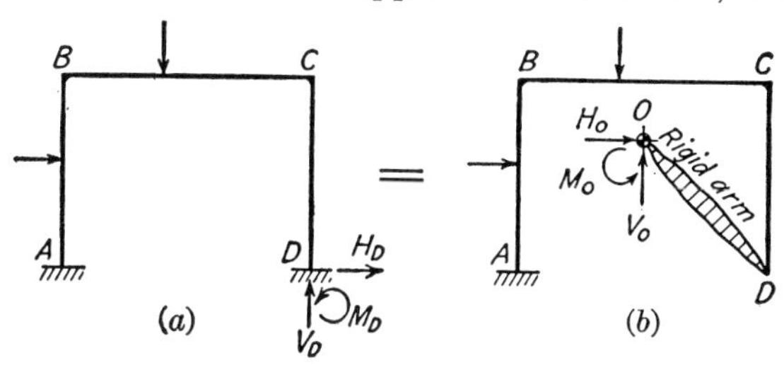

FIG. 263.

structure of Fig. 263*b* will be obtained. Since arm *OD* is rigid and cannot be deformed, the rotation and horizontal and vertical deflections at point *O* are also equal to zero. Let M_o, H_o, and V_o be the three unknown redundant reaction components to act at the elastic center. If the letter *O* is substituted for the letter *D* in Eqs. (123),

$$\text{Total } \Delta_\phi \text{ at } O = +\sum \frac{M_s\,ds}{EI} + M_O \sum \frac{ds}{EI} + H_O \sum \frac{y\,ds}{EI} - V_O \sum \frac{x\,ds}{EI} = 0$$

$$\text{Total } \Delta_H \text{ at } O = +\sum \frac{M_s y\,ds}{EI} + M_O \sum \frac{y\,ds}{EI} + H_O \sum \frac{y^2\,ds}{EI} - V_O \sum \frac{xy\,ds}{EI} = 0 \qquad (125)$$

$$\text{Total } \Delta_V \text{ at } O = -\sum \frac{M_s x\,ds}{EI} - M_O \sum \frac{x\,ds}{EI} - H_O \sum \frac{xy\,ds}{EI} + V_O \sum \frac{x^2\,ds}{EI} = 0$$

* True only if *y-y* is an axis of symmetry.

in which x and y are the coordinates referred to the elastic center, point O, as origin. Substituting Eqs. (124) into Eqs. (125),

$$
\begin{aligned}
0 &= \sum \frac{M_s\,ds}{EI} + M_O A &\quad\text{or}\quad M_O &= -\frac{\Sigma(M_s\,ds/EI)}{A} \\
0 &= \sum \frac{M_s y\,ds}{EI} + H_O I_x &\quad\text{or}\quad H_O &= -\frac{\Sigma(M_s y\,ds/EI)}{I_x} \\
0 &= -\sum \frac{M_s x\,ds}{EI} + V_O I_y &\quad\text{or}\quad V_O &= +\frac{\Sigma(M_s x\,ds/EI)}{I_y}
\end{aligned}
\qquad (126)
$$

The total moment, M, at any point of the given structure is equal to

$$M = M_s + M_O + H_O y - V_O x \qquad (127)$$

where x and y are coordinates referred to the elastic center as origin. Substituting Eqs. (126) into Eq. (127),

$$M = M_s - \left[\frac{\sum \frac{M_s\,ds}{EI}}{A} + \frac{\left(\sum \frac{M_s y\,ds}{EI}\right)(y)}{I_x} + \frac{\left(\sum \frac{M_s x\,ds}{EI}\right)(x)}{I_y} \right] \qquad (128)$$

If the M_s/EI diagram is considered to act as loads on the top of the analogous column, the sum of the three terms in the brackets of Eq. (128) represents the pressure at any point (x,y). This pressure will be called M_i. Thus

$$M = M_s - M_i \qquad (129)$$

It has been pointed out before that in the application of the method of consistent deformation there are in general several ways of choosing the basic determinate structure. Another way of deriving a basic determinate structure from the quadrangular frame with two fixed supports is shown as Fig. 264a to c; in such a case the redundants are M_A, M_D, and H_D, and the conditions of geometry required are $\theta_A = 0$, $\theta_D = 0$, and

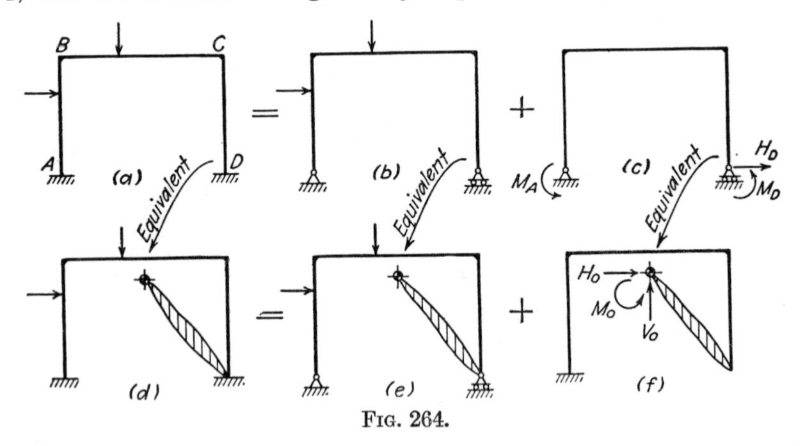

FIG. 264.

Δ_H at $D = 0$. When a rigid arm joining the elastic center and the fixed support D is attached, it is seen that Fig. 264*d* is equivalent to Fig. 264*a*, Fig. 264*e* to Fig. 264*b*, and Fig. 264*f* to Fig. 264*c* because there is certainly a set of values for M_O, H_O, and V_O which can be found such that the moment at any point in the frame of Fig. 264*f* is the same as that of Fig. 264*c*. Thus the moment M at any point in the frame of Fig. 264*d* is the sum of the M_s of Fig. 264*e* and the $(M_O + H_Oy - V_Ox)$ of Fig. 264*f*. Thus, referring to Fig. 264*d* to *f*,

$$M = M_s + M_O + H_Oy - V_Ox$$

in which x and y are the coordinates referred to the elastic center as origin. Therefore it can be concluded that, in applying the method of column analogy, M_s may be the statical moment in any basic determinate structure which can be derived from the given structure with two fixed supports.

As a summary of the method of column analogy in analyzing quadrangular frames with one axis of symmetry and two fixed supports, the following items should be noted:

1. The loading on the top of the analogous column is equal to the M_s/EI diagram, where M_s is the statical moment in *any* basic determinate structure derived from the given frame. The loading is in the downward direction if M_s is positive, which means that it causes compression on the outside.

2. The cross section of the analogous column consists of an area, the shape of which is the same as that of the given frame and the thickness of which is equal to $1/EI$.

3. The moment at any point of the given frame, M, is equal to

$$(M_s - M_i)$$

where M_i is the pressure on the bottom of the analogous column at the point under consideration.

Example 90. Analyze the quadrangular frame shown in Fig. 265 by the method of column analogy.

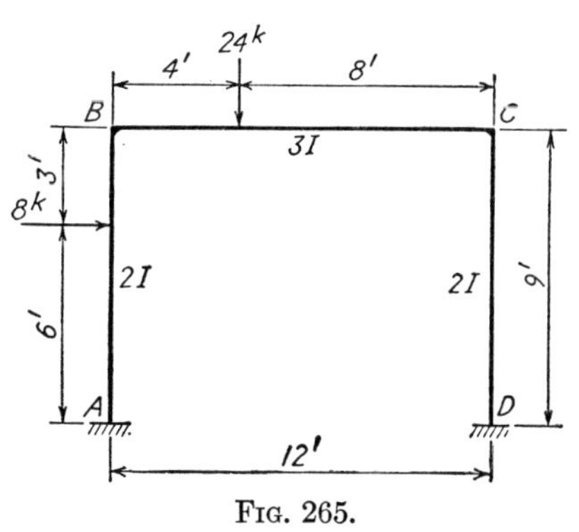

FIG. 265.

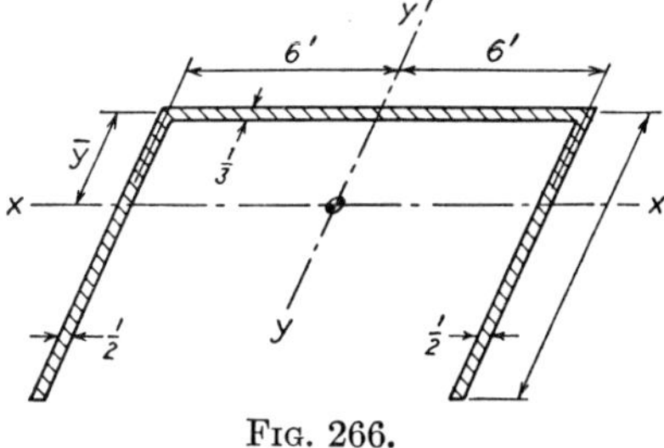

FIG. 266.

Solution. The properties of the analogous-column section will first be determined. The analogous-column section by considering EI to be unity is shown in Fig. 266. It is to be noted that the width, being equal to $1/EI$, is in effect very small, so that the area is what can be mathematically considered as a *line area.* Thus the moment of inertia of the line area about its own centroidal axis in the direction of the "line" is equal to zero.

Properties of the Analogous-column Section (Fig. 266)

$$A = (12)(\tfrac{1}{3}) + (2)(9)(\tfrac{1}{2}) = 13$$

$$\bar{y} = \frac{(4)(0) + (9)(4.5)}{13} = 3.115 \text{ ft}$$

$$I_x = (4)(3.115)^2 + 2\left[\frac{(\frac{1}{2})(9)^3}{12} + (4.5)(1.385)^2\right] = 116.83$$

$$I_y = \frac{(\frac{1}{3})(12)^3}{12} + (2)(4.5)(6)^2 = 372$$

For purposes of illustration, three different cases of loading on the analogous column by choosing three different basic determinate structures will be fully worked out (see Fig. 267).

Case I (see Table 90-1 and Fig. 267)

$$P = 64 + 432 + 72 = 568 \text{ upward}$$

$$M_x = (432)(1.385) + (72)(3.885) - (64)(3.115)$$
$$= 678.7 \text{ clockwise}$$

$$M_y = (64)(4.667) + (432 + 72)(6)$$
$$= 3{,}322.7 \text{ clockwise}$$

TABLE 90-1

Point	M_s	$\left[\frac{P}{A}\right]$	$\left[\frac{M_x y}{I_x}\right]$	$\left[\frac{M_y x}{I_y}\right]$	M_i	M
A....	−144	$-568/13 = -43.69$	$-\frac{678.7}{116.83}(5.885) = -34.19$	$-\frac{3{,}322.7}{372}(6) = -53.59$	−131.47	−12.53
B....	− 96	−43.69	$+\frac{678.7}{116.83}(3.115) = +18.10$	$-\frac{3{,}322.7}{372}(6) = -53.59$	− 79.18	−16.82
C....	0	−43.69	$+\frac{678.7}{116.83}(3.115) = +18.10$	$+\frac{3{,}322.7}{372}(6) = +53.59$	+ 28.00	−28.00
D....	0	−43.69	$-\frac{678.7}{116.83}(5.885) = -34.19$	$+\frac{3{,}322.7}{372}(6) = +53.59$	− 24.29	+24.29

In Table 90-1 the signs for the values in the $[P/A]$, $[M_x y/I_x]$, and $[M_y x/I_y]$ columns are determined by inspection. For instance, the total

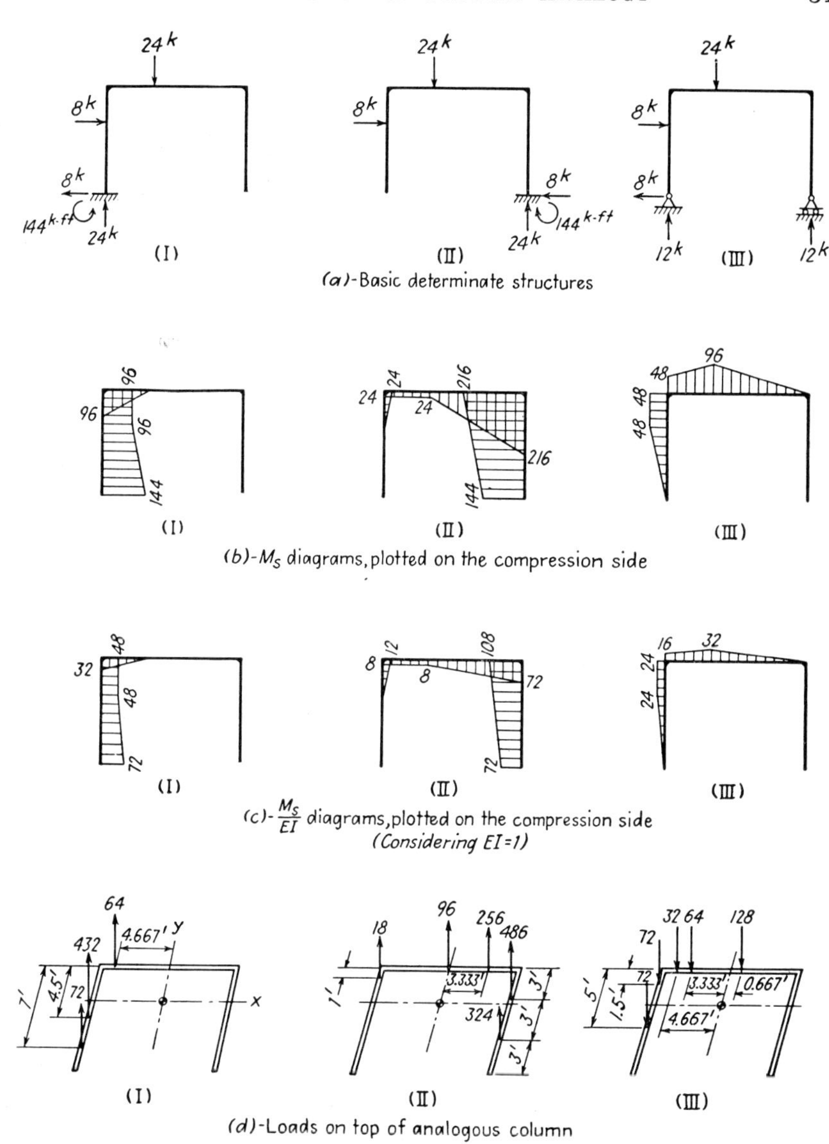

FIG. 267.

load P is upward; so it causes tension, which is considered negative, at all points in the column. M_x is observed to be clockwise when viewed from the right side; so it causes negative pressure at A and D and positive pressure at B and C. M_y is clockwise when viewed from front; so it causes negative pressure at A and B and positive pressure at C and D.

TABLE 90-2

Point	M_s	$\left[\frac{P}{A}\right]$	$\left[\frac{M_x y}{I_x}\right]$	$\left[\frac{M_y x}{I_y}\right]$	M_i	M
A	0	$-\frac{1,180}{13} = -90.77$	$+\frac{255.70}{116.83}(5.885) = +12.88$	$+\frac{5,605.3}{372}(6) = +90.41$	+ 12.52	−12.52
B	− 24	−90.77	$-\frac{255.70}{116.83}(3.115) = -\ 6.82$	$+\frac{5,605.3}{372}(6) = +90.41$	− 7.18	−16.82
C	−216	−90.77	$-\frac{255.70}{116.83}(3.115) = -\ 6.82$	$-\frac{5,605.3}{372}(6) = -90.41$	−188.00	−28.00
D	−144	−90.77	$+\frac{255.70}{116.83}(5.885) = +12.88$	$-\frac{5,605.3}{372}(6) = -90.41$	−168.30	+24.30

Case II (see Table 90-2 and Fig. 267)

$$\begin{aligned}
P &= 18 + 96 + 256 + 486 + 324 \\
&= 1{,}180 \text{ upward} \\
M_x &= (18)(2.115) + (96 + 256)(3.115) + (486)(0.115) - (324)(2.885) \\
&= 255.70 \text{ counterclockwise} \\
M_y &= (256)(3.333) + (486 + 324)(6) - (18)(6) \\
&= 5{,}605.3 \text{ counterclockwise}
\end{aligned}$$

Case III (see Table 90-3 and Fig. 267)

$$\begin{aligned}
P &= 72 + 72 + 32 + 64 + 128 \\
&= 368 \text{ downward} \\
M_x &= (72)(1.615) + (32 + 64 + 128)(3.115) - (72)(1.885) \\
&= 678.32 \text{ clockwise} \\
M_y &= (72 + 72)(6) + (32)(4.667) + (64)(3.333) - (128)(0.667) \\
&= 1{,}141.3 \text{ counterclockwise}
\end{aligned}$$

TABLE 90-3

Point	M_s	$\left[\frac{P}{A}\right]$	$\left[\frac{M_x y}{I_x}\right]$	$\left[\frac{M_y x}{I_y}\right]$	M_i	M
A....	0	$+\frac{368}{13} = +28.31$	$-\frac{678.32}{116.83}(5.885) = -34.17$	$+\frac{1,141.3}{372}(6) = +18.41$	+12.55	−12.55
B....	+48	+28.31	$+\frac{678.32}{116.83}(3.115) = +18.09$	$+\frac{1,141.3}{372}(6) = +18.41$	+64.81	−16.81
C....	0	+28.31	$+\frac{678.32}{116.83}(3.115) = +18.09$	$-\frac{1,141.3}{372}(6) = -18.41$	+27.99	−27.99
D....	0	+28.31	$-\frac{678.32}{116.83}(5.885) = -34.17$	$-\frac{1,141.3}{372}(6) = -18.41$	−24.27	+24.27

The results obtained from the three different cases check reasonably well within the limits of accuracy. The answers are

$$\begin{aligned}
M_A &= -12.53 \text{ kip-ft compression inside} \\
M_B &= -16.82 \text{ kip-ft compression inside} \\
M_C &= -28.00 \text{ kip-ft compression inside} \\
M_D &= +24.29 \text{ kip-ft compression outside}
\end{aligned}$$

The free-body diagrams, moment diagrams, and the deformed structure are shown in Fig. 268.

It is to be noted that a basic determinate structure can also be derived by placing three hinges (external or internal as the case may be) at any three of the four locations A, B, C, or D.

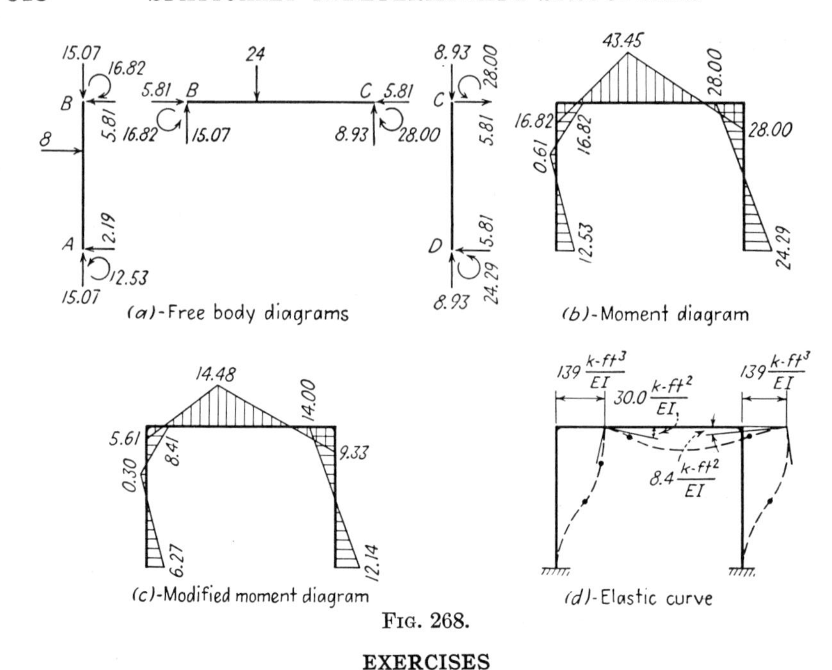

Fig. 268.

EXERCISES

153 to 155. Analyze the quadrangular frame shown by the method of column analogy. Check the solution by using a different basic determinate structure.

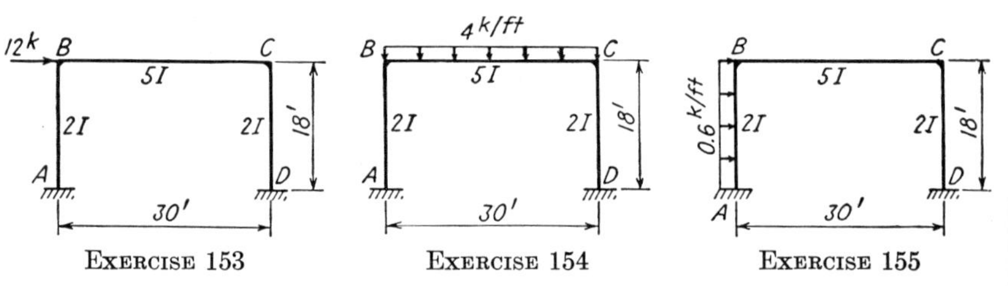

Exercise 153 Exercise 154 Exercise 155

61. The Method of Column Analogy Applied to Closed Frames with One Axis of Symmetry. The method of column analogy can be used to analyze closed frames with one axis of symmetry, when such frames are subjected to self-balancing loads or may be supported by external reactions. Two typical closed frames fulfilling the requirements mentioned above are shown in Fig. 269. It is to be noted again that the axis of symmetry refers only to the properties of the frame, and not to the applied loads, although in the case of Fig. 269*b* the self-balancing loads are shown as being also symmetrical with respect to the axis of symmetry. By the method of column analogy the moment at any point of the frame can be determined by a *direct* procedure.

Consider the closed frame of Fig. 270*a*. If this frame is cut at any point A and the points on each side of the cut be called A_1 and A_2, respectively, the thrust, shear, and moment acting on points A_1 and A_2 are as shown in Fig. 270*b*, wherein, by reason of statics,

$$M_{A1} = M_{A2}$$
$$H_{A1} = H_{A2}$$
$$V_{A1} = V_{A2}$$

Now if the position of the point A_1 and the direction of the tangent at A_1 are held fixed (actually these are not fixed when the deformed frame rests on the two external supports), it is obvious that the Δ_ϕ, Δ_H, and Δ_V at

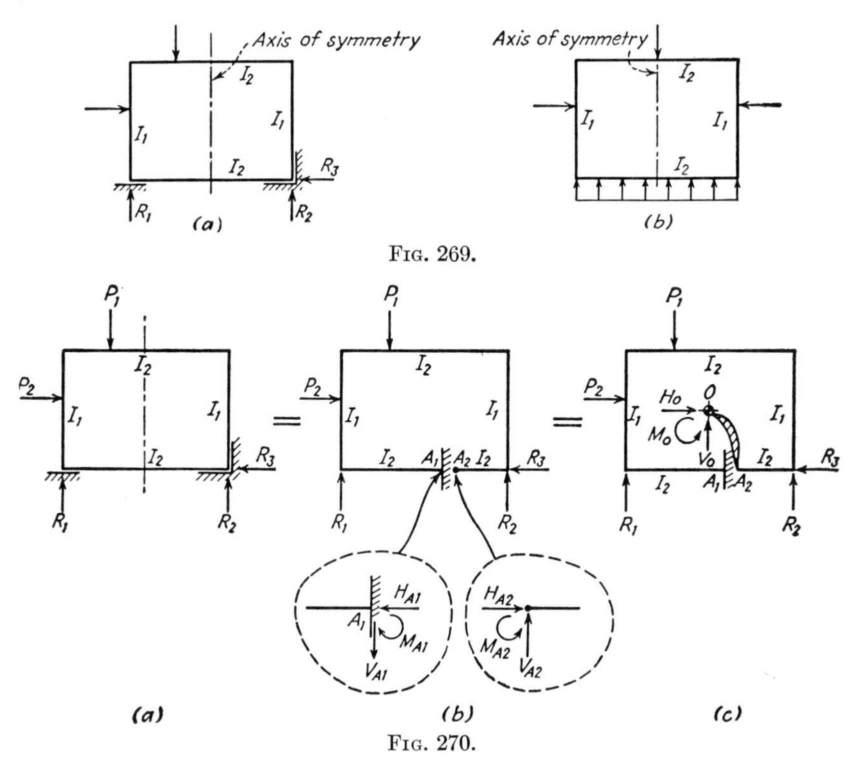

FIG. 269.

FIG. 270.

A_2 relative to A_1 should all be zero. Thus, when the method of consistent deformation is used, the redundants M_{A2}, H_{A2}, and V_{A2} can be determined by solving the three simultaneous equations similar to Eqs. (123), only that point D in Eqs. (123) is now replaced by point A_2 and M_s is now the statical moment due to the applied loadings in the cantilever frame fixed at A_1 and free at A_2. If a rigid arm joining the point A_2 and the elastic center O is attached to the frame of Fig. 270*b*, the equivalent frame of Fig. 270*c* is obtained. Since OA_2 is a rigid arm and cannot be deformed, the Δ_ϕ, Δ_H, and Δ_V of point O relative to point A_1 are also all zero. The redundants M_O, H_O, and V_O can now be determined by solving the equations similar to Eqs. (125) or (126). It follows that Eq. (128) can be used to determine the moment at any point of the frame of Fig. 270*a*, and thus the method of column analogy as discussed in Art. 60 can be equally applied to the analysis of closed frames with one axis of symmetry.

The discussion relating to the sign convention and the possibility of choosing different basic determinate structures as given in Art. 60 can be applied to closed frames in the same manner.

Example 91. Analyze the closed rectangular frame shown in Fig. 271*a* by the method of column analogy.

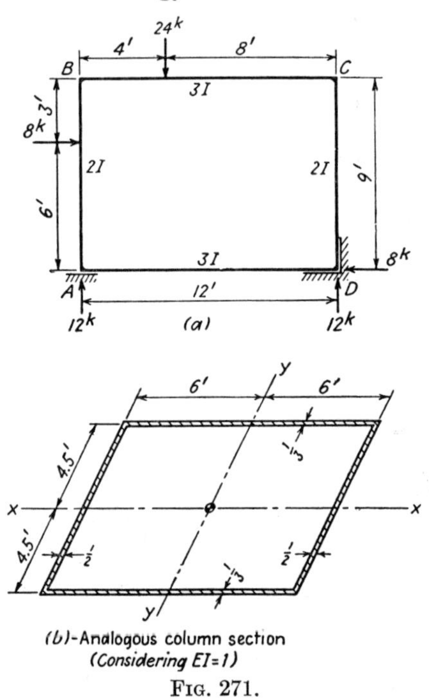

FIG. 271.

Solution. *Properties of the Analogous-column Section* (Fig. 271*b*)

$$A = (2)(12)(\tfrac{1}{3}) + (2)(9)(\tfrac{1}{2}) = 17$$

$$I_x = (2)(4)(4.5)^2 + (2)\frac{(\frac{1}{2})(9)^3}{12} = 222.75$$

$$I_y = (2)(4.5)(6)^2 + (2)\frac{(\frac{1}{3})(12)^3}{12} = 420$$

For purpose of illustration, two different cases of loading on the analogous column by choosing two different basic determinate structures will be fully worked out (see Fig. 272).

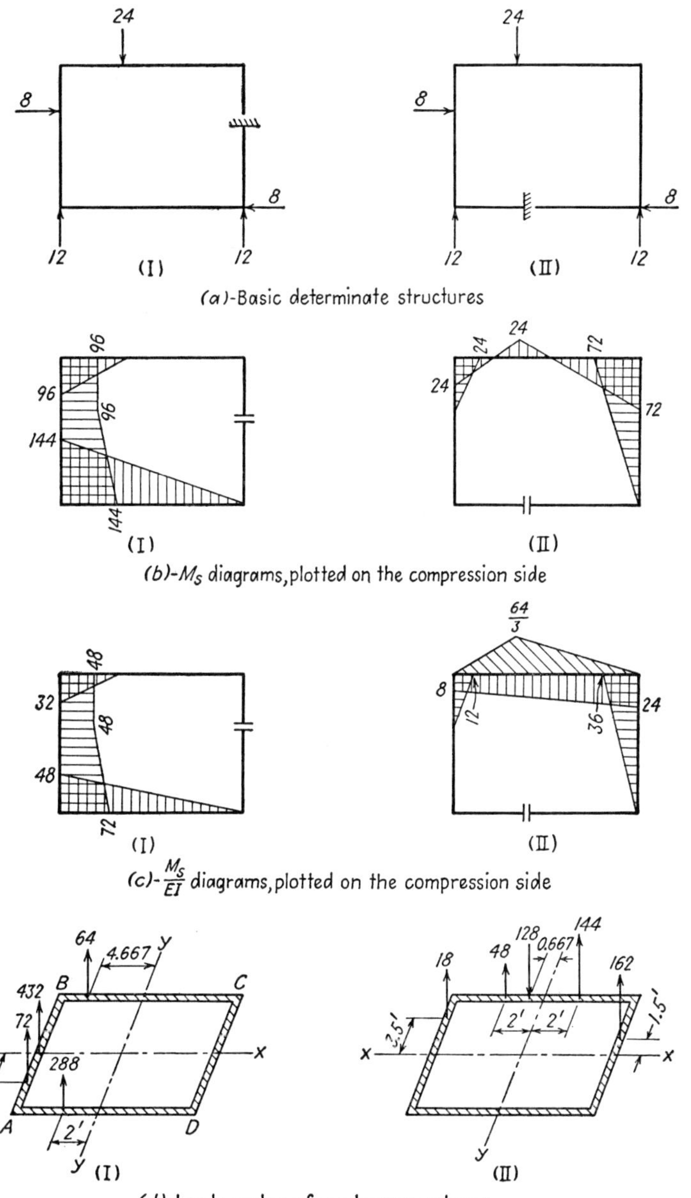
24
8
8
12
12
(I)
24
8
8
12
12
(II)
(a)-Basic determinate structures
96
96
96
144
144
(I)
24
24
24
72
72
(II)
(b)-M_S diagrams, plotted on the compression side
48
32
48
48
72
(I)
64/3
8
12
36
24
(II)
(c)-$\frac{M_S}{EI}$ diagrams, plotted on the compression side
64
4.667
432
72
288
2.5'
2'
A
B
C
D
x
y
(I)
18
48
128
0.667
144
162
2'
2'
2.5'
1.5'
x
y
(II)
(d)-Loads on top of analogous column

Fig. 272.

TABLE 91-1

Point	M_s	$\left[\frac{P}{A}\right]$	$\left[\frac{M_x y}{I_x}\right]$	$\left[\frac{M_y x}{I_y}\right]$	M_i	M
A	−144	$-{}^{856}/_{17} = -50.35$	$-\frac{(1,188)(4.5)}{222.75} = -24.00$	$-\frac{(3,898.7)(6)}{420} = -55.70$	−130.05	−13.95
B	− 96	−50.35	+24.00	−55.70	− 82.05	−13.95
C	0	−50.35	+24.00	+55.70	+ 29.35	−29.35
D	0	−50.35	−24.00	+55.70	− 18.65	+18.65

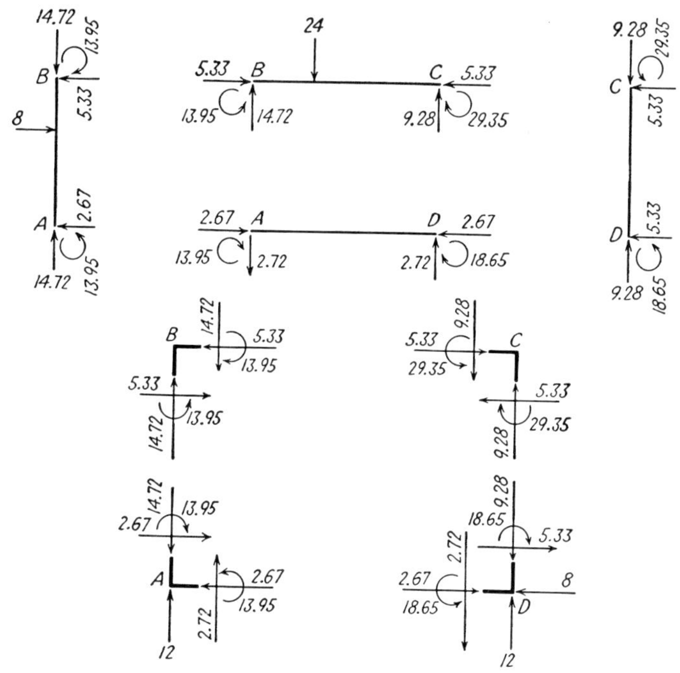

(a)-Free body diagrams of all members and joints

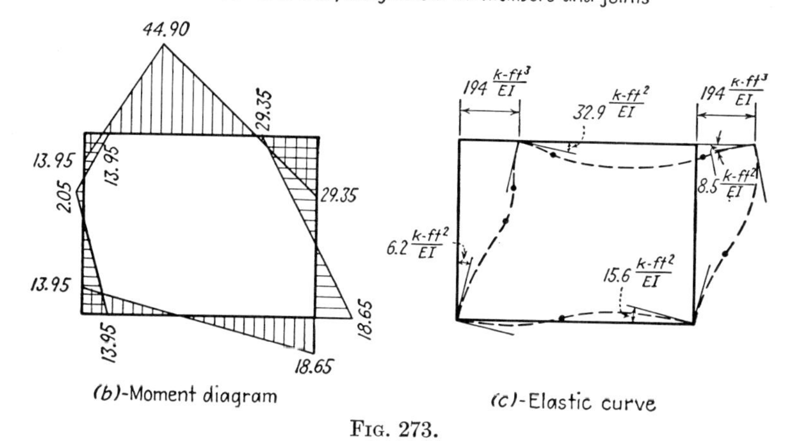

(b)-Moment diagram

(c)-Elastic curve

FIG. 273.

Case I (see Table 91-1 and Fig. 272)

$$
\begin{aligned}
P &= 72 + 432 + 64 + 288 \\
&= 856 \text{ upward} \\
M_x &= (288)(4.5) + (72)(2.5) - (64)(4.5) \\
&= 1{,}188 \text{ clockwise} \\
M_y &= (64)(4.667) + (288)(2) + (432 + 72)(6) \\
&= 3{,}898.7 \text{ clockwise}
\end{aligned}
$$

Case II (see Table 91-2 and Fig. 272)

$$\begin{aligned} P &= 18 + 48 + 144 + 162 - 128 \\ &= 244 \text{ (upward)} \\ M_x &= (18)(3.5) + (162)(1.5) + (144 + 48 - 128)(4.5) \\ &= 594 \text{ (counterclockwise)} \\ M_y &= (162)(6) + (144)(2) + (128)(0.667) - (48)(2) - (18)(6) \\ &= 1{,}141.4 \text{ (counterclockwise)} \end{aligned}$$

TABLE 91-2

Point	M_s	$\left[\frac{P}{A}\right]$	$\left[\frac{M_x y}{I_x}\right]$	$\left[\frac{M_y x}{I_y}\right]$	M_i	M
A......	0	$-\,244/17 = -14.35$	$+\,\frac{(594)(4.5)}{222.75} = +12.00$	$+\,\frac{(1,141.4)(6)}{420} = +16.31$	+13.96	−13.96
B......	−24	−14.35	−12.00	+16.31	−10.04	−13.96
C......	−72	−14.35	−12.00	−16.31	−42.66	−29.34
D......	0	−14.35	+12.00	−16.31	−18.66	+18.66

The answers are

$$\begin{aligned} M_A &= -13.95 \text{ kip-ft compression inside} \\ M_B &= -13.95 \text{ kip-ft compression inside} \\ M_C &= -29.35 \text{ kip-ft compression inside} \\ M_D &= +18.65 \text{ kip-ft compression outside} \end{aligned}$$

The free-body diagrams of all members and joints, the moment diagram, and the deformed structure are shown in Fig. 273.

EXERCISES

156 and 157. Analyze the closed rectangular frame shown by the method of column analogy. Check the solution by using a different basic determinate structure.

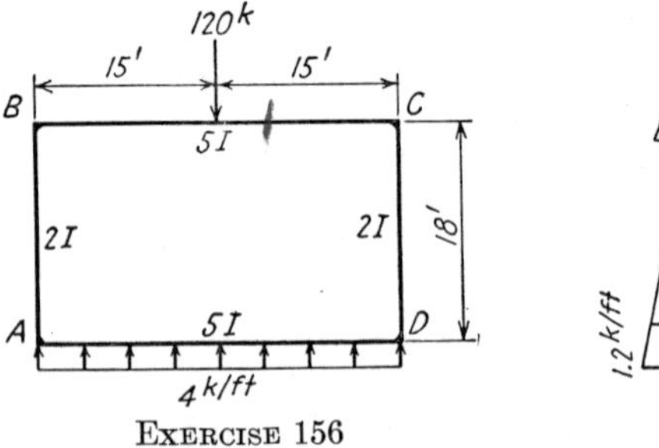

EXERCISE 156

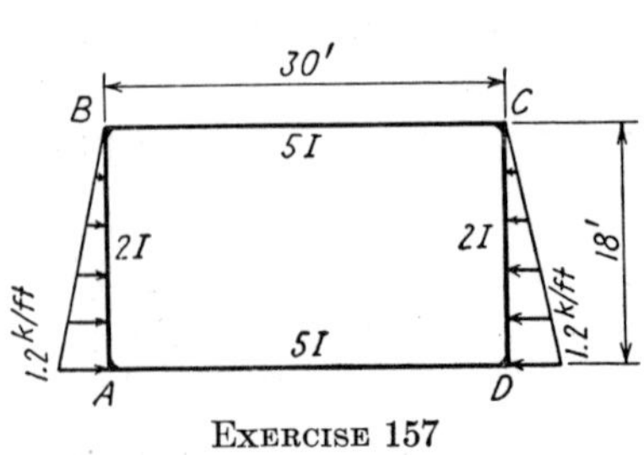

EXERCISE 157

62. Analysis of Gable Frames by the Method of Column Analogy. A gable frame with one axis of symmetry such as the one shown in Fig. 274

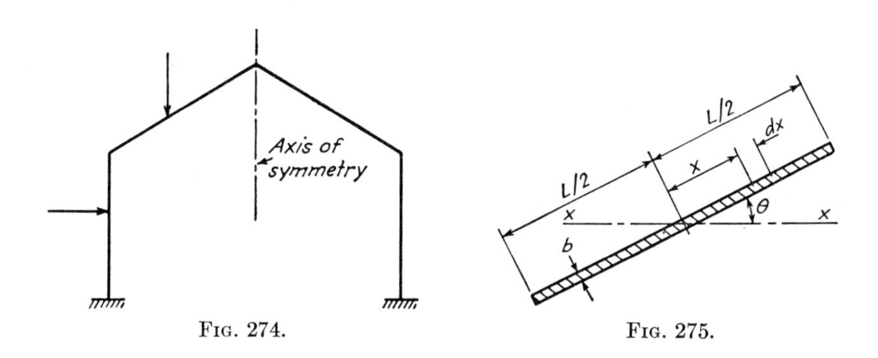

FIG. 274. FIG. 275.

can be analyzed by the method of column analogy in the same manner as the quadrangular frame with two fixed supports. In determining the moments of inertia of the analogous-column section for the gable frame, it is necessary to find the moment of inertia of a line area about a centroidal axis which is at an angle θ with the direction of the line. From Fig. 275,

$$I_{xx} = 2\int_0^{L/2} (b\,dx)(x\sin\theta)^2 = [\tfrac{2}{3}bx^3]_0^{L/2}(\sin^2\theta)$$

$$= \frac{bL^3}{12}\sin^2\theta \tag{130}$$

Example 92. Analyze the gable frame shown in Fig. 276*a* by the method of column analogy.

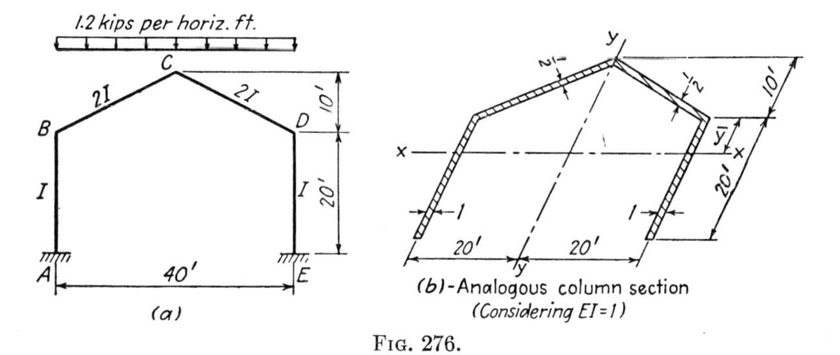

FIG. 276.

Solution. *Properties of the Analogous-column Section* (Fig. 276*b*)

$$A = (2)(20)(1) + (2)(22.36)(\tfrac{1}{2}) = 62.36$$

$$\bar{y} = \frac{(40)(10) - (22.36)(5)}{62.36} = 4.62 \text{ ft}$$

$$I_x = (2)\left[\frac{(1)(20)^3}{12} + (20)(5.38)^2\right] + (2)\left[\frac{(\frac{1}{2})(22.36)^3}{12}\left(\frac{1}{\sqrt{5}}\right)^2 + (11.18)(9.62)^2\right] = 4{,}746.7$$

$$I_y = (2)(20)(20)^2 + 2\left[\frac{(\frac{1}{2})(22.36)^3}{12}\left(\frac{2}{\sqrt{5}}\right)^2 + (11.18)(10)^2\right] = 18{,}981$$

Loads on Top of the Analogous Column (Fig. 277)

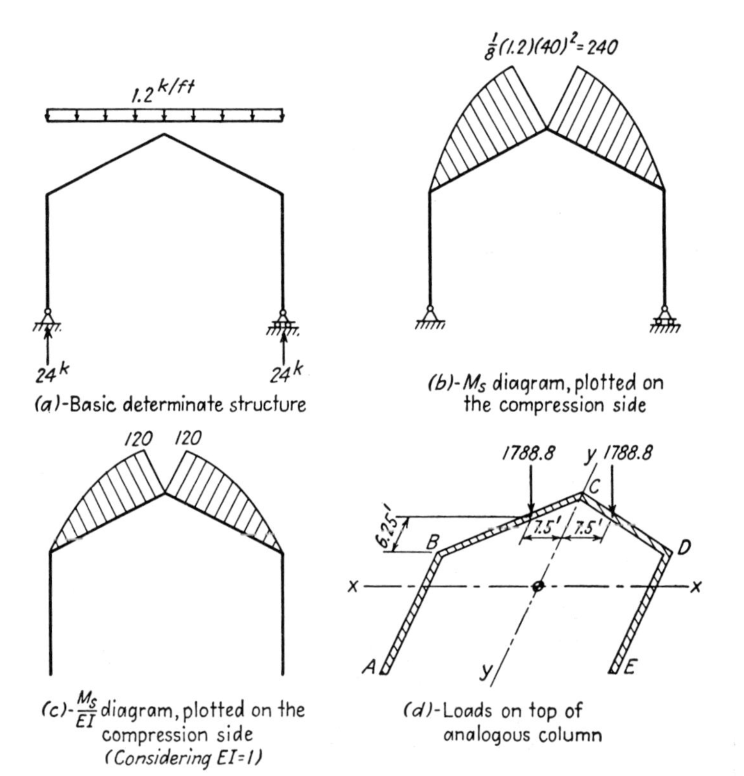

FIG. 277.

$$P = (2)(1{,}788.8) = 3{,}577.6 \text{ downward}$$

$$M_x = (2)(1{,}788.8)(10.87) = 38{,}888 \text{ clockwise}$$

$$M_y = 0$$

TABLE 92-1

Point	M_s	$\left[\frac{P}{A}\right]$	$\left[\frac{M_x y}{I_x}\right]$	M_i	M
A.....	0	$+\frac{3,577.6}{62.36} = +57.37$	$-\frac{38,888}{4,746.7}(15.38) = -126.00$	-68.63	$+68.63$
B.....	0	$+57.37$	$+\frac{38,888}{4,746.7}(4.62) = +37.85$	$+95.22$	-95.22
C.....	+240	$+57.37$	$+\frac{38,888}{4,746.7}(14.62) = +119.78$	$+177.15$	$+62.85$
D.....	0	$+57.37$	$+\frac{38,888}{4,746.7}(4.62) = +37.85$	$+95.22$	-95.22
E.....	0	$+57.37$	$-\frac{38,888}{4,746.7}(15.38) = -126.00$	-68.63	$+68.63$

The answers are (see Table 92-1)

$$M_A = M_E = +68.63 \text{ kip-ft compression outside}$$
$$M_B = M_D = -95.22 \text{ kip-ft compression inside}$$
$$M_C = +62.85 \text{ kip-ft compression outside}$$

For reactions, shear and moment diagrams, and the deformed structure, see Example 66.

Example 93. Analyze the gable frame shown in Fig. 278 by the method of column analogy.

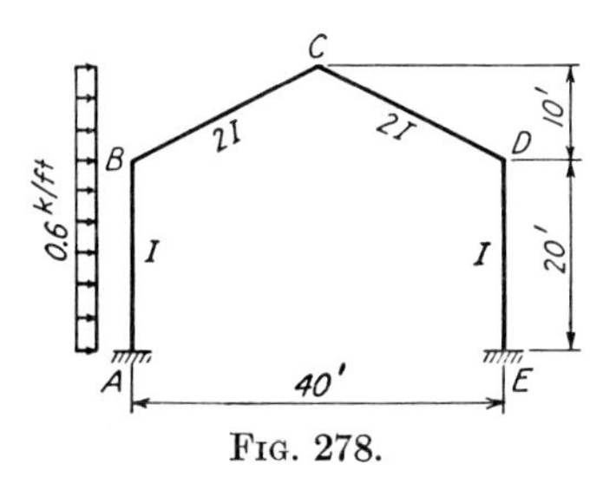

FIG. 278.

Solution. The properties of the analogous-column section have been determined in Example 92.

Loads on Top of the Analogous Column (Fig. 279). Referring to Fig. 279c,

$$A_2 = (\tfrac{1}{3})(15)(22.36) = 111.8$$
$$A_1 = (\tfrac{1}{3})(270)(30) - (\tfrac{1}{3})(30)(10) = 2,600$$

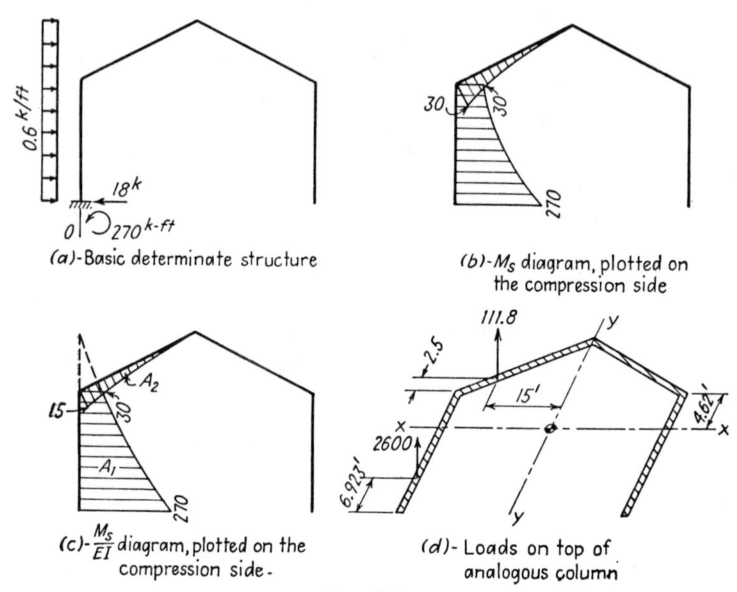

FIG. 279.

Distance from centroid of A_1 to the left support

$$= \frac{(2{,}700)(7.5) - (100)(22.5)}{2{,}600} = 6.923 \text{ ft}$$

Thus, on the analogous column,

$$P = 2{,}600 + 111.8 = 2{,}711.8 \text{ upward}$$
$$M_x = (2{,}600)(8.457) - (111.8)(7.12) = 21{,}192 \text{ clockwise}$$
$$M_y = (2{,}600)(20) + (111.8)(15) = 53{,}677 \text{ clockwise}$$

The answers are (see Table 93-1)

$$M_A = -101.29 \text{ kip-ft compression inside}$$
$$M_B = +49.42 \text{ kip-ft compression outside}$$
$$M_C = -21.78 \text{ kip-ft compression inside}$$
$$M_D = -33.70 \text{ kip-ft compression inside}$$
$$M_E = +55.59 \text{ kip-ft compression outside}$$

For reactions, shear and moment diagrams, and the deformed structure, see Example 67.

TABLE 93-1

Point	M_s	$\left[\frac{P}{A}\right]$	$\left[\frac{M_x y}{I_x}\right]$	$\left[\frac{M_y x}{I_y}\right]$	M_i	M
A..........	−270	$-\frac{2,711.8}{62.36} = -43.49$	$-\frac{21,192}{4,746.7}(15.38) = -68.66$	$-\frac{53,677}{18,981}(20) = -56.56$	−168.71	−101.29
B..........	− 30	−43.69	$+\frac{21,192}{4,746.7}(4.62) = +20.63$	−56.56	− 79.42	+ 49.42
C..........	0	−43.49	$+\frac{21,192}{4,746.7}(14.62) = +65.27$	0	+ 21.78	− 21.78
D..........	0	−43.49	$+\frac{21,192}{4,746.7}(4.62) = +20.63$	+56.56	+ 33.70	− 33.70
E..........	0	−43.49	$-\frac{21,192}{4,746.7}(15.38) = -68.66$	+56.56	− 55.59	+ 55.59

EXERCISES

158 to 161. Analyze the gable frame shown by the method of column analogy. Check the solution by using a different basic determinate structure.

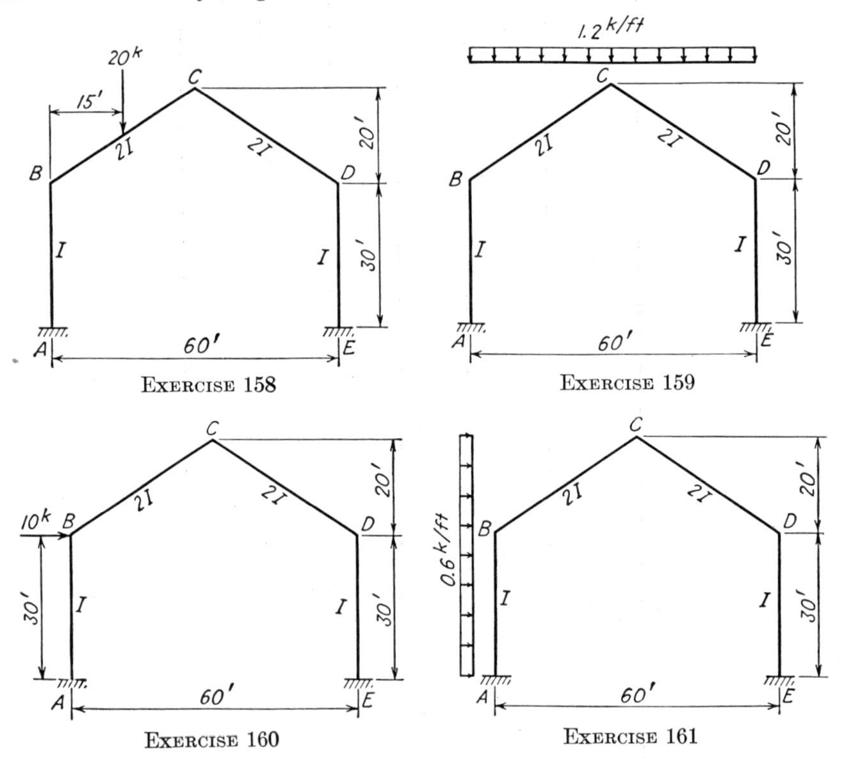

EXERCISE 158 EXERCISE 159

EXERCISE 160 EXERCISE 161

63. The Method of Column Analogy Applied to Unsymmetrical Quadrangular Frames. The method of column analogy can be used to analyze unsymmetrical quadrangular frames when such frames have two fixed supports and are subjected to the applied loadings. The analysis of an unsymmetrical quadrangular frame as shown in Fig. 280 by the method

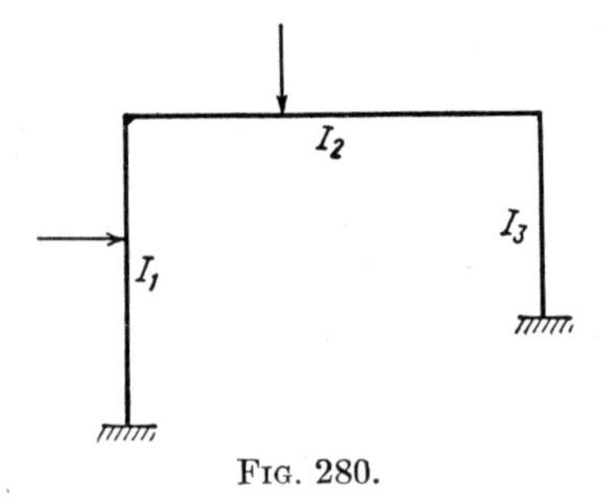

FIG. 280.

of consistent deformation requires the solution of three simultaneous equations. In the method of column analogy, however, the moment at any point of the frame can be determined by a *direct* procedure.

The problem of determining the pressure at any point in a column with an unsymmetrical cross section due to the combined action of a direct load at the centroid and bending moments about a pair of mutually perpendicular axes through the centroid will be discussed first. Consider the short column shown in Fig. 281a. It is subjected to the downward

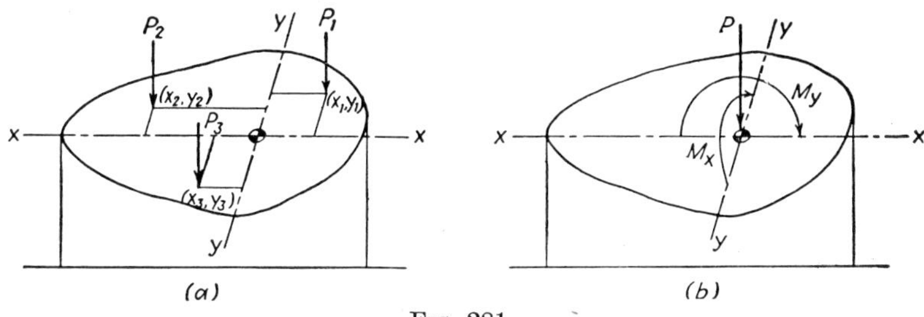

FIG. 281.

loads P_1, P_2, P_3, etc., at the points (x_1,y_1), (x_2,y_2), (x_3,y_3), etc. By the principles of statics, the loads P_1, P_2, P_3, etc., shown in Fig. 281a can be replaced by P, M_x, and M_y shown in Fig. 281b. Thus,

$$\begin{aligned} P &= P_1 + P_2 + P_3 + \cdots = \Sigma P \text{ downward} \\ M_x &= P_1y_1 + P_2y_2 + P_3y_3 + \cdots = \Sigma Py \text{ clockwise} \\ M_y &= P_1x_1 + P_2x_2 + P_3x_3 + \cdots = \Sigma Px \text{ clockwise} \end{aligned} \tag{131}$$

in which P_1, P_2, P_3, etc., are considered positive when acting downward and (x_1,y_1), (x_2,y_2), (x_3,y_3), etc., are coordinates of the points of application of P_1, P_2, P_3, etc., referred to the centroid as origin.

From the assumption of planar distribution of pressure over the cross section, the pressure p at any point (x,y) can be expressed by

$$p = a + bx + cy \tag{132}$$

where the constants a, b, and c can be determined from the three equations of statics, namely,

$$\begin{aligned} \int_0^A p\,dA &= P \\ \int_0^A (p\,dA)(y) &= M_x \\ \int_0^A (p\,dA)(x) &= M_y \end{aligned} \tag{133}$$

By substituting Eq. (132) into Eqs. (133), and noting that $\int_0^A dA = A$, $\int_0^A x\,dA = 0$, $\int_0^A y\,dA = 0$, $\int_0^A y^2\,dA = I_x$, $\int_0^A xy\,dA = I_{xy}$, and $\int_0^A x^2\,dA = I_y$, the following three equations are obtained:

$$\begin{aligned} P &= \int_0^A p\,dA = \int_0^A (a + bx + cy)\,dA \\ &= a\int_0^A dA + b\int_0^A x\,dA + c\int_0^A y\,dA \\ &= aA + 0 + 0 = aA \end{aligned} \tag{134}$$

$$\begin{aligned} M_x &= \int_0^A (p\,dA)(y) = \int_0^A (ay + bxy + cy^2)\,dA \\ &= a\int_0^A y\,dA + b\int_0^A xy\,dA + c\int_0^A y^2\,dA \\ &= 0 + bI_{xy} + cI_x = bI_{xy} + cI_x \end{aligned} \tag{135}$$

$$\begin{aligned} M_y &= \int_0^A (p\,dA)(x) = \int_0^A (ax + bx^2 + cxy)\,dA \\ &= a\int_0^A x\,dA + b\int_0^A x^2\,dA + c\int_0^A xy\,dA \\ &= 0 + bI_y + cI_{xy} = bI_y + cI_{xy} \end{aligned} \tag{136}$$

Solving Eqs. (134), (135), and (136) for a, b, and c,

$$\begin{aligned} a &= \frac{P}{A} \\ b &= \frac{\left(M_y - M_x\dfrac{I_{xy}}{I_x}\right)}{I_y\left(1 - \dfrac{I_{xy}^2}{I_xI_y}\right)} \\ c &= \frac{\left(M_x - M_y\dfrac{I_{xy}}{I_y}\right)}{I_x\left(1 - \dfrac{I_{xy}^2}{I_xI_y}\right)} \end{aligned} \tag{137}$$

Substituting Eqs. (137) into Eq. (132),

$$\begin{aligned} p &= \frac{P}{A} + \frac{\left(M_y - M_x\dfrac{I_{xy}}{I_x}\right)}{I_y\left(1 - \dfrac{I_{xy}^2}{I_xI_y}\right)}x + \frac{\left(M_x - M_y\dfrac{I_{xy}}{I_y}\right)}{I_x\left(1 - \dfrac{I_{xy}^2}{I_xI_y}\right)}y \\ &= \frac{P}{A} + \frac{M_y'x}{I_y'} + \frac{M_x'y}{I_x'} \end{aligned} \tag{138}$$

in which

$$M'_x = M_x - M_y\left(\frac{I_{xy}}{I_y}\right) \qquad M'_y = M_y - M_x\left(\frac{I_{xy}}{I_x}\right)$$

and

$$I'_x = I_x\left(1 - \frac{I_{xy}^2}{I_xI_y}\right) \qquad I'_y = I_y\left(1 - \frac{I_{xy}^2}{I_xI_y}\right)$$

The analysis of the quadrangular frame shown in Fig. 282*a* by the method of column analogy will now be taken up.

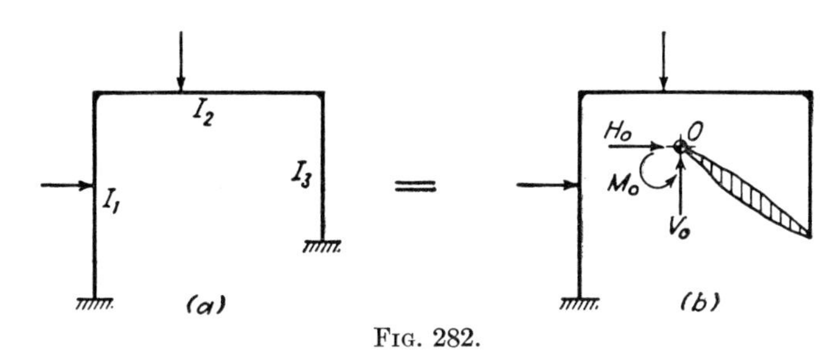

FIG. 282.

The discussion in regard to quadrangular frames with one axis of symmetry in Art. 60 applies equally well to the present problem of unsymmetrical quadrangular frames except that I_{xy} is now *not* equal to zero. Restating Eqs. (125),

$$\begin{aligned}
\text{Total } \Delta_\phi \text{ at } O &= +\sum \frac{M_s\,ds}{EI} + M_O \sum \frac{ds}{EI} + H_O \sum \frac{y\,ds}{EI} \\
&\qquad - V_O \sum \frac{x\,ds}{EI} = 0 \\
\text{Total } \Delta_H \text{ at } O &= +\sum \frac{M_s y\,ds}{EI} + M_O \sum \frac{y\,ds}{EI} + H_O \sum \frac{y^2\,ds}{EI} \\
&\qquad - V_O \sum \frac{xy\,ds}{EI} = 0 \qquad (125) \\
\text{Total } \Delta_V \text{ at } O &= -\sum \frac{M_s x\,ds}{EI} - M_O \sum \frac{x\,ds}{EI} - H_O \sum \frac{xy\,ds}{EI} \\
&\qquad + V_O \sum \frac{x^2\,ds}{EI} = 0
\end{aligned}$$

Substituting the loads on the top of the analogous column,

$$P = \sum \frac{M_s\,ds}{EI} \qquad M_x = \sum \frac{M_s y\,ds}{EI} \qquad M_y = \sum \frac{M_s x\,ds}{EI}$$

and the properties of the analogous-column section,

$$\sum \frac{ds}{EI} = A \qquad \sum \frac{x\,ds}{EI} = 0 \qquad \sum \frac{y\,ds}{EI} = 0$$
$$\sum \frac{xy\,ds}{EI} = I_{xy} \qquad \sum \frac{x^2\,ds}{EI} = I_y \qquad \sum \frac{y^2\,ds}{EI} = I_x$$

into Eqs. (125),

$$\begin{aligned} P + M_O A &= 0 \\ M_x + H_O I_x - V_O I_{xy} &= 0 \\ -M_y - H_O I_{xy} + V_O I_y &= 0 \end{aligned} \tag{139}$$

Solving Eqs. (139) for M_O, H_O, and V_O,

$$\begin{aligned} M_O &= -\frac{P}{A} \\ H_O &= -\frac{\left(M_x - M_y \dfrac{I_{xy}}{I_y}\right)}{I_x\left(1 - \dfrac{I_{xy}^2}{I_x I_y}\right)} = -\frac{M'_x}{I'_x} \\ V_O &= +\frac{\left(M_y - M_x \dfrac{I_{xy}}{I_x}\right)}{I_y\left(1 - \dfrac{I_{xy}^2}{I_x I_y}\right)} = +\frac{M'_y}{I'_y} \end{aligned} \tag{140}$$

Substituting Eqs. (140) into Eq. (127),

$$M = M_s + M_O + H_O y - V_O x = M_s - \left[\frac{P}{A} + \frac{M'_x y}{I'_x} + \frac{M'_y x}{I'_y}\right] \tag{141}$$

The sum of the three terms in the brackets of Eq. (141) is seen to be the same as the formula [Eq. (138)] derived for the pressure in a column when subjected to the loads P, M_x, and M_y. When $P = \Sigma(M_s\,ds/EI)$, $M_x = \Sigma(M_s y\,ds/EI)$, and $M_y = \Sigma(M_s x\,ds/EI)$, the pressure in the analogous column will be designated as M_i. Thus

$$M = M_s - M_i$$

As a summary of the method of column analogy in analyzing unsymmetrical quadrangular frames with two fixed supports, the following items should be noted:

1. The loading on the top of the analogous column is equal to the M_s/EI diagram. The loading is in the downward direction if M_s is positive, which means that it causes compression on the outside.

2. The properties of the analogous-column section may be found by referring to any pair of convenient x and y axes passing through the centroid, or the elastic center. The properties are A, I_x, I_{xy}, I_y, $I'_x = I_x(1 - I^2_{xy}/I_xI_y)$, and $I'_y = I_y(1 - I^2_{xy}/I_xI_y)$.

3. The moment at any point in the frame, M, is equal to $M_s - M_i$, where $M_i = \left[\frac{P}{A} + \frac{M'_x y}{I'_x} + \frac{M'_y x}{I'_y}\right]$, $M'_x = M_x - M_y(I_{xy}/I_y)$, $M'_y = M_y - M_x(I_{xy}/I_x)$, $P = \Sigma(M_s\,ds/EI)$, $M_x = \Sigma(M_s y\,ds/EI)$, and $M_y = \Sigma(M_s x\,ds/EI)$. In computing values of M'_x and M'_y, care must be exercised in substituting the correct signs for I_{xy}, M_x, and M_y. Clockwise M_x and M_y are positive.

Example 94. Analyze the unsymmetrical quadrangular frame shown in Fig. 283a by the method of column analogy.

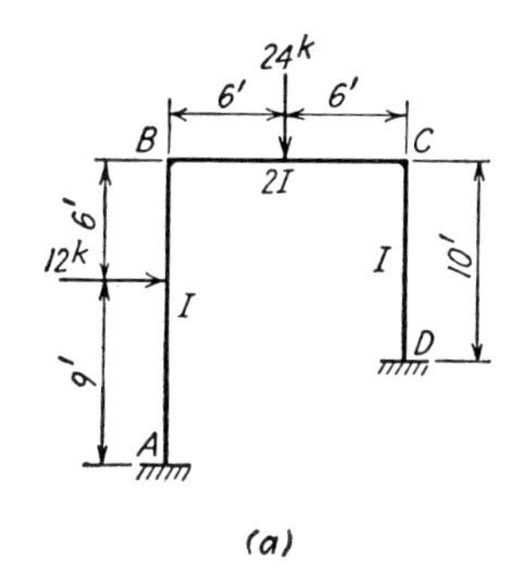

(a)

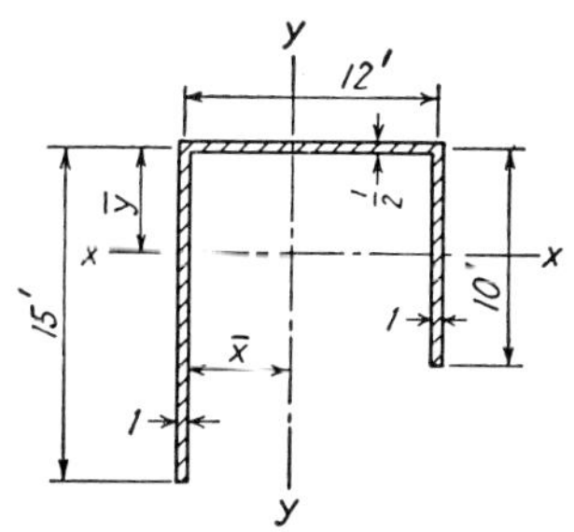

(b)-Analogous column section (Considering $EI=1$)

FIG. 283.

Solution. *Properties of the Analogous-column Section* (Fig. 283b)

$$A = (15)(1) + (12)(\tfrac{1}{2}) + (10)(1) = 31$$

$$\bar{x} = \frac{(15)(0) + (6)(6) + (10)(12)}{31} = 5.032 \text{ ft}$$

$$\bar{y} = \frac{(15)(7.5) + (6)(0) + (10)(5)}{31} = 5.242 \text{ ft}$$

$$I_x = \frac{(1)(15)^3}{12} + (15)(2.258)^2 + (6)(5.242)^2 + \frac{(1)(10)^3}{12} + (10)(0.242)^2$$
$$= 606.52$$

$$I_y = (15)(5.032)^2 + \frac{(\tfrac{1}{2})(12)^3}{12} + (6)(0.968)^2 + (10)(6.968)^2$$
$$= 942.97$$

$$I_{xy} = (15)(-5.032)(-2.258) + (6)(+0.968)(+5.242) + (10)(+6.968)(+0.242)$$
$$= +217.74$$

$$I'_x = I_x\left(1 - \frac{I_{xy}^2}{I_x I_y}\right) = (606.52)\left[1 - \frac{(+217.74)^2}{(606.52)(942.97)}\right]$$
$$= (606.52)(0.9171)$$
$$= 556.24$$

$$I'_y = I_y\left(1 - \frac{I_{xy}^2}{I_x I_y}\right) = (942.97)(0.9171) = 864.80$$

Loads on Top of Analogous Column (Fig. 284)

Case I (see Tables 94-1 and 94-2)

TABLE 94-1

P	x	y	$M_y = Px$	$M_x = Py$
− 486	−5.032	−6.758	+ 2,445.6	+3,284.4
−2,160	−5.032	−2.258	+10,869.1	+4,877.3
− 216	−3.032	+5.242	+ 654.9	−1,132.3
−2,862			+13,970	+7,029.4

$$M'_x = M_x - M_y\frac{I_{xy}}{I_y} = (+7{,}029.4) - (+13{,}970)\frac{+217.74}{+942.97} = +3{,}803.6$$

$$M'_y = M_y - M_x\frac{I_{xy}}{I_x} = (+13{,}970) - (+7{,}029.4)\frac{+217.74}{+606.52} = +11{,}447$$

It is to be noted that in this problem the signs for $[P/A]$, $[M'_y x/I'_y]$, and $[M'_x y/I'_x]$ develop automatically when the sign conventions for P, M'_x, M'_y, and the coordinates (x,y) are strictly followed.

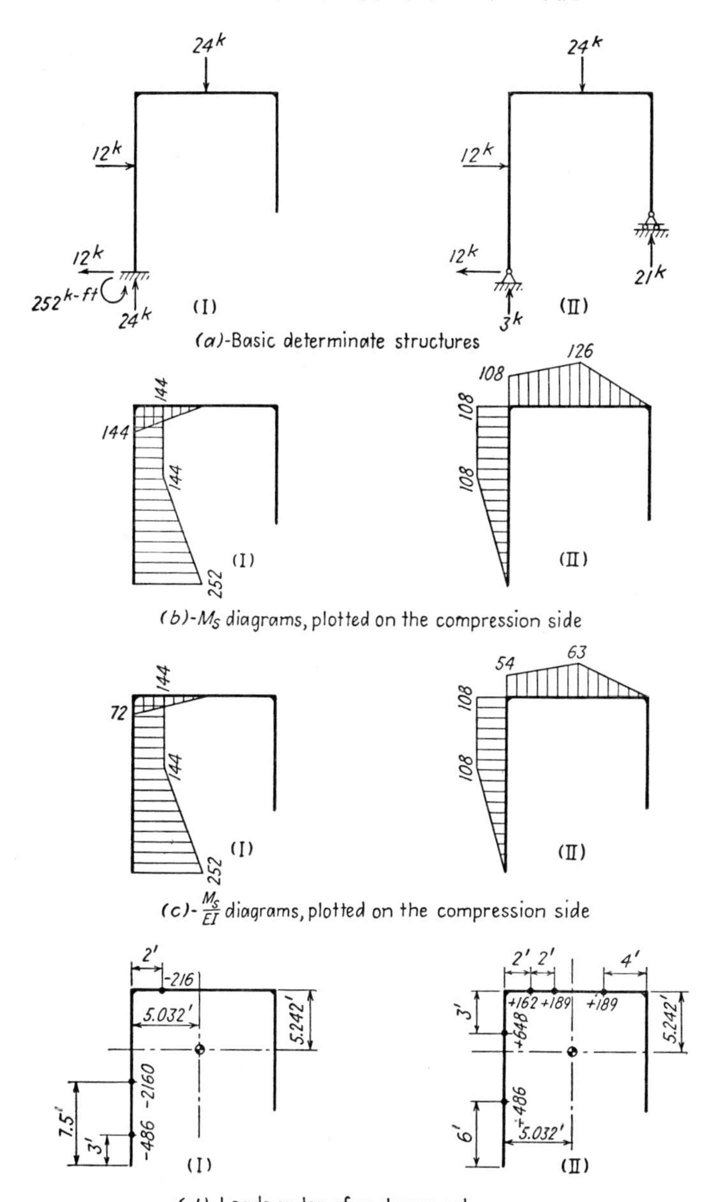

24k
12k
12k
252k-ft
24k
(I)
24k
12k
12k
3k
21k
(II)
(a)-Basic determinate structures
144
144
144
252
(I)
108
126
108
108
(II)
(b)-Ms diagrams, plotted on the compression side
72
144
144
252
(I)
54
63
108
108
(II)
(c)-Ms/EI diagrams, plotted on the compression side
2'
-216
5.032'
5.242'
7.5'
3'
-2160
-486
(I)
2' 2'
4'
+162 +189
+189
3'
+648
5.242'
+486
6'
5.032'
(II)
(d)-Loads on top of analogous column

Fig. 284.

TABLE 94-2

Point	x	y	M_s	$\left[\frac{P}{A}\right]$	$\left[\frac{M'_y x}{I'_y}\right]$	$\left[\frac{M'_x y}{I'_x}\right]$	M_i	M
A....	−5.032	−9.758	−252	$\frac{-2,862}{31} = -92.32$	$\frac{+11,447}{864.80}(-5.032) = -66.61$	$\frac{+3,803.6}{556.24}(-9.758) = -66.73$	−225.66	−26.34
B....	−5.032	+5.242	−144	−92.32	$\frac{+11,447}{864.80}(-5.032) = -66.61$	$\frac{+3,803.6}{556.24}(+5.242) = +35.85$	−123.08	−20.92
C.....	+6.968	+5.242	0	−92.32	$\frac{+11,447}{864.80}(+6.968) = +92.23$	$\frac{+3,803.6}{556.24}(+5.242) = +35.85$	+ 35.76	−35.76
D....	+6.968	−4.758	0	−92.32	$\frac{+11,447}{864.80}(+6.968) = +92.23$	$\frac{+3,803.6}{556.24}(-4.758) = -32.54$	− 32.63	+32.63

TABLE 94-4

Point	x	y	M_s	$\left[\frac{P}{A}\right]$	$\left[\frac{M'_y x}{I'_y}\right]$	$\left[\frac{M'_x y}{I'_x}\right]$	M_i	M
A...	−5.032	−9.758	0	$\frac{+1,374}{31} = +54.00$	$\frac{-6,713.6}{864.80}(-5.032) = +39.06$	$\frac{+3,803.5}{556.24}(-9.758) = -66.72$	+ 26.34	−26.34
B...	−5.032	+5.242	+108	+54.00	$\frac{-6,713.6}{864.80}(-5.032) = -39.06$	$\frac{+3,803.5}{556.24}(+5.242) = +35.84$	−128.90	−20.90
C....	+6.968	+5.242	0	+54.00	$\frac{-6,713.6}{864.80}(+6.968) = -54.09$	$\frac{+3,803.5}{556.24}(+5.242) = +35.84$	+ 35.75	−35.75
D...	+6.968	−4.758	0	+54.00	$\frac{-6,713.6}{864.80}(+6.968) = -54.09$	$\frac{+3,803.5}{556.24}(-4.758) = -32.53$	− 32.62	+32.62

Case II (see Tables 94-3 and 94-4)

TABLE 94-3

P	x	y	$M_y = Px$	$M_x = Py$
+ 486	−5.032	−3.758	−2,445.6	−1,826.4
+ 648	−5.032	+2.242	−3,260.7	+1,452.8
+ 162	−3.032	+5.242	− 491.2	+ 849.2
+ 189	−1.032	+5.242	− 195.0	+ 990.7
+ 189	+2.968	+5.242	+ 561.0	+ 990.7
+1,674			−5,831.5	+2,457.0

$$M'_x = M_x - M_y \frac{I_{xy}}{I_y} = (+2{,}457.0) - (-5{,}831.5)\frac{+217.74}{+942.97} = +3{,}803.5$$

$$M'_y = M_y - M_x \frac{I_{xy}}{I_x} = (-5{,}831.5) - (+2{,}457.0)\frac{+217.74}{+606.52} = -6{,}713.6$$

For reactions, shear and moment diagrams, and the deformed structure, see Example 62.

EXERCISES

162 and 163. Analyze the unsymmetrical quadrangular frame shown by the method of column analogy. Check the solution by using a different basic determinate structure.

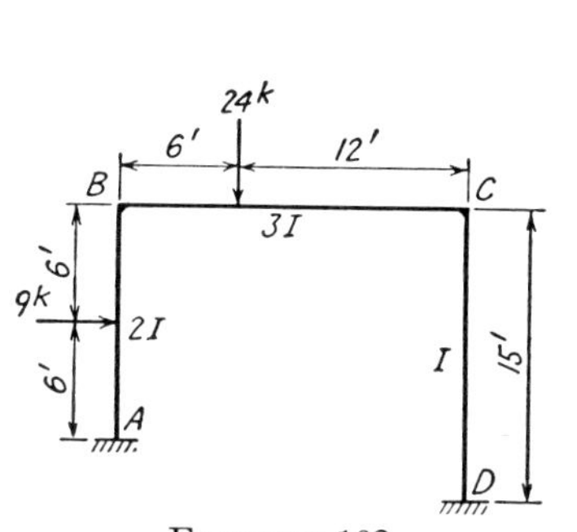

EXERCISE 162

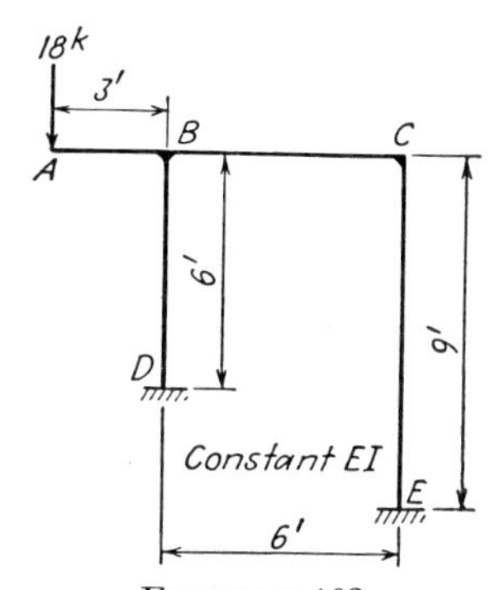

EXERCISE 163

64. The Method of Column Analogy Applied to Unsymmetrical Closed Frames. The method of column analogy can be used to analyze unsymmetrical closed frames when such frames are subjected to the applied loadings. It has been stated that the procedure of analyzing closed frames with one axis of symmetry is similar to that for quadrangular frames with one axis of symmetry and two fixed supports. Thus the procedure of analyzing unsymmetrical quadrangular frames with two fixed supports as summarized at the end of the preceding article can be similarly applied to the analysis of unsymmetrical closed frames. The following example will illustrate the detailed procedure.

Example 95. Analyze the unsymmetrical closed frame shown in Fig. 285*a* by the method of column analogy.

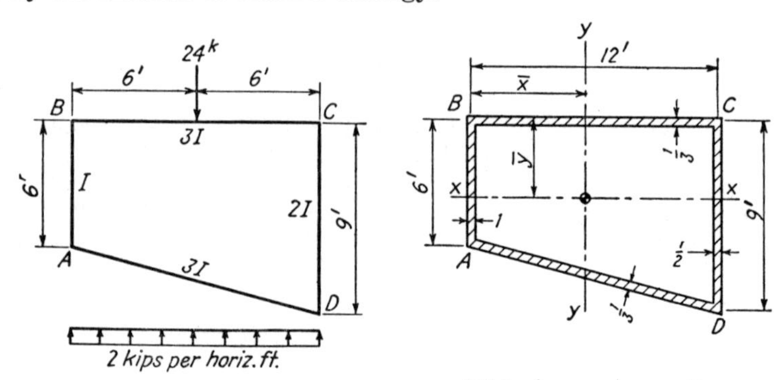

FIG. 285.

Solution. *Properties of the Analogous-column Section* (Fig. 285*b*)

$$A = (12)(\tfrac{1}{3}) + (12.369)(\tfrac{1}{3}) + (6)(1) + (9)(\tfrac{1}{2})$$
$$= 4 + 4.123 + 6 + 4.5 = 18.623$$

$$\bar{x} = \frac{(6)(0) + (4)(6) + (4.123)(6) + (4.5)(12)}{18.623}$$
$$= \frac{102.738}{18.623} = 5.5167$$

$$\bar{y} = \frac{(4)(0) + (6)(3) + (4.5)(4.5) + (4.123)(7.5)}{18.623}$$
$$= \frac{69.1725}{18.623} = 3.7144$$

$$I_x = (4)(3.7144)^2 + \frac{(1)(6)^3}{12} + (6)(0.7144)^2 + \frac{(\frac{1}{2})(9)^3}{12} + (4.5)(0.7856)^2 + \frac{(\frac{1}{3})(12.369)^3}{12}\left(\frac{1}{\sqrt{17}}\right)^2 + (4.123)(3.7856)^2$$
$$= 171.58$$

$$I_y = (6)(5.5167)^2 + (4.5)(6.4833)^2 + \frac{(\frac{1}{3})(12)^3}{12} + (4)(0.4833)^2 + \frac{(\frac{1}{3})(12.369)^3}{12}\left(\frac{4}{\sqrt{17}}\right)^2 + (4.123)(0.4833)^2$$
$$= 471.13$$

$$I_{xy} = (4)(+0.4833)(+3.7144) - \frac{(\frac{1}{3})(12.369)^3}{12}\left(\frac{1}{\sqrt{17}}\right)\left(\frac{4}{\sqrt{17}}\right) + (4.123)(+0.4833)(-3.7856) + (6)(-5.5167)(+0.7144) + (4.5)(+6.4833)(-0.7856)$$
$$= -59.298$$

$$I'_x = I_x\left(1 - \frac{I_{xy}^2}{I_xI_y}\right)$$
$$= (171.58)\left[1 - \frac{(-59.298)^2}{(171.58)(471.13)}\right]$$
$$= (171.58)(0.9565) = 164.12$$
$$I'_y = I_y\left(1 - \frac{I_{xy}^2}{I_xI_y}\right) = (471.13)(0.9565) = 450.64$$

Loads on Top of the Analogous Column (Fig. 286)

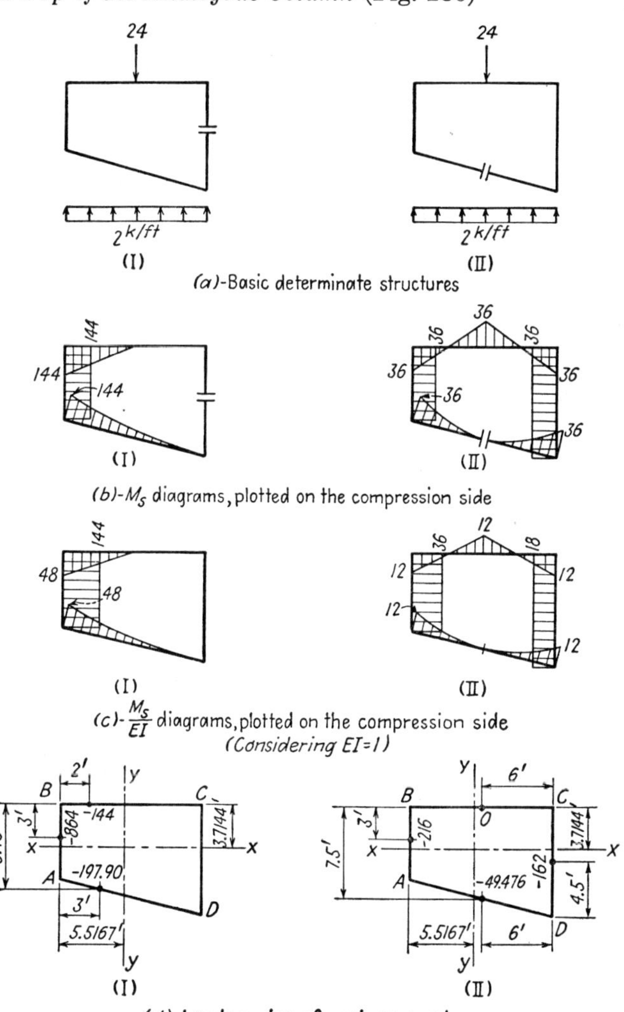

Fig. 286.

TABLE 95-1

Point	x	y	M_s	$\left[\frac{P}{A}\right]$	$\left[\frac{M_y'x}{I_y'}\right]$	$\left[\frac{M_x'y}{I_x'}\right]$	M_i	M
A.......	−5.5167	−2.2856	−144	$\frac{-1,205.90}{18.623} = -64.75$	$\frac{+5,580.3}{450.64}(-5.5167) = -68.31$	$\frac{+174.98}{164.12}(-2.2856) = -2.44$	−135.50	− 8.50
B.......	−5.5167	+3.7144	−144	−64.75	$\frac{+5,580.3}{450.64}(-5.5167) = -68.31$	$\frac{+174.98}{164.12}(+3.7144) = +3.96$	−129.10	−14.90
C.......	+6.4833	+3.7144	0	−64.75	$\frac{+5,580.3}{450.64}(+6.4833) = +80.28$	$\frac{+174.98}{164.12}(+3.7144) = +3.96$	+ 19.49	−19.49
D.......	+6.4833	−5.2856	0	−64.75	$\frac{+5,580.3}{450.64}(+6.4833) = +80.28$	$\frac{+174.98}{164.12}(-5.2856) = -5.64$	+ 9.89	− 9.89

TABLE 95-2

Point	x	y	M_s	$\left[\frac{P}{A}\right]$	$\left[\frac{M'_yx}{I'_y}\right]$	$\left[\frac{M'_xy}{I'_x}\right]$	M_i	M
A....	−5.5167	−2.2856	−36	$\frac{-427.48}{18.623} = -22.95$	$\frac{+172.79}{450.64}(-5.5167) = -2.11$	$\frac{+175.04}{164.12}(-2.2856) = -2.44$	−27.50	− 8.50
B....	−5.5167	+3.7144	−36	−22.95	$\frac{+172.79}{450.64}(-5.5167) = -2.11$	$\frac{+175.04}{164.12}(+3.7144) = +3.96$	−21.10	−14.90
C.....	+6.4833	+3.7144	−36	−22.95	$\frac{+172.79}{450.64}(+6.4833) = +2.48$	$\frac{+175.04}{164.12}(+3.7144) = +3.96$	−16.51	−19.49
D....	+6.4833	−5.2856	−36	−22.95	$\frac{+172.79}{450.64}(+6.4833) = +2.48$	$\frac{+175.04}{164.12}(-5.2856) = -5.64$	−26.11	− 9.89

Case I (see Table 95-1)

$$P = -144 - 864 - 197.90 = -1{,}205.90$$
$$M_x = (-144)(+3.7144) + (-864)(+0.7144) + (-197.90)(-3.0356)$$
$$= -551.36$$
$$M_y = (-144)(-3.5167) + (-864)(-5.5167) + (-197.90)(-2.5167)$$
$$= +5{,}770.88$$
$$M'_x = M_x - M_y \frac{I_{xy}}{I_y}$$
$$= (-551.36) - (+5{,}770.88)\left(\frac{-59.298}{471.13}\right) = +174.98$$
$$M'_y = M_y - M_x \frac{I_{xy}}{I_x}$$
$$= (+5{,}770.88) - (-551.36)\left(\frac{-59.298}{171.58}\right) = +5{,}580.3$$

Case II (see Table 95-2)

$$P = -216 - 49.476 - 162 = -427.48$$
$$M_x = (-216)(+0.7144) + (-49.476)(-3.7856) + (-162)(-0.7856)$$
$$= +160.26$$
$$M_y = (-216)(-5.5167) + (-49.476)(+0.4833) + (-162)(+6.4833)$$
$$= +117.41$$
$$M'_x = M_x - M_y \frac{I_{xy}}{I_y}$$
$$= (+160.26) - (+117.41)\left(\frac{-59.298}{471.13}\right) = +175.04$$
$$M'_y = M_y - M_x \frac{I_{xy}}{I_x}$$
$$= (+117.41) - (+160.26)\left(\frac{-59.298}{171.58}\right) = +172.79$$

The free-body diagrams, moment diagram, and the deformed structure are shown in Fig. 287.

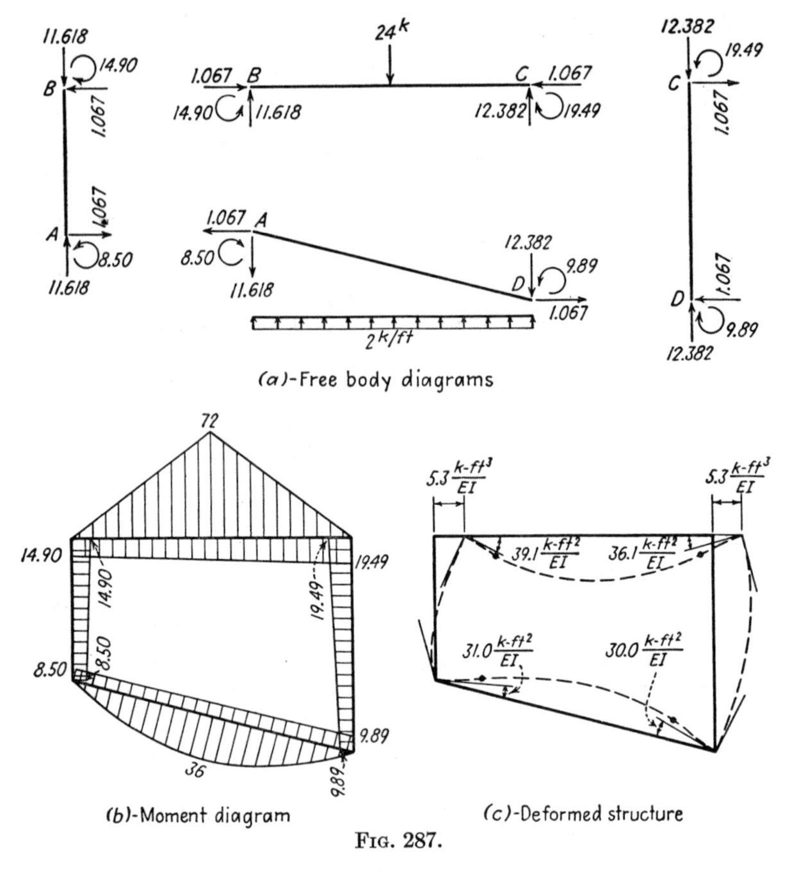

FIG. 287.

EXERCISE

164. Analyze the unsymmetrical closed frame shown by the method of column analogy. Check the solution by using a different basic determinate structure.

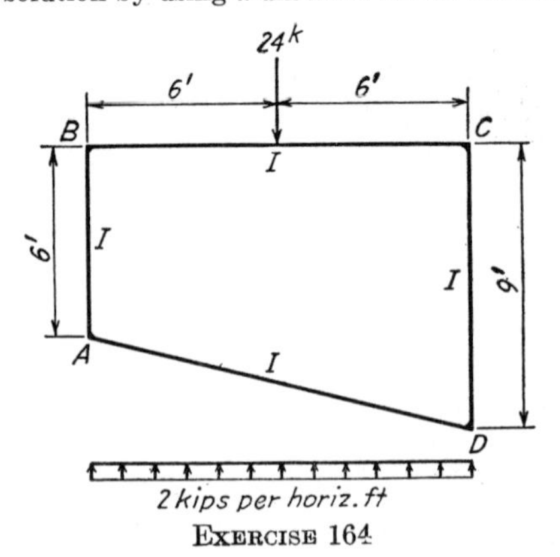

EXERCISE 164

CHAPTER X

ANALYSIS OF FIXED ARCHES

65. General Description. A fixed arch is a curved member with two fixed supports. The fixed arch may be symmetrical about a vertical axis (Fig. 288*a*) or unsymmetrical (Fig. 288*b*). A fixed arch is completely

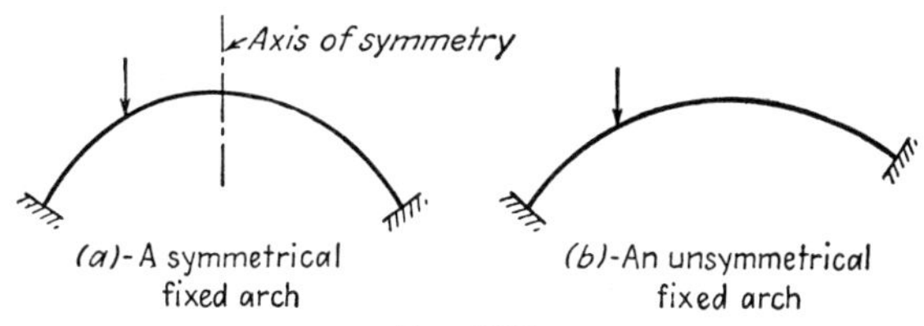

FIG. 288.

analyzed when the thrust, shear, and moment at any section perpendicular to the arch axis are known. The thrust, N, usually in the form of a push, is the total force acting in a direction perpendicular to the section at its centroid. The shear, V, is the total force acting in a direction parallel to the section. The moment, M, is the total moment about the horizontal centroidal axis in the cross section. In regard to the signs for N, V, and M, the directions as shown in Fig. 289 are positive.

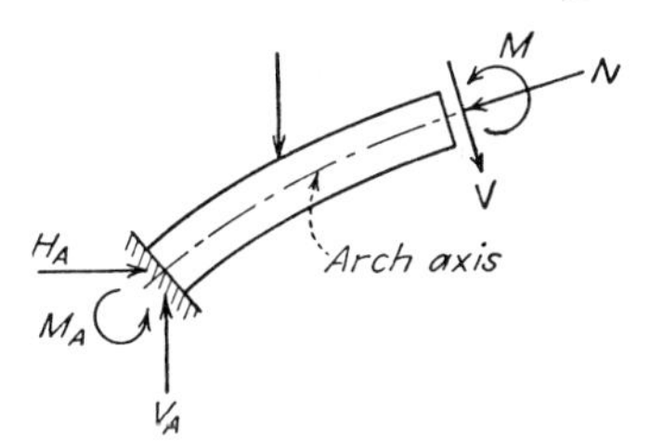

FIG. 289.

It is apparent that the thrust, shear, and moment at any section of a fixed arch can be readily determined by the laws of statics if the six reaction components at the two fixed supports are known. Since, by considering the whole arch as a free body, only three independent equations can be made by use of the principles of statics, the fixed arch is statically indeterminate to the third degree. The elastic-center and the column-analogy methods, as will be discussed in the next article, are the most convenient methods of analyzing the fixed arch.

As is true of any statically indeterminate structure, the properties of the fixed arch must be completely known before it can be analyzed. Thus before analyzing *an arch*, a design must be assumed. From the results of the ensuing analysis, the first assumed arch may be modified and, if necessary, reanalyzed. The procedure should be repeated until the last adopted arch can sustain at all sections the thrust, shear, and moment as found in the final analysis.

While a fixed arch may be used for roofs or bridges, its analysis can best be performed by constructing influence lines for the reaction components or the moments at certain evenly spaced sections. Usually the position of live load for maximum combined stress at any section is determined by the criterion for maximum moment, and the corresponding shear and thrust are then found from this loading position. In this chapter the discussion will be limited to the determination of the influence lines for the various functions for a fixed arch when its properties are given. In fact, for this purpose, only the shape of the arch axis and the *relative* sizes of the moments of inertia at different sections are required. Then the effect of temperature, shrinkage, rib shortening, and foundation yielding will be discussed. In the latter case, the *absolute* sizes of the moments of inertia are required.

66. The Elastic-center Method vs. the Column-analogy Method. The influence line for a certain function is defined as the graph showing the variation of this function when a unit load moves across the span. Thus the problem of analyzing a fixed arch reduces to that of first finding the six reaction components when the unit load takes various positions. Usually the shape of the arch axis and the way in which the moment of inertia varies along the span do not fit into any mathematical equations. The curved member, however, can be considered to be made of a finite number of straight segments, each of which has a *constant* moment of inertia. Thus the rigid frame *ABCDEFG* as shown in Fig. 290 is in

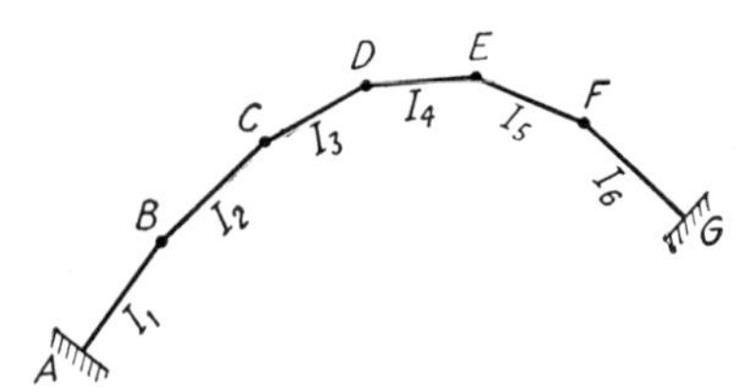

FIG. 290.

effect a fixed arch. Of course, the more divisions that are made in the arch, the more accurate will be the results. The quadrangular frames with two fixed supports can be considered as a "simplified arch" with only three finite members. Thus the method of column analogy as explained in the preceding chapter can be directly applied to the analysis of fixed arches.

It has sometimes been considered that there are two methods of analyzing the fixed arch, namely, the elastic-center method and the column-analogy method. In reality, the two methods are *identical* in every detail of arithmetic and only slightly different in the final step when the moment at any section is found. In the elastic-center method the redundants M_O, H_O, and V_O are first found, and the moment M at any section is determined as the bending moment in a cantilever structure $ABCDEFGO$ fixed at A and free at the elastic center O (Fig. 291); or

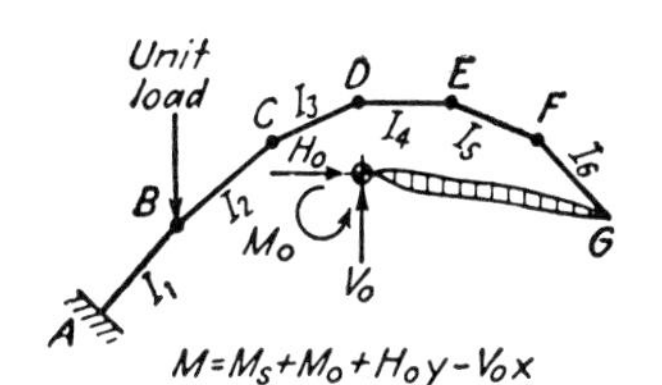

The elastic center method

FIG. 291.

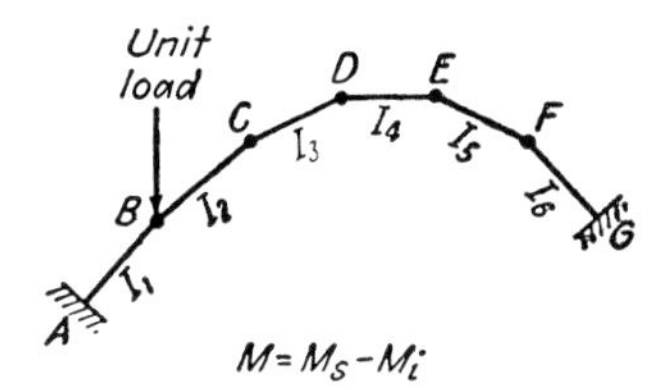

The column analogy method

FIG. 292.

$M = M_s + M_O + H_O y - V_O x$. In the column-analogy method the moment M at any section is directly determined from the relation $M = M_s - M_i$, where M_i is the pressure in the analogous column (Fig. 292). A mixed procedure will be suggested for use in the analysis of a fixed arch wherein values of M_O, H_O, and V_O are found for any loading position as in the elastic-center method; and the moments at the left support (or left springing), at the crown, and at the right support (or right springing) are found as in the column-analogy method. Of course the results from both methods should agree, a fact which sometimes serves as a good check for arithmetical mistakes, which so often happen in the last operation of computing the moment.

For convenience, a résumé of the formulas derived in the preceding chapter is now stated.

For both symmetrical and unsymmetrical arches,

$$P = \sum \frac{M_s\, ds}{EI} \qquad M_x = \sum \frac{M_s y\, ds}{EI} \qquad M_y = \sum \frac{M_s x\, ds}{EI}$$

$$A = \int_0^A \frac{ds}{EI} \qquad I_x = \int_0^A \frac{y^2\, ds}{EI} \qquad I_{xy} = \int_0^A \frac{xy\, ds}{EI} \qquad I_y = \int_0^A \frac{x^2\, ds}{EI}$$

x, y = coordinates referred to the elastic center as origin.

$$\int_0^A \frac{x\, ds}{EI} = 0 \qquad \int_0^A \frac{y\, ds}{EI} = 0$$

For symmetrical arches,

$$M_O = -\frac{P}{A} \qquad H_O = -\frac{M_x}{I_x} \qquad V_O = +\frac{M_y}{I_y}$$

$$M = M_s + M_O + H_O y - V_O x \text{ (elastic-center method)}$$

or

$$M = M_s - M_i \text{ (column-analogy method)}$$

in which

$$M_i = \left[\frac{P}{A}\right] + \left[\frac{M_x y}{I_x}\right] + \left[\frac{M_y x}{I_y}\right]$$

For unsymmetrical arches,

$$M'_x = M_x - M_y\left(\frac{I_{xy}}{I_y}\right) \qquad M'_y = M_y - M_x\left(\frac{I_{xy}}{I_x}\right)$$

$$I'_x = I_x\left(1 - \frac{I_{xy}^2}{I_x I_y}\right) \qquad I'_y = I_y\left(1 - \frac{I_{xy}^2}{I_x I_y}\right)$$

$$M_O = -\frac{P}{A} \qquad H_O = -\frac{M'_x}{I'_x} \qquad V_O = +\frac{M'_y}{I'_y}$$

$$M = M_s + M_O + H_O y - V_O x \text{ (elastic-center method)}$$

or

$$M = M_s - M_i \text{ (column-analogy method)}$$

in which

$$M_i = \left[\frac{P}{A}\right] + \left[\frac{M'_x y}{I'_x}\right] + \left[\frac{M'_y x}{I'_y}\right]$$

For explanation of notations and derivation of the above formulas, the reader is referred to Arts. 60 and 63.

For the purpose of showing the detailed procedure in applying the above formulas, some simple form of the arch axis and way of variation of the cross section will be chosen in the examples which follow. For the usual types of arch axis and ways of variation of the cross section, the reader is referred to treatments elsewhere on the design aspects of fixed arches.[1]

67. Influence Lines for a Symmetrical Fixed Arch

Example 96. For the symmetrical parabolic fixed arch shown in Fig. 293, draw influence lines for (a) the horizontal reaction at the left spring-

[1] Hale Sutherland and Raymond C. Reese, "Reinforced Concrete Design," 2d ed., Chap. XX, John Wiley & Sons, Inc., New York, 1943.

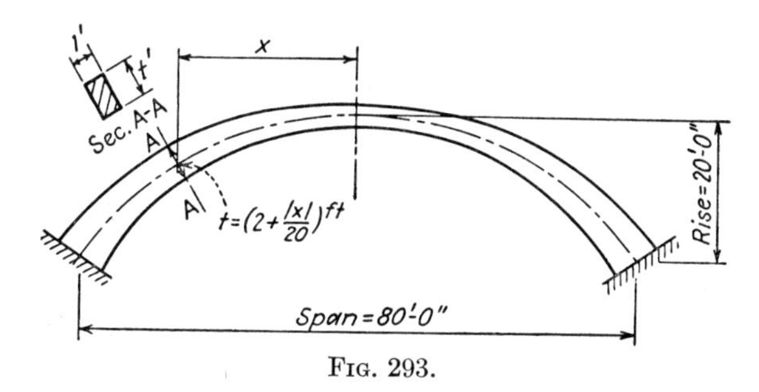

FIG. 293.

ing, (*b*) the vertical reaction at the left springing, (*c*) the vertical shear at the crown, (*d*) the moment at the left springing, (*e*) the moment at the quarter point, and (*f*) the moment at the crown.

Solution. The arch rib will be replaced by eight finite segments, as represented by *AB*, *BC*, *CD*, etc. (Fig. 294). The moment of inertia

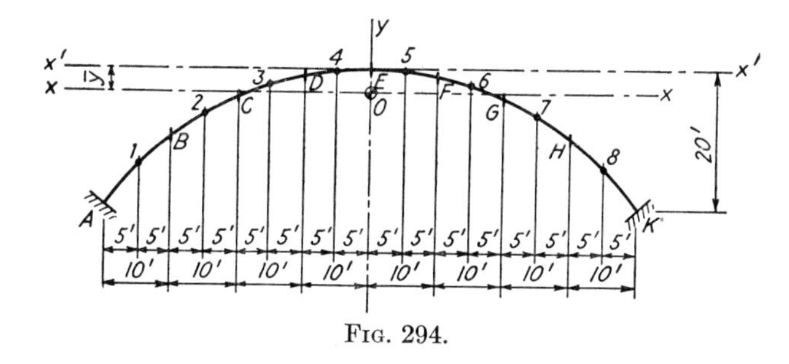

FIG. 294.

of segment *AB* will be considered to be constant and equal to that of the section perpendicular to the arch axis at the point 1. Also, point 1 is considered to be the centroid of segment *AB*, and the moment or product of inertia of segment *AB* about any axes through point 1 will be neglected. The same assumptions are made for all other segments. The moment of inertia of the section through the crown, *E*, will be called I_c, and the moments of inertia at sections through points 1, 2, 3, etc., will be expressed in terms of I_c. In computing properties of the analogous-column section EI_c will be considered equal to unity. The points on the influence lines will be plotted as the unit load takes successively the positions *B*, *C*, *D*, *E*, *F*, *G*, and *H*. It can be shown that, on account of symmetry, computations of redundants for loading positions *F*, *G*, and *H* are not necessary; for instance, the reaction components at the right springing when the unit load is at *C* will be the same as those at the left springing when the unit load is at *G*.

Properties of the Analogous-column Section (Fig. 294) (see Table 96-1)

TABLE 96-1. PROPERTIES OF THE ANALOGOUS-COLUMN SECTION

Segment	Centroid	ds, ft	t, ft	n ($I = nI_c$)	$\frac{ds}{EI} = \frac{ds}{E(nI_c)} = \frac{ds}{n}$ (Let $EI_c = 1$)	y'	$y' \frac{ds}{EI}$	x	y	$x^2 \frac{ds}{EI}$	$y^2 \frac{ds}{EI}$
AB.........	1	13.29	3.75	6.592	2.016	−15.31	−30.86	−35	−11.17	2,470	251.5
BC.........	2	11.79	3.25	4.291	2.748	− 7.81	−21.46	−25	− 3.67	1,718	37.0
CD.........	3	10.68	2.75	2.600	4.108	− 2.81	−11.54	−15	+ 1.33	924	7.3
DE.........	4	10.08	2.25	1.424	7.079	− 0.31	− 2.19	− 5	+ 3.83	177	103.8
					15.951	$\bar{y}' = -4.14$	−66.05			5,289	399.6
					$A = 31.902$					$I_y = 10,578$	$I_x = 799.2$

Sample Calculation for Segment CD (Fig. 295)

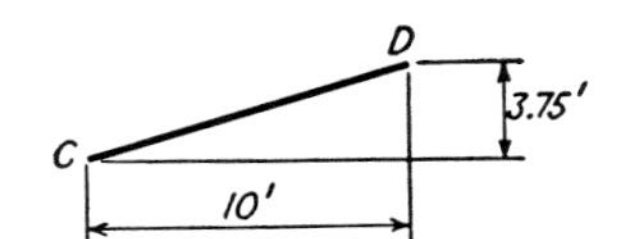

FIG. 295.

y' of point 3 $= -(\frac{15}{40})^2(20) = -(\frac{3}{8})^2(20) = -2.81$ ft

$ds = \sqrt{(10)^2 + (3.75)^2} = 10.68$ ft

$t = 2 + \frac{15}{20} = 2.75$ ft

$$n = \frac{I}{I_c} = \frac{(1)(t)^3/12}{(1)(2)^3/12} = \left(\frac{t}{2}\right)^3 = \left(\frac{2.75}{2}\right)^3 = 2.600$$

$$\bar{y}' = -\frac{66.05}{15.951} = -4.14 \text{ ft}$$

y of point 3 $= 4.14 - 2.81 = +1.33$ ft

In Table 96-2 are shown the computations for the values of the redundants M_O, H_O, and V_O as in the elastic-center method, and also for the values of the moments at the points A, E, and K as in the column-analogy method.

In Fig. 296 are shown the free-body diagrams for the equivalent cantilever structure with the rigid arm attached and also for the given fixed arch, when the unit loads are at B, C, D, and E, respectively.

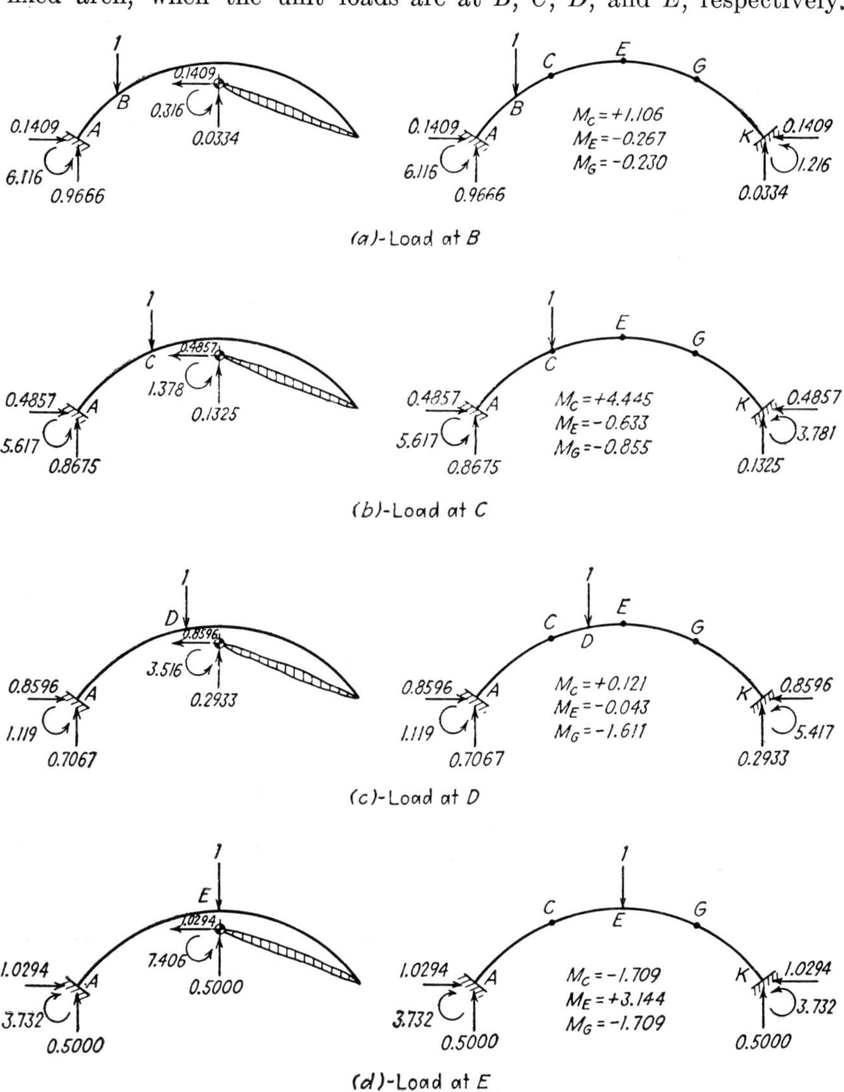

FIG. 296.

TABLE 96-2

Properties of analogous-column section					Load at B				Load at C			
Segment	Centroid	x	y	$\frac{ds}{EI}$	M_s	$\frac{M_s\,ds}{EI}$	$\frac{M_sx\,ds}{EI}$	$\frac{M_sy\,ds}{EI}$	M_s	$\frac{M_s\,ds}{EI}$	$\frac{M_sx\,ds}{EI}$	$\frac{M_sy\,ds}{EI}$
AB........	1	−35	−11.17	2.016	−5	−10.08	+352.8	+112.6	−15	−30.24	+1,058.4	+337.7
BC........	2	−25	− 3.67	2.748					− 5	−13.74	+ 343.5	+ 50.4
CD........	3	−15	+ 1.33	4.108								
DE........	4	− 5	+ 3.83	7.079								
	$A = 31.902$ $I_x = 799.2$ $I_y = 10{,}578$					−10.08	+352.8	+112.6		−43.98	+1,401.9	+388.1
						$P = -10.08$	$M_y = +352.8$	$M_x = +112.6$		$P = -43.98$	$M_y = +1.401.9$	$M_x = +338.1$
					$Mo = -\frac{P}{A} = -\frac{(-10.08)}{31.902} = +0.316$				$Mo = -\frac{P}{A} = -\frac{(-43.98)}{31.902} = +1.378$			
					$Ho = -\frac{M_x}{I_x} = -\frac{(+112.6)}{799.2} = -0.1409$				$Ho = -\frac{M_x}{I_x} = -\frac{(+388.1)}{799.2} = -0.4856$			
					$V_o = +\frac{M_y}{I_y} = +\frac{(+352.8)}{10{,}578} = +0.0334$				$Vo = +\frac{M_y}{I_y} = +\frac{(+1{,}401.9)}{10{,}578} = +0.1325$			

Load at B

Point	x	y	M_s	$\left[\frac{P}{A}\right]$	$\left[\frac{M_xy}{I_x}\right]$	$\left[\frac{M_yx}{I_y}\right]$	M_i	M
A...........	−40	−15.86	−10	$\frac{(-10.08)}{31.902} = -0.316$	$\frac{(+112.6)}{799.2}(-15.86) = -2.234$	$\frac{(+352.8)}{10{,}578}(-40) = -1.334$	−3.884	−6.116
E...........	0	+ 4.14	0	−0.316	$\frac{(+112.6)}{799.2}(+4.14) = +0.583$	$\frac{(+352.8)}{10{,}578}(0) = 0$	+0.267	−0.267
K...........	+40	−15.86	0	−0.316	$\frac{(+112.6)}{799.2}(-15.86) = -2.234$	$\frac{(+352.8)}{10{,}578}(+40) = +1.334$	−1.216	+1.216

Load at C

Point	x	y	M_s	$\left[\frac{P}{A}\right]$	$\left[\frac{M_xy}{I_x}\right]$	$\left[\frac{M_yx}{I_y}\right]$	M_i	M
A...........	−40	−15.86	−20	$\frac{(-43.98)}{31.902} = -1.378$	$\frac{(+388.1)}{799.2}(-15.86) = -7.704$	$\frac{(+1{,}401.9)}{10{,}578}(-40) = -5.301$	−14.383	−5.617
E...........	0	+ 4.14	0	−1.378	$\frac{(+388.1)}{799.2}(+4.14) = +2.011$	$\frac{(+1{,}401.9)}{10{,}578}(0) = 0$	+ 0.633	−0.633
K...........	+40	−15.86	0	−1.378	$\frac{(+388.1)}{799.2}(-15.86) = -7.704$	$\frac{(+1{,}401.9)}{10{,}578}(+40) = +5.301$	− 3.781	+3.781

TABLE 96-2.—(*Continued*)

Properties of analogous-column section					Load at D				Load at E			
Segment	Centroid	x	y	$\frac{ds}{EI}$	M_s	$\frac{M_s\,ds}{EI}$	$\frac{M_sx\,ds}{EI}$	$\frac{M_sy\,ds}{EI}$	M_s	$\frac{M_s\,ds}{EI}$	$\frac{M_sx\,ds}{EI}$	$\frac{M_sy\,ds}{EI}$
AB.......	1	−35	−11.17	2.016	−25	−50.40	+1,764.0	+563.0	−35	−70.56	+2,469.6	+788.2
BC.......	2	−25	−3.67	2.748	−15	−41.22	+1,030.5	+151.3	−25	−68.70	+1,717.5	+252.1
CD.......	3	−15	+1.33	4.108	−5	−20.54	+308.1	−27.3	−15	−61.62	+924.3	−82.0
DE.......	4	−5	+3.83	7.079					−5	−35.40	+177.0	−135.6
	$A = 31.902$ $I_x = 799.2$ $I_y = 10{,}578$					−112.16	+3,102.6	+687.0		−236.28	+5,288.4	+822.7
						$P = -112.16$	$M_y = +3{,}102.6$	$M_x = +687.0$		$P = -236.28$	$M_y = +5{,}288.4$	$M_x = +822.7$

Load at D:

$$Mo = -\frac{P}{A} = -\frac{(-112.16)}{31.902} = +3.516$$

$$Ho = -\frac{M_x}{I_x} = -\frac{(+687.0)}{799.2} = -0.8596$$

$$Vo = +\frac{M_y}{I_y} = +\frac{(+3{,}102.6)}{10{,}578} = +0.2933$$

Load at E:

$$Mo = -\frac{P}{A} = -\frac{(-236.28)}{31.902} = +7.406$$

$$Ho = -\frac{M_x}{I_x} = -\frac{(+822.7)}{799.2} = -1.0294$$

$$Vo = +\frac{M_y}{I_y} = +\frac{(+5{,}288.4)}{10{,}578} = +0.5000$$

Load at D

Point	x	y	M_s	$\left[\frac{P}{A}\right]$	$\left[\frac{M_xy}{I_x}\right]$	$\left[\frac{M_yx}{I_y}\right]$	M_i	M
A...........	−40	−15.86	−30	$\frac{-112.16}{31.902} = -3.516$	$\frac{+687.0}{799.2}(-15.86) = -13.633$	$\frac{+3{,}102.6}{10{,}578}(-40) = -11.732$	−28.881	−1.119
E...........	0	+4.14	0	−3.516	$\frac{+687.0}{799.2}(+4.14) = +3.559$	$\frac{+3{,}102.6}{10{,}578}(0) = 0$	+0.043	−0.043
K...........	+40	−15.86	0	−3.516	$\frac{+687.0}{799.2}(-15.86) = -13.633$	$\frac{+3{,}102.6}{10{,}578}(+40) = +11.732$	−5.417	+5.417

Load at E

Point	x	y	M_s	$\left[\frac{P}{A}\right]$	$\left[\frac{M_xy}{I_x}\right]$	$\left[\frac{M_yx}{I_y}\right]$	M_i	M
A...........	−40	−15.86	−40	$\frac{-236.28}{31.902} = -7.406$	$\frac{+822.7}{799.2}(-15.86) = -16.326$	$\frac{+5{,}288.4}{10{,}578}(-40) = -20.000$	−43.732	+3.732
E...........	0	+4.14	0	−7.406	$\frac{+822.7}{799.2}(+4.14) = +4.262$	$\frac{+5{,}288.4}{10{,}578}(0) = 0$	−3.144	+3.144
K...........	+40	−15.86	0	−7.406	$\frac{+822.7}{799.2}(-15.86) = -16.326$	$\frac{+5{,}288.4}{10{,}578}(+40) = +20.000$	−3.732	+3.732

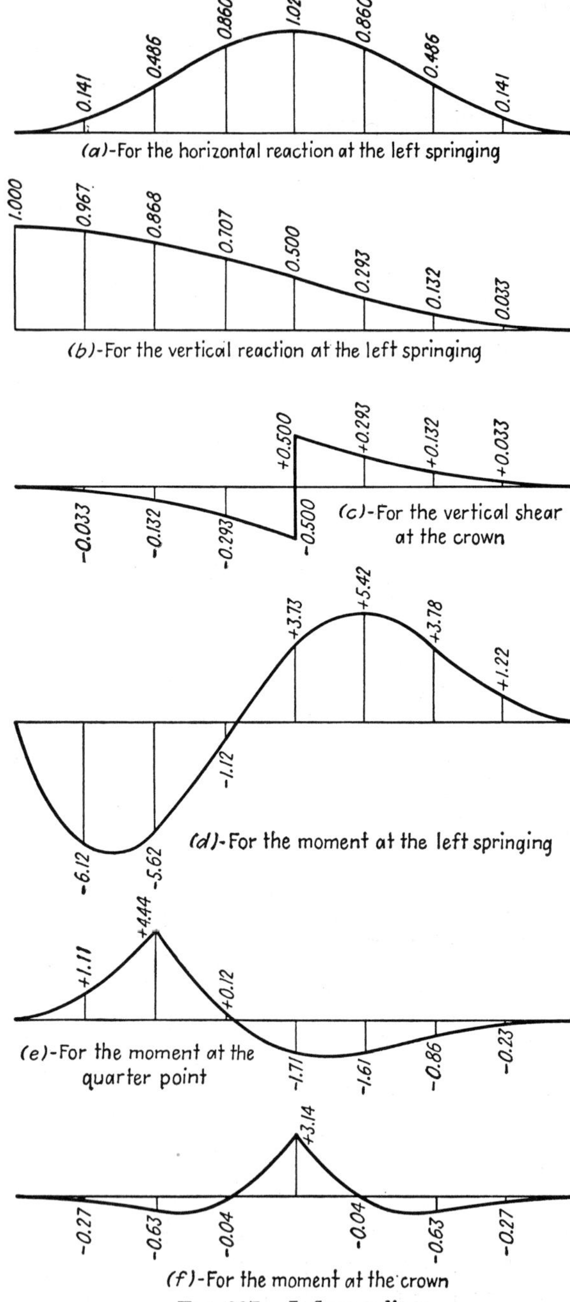

FIG. 297. Influence lines.

It is to be noted that the moment at A, E, and G can be found as the bending moment on the cantilever structure and the values thus found should check with those obtained by $(M_s - M_i)$ in the method of column analogy.

Before plotting the influence lines, it is advisable to make the accompanying influence table (Table 96-3). Note how the influence values for unit loads at F, G, and H can be picked up by inspection from Fig. 296.

TABLE 96-3. INFLUENCE TABLE

Load at	Distance from A	H_A	V_A	V_E	M_A	M_C	M_E
A	0	0	1.000	0	0	0	0
B	10	0.141	0.967	−0.033	−6.12	+1.11	−0.27
C	20	0.486	0.868	−0.132	−5.62	+4.44	−0.63
D	30	0.860	0.707	−0.293	−1.12	+0.12	−0.04
E	40	1.029	0.500	−0.500 +0.500	+3.73	−1.71	+3.14
F	50	0.860	0.293	+0.293	+5.42	−1.61	−0.04
G	60	0.486	0.132	+0.132	+3.78	−0.86	−0.63
H	70	0.141	0.033	+0.033	+1.22	−0.23	−0.27
K	80	0	0	0	0	0	0

The required influence lines are drawn as in Fig. 297.

EXERCISE

165. For the symmetrical circular fixed arch shown, draw influence lines for (*a*) the horizontal reaction at the left springing, (*b*) the vertical reaction at the left springing,

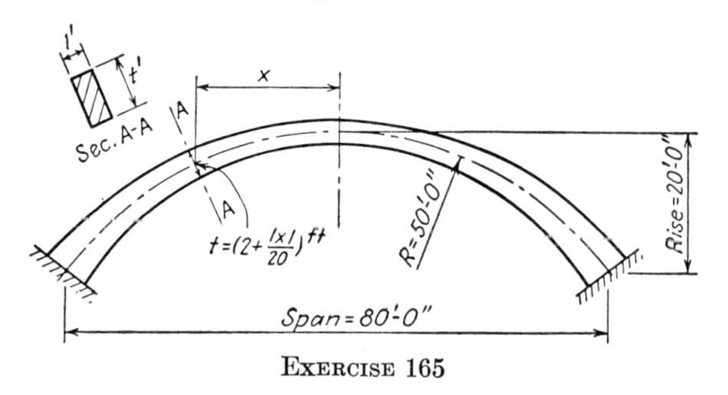

EXERCISE 165

(*c*) the vertical shear at the crown, (*d*) the moment at the left springing, (*e*) the moment at the quarter point, and (*f*) the moment at the crown.

68. Influence Lines for an Unsymmetrical Fixed Arch

Example 97. For the unsymmetrical parabolic fixed arch shown in Fig. 298, draw influence lines for (*a*) the horizontal reaction at the left

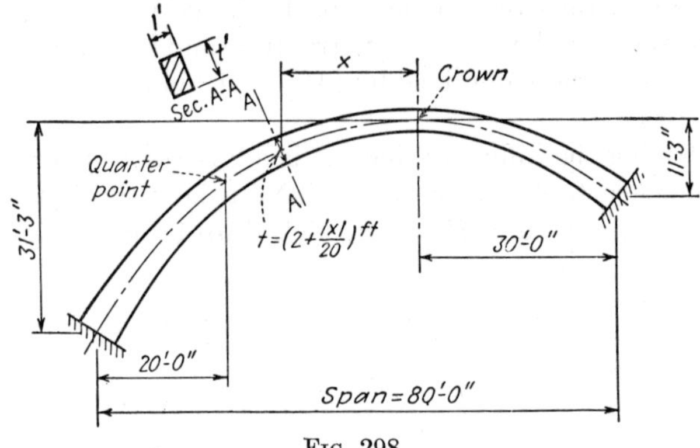

FIG. 298.

springing, (*b*) the vertical reaction at the left springing, (*c*) the vertical shear at the crown, (*d*) the moment at the left springing, (*e*) the moment at the quarter point, (*f*) the moment at the crown, and (*g*) the moment at the right springing.

Solution. The arch rib will be replaced by eight finite segments, as represented by *AB*, *BC*, *CD*, etc. (Fig. 299). Each segment is assumed

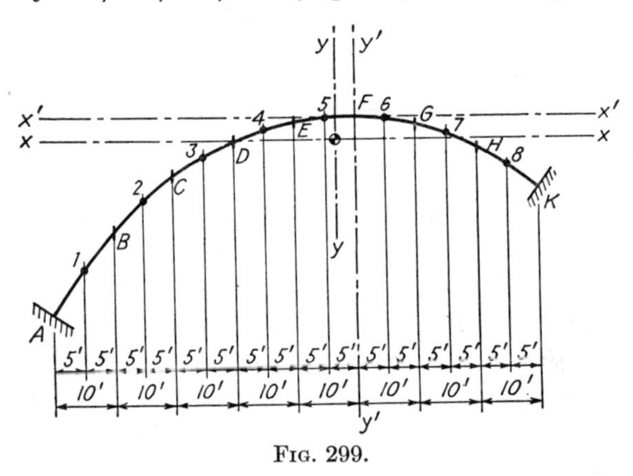

FIG. 299.

to have a constant moment of inertia equal to that at its mid-point, which is considered to be the centroid of the segment. I_c is the moment of inertia of the vertical section at the crown; EI_c is considered equal to unity. Moments of inertia of the segment about the axes through its own centroid are neglected.

It is to be noted that in this example all computations are carried to an unusual degree of accuracy. This is done so that a check can be shown at the end. In an actual design problem the usual practice of accuracy to three or four significant figures in all numbers is recommended.

Properties of the Analogous-column Section (Fig. 299) (see Table 97-1)

TABLE 97-1. PROPERTIES OF THE ANALOGOUS-COLUMN SECTION

Segment	Centroid	ds, ft	t, ft	n $(I = nI_c)$	$\frac{ds}{EI} = \frac{ds}{n}$ $(EI_c = 1)$	x'	y'	$x' \frac{ds}{EI}$	$y' \frac{ds}{EI}$	x	y	$x^2 \frac{ds}{EI}$	$xy \frac{ds}{EI}$	$y^2 \frac{ds}{EI}$
AB.........	1	15.052	4.25	9.5957	1.5686	−45	−25.3125	− 70.587	− 39.7052	−40.5128	−20.8283	2,574.52	+1,323.605	680.488
BC.........	2	13.288	3.75	6.5918	2.0158	−35	−15.3125	− 70.553	− 30.8669	−30.5128	−10.8283	1,876.77	+ 666.024	236.357
CD.........	3	11.792	3.25	4.2910	2.7481	−25	− 7.8125	− 68.702	− 21.4695	−20.5128	− 3.3283	1,156.33	+ 187.620	30.442
DE.........	4	10.680	2.75	2.5996	4.1083	−15	− 2.8125	− 61.625	− 11.5546	−10.5128	+ 1.6717	454.04	− 72.200	11.481
EF.........	5	10.078	2.25	1.4238	7.0782	− 5	− 0.3125	− 35.391	− 2.2119	− 0.5128	+ 4.1717	1.86	− 15.142	123.182
FG.........	6	10.078	2.25	1.4238	7.0782	+ 5	− 0.3125	+ 35.391	− 2.2119	+ 9.4872	+ 4.1717	637.09	+ 280.139	123.182
GH.........	7	10.680	2.75	2.5996	4.1083	+15	− 2.8125	+ 61.625	− 11.5546	+19.4872	+ 1.6717	1,560.13	+ 133.835	11.481
HK.........	8	11.792	3.25	4.2910	2.7481	+25	− 7.8125	+ 68.702	− 21.4695	+29.4872	− 3.3283	2,389.46	− 269.705	30.442
					31.4536	−4.4872	− 4.4842	−141.140	−141.0441			10,650.20	+2,234.176	1,247.005
					A	$\bar{x}'$	$\bar{y}'$					I_y	I_{xy}	I_x

$A = 31.4536$

$$I_x' = I_x\left(1 - \frac{I_{xy}^2}{I_xI_y}\right) = (1,247.055)\left[1 - \frac{(+2,234.176)^2}{(1,247.055)(10,650.20)}\right] = (1,247.055)(0.62417) = 778.374$$

$$I_y' = I_y\left(1 - \frac{I_{xy}^2}{I_xI_y}\right) = (10,650.20)(0.62417) = 6,647.54$$

TABLE 97-2

Properties of analogous-column section					Load at B				Load at C			
Segment	Centroid	x	y	$\frac{ds}{EI}$	M_s	$\frac{M_s\,ds}{EI}$	$\frac{M_sx\,ds}{EI}$	$\frac{M_sy\,ds}{EI}$	M_s	$\frac{M_s\,ds}{EI}$	$\frac{M_sx\,ds}{EI}$	$\frac{M_sy\,ds}{EI}$
AB...	1	−40.5128	−20.8283	1.5686	−5	−7.8430	+317.742	+163.356	−15	−23.529	+953.226	+490.069
BC...	2	−30.5128	−10.8283	2.0158					− 5	−10.079	+307.538	+109.138
CD...	3	−20.5128	− 3.3283	2.7481								
DE...	4	−10.5128	+ 1.6717	4.1083								
EF...	5	− 0.5128	+ 4.1717	7.0782								
FG...	6	+ 9.4872	+ 4.1717	7.0782								
GH...	7	+19.4872	+ 1.6717	4.1083								
HK..	8	+29.4872	− 3.3283	2.7481								
						−7.8430	+317.742	+163.356		−33.608	+1,260.764	+599.207
						P	M_y	M_x		P	M_y	M_x

$A = 31.4536$

$I_x' = 778.374$

$I_y' = 6{,}647.54$

$$\frac{I_{xy}}{I_x} = +1.79218$$

$$\frac{I_{xy}}{I_y} = +0.20978$$

Load at B:

$$M_x' = (+163.356) - (+317.742)(0.20978) = +96.700$$
$$M_y' = (+317.742) - (+163.356)(1.79218) = +24.979$$

$$M_o = -\frac{P}{A} = -\frac{-7.8430}{31.4536} = +0.249$$

$$H_o = -\frac{M_x'}{I_x'} = -\frac{+96.700}{778.374} = -0.1242$$

$$V_o = +\frac{M_y'}{I_y'} = +\frac{+24.979}{6{,}647.54} = +0.0038$$

Load at C:

$$M_x' = 599.207 - (1{,}260.764)(0.20978) = +334.72$$
$$M_y' = 1{,}260.764 - (599.207)(1.79218) = +186.88$$

$$M_o = -\frac{P}{A} = -\frac{-33.608}{31.4536} = +1.068$$

$$H_o = -\frac{M_x'}{I_x'} = -\frac{+334.72}{778.374} = -0.4300$$

$$V_o = +\frac{M_y'}{I_y'} = +\frac{+186.88}{6{,}647.54} = +0.0281$$

TABLE 97-2.—(*Continued*)

Load at *B*

Point	x	y	M_s	$\left[\frac{P}{A}\right]$	$\left[\frac{M_x'y}{I_x'}\right]$	$\left[\frac{M_y'x}{I_y'}\right]$	M_i	M
A......	−45.5128	−26.7658	−10	$\frac{-7.8430}{31.4536} = -0.249$	$\frac{+96.700}{778.374}(-26.7658) = -3.325$	$\frac{+24,979}{6,647.54}(-45.5128) = -0.171$	−3.745	−6.255
C.......	−25.5128	− 6.7658	0	−0.249	$\frac{+96.700}{778.374}(-6.7658) = -0.840$	$\frac{+24.979}{6,647.54}(-25.5128) = -0.096$	−1.185	+1.185
F.......	+ 4.4872	+ 4.4842	0	−0.249	$\frac{+96.700}{778.374}(+4.4842) = +0.557$	$\frac{+24.979}{6,647.54}(+4.4872) = +0.017$	+0.325	−0.325
K......	+34.4872	− 6.7658	0	−0.249	$\frac{+96.700}{778.374}(-6.7658) = -0.840$	$\frac{+24.979}{6,647.54}(+34.4872) = +0.130$	−0.959	+0.959

Load at *C*

Point	x	y	M_s	$\left[\frac{P}{A}\right]$	$\left[\frac{M_x'y}{I_x'}\right]$	$\left[\frac{M_y'x}{I_y'}\right]$	M_i	M
A......	−45.5128	−26.7658	−20	$\frac{-33.608}{31.4536} = -1.068$	$\frac{+334.72}{778.374}(-26.7658) = -11.510$	$\frac{+186.88}{6,647.54}(-45.5128) = -1.279$	−13.857	−6.143
C.......	−25.5128	− 6.7658	0	−1.068	$\frac{+334.72}{778.374}(-6.7658) = -2.909$	$\frac{+186.88}{6,647.54}(-25.5128) = -0.717$	− 4.694	+4.694
F.......	+ 4.4872	+ 4.4842	0	−1.068	$\frac{+334.72}{778.374}(+4.4842) = +1.928$	$\frac{+186.88}{6,647.54}(+4.4872) = +0.126$	+ 0.986	−0.986
K......	+34.4872	− 6.7658	0	−1.068	$\frac{+334.72}{778.374}(-6.7658) = -2.909$	$\frac{+186.88}{6,647.54}(+34.4872) = +0.970$	− 3.007	+3.007

TABLE 97-2.—(*Continued*)

Properties of analogous-column section					Load at D				Load at E			
Segment	Centroid	x	y	$\frac{ds}{EI}$	M_s	$\frac{M_s\,ds}{EI}$	$\frac{M_sx\,ds}{EI}$	$\frac{M_sy\,ds}{EI}$	M_s	$\frac{M_s\,ds}{EI}$	$\frac{M_sx\,ds}{EI}$	$\frac{M_sy\,ds}{EI}$
AB...	1	−40.5128	−20.8283	1.5686	−25	−39.215	+1,588.71	+816.78	−35	−54.901	+2,224.19	+1,143.49
BC...	2	−30.5128	−10.8283	2.0158	−15	−30.237	+ 922.62	+327.42	−25	−50.395	+1,537.69	+ 545.69
CD...	3	−20.5128	− 3.3283	2.7481	− 5	−13.740	+ 281.84	+ 45.73	−15	−41.222	+ 845.57	+ 137.20
DE...	4	−10.5128	+ 1.6717	4.1083					− 5	−20.542	+ 215.95	− 34.34
EF...	5	− 0.5128	+ 4.1717	7.0782								
FG...	6	+ 9.4872	+ 4.1717	7.0782								
GH...	7	+19.4872	+ 1.6717	4.1083								
HK..	8	+29.4872	− 3.3283	2.7481								
						−83.192	+2,793.17	+1,189.93		−167.060	+4,823.40	+1,792.04
						P	M_y	M_x		P	M_y	M_x

$A = 31.4536$
$I_x' = 778.374$
$I_y' = 6{,}647.54$

$\frac{I_{xy}}{I_x} = +1.79218$

$\frac{I_{xy}}{I_y} = +0.20978$

Load at D:

$M_x' = 1{,}189.93 - (2{,}793.17)(0.20978) = +603.98$
$M_y' = 2{,}793.17 - (1{,}189.93)(1.79218) = +660.60$

$$M_o = -\frac{P}{A} = -\frac{-83.192}{31.4536} = +2.645$$

$$H_o = -\frac{M_x'}{I_x'} = -\frac{603.98}{778.374} = -0.7760$$

$$V_o = +\frac{M_y'}{I_y'} = +\frac{660.60}{6{,}647.54} = +0.0994$$

Load at E:

$M_x' = 1{,}792.04 - (4{,}823.40)(0.20978) = 780.19$
$M_y = 4{,}823.40 - (1{,}792.04)(1.79218) = 1{,}611.74$

$$M_o = -\frac{P}{A} = -\frac{-167.060}{31.4536} = +5.311$$

$$H_o = -\frac{M_x'}{I_x'} = -\frac{780.19}{778.374} = -1.0023$$

$$V_o = +\frac{M_y'}{I_y'} = +\frac{1{,}611.74}{6{,}647.54} = +0.2425$$

TABLE 97-2.—(*Continued*)

Load at D

Point	x	y	M_s	$\left[\frac{P}{A}\right]$	$\left[\frac{M_x'y}{I_x'}\right]$	$\left[\frac{M_y'x}{I_y'}\right]$	M_i	M
A......	−45.5128	−26.7658	−30	$\frac{-83.192}{31.4536} = -2.645$	$\frac{603.98}{778.374}(-26.7658) = -20.769$	$\frac{660.60}{6,647.54}(-45.5128) = -4.523$	−27.937	−2.063
C......	−25.5128	− 6.7658	−10	−2.645	$\frac{603.98}{778.374}(-6.7658) = -5.250$	$\frac{660.60}{6,647.54}(-25.5128) = -2.535$	−10.430	+0.430
F......	+ 4.4872	+ 4.4842	0	−2.645	$\frac{603.98}{778.374}(+4.4842) = +3.480$	$\frac{660.60}{6,647.54}(+4.4872) = +0.446$	+ 1.281	−1.281
K......	+34.4872	− 6.7658	0	−2.645	$\frac{603.98}{778.374}(-6.7658) = -5.250$	$\frac{660.60}{6,647.54}(+34.4872) = +3.427$	− 4.468	+4.468

Load at E

Point	x	y	M_s	$\left[\frac{P}{A}\right]$	$\left[\frac{M_x'y}{I_x'}\right]$	$\left[\frac{M_y'x}{I_y'}\right]$	M_i	M
A......	−45.5128	−26.7658	−40	$\frac{-167.060}{31.4536} = -5.311$	$\frac{780.19}{778.374}(-26.7658) = -26.828$	$\frac{1,611.74}{6,647.54}(-45.5128) = -11.035$	−43.174	+3.174
C......	−25.5128	− 6.7658	−20	−5.311	$\frac{780.19}{778.374}(-6.7658) = -6.782$	$\frac{1,611.74}{6,647.54}(-25.5128) = -6.186$	−18.279	−1.721
F......	+ 4.4872	+ 4.4842	0	−5.311	$\frac{780.19}{778.374}(+4.4842) = +4.495$	$\frac{1,611.74}{6,647.54}(+4.4872) = +1.088$	+ 0.272	−0.272
K......	+34.4872	− 6.7658	0	−5.311	$\frac{780.19}{778.374}(-6.7658) = -6.782$	$\frac{1,611.74}{6,647.54}(+34.4872) = +8.362$	− 3.731	+3.731

TABLE 97-2.—(*Continued*)

Properties of analogous-column section					Load at F				Load at G			
Segment	Centroid	x	y	$\frac{ds}{EI}$	M_s	$\frac{M_s\,ds}{EI}$	$\frac{M_s x\,ds}{EI}$	$\frac{M_s y\,ds}{EI}$	M_s	$\frac{M_s\,ds}{EI}$	$\frac{M_s x\,ds}{EI}$	$\frac{M_s y\,ds}{EI}$
AB...	1	−40.5128	−20.8283	1.5686								
BC...	2	−30.5128	−10.8283	2.0158								
CD...	3	−20.5128	− 3.3283	2.7481								
DE...	4	−10.5128	+ 1.6717	4.1083								
EF...	5	− 0.5128	+ 4.1717	7.0782								
FG...	6	+ 9.4872	+ 4.1717	7.0782	− 5	− 35.391	− 335.76	−147.641				
GH...	7	+19.4872	+ 1.6717	4.1083	−15	− 61.624	−1,200.88	−103.017	− 5	−20.542	− 400.31	− 34.340
HK..	8	+29.4872	− 3.3283	2.7481	−25	− 68.702	−2,025.83	+228.661	−15	−41.222	−1,215.52	+137.199
						−165.717	−3,562.47	− 21.997		−61.764	−1,615.83	+102.859
						P	M_y	M_x		P	M_y	M_x

$A = 31.4536$
$I_x' = 778.374$
$I_y' = 6{,}647.54$

$$\frac{I_{xy}}{I_x} = +1.79218$$

$$\frac{I_{xy}}{I_y} = +0.20978$$

Load at F:

$$M_x' = (-21.997) - (-3{,}562.47)(0.20978) = +725.34$$
$$M_y' = (-3{,}562.47) - (-21.997)(1.79218) = -3{,}523.05$$

$$M_o = -\frac{P}{A} = -\frac{-165.717}{31.4536} = +5.269$$

$$H_o = -\frac{M_x'}{I_x'} = -\frac{+725.34}{778.374} = -0.9319$$

$$V_o = +\frac{M_y'}{I_y'} = +\frac{-3{,}523.05}{6{,}647.54} = -0.5300$$

Load at G:

$$M_x' = (+102.859) - (-1{,}615.83)(0.20978) = +441.83$$
$$M_y' = (-1{,}615.83) - (+102.859)(1.79218) = -1{,}800.17$$

$$M_o = -\frac{P}{A} = -\frac{-61.764}{31.4536} = +1.964$$

$$H_o = -\frac{M_x'}{I_x'} = -\frac{441.83}{778.374} = -0.5676$$

$$V_o = +\frac{M_y'}{I_y'} = +\frac{-1{,}800.17}{6{,}647.54} = -0.2708$$

TABLE 97-2.—(*Continued*)

Load at F

Point	x	y	M_s	$\left[\frac{P}{A}\right]$	$\left[\frac{M_x'y}{I_x'}\right]$	$\left[\frac{M_y'x}{I_y'}\right]$	M_i	M
A......	−45.5128	−26.7658	0	$\frac{-165.717}{31.4536} = -5.269$	$\frac{+725.34}{778.374}(-26.7658) = -24.942$	$\frac{-3,523.05}{6,647.54}(-45.5128) = +24.121$	− 6.090	+6.090
C.......	−25.5128	− 6.7658	0	−5.269	$\frac{+725.34}{778.374}(-\ 6.7658) = -\ 6.305$	$\frac{-3,523.05}{6,647.54}(-25.5128) = +13.521$	+ 1.947	−1.947
F.......	+ 4.4872	+ 4.4842	0	−5.269	$\frac{+725.34}{778.374}(+\ 4.4842) = +\ 4.179$	$\frac{-3,523.05}{6,647.54}(+\ 4.4872) = -\ 2.378$	− 3.468	+3.468
K......	+34.4872	− 6.7658	−30	−5.269	$\frac{+725.34}{778.374}(-\ 6.7658) = -\ 6.305$	$\frac{-3,523.05}{6,647.54}(+34.4872) = -18.277$	−29.851	−0.149

Load at G

Point	x	y	M_s	$\left[\frac{P}{A}\right]$	$\left[\frac{M_x'y}{I_x'}\right]$	$\left[\frac{M_y'x}{I_y'}\right]$	M_i	M
A......	−45.5128	−26.7658	0	$\frac{-61.764}{31.4536} = -1.964$	$\frac{441.83}{778.374}(-26.7658) = -15.193$	$\frac{-1,800.17}{6,647.54}(-45.5128) = +12.325$	− 4.832	+4.832
C.......	−25.5128	− 6.7658	0	−1.964	$\frac{441.83}{778.374}(-\ 6.7658) = -\ 3.840$	$\frac{-1,800.17}{6,647.54}(-25.5128) = +\ 6.909$	+ 1.105	−1.105
F.......	+ 4.4872	+ 4.4842	0	−1.964	$\frac{441.83}{778.374}(+\ 4.4842) = +\ 2.545$	$\frac{-1,800.17}{6,647.54}(+\ 4.4872) = -\ 1.215$	− 0.634	+0.634
K......	+34.4872	− 6.7658	−20	−1.964	$\frac{441.83}{778.374}(-\ 6.7658) = -\ 3.840$	$\frac{-1,800.17}{6,647.54}(+34.4872) = -\ 9.339$	−15.143	−4.857

TABLE 97-2.—(*Continued*)

Properties of analogous-column section					Load at H				Loads at B, C, D, E, F, G, and H			
Segment	Centroid	x	y	$\frac{ds}{EI}$	M_s	$\frac{M_s\,ds}{EI}$	$\frac{M_sx\,ds}{EI}$	$\frac{M_sy\,ds}{EI}$	M_s	$\frac{M_s\,ds}{EI}$	$\frac{M_sx\,ds}{EI}$	$\frac{M_sy\,ds}{EI}$
AB...	1	−40.5128	−20.8283	1.5686					+17.5	+ 27.450	−1,112.076	− 571.737
BC...	2	−30.5128	−10.8283	2.0158					+47.5	+ 95.750	−2,921.601	−1,036.810
CD...	3	−20.5128	− 3.3283	2.7481					+67.5	+ 185.497	−3,805.063	− 617.390
DE...	4	−10.5128	+ 1.6717	4.1083					+77.5	+ 318.393	−3,347.202	+ 532.258
EF...	5	− 0.5128	+ 4.1717	7.0782					+77.5	+ 548.561	− 281.302	+2,288.432
FG...	6	+ 9.4872	+ 4.1717	7.0782					+67.5	+ 477.778	+4,532.775	+1,993.146
GH...	7	+19.4872	+ 1.6717	4.1083					+47.5	+ 195.144	+3,802.810	+ 326.222
HK..	8	+29.4872	− 3.3283	2.7481	−5	−13.740	−405.169	+45.732	+17.5	+ 48.092	+1,418.098	− 160.065
						−13.740	−405.169	+45.732		+1,896.665	−1,713.561	+2,754.056
						P	M_y	M_x		P	M_y	M_x

$A = 31.4536$
$I_x' = 778.374$
$I_y' = 6{,}647.54$

$$\frac{I_{xy}}{I_x} = +1.79218$$

$$\frac{I_{xy}}{I_y} = +0.20978$$

Load at H:

$$M_x' = (+45.732) - (-405.169)(0.20978) = +130.728$$
$$M_y' = (-405.169) - (+45.732)(1.79218) = -487.13$$

$$M_o = -\frac{P}{A} = -\frac{-13.740}{31.4536} = +0.437$$

$$H_o = -\frac{M_x'}{I_x'} = -\frac{130.728}{778.374} = -0.1680$$

$$V_o = +\frac{M_y'}{I_y'} = +\frac{-487.13}{6{,}647.54} = -0.0733$$

Loads at B, C, D, E, F, G, and H:

$$M_x' = 2{,}754.056 - (-1{,}713.561)(0.20978) = +3{,}113.527$$
$$M_y' = -1{,}713.561 - (2{,}754.056)(1.79218) = -6{,}649.325$$

$$M_o = -\frac{P}{A} = -\frac{1{,}896.665}{31.4536} = -60.300$$

$$H_o = -\frac{M_x'}{I_x'} = -\frac{3{,}113.527}{778.374} = -4.0000$$

$$V_o = +\frac{M_y'}{I_y'} = \frac{-6{,}649.325}{6{,}647.54} = -1.0003$$

TABLE 97-2.—(*Continued*)

Load at H

Point	x	y	M_s	$\left[\frac{P}{A}\right]$	$\left[\frac{M_x'y}{I_x'}\right]$	$\left[\frac{M_y'x}{I_y'}\right]$	M_i	M
A......	−45.5128	−26.7658	0	$\frac{-13.740}{31.4536} = -0.437$	$\frac{130.728}{778.374}(-26.7658) = -4.495$	$\frac{-487.13}{6,647.54}(-45.5128) = +3.335$	−1.597	+1.597
C......	−25.5128	− 6.7658	0	−0.437	$\frac{130.728}{778.374}(-6.7658) = -1.136$	$\frac{-487.13}{6,647.54}(-25.5128) = +1.870$	+0.297	−0.297
F......	+ 4.4872	+ 4.4842	0	−0.437	$\frac{130.728}{778.374}(+4.4842) = +0.753$	$\frac{-487.13}{6,647.54}(+4.4872) = -0.329$	−0.013	+0.013
K......	+34.4872	− 6.7658	−10	−0.437	$\frac{130.728}{778.374}(-6.7658) = -1.136$	$\frac{-487.13}{6,647.54}(+34.4872) = -2.527$	−4.100	−5.900

Loads at B, C, D, E, F, G, and H

Point	x	y	M_s	$\left[\frac{P}{A}\right]$	$\left[\frac{M_x'y}{I_x'}\right]$	$\left[\frac{M_y'x}{I_y'}\right]$	M_i	M
A......	−45.5128	−26.7658	0	$\frac{1,896.665}{31.4536} = +60.300$	$\frac{3,113.527}{778.374}(-26.7658) = -107.064$	$\frac{-6,649.325}{6,647.54}(-45.5128) = +45.525$	− 1.239	+1.239
C......	−25.5128	− 6.7658	+60	+60.300	$\frac{3,113.527}{778.374}(-6.7658) = -27.063$	$\frac{-6,649.325}{6,647.54}(-25.5128) = +25.520$	+58.757	+1.243
F......	+ 4.4872	+ 4.4842	+75	+60.300	$\frac{3,113.527}{778.374}(+4.4842) = +17.937$	$\frac{-6,649.325}{6,647.54}(+4.4872) = -4.488$	+73.749	+1.251
K......	+34.4872	− 6.7658	0	+60.300	$\frac{3,113.527}{778.374}(-6.7658) = -27.063$	$\frac{-6,649.325}{6,647.54}(+34.4872) = -34.496$	− 1.259	+1.259

The fixed arch is analyzed for eight loading conditions: unit loads at B, C, D, E, F, G, or H applied singly and then unit loads at all seven points jointly (see Table 97-2). For unit loads applied singly at B, C, D, or E, the basic determinate structure from which M_s is found is the simple cantilever structure fixed at A and free at K, but for unit loads applied singly at F, G, or H, the simple cantilever structure fixed at K and free at A is chosen as the basic determinate structure. Again for

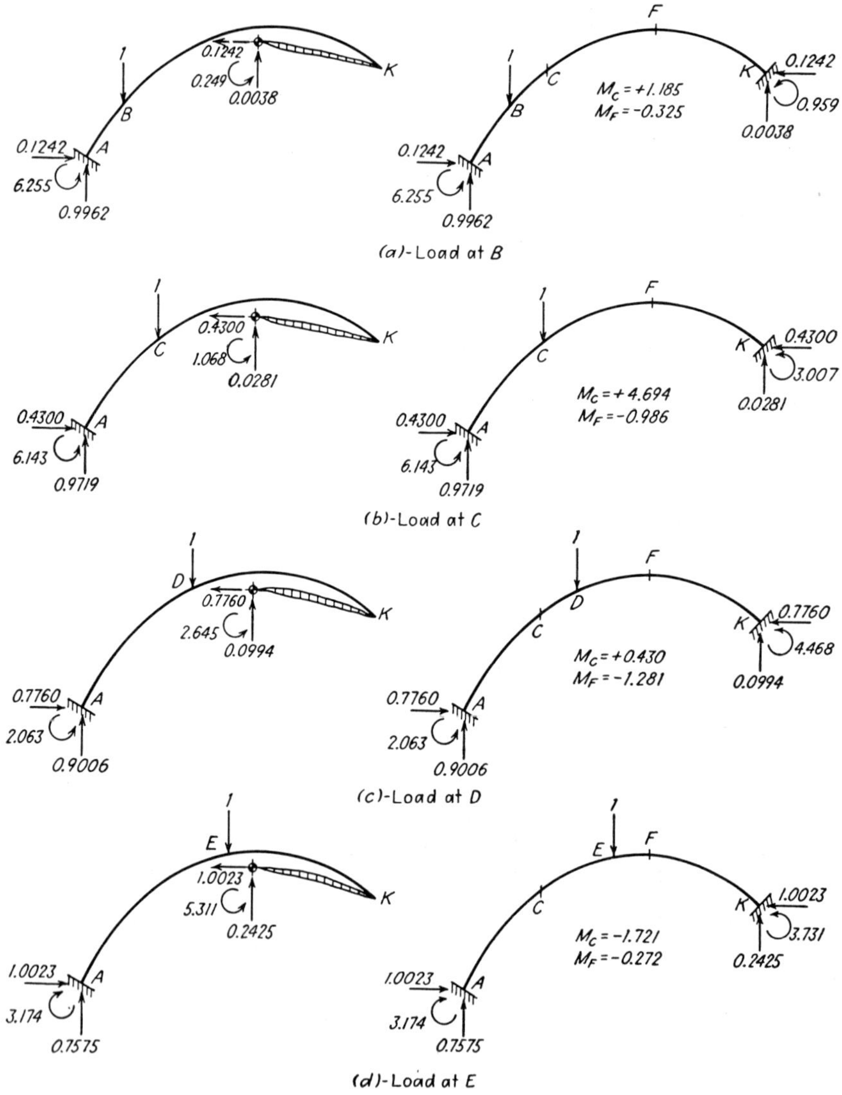

(a)-Load at B

(b)-Load at C

(c)-Load at D

(d)-Load at E

FIG. 300.

unit loads applied jointly at all seven points, the basic determinate structure wherein the arch is simply supported at A and K is used.

The free-body diagrams of all cantilever structures with the rigid arm attached and of the given fixed arches under the various loading conditions are shown in Fig. 300. Note particularly the free-body diagrams

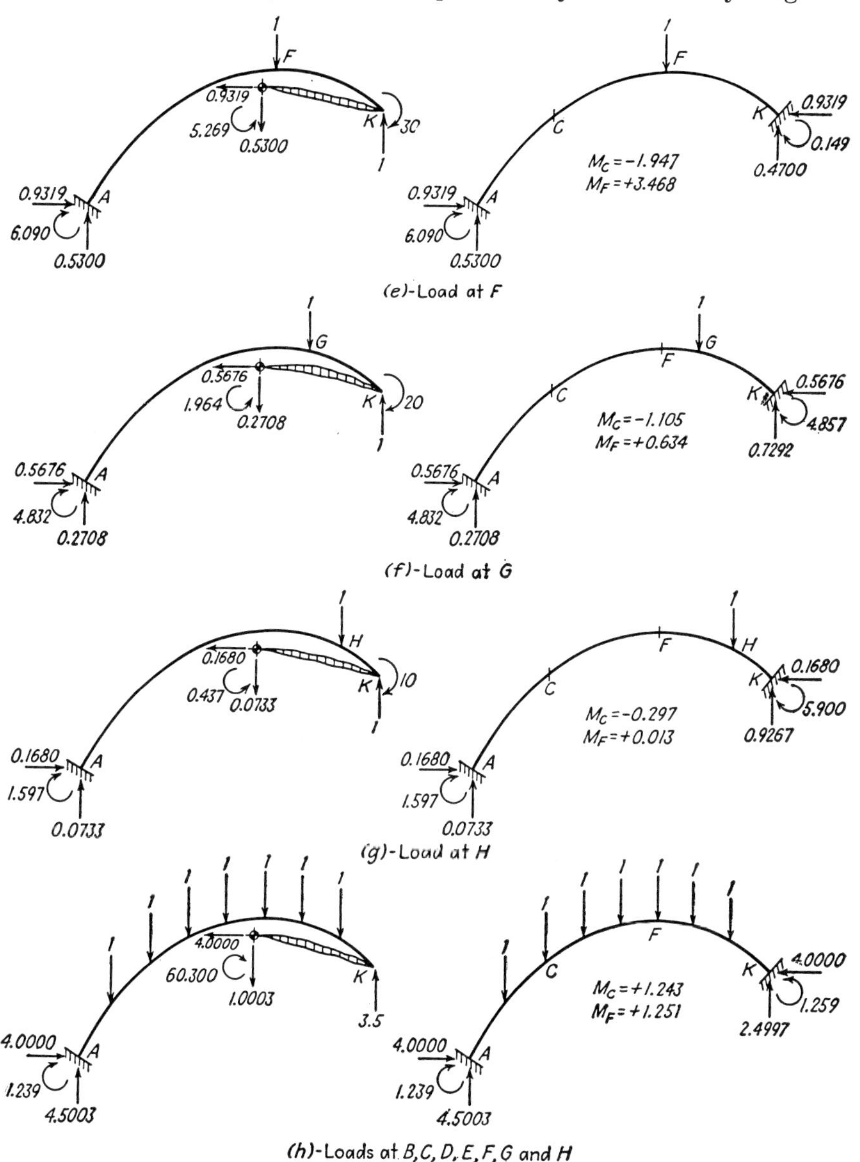

FIG. 300 (*continued*).

of the cantilever structures shown in Fig. 300*e* to *h*. These cantilever structures fixed at A and free at the elastic center are subjected to the self-balancing forces and moments which are comprised of all those which act on the *chosen* basic determinate structures and the (M_O, H_O, V_O)'s acting at the elastic center. The moments at A, C, F, K as determined by $M = M_s - M_i$ in the method of column analogy should check very closely with those determined as the bending moments in the cantilever structures.

The total effects on the fixed arch owing to the singly applied loads at B, C, D, E, F, G, or H must be the same as the effect owing to the jointly applied loads at all seven points. The verification of this fact serves as a check on the correctness of the solution. The comparisons for H_A, V_A, M_A, M_C, M_F, and M_K are listed in Table 97-3. The check is close enough to be considered satisfactory.

TABLE 97-3

	Sum of values due to unit loads applied singly at B, C, D, E, F, G, or H	Combined value due to unit loads applied jointly at B, C, D, E, F, G, and H
H_A	4.0000	4.0000
V_A	4.5003	4.5003
M_A	1.232	1.239
M_C	1.239	1.243
M_F	1.251	1.251
M_K	1.259	1.259

The influence table in which all necessary values for plotting the required influence lines are recorded is given in Table 97–4. The required influence lines are drawn as in Fig. 301.

TABLE 97-4. INFLUENCE TABLE

Load at	Distance from A	H_A	V_A	V_F	M_A	M_C	M_F	M_K
A	0	0	1.000	0	0	0	0	0
B	10	0.124	0.996	−0.004	−6.26	+1.18	−0.32	+0.96
C	20	0.430	0.972	−0.028	−6.14	+4.69	−0.99	+3.01
D	30	0.776	0.901	−0.099	−2.06	+0.43	−1.28	+4.47
E	40	1.002	0.758	−0.242	+3.17	−1.72	−0.27	+3.73
F	50	0.932	0.530	−0.470 +0.530	+6.09	−1.95	+3.47	−0.15
G	60	0.568	0.271	+0.271	+4.83	−1.10	+0.63	−4.86
H	70	0.168	0.073	+0.073	+1.60	−0.30	+0.01	−5.90
K	80	0	0	0	0	0	0	0

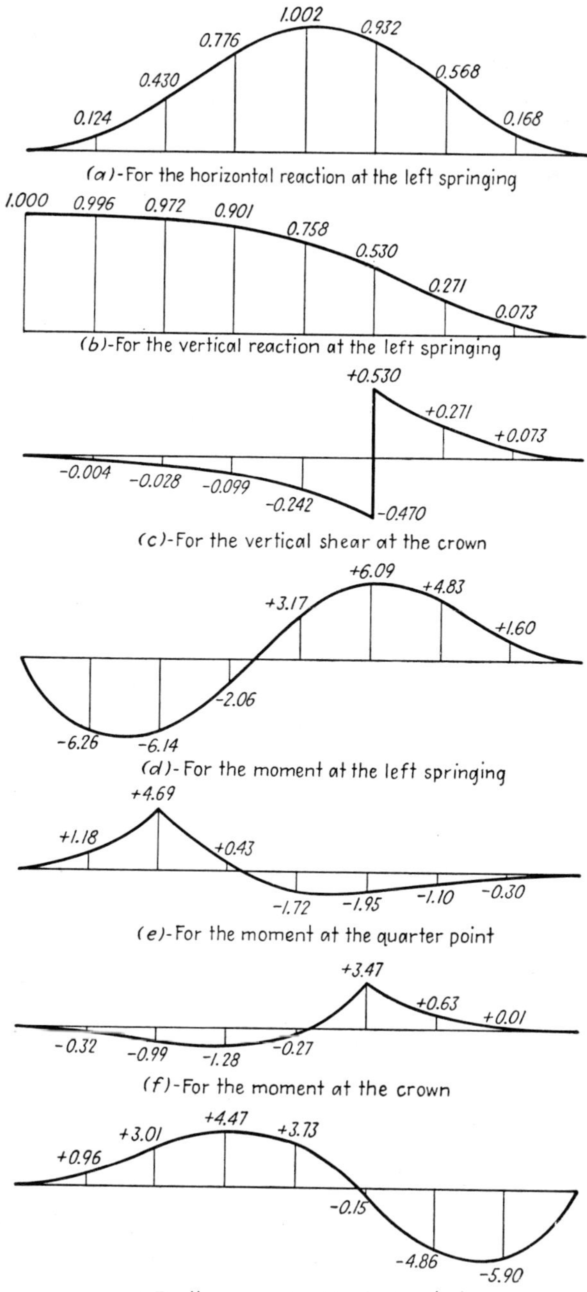

FIG. 301. Influence lines.

EXERCISE

166. For the unsymmetrical circular fixed arch shown, draw influence lines for (*a*) the horizontal reaction at the left springing, (*b*) the vertical reaction at the left

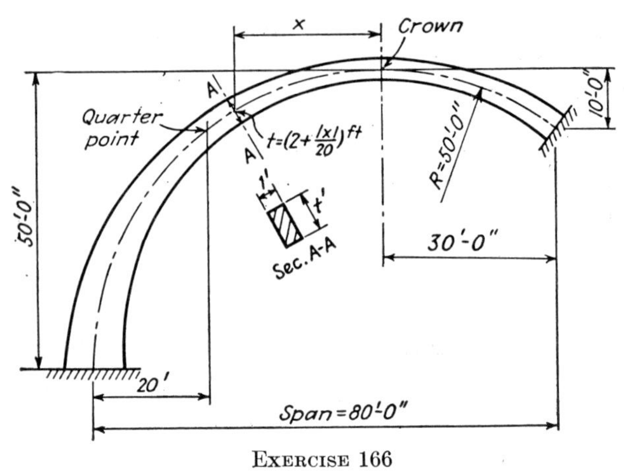

EXERCISE 166

springing, (*c*) the vertical shear at the crown, (*d*) the moment at the left springing, (*e*) the moment at the quarter point, (*f*) the moment at the crown, and (*g*) the moment at the right springing.

69. Effects of Temperature, Shrinkage, Rib Shortening, and Foundation Yielding. Any structure, whether statically determinate or indeterminate, is stressed when loads are applied. An unloaded structure, if it is statically determinate, is not stressed as a result of change in temperature or yielding of supports, but a statically indeterminate structure is stressed by these phenomena. Inasmuch as the fixed arch is statically indeterminate to the third degree, it will be subjected to self-balancing reaction components due to a change in temperature or yielding of supports, which may be a rotational slip, a horizontal displacement, or a vertical settlement.

If the fixed arch is built of concrete, it is necessary to investigate the effect of shrinkage. It is conceivable that the effect of shrinkage is equivalent to that of a temperature drop. Usually the amount of shrinkage is considered to be the same as that due to a drop in temperature of 15°F or thereabouts, depending on the properties of the concrete mix.

Whether the elastic-center method or the column-analogy method is used in analyzing the fixed arch, the basis for analysis is the satisfaction of the conditions of geometry that Δ_ϕ, Δ_H, and Δ_V at the right support are all equal to zero when the left support is considered fixed and the right support free. In figuring the rotation of the tangent and the horizontal or vertical deflections at the right support, only the effect of moment in the arch rib is considered. In fact, the direct stresses in the arch rib would sometimes cause an appreciable amount of shortening or a consequent decrease in the span if the span were not fixed by the supports at each end. The amount of this rib shortening is usually estimated or assumed, and its effect can be investigated in the same way as that of a temperature drop.

The effects of temperature, shrinkage, rib shortening, or foundation yielding on a fixed arch can be found pending the answers to the following problem (Fig. 302*a*):

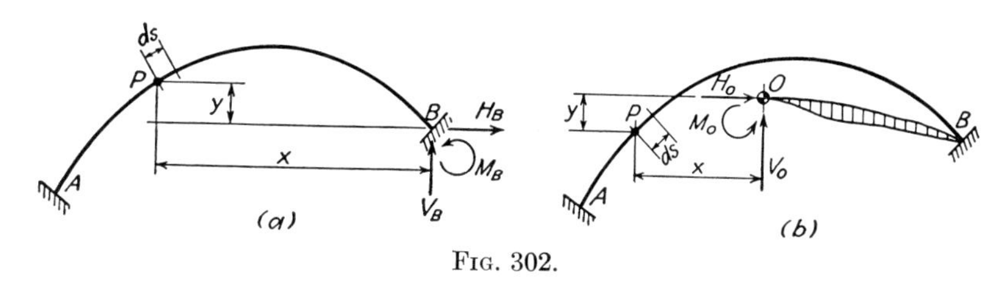

FIG. 302.

Determine all reaction components induced to act on the fixed arch due to the combined effect of

1. A rotation of α in the counterclockwise direction of the tangent at the support B.
2. A horizontal displacement, a, toward the right at the support B.
3. A vertical settlement, b, in the downward direction at the support B.

If a rigid, inelastic arm joining the support B and the elastic center O is attached (Fig. 302*b*), the corresponding displacement components at O, due to the displacements of α counterclockwise, a to the right, and b downward at the support B, will be

$$\begin{aligned} \Delta_\phi \text{ at } O &= \alpha \text{ counterclockwise} \\ \Delta_H \text{ at } O &= a + y_B\alpha \text{ to the right} \\ \Delta_V \text{ at } O &= b + x_B\alpha \text{ downward} \end{aligned} \qquad (142)$$

in which (x_B, y_B) are the coordinates of B referred to O as origin. Note that, in ordinary cases, x_B is a positive quantity and y_B is a negative quantity.

Let M = moment at any point P owing to the action of the self-balancing reaction components

x, y = coordinates of P referred to O as origin.

Following the reasoning of Eq. (120) in Art. 60, the deflection components at O due to the action of M on the whole arch rib are

$$
\begin{aligned}
\Delta_\phi \text{ at } O &= +\sum \frac{M\,ds}{EI} \text{ counterclockwise} \\
\Delta_H \text{ at } O &= +\sum \frac{My\,ds}{EI} \text{ to the right} \\
\Delta_V \text{ at } O &= -\sum \frac{Mx\,ds}{EI} \text{ upward}
\end{aligned}
\tag{143}
$$

Equating Eqs. (142) to Eqs. (143), and substituting

$$M = M_O + H_O y - V_O x$$

$$
\begin{aligned}
+\sum \frac{M\,ds}{EI} &= +\frac{(M_O + H_O y - V_O x)\,ds}{EI} = \alpha \text{ counterclockwise} \\
+\sum \frac{My\,ds}{EI} &= +\frac{(M_O + H_O y - V_O x)y\,ds}{EI} = a + y_B\alpha \text{ to the right} \\
+\sum \frac{Mx\,ds}{EI} &= +\frac{(M_O + H_O y - V_O x)x\,ds}{EI} = b + x_B\alpha \text{ downward}
\end{aligned}
\tag{144}
$$

Substituting

$$
\sum \frac{ds}{EI} = A \qquad \sum \frac{x\,ds}{EI} = 0 \qquad \sum \frac{y\,ds}{EI} = 0
$$

$$
\sum \frac{x^2\,ds}{EI} = I_y \qquad \sum \frac{xy\,ds}{EI} = I_{xy} \qquad \sum \frac{y^2\,ds}{EI} = I_x
$$

into Eqs. (144),

$$
\begin{aligned}
M_O A &= \alpha \\
H_O I_x - V_O I_{xy} &= a + y_B\alpha \\
H_O I_{xy} - V_O I_y &= b + x_B\alpha
\end{aligned}
\tag{145}
$$

Solving Eqs. (145) for M_O, H_O, and V_O,

$$
\begin{aligned}
M_O &= +\frac{\alpha}{A} \\
H_O &= +\frac{(a + y_B\alpha) - (b + x_B\alpha)\dfrac{I_{xy}}{I_y}}{I_x\left(1 - \dfrac{I_{xy}^2}{I_x I_y}\right)} \\
V_O &= -\frac{(b + x_B\alpha) - (a + y_B\alpha)\dfrac{I_{xy}}{I_y}}{I_y\left(1 - \dfrac{I_{xy}^2}{I_x I_y}\right)}
\end{aligned}
\tag{146}
$$

Equations (146) are the general formulas for an unsymmetrical fixed arch when α, a, and b occur simultaneously. For symmetrical fixed arches, I_{xy} is zero in Eqs. (146). From these formulas, the following table is prepared.

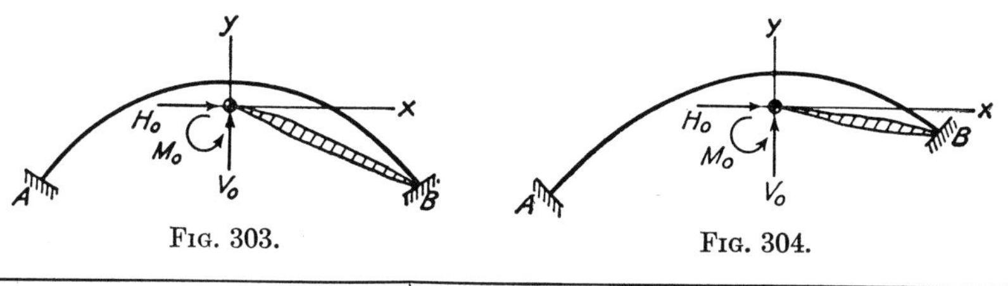

FIG. 303. FIG. 304.

Fixed arches	Symmetrical, Fig. 303	Unsymmetrical, Fig. 304
Due to rotational slip of α (counterclockwise) at support B	$M_O = +\frac{\alpha}{A}$ $H_O = +\frac{y_B\alpha}{I_x}$ $V_O = -\frac{x_B\alpha}{I_y}$	$M_O = +\frac{\alpha}{A}$ $H_O = +\frac{y_B\alpha - x_B\alpha(I_{xy}/I_y)}{I'_x}$ $V_O = -\frac{x_B\alpha - y_B\alpha(I_{xy}/I_x)}{I'_y}$
Due to horizontal displacement of a (to the right) at support B	$M_O = 0$ $H_O = +\frac{a}{I_x}$ $V_O = 0$	$M_O = 0$ $H_O = +\frac{a}{I'_x}$ $V_O = +\frac{a(I_{xy}/I_x)}{I'_y}$
Due to vertical settlement of b (downward) at support B	$M_O = 0$ $H_O = 0$ $V_O = -\frac{b}{I_y}$	$M_O = 0$ $H_O = -\frac{b(I_{xy}/I_y)}{I'_x}$ $V_O = -\frac{b}{I'_y}$

Note that, in ordinary cases, x_B is a positive quantity and y_B is a negative quantity.

When the yielding occurs at support A, perhaps a convenient procedure would be to draw an opposite-handed picture of the arch such that the yielding may then happen at the right support and the formulas in the accompanying table can thus be used. When the reaction components acting on the opposite-handed arch are found, those acting on the given arch can be determined by inspection.

Example 98. Determine all reaction components induced to act on the symmetrical parabolic fixed arch shown in Fig. 305 due to each of the following causes: (*a*) a temperature rise of 50°F; (*b*) a temperature drop of 80°F; (*c*) shrinkage equivalent to a temperature drop of 15°F; (*d*) rib shortening equivalent to an axial stress of 300 psi throughout the arch rib; (*e*) a rotational slip of 0.001 rad counterclockwise at support B; (*f*) a horizontal displacement of 0.200 in. to the right at support B; (*g*) a vertical settlement of 0.250 in. downward at support B.

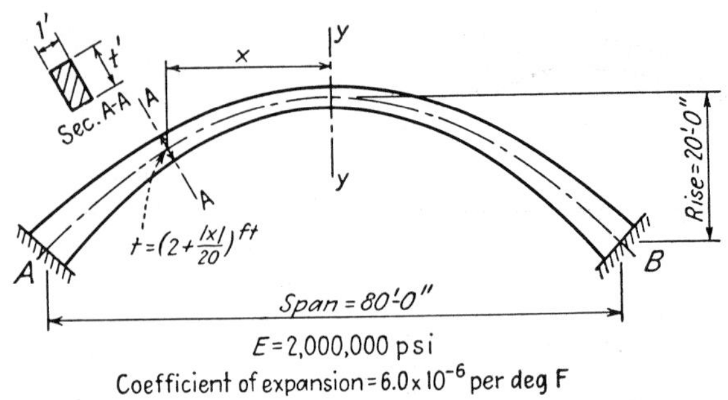

FIG. 305.

Solution. The properties of the analogous-column section are taken from Example 96 and summarized in Fig. 306. However, in the present

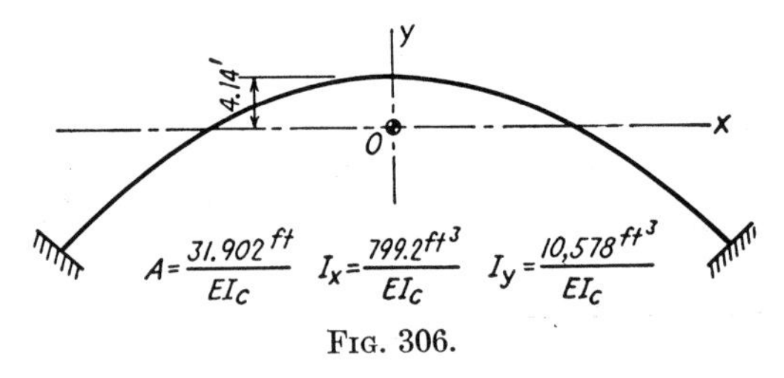

FIG. 306.

problem the true value of EI_c must be used in computing the values of A, I_x, and I_y. Thus

$$A = \frac{31.902\text{ ft}}{EI_c} = \frac{31.902\text{ ft}}{(2{,}000 \times 144)\dfrac{\text{kips}}{\text{ft}^2}\dfrac{(1)(2)^3}{12}\text{ ft}^4} = 0.16616 \times 10^{-3}\,\frac{1}{\text{kip-ft}}$$

$$I_x = \frac{799.2\text{ ft}^3}{EI_c} = \frac{799.2\text{ ft}^3}{(2{,}000 \times 144)\dfrac{\text{kips}}{\text{ft}^2}\frac{2}{3}\text{ ft}^4} = 4.1625 \times 10^{-3}\,\frac{\text{ft}}{\text{kips}}$$

$$I_y = \frac{10{,}578\text{ ft}^3}{EI_c} = \frac{10{,}578\text{ ft}^3}{(2{,}000 \times 144)\dfrac{\text{kips}}{\text{ft}^2}\frac{2}{3}\text{ ft}^4} = 55.094 \times 10^{-3}\,\frac{\text{ft}}{\text{kips}}$$

a. A Temperature Rise of 50°*F.* If the arch rib is free to expand owing to a temperature rise, it will change from the shape of the solid curve in Fig. 307*a* to that of the dotted curve, which is *geometrically* similar to the solid curve. The tangent at A (or B) is parallel to that at A' (or B').

In Fig. 307*a*, $AA' = Lct/2 = BB'$, in which L is the span, c is the coefficient of expansion, and t is the temperature rise. Now if the tangent at A' is made to coincide with that at A, as is done in Fig. 307*b*, $BB' = Lct$. AB' in Fig. 307*b* is the unstressed arch. It is necessary to give the right

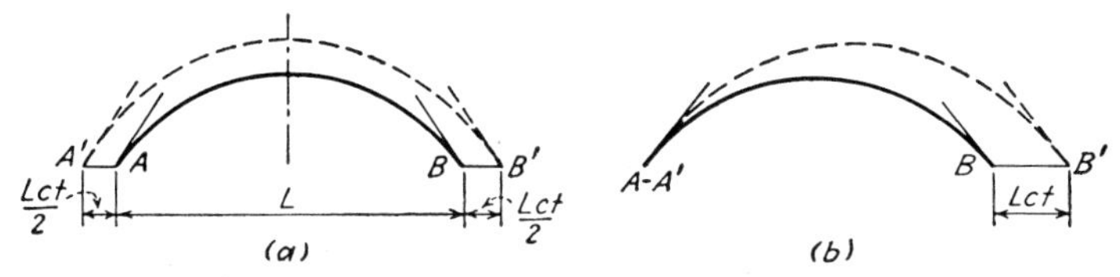

Expansion of arch rib due to temperature rise

FIG. 307.

support a horizontal displacement to the left equal to Lct in order to maintain the original span length, L, between the two fixed supports. Thus

$$a = -Lct = -(80)(6.0 \times 10^{-6})(50) = -24 \times 10^{-3} \text{ ft}$$

Applying the formulas in the table above,

$$M_O = 0$$

$$H_O = +\frac{a}{I_x} = +\frac{-24 \times 10^{-3} \text{ ft}}{4.1625 \times 10^{-3} \dfrac{\text{ft}}{\text{kips}}} = -5.766 \text{ kips}$$

$$V_O = 0$$

The reaction components are shown in Fig. 308.

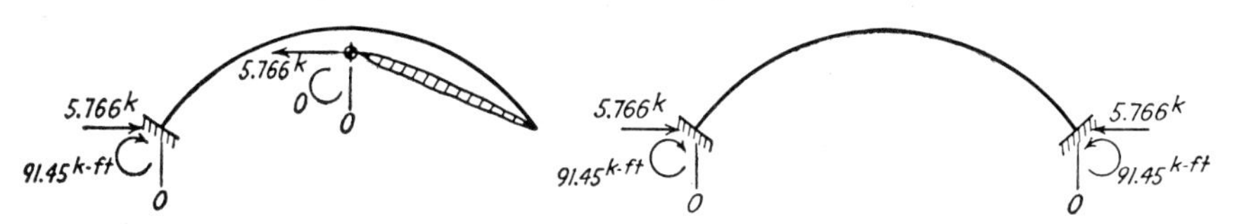

Induced reaction components due to a temperature rise of 50°F

FIG. 308.

b. A Temperature Drop of 80°*F* (Fig. 309). The arch span would become shorter if the rib were free to contract, or the right support should be pulled out by an amount Lct.

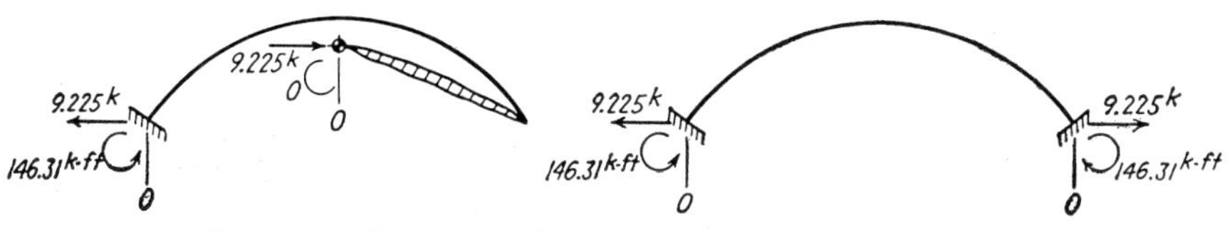

Induced reaction components due to a temperature drop of 80°F

FIG. 309.

$$a = +(80)(6.0 \times 10^{-6})(80) = +38.4 \times 10^{-3} \text{ ft}$$
$$M_O = 0$$
$$H_O = +\frac{a}{I_x} = +\frac{38.4 \times 10^{-3} \text{ ft}}{4.1625 \times 10^{-3} \dfrac{\text{ft}}{\text{kips}}} = +9.225 \text{ kips}$$
$$V_O = 0$$

c. Shrinkage Equivalent to a Temperature Drop of 15°F (Fig. 310)

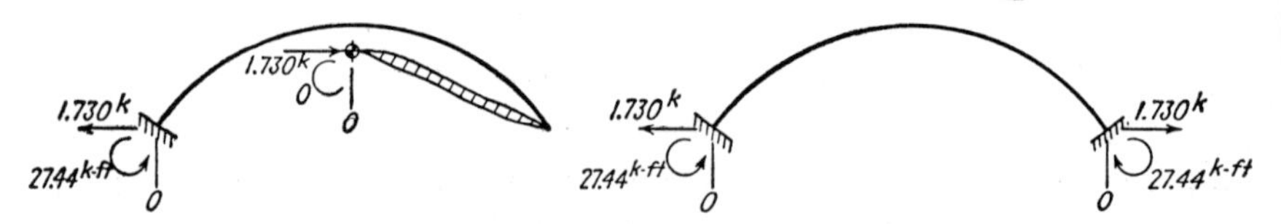

Induced reaction components due to shrinkage equivalent to a temperature drop of 15°F

FIG. 310.

$$a = +(15)(6.0 \times 10^{-6})(80) = +7.2 \times 10^{-3} \text{ ft}$$
$$M_O = 0$$
$$H_O = +\frac{a}{I_x} = +\frac{+7.2 \times 10^{-3} \text{ ft}}{4.1625 \times 10^{-3} \dfrac{\text{ft}}{\text{kips}}} = +1.730 \text{ kips}$$
$$V_O = 0$$

d. Rib Shortening Equivalent to an Axial Stress of 300 Psi throughout the Arch Rib (Fig. 311)

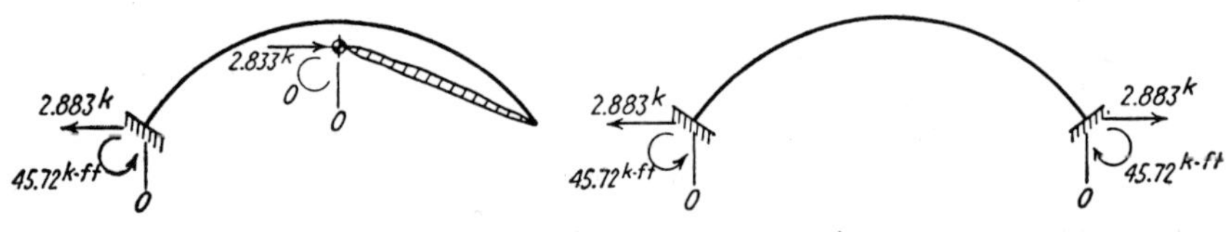

Induced reaction components due to rib shortening

FIG. 311.

$$a = +\frac{300}{2{,}000{,}000} \times 80 = +12 \times 10^{-3} \text{ ft}$$
$$M_O = 0$$
$$H_O = +\frac{a}{I_x} = +\frac{+12 \times 10^{-3} \text{ ft}}{4.1625 \times 10^{-3} \dfrac{\text{ft}}{\text{kips}}} = +2.883 \text{ kips}$$
$$V_O = 0$$

e. A Rotational Slip of 0.001 *Rad Counterclockwise at Support B* (Fig. 312)

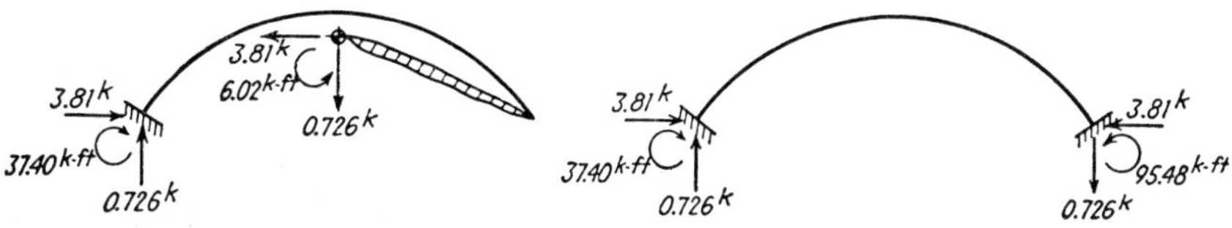

Induced reaction components due to a rotational slip of 0.001 rad. counterclockwise at support B

FIG. 312.

$$\alpha = +0.001 \text{ rad}$$

$$M_O = +\frac{\alpha}{A} = +\frac{0.001}{0.16616 \times 10^{-3}} = +6.02 \text{ kip-ft}$$

$$H_O = +\frac{y_B \alpha}{I_x} = +\frac{(-15.86)(+0.001)}{4.1625 \times 10^{-3}} = -3.81 \text{ kips}$$

$$V_O = -\frac{x_B \alpha}{I_y} = -\frac{(+40)(+0.001)}{55.094 \times 10^{-3}} = -0.726 \text{ kip}$$

f. A Horizontal Displacement of 0.200 *In. to the Right at Support B* (Fig. 313)

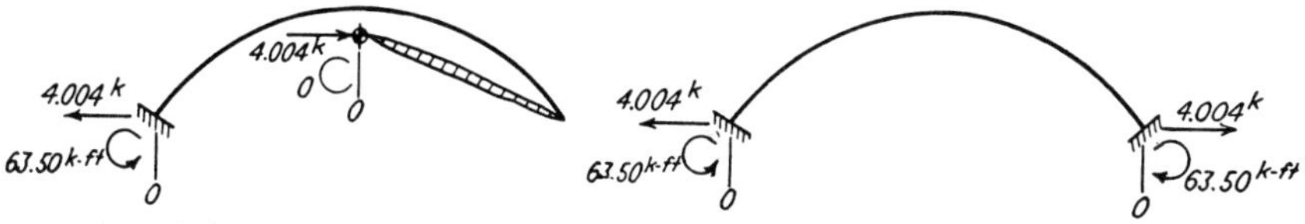

Induced reaction components due to a horizontal displacement of 0.200 in. to the right at support B

FIG. 313.

$$a = +0.200 \text{ in.}$$

$$M_O = 0$$

$$H_O = +\frac{a}{I_x} = +\frac{(+0.200/12) \text{ ft}}{4.1625 \times 10^{-3} \dfrac{\text{ft}}{\text{kips}}} = +4.004 \text{ kips}$$

$$V_O = 0$$

g. A Vertical Settlement of 0.250 *In. Downward at Support B* (Fig. 314)

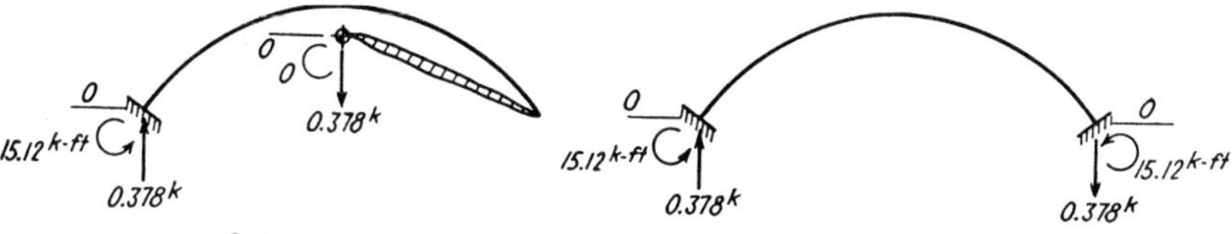

Induced reaction components due to a vertical settlement of 0.250 in. downward at support B

FIG. 314.

$$b = +0.250 \text{ in.}$$
$$M_O = 0$$
$$H_O = 0$$
$$V_O = -\frac{b}{I_y} = -\frac{(+0.250/12) \text{ ft}}{55.094 \times 10^{-3} \dfrac{\text{ft}}{\text{kips}}} = -0.378 \text{ kip}$$

Example 99. Determine all reaction components induced to act on the unsymmetrical parabolic fixed arch shown in Fig. 315 due to each of the

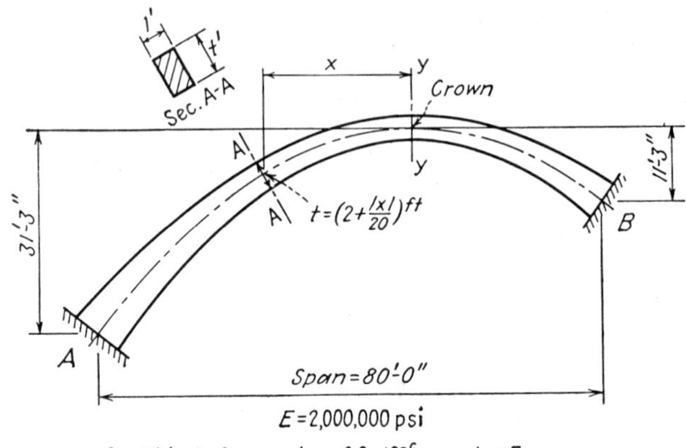

FIG. 315.

following causes: (a) a temperature rise of 50°F; (b) a temperature drop of 80°F; (c) shrinkage equivalent to a temperature drop of 15°F; (d) rib shortening equivalent to an axial stress of 300 psi throughout the arch rib; (e) a rotational slip of 0.001 rad counterclockwise at support B; (f) a horizontal displacement of 0.200 in. to the right at support B; (g) a vertical settlement of 0.250 in. downward at support B.

Solution. The properties of the analogous-column section are taken from Example 97 and summarized in Fig. 316. In the present problem

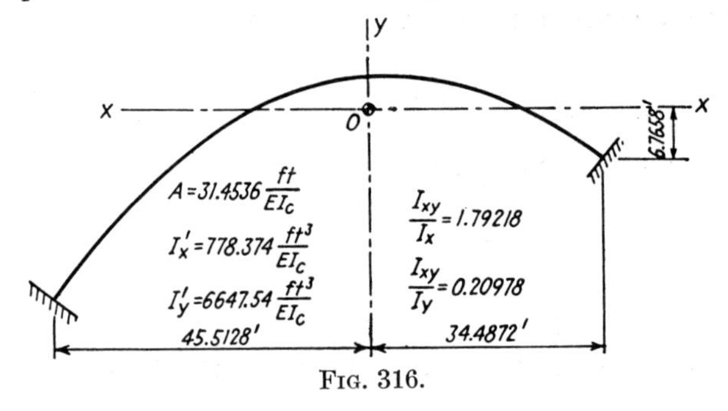

FIG. 316.

the true value of EI_c must be used in computing the values of A, I'_x, and I'_y. Thus

$$A = \frac{31.4536 \text{ ft}}{EI_c} = \frac{31.4536 \text{ ft}}{(2{,}000 \times 144) \frac{\text{kips}}{\text{ft}^2} \frac{2}{3} \text{ ft}^4} = 0.16382 \times 10^{-3} \frac{1}{\text{kip-ft}}$$

$$I'_x = \frac{778.374 \text{ ft}^3}{EI_c} = \frac{778.374 \text{ ft}^3}{(2{,}000 \times 144) \frac{\text{kips}}{\text{ft}^2} \frac{2}{3} \text{ ft}^4} = 4.0540 \times 10^{-3} \frac{\text{ft}}{\text{kips}}$$

$$I'_y = \frac{6{,}647.54 \text{ ft}^3}{EI_c} = \frac{6{,}647.54 \text{ ft}^3}{(2{,}000 \times 144) \frac{\text{kips}}{\text{ft}^2} \frac{2}{3} \text{ ft}^4} = 34.623 \times 10^{-3} \frac{\text{ft}}{\text{kips}}$$

a. A Temperature Rise of 50°*F*. If the arch rib is free to expand owing to a temperature rise, it will change from the shape of the solid curve in Fig. 317*a* to that of the dotted curve, which is *geometrically* similar to the

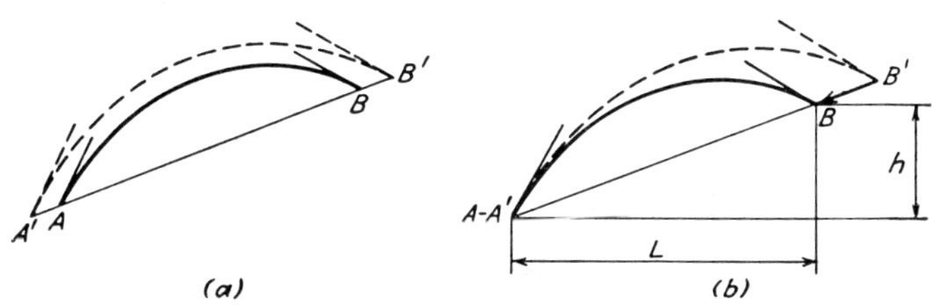

Expansion of arch rib due to temperature rise

FIG. 317.

solid curve. The tangent at A (or B) is parallel to that at A' (or B'). In Fig. 317*a*, $(AA') + (BB') = (AB)(c)(t)$, in which c is the coefficient of expansion and t is the temperature rise. Now if the tangent at A' is made to coincide with that at A, as is done in Fig. 317*b*, then BB' in Fig. 317*b* $= [(AA') + (BB')]$ in Fig. 317*a* $= (AB)(c)(t)$. AB' in Fig. 317*b* is the unstressed arch. It is necessary to give a displacement at the right support from B' to B to restore the position of the fixed support. Thus the effect of a temperature rise is the same as the combined effect of:

1. A horizontal displacement to the left of

$$(AB)(c)(t)\left(\frac{L}{AB}\right) = Lct$$

2. A downward vertical displacement of

$$(AB)(c)(t)\left(\frac{h}{AB}\right) = hct$$

Thus

$$a = -(80)(6.0 \times 10^{-6})(50) = -24 \times 10^{-3} \text{ ft}$$
$$b = +(20)(6.0 \times 10^{-6})(50) = +6 \times 10^{-3} \text{ ft}$$

Due to $a = -24 \times 10^{-3}$ ft

$$M_O = 0$$

$$H_O = +\frac{a}{I'_x} = +\frac{-24 \times 10^{-3} \text{ ft}}{4.0540 \times 10^{-3} \dfrac{\text{ft}}{\text{kips}}} = -5.920 \text{ kips}$$

$$V_O = +\frac{a(I_{xy}/I_x)}{I'_y} = +\frac{(-24 \times 10^{-3})(1.79218)}{34.623 \times 10^{-3}} = -1.242 \text{ kips}$$

Due to $b = +6 \times 10^{-3}$ ft,

$$M_O = 0$$

$$H_O = -\frac{b(I_{xy}/I_y)}{I'_x} = -\frac{(6 \times 10^{-3})(0.20978)}{4.0540 \times 10^{-3}} = -0.310 \text{ kip}$$

$$V_O = -\frac{b}{I'_y} = -\frac{6 \times 10^{-3}}{34.623 \times 10^{-3}} = -0.173 \text{ kip}$$

Due to a temperature rise of 50°F,

$$M_O = 0$$
$$H_O = -5.920 - 0.310 = -6.230 \text{ kips}$$
$$V_O = -1.242 - 0.173 = -1.415 \text{ kips}$$

The reaction components are shown in Fig. 318.

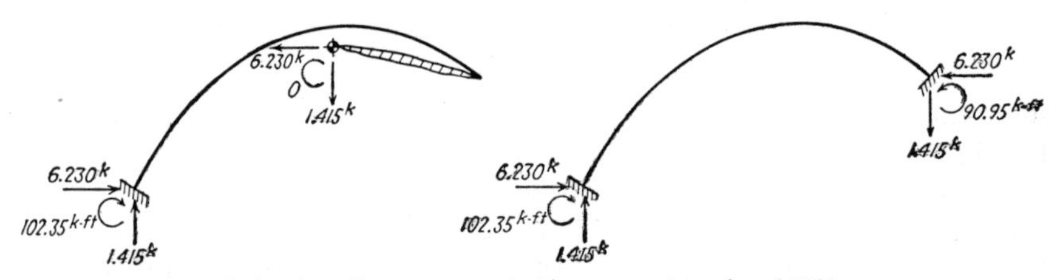

Induced reaction components due to a temperature rise of 50°F

FIG. 318.

b. A Temperature Drop of 80°*F* (Fig. 319). By multiplying the results in (*a*) by $-\frac{80}{50} = -1.6$, the reaction components due to a temperature drop of 80°F are found.

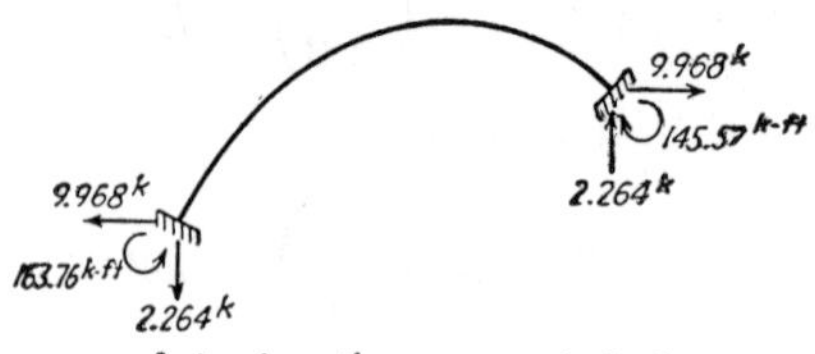

Induced reaction components due to a temperature drop of 80°F

FIG. 319.

c. Shrinkage Equivalent to a Temperature Drop of 15°*F* (Fig. 320). By multiplying the results in (*a*) by $-\frac{15}{50} = -0.3$, the reaction components due to shrinkage equivalent to a temperature drop of 15°F are found.

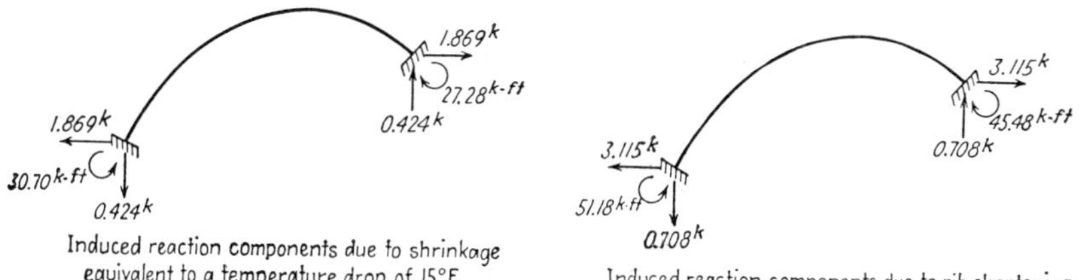

Fig. 320. Fig. 321.

d. Rib Shortening Equivalent to an Axial Stress of 300 *Psi throughout the Arch Rib* (Fig. 321). By multiplying the results in (*a*) by

$$-\left(\frac{300}{2 \times 10^6}\right) \div \left(50 \times 6.0 \times 10^{-6}\right) = -\tfrac{1}{2},$$

the reaction components due to rib shortening are found.

e. A Rotational Slip of 0.001 *Rad Counterclockwise at Support B* (Fig. 322)

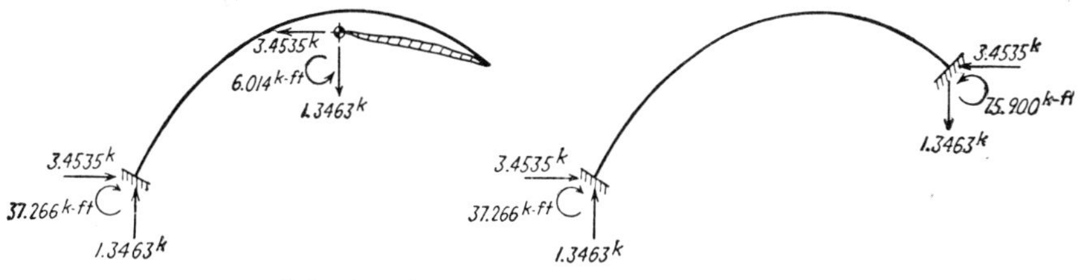

Fig. 322.

$$\alpha = +0.001 \text{ rad}$$

$$M_O = +\frac{\alpha}{A} = +\frac{0.001}{0.16382 \times 10^{-3}} = +6.104 \text{ kip-ft}$$

$$H_O = +\frac{y_B\alpha - x_B\alpha(I_{xy}/I_y)}{I'_x}$$

$$= \frac{(-6.7658)(+0.001) - (+34.4872)(+0.001)(0.20978)}{4.0540 \times 10^{-3}}$$

$$= -3.4535 \text{ kips}$$

$$V_O = -\frac{x_B\alpha - y_B\alpha(I_{xy}/I_x)}{I'_y}$$

$$= -\frac{(+34.4872)(+0.001) - (-6.7658)(+0.001)(1.79218)}{34.623 \times 10^{-3}}$$

$$= -1.3463 \text{ kips}$$

f. A Horizontal Displacement of 0.200 *In. to the Right at Support B* (Fig. 323)

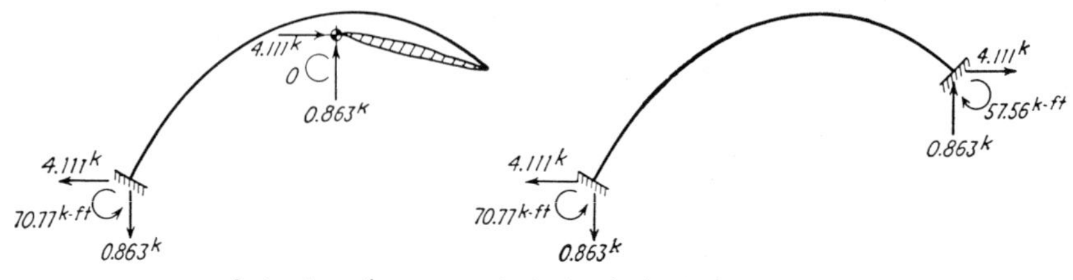

Induced reaction components due to a horizontal displacement of 0.200 in. to the right at support *B*

FIG. 323.

$$a = +0.200 \text{ in.}$$

$$M_O = 0$$

$$H_O = +\frac{a}{I'_x} = +\frac{(0.200/12) \text{ ft}}{4.0540 \times 10^{-3} \dfrac{\text{ft}}{\text{kips}}} = +4.111 \text{ kips}$$

$$V_O = +\frac{a(I_{xy}/I_x)}{I'_y} = +\frac{(0.200/12)(1.79218)}{34.623 \times 10^{-3}} = +0.863 \text{ kip}$$

g. A Vertical Settlement of 0.250 *In. Downward at Support B* (Fig. 324)

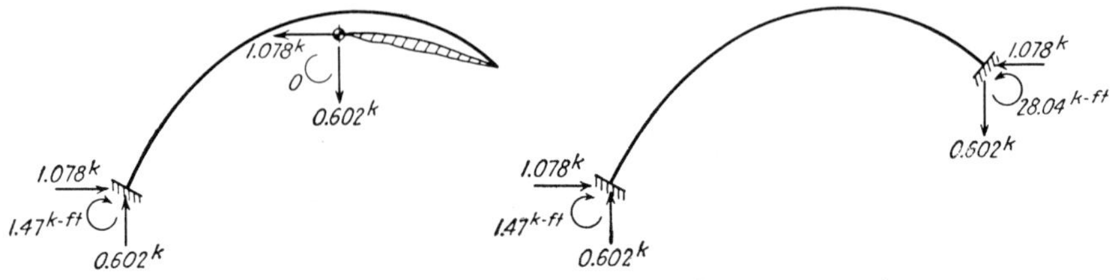

Induced reaction components due to a vertical settlement of 0.250 in. downward at support *B*

FIG. 324.

$$b = +0.250 \text{ in.}$$

$$M_O = 0$$

$$H_O = -\frac{b(I_{xy}/I_y)}{I'_x} = -\frac{(0.250/12)(0.20978)}{4.0540 \times 10^{-3}} = -1.078 \text{ kips}$$

$$V_O = -\frac{b}{I'_y} = -\frac{0.250/12}{34.623 \times 10^{-3}} = -0.602 \text{ kip}$$

EXERCISES

167. Determine all reaction components induced to act on the symmetrical circular fixed arch shown below, due to each of the following causes: (*a*) a temperature rise of 50°F; (*b*) a temperature drop of 80°F; (*c*) shrinkage equivalent to a temperature drop of 15°F; (*d*) rib shortening equivalent to an axial stress of 300 psi throughout the arch rib; (*e*) a rotational slip of 0.001 rad counterclockwise at support *B*, (*f*) a horizontal displacement of 0.200 in. to the right at support *B*; (*g*) a vertical settlement of 0.250 in. downward at support *B*.

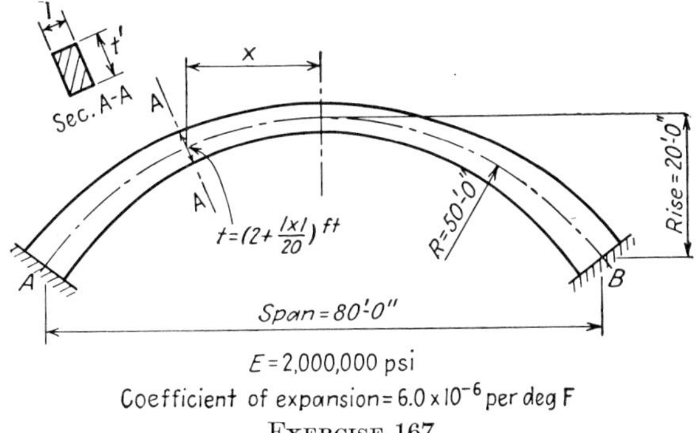

EXERCISE 167

168. Determine all reaction components induced to act on the unsymmetrical circular fixed arch shown below, due to each of the following causes: (*a*) a temperature rise of 50°F; (*b*) a temperature drop of 80°F; (*c*) shrinkage equivalent to a temperature

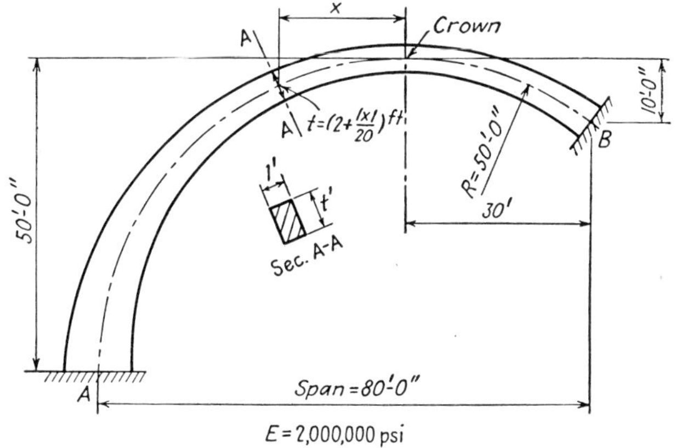

EXERCISE 168

drop of 15°F; (*d*) rib shortening equivalent to an axial stress of 300 psi throughout the arch rib; (*e*) a rotational slip of 0.001 rad counterclockwise at support *B*; (*f*) a horizontal displacement of 0.200 in. to the right at support *B*; (*g*) a vertical settlement of 0.250 in. downward at support *B*.

CHAPTER XI

SECONDARY STRESSES IN TRUSSES WITH RIGID JOINTS

70. General Description. The usual steel truss is a structure composed of individual members so joined together as to form a series of triangles. The joints may be pin-connected, riveted, or welded. In the first stage of stress analysis, however, the joints are assumed to act as smooth hinges. If the truss were so built, it follows that the members are subjected to direct stress of tension or compression only and not subjected to bending. The direct stress in each member computed by the methods of statics is called the *primary stress.* When the lengths of members change owing to the direct stresses, the joints or hinges must move accordingly to some new position so as to accommodate themselves to the new lengths. These movements, or deflections, of joints must be accompanied by changes in the angles between members. This may easily happen if the *straight* members can freely rotate around the joints or if the joints can act as perfectly smooth hinges. In actual cases, however, the joints are far from the condition of frictionless hinges. For pin-connected trusses, the members may not be free to turn around the pins because of the friction which may be developed. In riveted trusses, the members are almost impossible to rotate at the joints except for a certain amount of yielding, or "play," owing to the deformations of the rivet pins or to the somewhat loose fitting of the rivets in the holes. In welded trusses, the joints are considered rigid; *i.e.*, the angles between members cannot change at all. In all cases except that of smooth hinges, when the members tend to rotate around the joints but are restrained from attaining the full amounts as in the case of semirigid joints or are entirely prevented from any rotation as in the case of rigid joints, the members will have to bend and thus bending stresses, or secondary stresses, will be developed.

Consider the triangular truss ABC of Fig. 325. Owing to the changes in the lengths of the three members, the joints A, B, C are displaced to take the positions of A', B', C' (A and A' happen to coincide). If the three members $A'B'$, $B'C'$, and $C'A'$ are to remain straight, apparently angles A' and B' should be, respectively, smaller than angles A and B, while angle C' should be larger than angle C. If, however, the joints are rigid, or the sizes of angles cannot change, the members $A'B'$, $B'C'$, and $C'A'$ will have to bend as shown so that the angles between the tangents to the bent members remain equal to the original angles. Note that the joints as a whole may rotate as long as the angle between tangents is not changed. In Fig. 325, the amount of rotation of the joints as a whole is shown by a short arrow starting from the original direction of the member (dotted line) and ending at the tangent to the bent member (solid line). In this chapter, the discussion will be limited to the evaluation of the bending stresses, or secondary stresses, in trusses with perfectly rigid joints only.

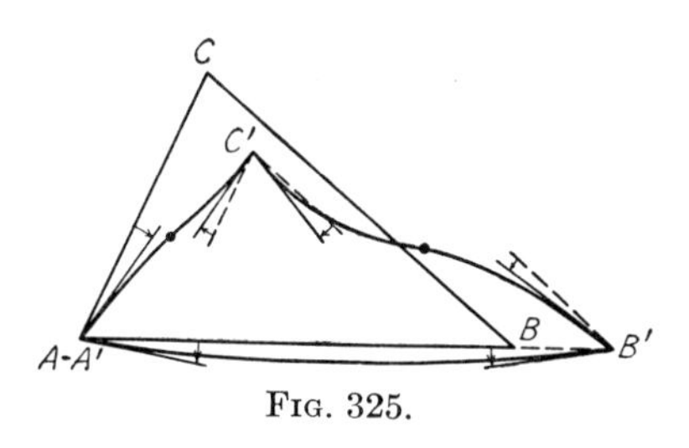

FIG. 325.

71. Procedure of Determining Secondary Stresses in Trusses with Rigid Joints. Since the truss is considered to have rigid joints, it is in fact a *rigid frame.* The slope-deflection method or the moment-distribution method can thus be applied to analyze such a frame. When the slope-deflection method is used, the joint rotations are treated as unknowns, the values of R for all members (for instance, R for member BC in Fig. 326 is the angle between BC and $B'C'$) can be computed, and the fixed-end moments are zero since there are no loads applied between the joints. When the moment-distribution method is used, the fixed-end moments due to R are first computed by the formula,

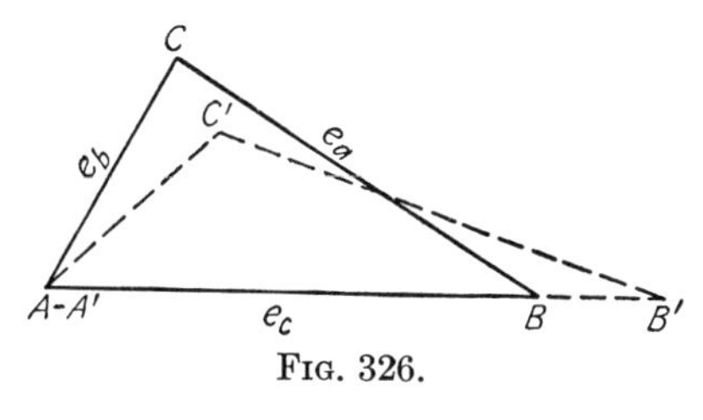

FIG. 326.

$$M_{FAB} = M_{FBA} = \frac{6EIR}{L}$$

The procedure of determining secondary stresses in trusses with rigid joints is described below:

1. Determine the change in the direction of each member, or the value of R, owing to the deflections of the joints. This can be done by first finding the changes in the sizes of all interior angles, using the following formulas (see Art. 19):

$$\begin{aligned} \Delta A &= (e_a - e_b) \cot C + (e_a - e_c) \cot B \\ \Delta B &= (e_b - e_c) \cot A + (e_b - e_a) \cot C \\ \Delta C &= (e_c - e_a) \cot B + (e_c - e_b) \cot A \end{aligned} \tag{31}$$

Then, referring to Fig. 326, if R_{AB} is known,

$$\begin{aligned} R_{BC} &= R_{AB} + \Delta B \\ R_{CA} &= R_{BC} + \Delta C \end{aligned} \tag{147}$$

Note that R is positive when the angle measured from the original direction of the member to the new direction is clockwise; and an increase in the size of any interior angle is considered positive. For any one triangle, if the value of R for one side is known, the values of R for the other two sides (starting from the side with known value of R and going around in the *counterclockwise* direction) can be computed by using Eqs. (147). The validity of this formula is obvious from the fact that when R_{AB} is zero, the increase, ΔB, in the size of angle B is equal to R_{BC} in the clockwise direction. To start this process, any member may be chosen as the reference member, or its value of R is assumed to be zero. For instance, in Fig. 327, R for member U_2L_3 may be made equal to zero.

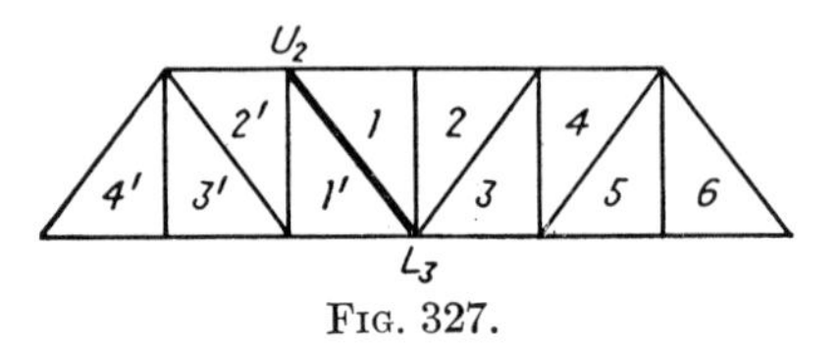

FIG. 327.

By proceeding through triangles 1, 2, 3, 4, 5, and 6 toward the right and then through triangles 1′, 2′, 3′, and 4′ toward the left, the values of R for all members can be determined.

2. With the values of R for all members computed, the end moments acting on each member can be found either by the slope-deflection method or by the moment-distribution method.

3. Compute the secondary stress for each member, which is the flexural stress caused by the larger end moment.

Example 100. Compute the secondary stresses in all members of the truss with rigid joints shown in Fig. 328 due to the applied loads. The

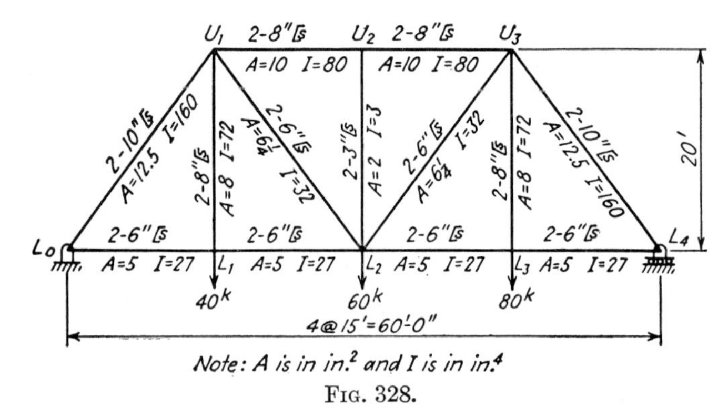

FIG. 328.

properties of the channel sections as given are arbitrary. Use $E = 30{,}000$ kips/in.²

Solution. The total direct stress, the primary stress (total direct stress divided by the cross-sectional area), and the unit axial deformation for each member are computed and shown in Fig. 329.

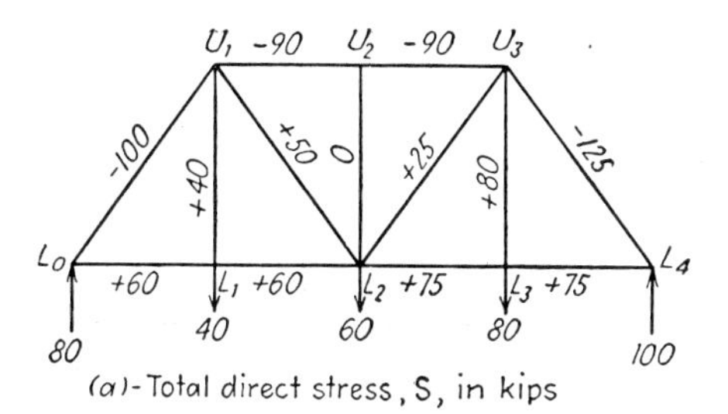

(a)-Total direct stress, S, in kips

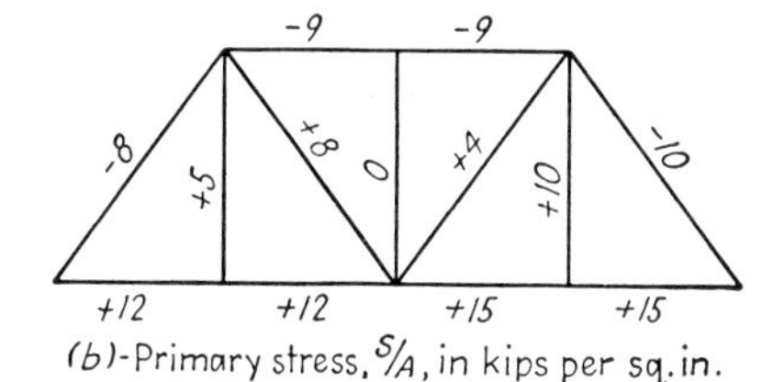

(b)-Primary stress, S/A, in kips per sq. in.

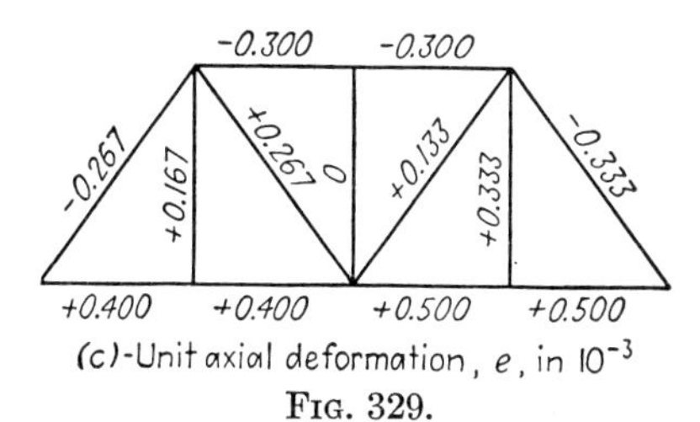

(c)-Unit axial deformation, e, in 10^{-3}

FIG. 329.

The changes in the sizes of all interior angles are computed in Table 100-1. Note that the sum of the changes in the three angles of any triangle should be equal to zero. Refer to Fig. 330 for the cotangent function of angles.

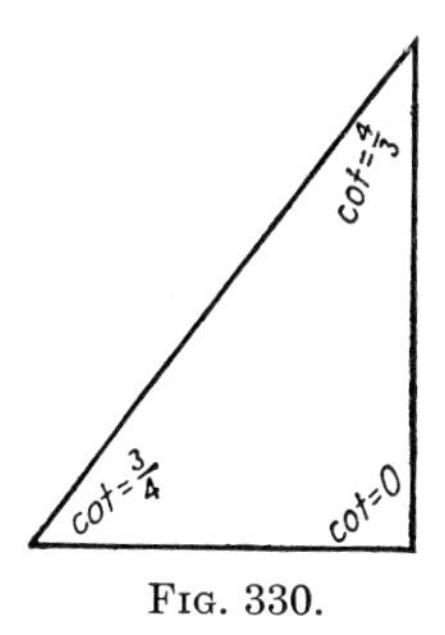

FIG. 330.

TABLE 100-1

Triangle	Angle	Change in size of angle, 10^{-3} rad
$L_0L_1U_1$	$L_0L_1U_1$	$(-0.267 - 0.400)(\frac{3}{4}) + (-0.267 - 0.167)(\frac{4}{3}) = -1.077$
	$L_1U_1L_0$	$(+0.400 - 0.167)(0) + (+0.400 + 0.267)(\frac{3}{4}) = +0.500$
	$U_1L_0L_1$	$(+0.167 - 0.400)(0) + (+0.167 + 0.267)(\frac{4}{3}) = +0.577$
$L_1L_2U_1$	$L_1L_2U_1$	$(+0.167 - 0.267)(\frac{4}{3}) + (+0.167 - 0.400)(0) = -0.133$
	$L_2U_1L_1$	$(+0.400 - 0.267)(\frac{3}{4}) + (+0.400 - 0.167)(0) = +0.100$
	$U_1L_1L_2$	$(+0.267 - 0.167)(\frac{4}{3}) + (+0.267 - 0.400)(\frac{3}{4}) = +0.033$
$L_2U_2U_1$	$L_2U_2U_1$	$(+0.267 - 0)(\frac{4}{3}) + (+0.267 + 0.300)(\frac{3}{4}) = +0.781$
	$U_2U_1L_2$	$(0 + 0.300)(0) + (0 - 0.267)(\frac{4}{3}) = -0.356$
	$U_1L_2U_2$	$(-0.300 - 0.267)(\frac{3}{4}) + (-0.300 - 0)(0) = -0.425$
$L_2U_3U_2$	$L_2U_3U_2$	$(0 + 0.300)(0) + (0 - 0.133)(\frac{4}{3}) = -0.178$
	$U_3U_2L_2$	$(+0.133 + 0.300)(\frac{3}{4}) + (+0.133 - 0)(\frac{4}{3}) = +0.503$
	$U_2L_2U_3$	$(-0.300 - 0)(0) + (-0.300 - 0.133)(\frac{3}{4}) = -0.325$
$L_2L_3U_3$	$L_2L_3U_3$	$(+0.133 - 0.500)(\frac{3}{4}) + (+0.133 - 0.333)(\frac{4}{3}) = -0.542$
	$L_3U_3L_2$	$(+0.500 - 0.333)(0) + (+0.500 - 0.133)(\frac{3}{4}) = +0.275$
	$U_3L_2L_3$	$(+0.333 - 0.133)(\frac{4}{3}) + (+0.333 - 0.500)(0) = +0.267$
$L_3L_4U_3$	$L_3L_4U_3$	$(+0.333 + 0.333)(\frac{4}{3}) + (+0.333 - 0.500)(0) = +0.889$
	$L_4U_3L_3$	$(+0.500 + 0.333)(\frac{3}{4}) + (+0.500 - 0.333)(0) = +0.625$
	$U_3L_3L_4$	$(-0.333 - 0.333)(\frac{4}{3}) + (-0.333 - 0.500)(\frac{3}{4}) = -1.514$

In order to demonstrate that the same secondary stresses will be obtained irrespective of which member is used as the reference member, two parallel solutions, R of $U_2L_2 = 0$ in one and R of $U_1L_2 = 0$ in the other, will be shown. The values of R for all members based on the two respective reference members are computed in Table 100-2.

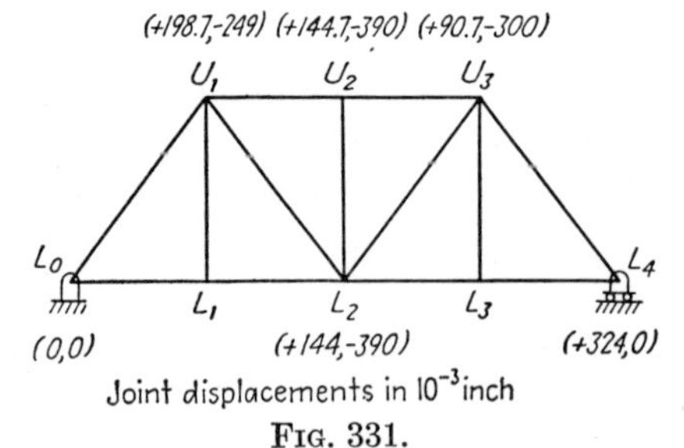

FIG. 331.

Inasmuch as the subsequent work depends entirely on the correctness of the values of R, it is advisable to make an independent check. One way of checking is to draw the Williot diagram, using the same reference member as is used in determining the R values. However, it is sometimes difficult to obtain the desired degree of accuracy by measurements from the Williot diagram.

The horizontal and vertical deflections at all joints of this truss have been solved by the method of joint displacements in Example 31 of Art. 20. The deflections of some of the joints are shown in Fig. 331.

TABLE 100-2. VALUES OF R (10^{-3} RAD)

U_2L_2 as reference member			U_1L_2 as reference member		
Triangle $U_2L_2U_3$	R of U_2L_2	0	Triangle $U_1L_2U_2$	R of U_1L_2	0
	$\Delta\theta$ at L_2	−0.325		$\Delta\theta$ at L_2	−0.425
	R of L_2U_3	−0.325		R of L_2U_2	−0.425
	$\Delta\theta$ at U_3	−0.178		$\Delta\theta$ at U_2	+0.781
	R of U_3U_2	−0.503		R of U_2U_1	+0.356
	$\Delta\theta$ at U_2	+0.503		$\Delta\theta$ at U_1	−0.356
	R of U_2L_2	0		R of U_1L_2	0
Triangle $U_3L_2L_3$	R of U_3L_2	−0.325	Triangle $U_2L_2U_3$	R of U_2L_2	−0.425
	$\Delta\theta$ at L_2	+0.267		$\Delta\theta$ at L_2	−0.325
	R of L_2L_3	−0.058		R of L_2U_3	−0.750
	$\Delta\theta$ at L_3	−0.542		$\Delta\theta$ at U_3	−0.178
	R of L_3U_3	−0.600		R of U_3U_2	−0.928
	$\Delta\theta$ at U_3	+0.275		$\Delta\theta$ at U_2	+0.503
	R of U_3L_2	−0.325		R of U_2L_2	−0.425
Triangle $U_3L_3L_4$	R of U_3L_3	−0.600	Triangle $U_3L_2L_3$	R of U_3L_2	−0.750
	$\Delta\theta$ at L_3	−1.514		$\Delta\theta$ at L_2	+0.267
	R of L_3L_4	−2.114		R of L_2L_3	−0.483
	$\Delta\theta$ at L_4	+0.889		$\Delta\theta$ at L_3	−0.542
	R of L_4U_3	−1.225		R of L_3U_3	−1.025
	$\Delta\theta$ at U_3	+0.625		$\Delta\theta$ at U_3	+0.275
	R of U_3L_3	−0.600		R of U_3L_2	−0.750
Triangle $U_2U_1L_2$	R of L_2U_2	0	Triangle $U_3L_3L_4$	R of U_3L_3	−1.025
	$\Delta\theta$ at U_2	+0.781		$\Delta\theta$ at L_3	−1.514
	R of U_2U_1	+0.781		R of L_3L_4	−2.539
	$\Delta\theta$ at U_1	−0.356		$\Delta\theta$ at L_4	+0.889
	R of U_1L_2	+0.425		R of L_4U_3	−1.650
	$\Delta\theta$ at L_2	−0.425		$\Delta\theta$ at U_3	+0.625
	R of L_2U_2	0		R of U_3L_3	−1.025
Triangle $U_1L_1L_2$	R of L_2U_1	+0.425	Triangle $L_2U_1L_1$	R of L_2U_1	0
	$\Delta\theta$ at U_1	+0.100		$\Delta\theta$ at U_1	+0.100
	R of U_1L_1	+0.525		R of U_1L_1	+0.100
	$\Delta\theta$ at L_1	+0.033		$\Delta\theta$ at L_1	+0.033
	R of L_1L_2	+0.558		R of L_1L_2	+0.133
	$\Delta\theta$ at L_2	−0.133		$\Delta\theta$ at L_2	−0.133
	R of L_2U_1	+0.425		R of L_2U_1	0
Triangle $U_1L_0L_1$	R of L_1U_1	+0.525	Triangle $L_1U_1L_0$	R of L_1U_1	+0.100
	$\Delta\theta$ at U_1	+0.500		$\Delta\theta$ at U_1	+0.500
	R of U_1L_0	+1.025		R of U_1L_0	+0.600
	$\Delta\theta$ at L_0	+0.577		$\Delta\theta$ at L_0	+0.577
	R of L_0L_1	+1.602		R of L_0L_1	+1.177
	$\Delta\theta$ at L_1	−1.077		$\Delta\theta$ at L_1	−1.077
	R of L_1U_1	+0.525		R of L_1U_1	+0.100

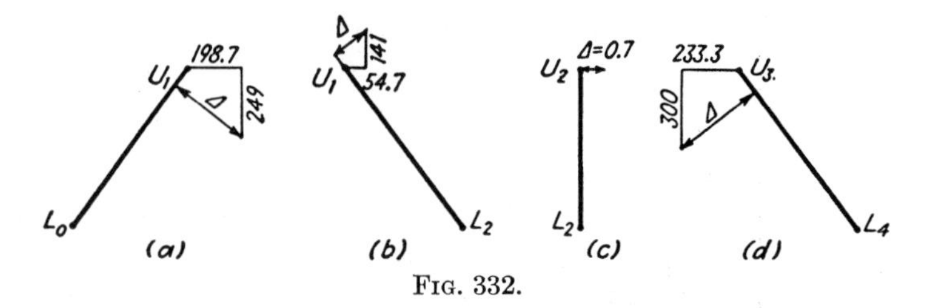

FIG. 332.

The rotation, or change in the direction of any member, can be computed from the displacements of its end joints. Thus, referring to Fig. 332,

$$R \text{ of } L_0U_1 \text{ (Fig. 332}a) = +\frac{\Delta}{300} = +\frac{(198.7 \times 0.8) + (249 \times 0.6)}{300} \times 10^{-3}$$
$$= +1.028 \times 10^{-3} \text{ rad}$$

$$R \text{ of } U_1L_2 \text{ (Fig. 332}b) = +\frac{\Delta}{300} = +\frac{(54.7 \times 0.8) + (141 \times 0.6)}{300} \times 10^{-3}$$
$$= +0.428 \times 10^{-3} \text{ rad}$$

$$R \text{ of } U_2L_2 \text{ (Fig. 332}c) = +\frac{\Delta}{240} = +\frac{0.7}{240} \times 10^{-3}$$
$$= +0.003 \times 10^{-3} \text{ rad}$$

$$R \text{ of } U_3L_4 \text{ (Fig. 332}d) = -\frac{\Delta}{300} = -\frac{(233.3 \times 0.8) + (300 \times 0.6)}{300} \times 10^{-3}$$
$$= -1.222 \times 10^{-3} \text{ rad}$$

Now if U_2L_2 is made the reference member, the R's of the above four members become

$$\begin{aligned}
R \text{ of } L_0U_1 &= +1.028 \times 10^{-3} - 0.003 \times 10^{-3} = +1.025 \times 10^{-3} \text{ rad}\\
R \text{ of } U_1L_2 &= +0.428 \times 10^{-3} - 0.003 \times 10^{-3} = +0.425 \times 10^{-3} \text{ rad}\\
R \text{ of } U_2L_2 &= +0.003 \times 10^{-3} - 0.003 \times 10^{-3} = 0\\
R \text{ of } U_3L_4 &= -1.222 \times 10^{-3} - 0.003 \times 10^{-3} = -1.225 \times 10^{-3} \text{ rad}
\end{aligned}$$

If U_1L_2 is made the reference member, the R's of the above four members become

$$\begin{aligned}
R \text{ of } L_0U_1 &= +1.028 \times 10^{-3} - 0.428 \times 10^{-3} = +0.600 \times 10^{-3} \text{ rad}\\
R \text{ of } U_1L_2 &= +0.428 \times 10^{-3} - 0.428 \times 10^{-3} = 0\\
R \text{ of } U_2L_2 &= +0.003 \times 10^{-3} - 0.428 \times 10^{-3} = -0.425 \times 10^{-3} \text{ rad}\\
R \text{ of } U_3L_4 &= -1.222 \times 10^{-3} - 0.428 \times 10^{-3} = -1.650 \times 10^{-3} \text{ rad}
\end{aligned}$$

It can be seen that the R values for the two end posts as computed above for both reference members check precisely with those found by the method of angle changes.

The values of $6EIR/L$ for all members are computed in Table 100-3.

TABLE 100-3

Member	L, ft	I in.[4]	R of $U_2L_2 = 0$		R of $U_1L_2 = 0$	
			R, 10^{-3} rad	$\frac{6EIR}{L}$, kip-ft	R, 10^{-3} rad	$\frac{6EIR}{L}$, kip-ft
U_1U_2	15	80	+0.781	+5.21	+0.356	+2.37
U_2U_3	15	80	−0.503	−3.35	−0.928	−6.19
L_0L_1	15	27	+1.602	+3.60	+1.177	+2.65
L_1L_2	15	27	+0.558	+1.26	+0.133	+0.30
L_2L_3	15	27	−0.058	−0.13	−0.483	−1.09
L_3L_4	15	27	−2.114	−4.76	−2.539	−5.71
L_0U_1	25	160	+1.025	+8.20	+0.600	+4.80
U_1L_2	25	32	+0.425	+0.68	0	0
L_2U_3	25	32	−0.325	−0.52	−0.750	−1.20
U_3L_4	25	160	−1.225	−9.80	−1.650	−13.20
U_1L_1	20	72	+0.525	+2.36	+0.100	+0.45
U_2L_2	20	3	0	0	−0.425	−0.08
U_3L_3	20	72	−0.600	−2.70	−1.025	−4.61

The solution for the end moments by the moment-distribution method is shown in Tables 100-4 and 100-5, with U_2L_2 and U_1L_2 as the respective reference members. In the check for moment distribution, when the "sum" of the "change" at the near end and $-\frac{1}{2}$ times the change at the far end is divided by $-3EI/L$ kip-ft, the true value of the angle between the tangent and the original direction of the member when the lower chord of the undeformed truss is oriented in the horizontal direction is obtained. However, whichever member is used as the reference member, the angle between the tangent to the elastic curve and the direction of the line joining the deflected points must be *constant*. Thus, in Fig. 333,

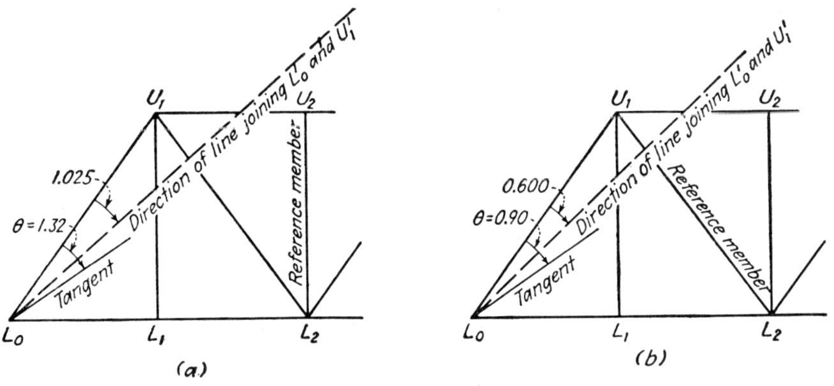

FIG. 333.

TABLE 100-4. MOMENT DISTRIBUTION (REFERENCE MEMBER: U_2L_2)

Joint		U_1				U_2			U_3				L_0	
Member		U_1L_0	U_1L_1	U_1L_2	U_1U_2	U_2U_1	U_2L_2	U_2U_3	U_3U_2	U_3L_2	U_3L_3	U_3L_4	L_0U_1	L_0L_1
$2EI/L$, kip-ft		2,667	1,500	533	2,222	2,222	62	2,222	2,222	533	1,500	2,667	2,667	750
Cycle	DF	0.385	0.217	0.077	0.321	0.493	0.014	0.493	0.321	0.077	0.217	0.385	0.780	0.220
1	FEM	+8.20	+2.36	+0.68	+5.21	+5.21		−3.35	−3.35	−0.52	−2.70	−9.80	+8.20	+3.60
	Bal.	−6.33	−3.57	−1.27	−5.28	−0.92	−0.02	−0.92	+5.25	+1.26	+3.56	+6.30	−9.20	−2.60
2	CO	−4.60	−1.80	−0.13	−0.46	−2.64	−0.01	+2.62	−0.46	−0.13	+1.90	+5 68	−3.16	−0.90
	Bal.	+2.69	+1.52	+0.54	+2.24	+0.01		+0.02	−2.24	−0.54	−1.52	−2.69	+3.17	+0.89
3	CO	+1.58	+0.82			+1.12		−1.12	+0.01		−0.80	−1.60	+1.34	+0.40
	Bal.	−0.92	−0.52	−0.19	−0.77				+0.77	+0.18	+0.52	+0.92	−1.36	−0.38
4	CO	−0.68	−0.30			−0.38		+0.38			+0.30	+0.68	−0.46	−0.15
	Bal.	+0.33	+0.21	+0.08	+0.31				−0.31	−0.08	−0.21	−0.38	+0.48	+0.13
5	CO	+0.24	+0.11			+0.16		−0.16			−0.11	−0.24	+0.19	+0.06
	Bal.	−0.13	−0.08	−0.03	−0.11				+0.11	+0.03	+0.08	+0.13	−0.20	−0.05
6	CO	−0.10	−0.04			−0.06		+0.06			+0.04	+0.10	−0.06	−0.02
	Bal.	+0.05	+0.03	+0.01	+0.05				−0.05	−0.01	−0.03	−0.05	+0.06	+0.02
7	CO	+0.03	+0.01			+0.02		−0.02			−0.01	−0.03	+0.03	+0.01
	Bal.	−0.02	−0.01		−0.01				+0.01		+0.01	+0.02	−0.03	−0.01
8	CO	−0.01										+0.01	−0.01	
	Bal.	+0.01										−0.01	+0.01	
Total		+0.39	−1.26	−0.31	+1.18	+2.52	−0.03	−2.49	−0.26	+0.19	+1.03	−0.96	−1.00	+1.00
Check:														
Change		−7.81	−3.62	−0.99	−4.03	−2.69	−0.03	+0.86	+3.09	+0.71	+3.73	+8.84	−9.20	−2.60
$-\frac{1}{2}$ (change)		+4.60	+1.81	+0.34	+1.34	+2.02	+0.02	−1.54	−0.43	−0.08	−1.92	−5.66	+3.90	+1.10
Sum		−3.21	−1.81	−0.65	−2.69	−0.67	−0.01	−0.68	+2.66	+0.63	+1.81	+3.18	−5.30	−1.50
$\theta = \frac{\text{sum}}{(-3EI/L)}$		+0.80	+0.80	+0.81	+0.81	+0.20	+0.10	+0.20	−0.80	−0.79	−0.80	−0.80	+1.32	+1.33
		$\theta = +0.81 \times 10^{-3}$				$\theta = +0.20 \times 10^{-3}$			$\theta = -0.80 \times 10^{-3}$				$\theta = +1.32 \times 10^{-3}$	

TABLE 100-4.—(*Continued*)

Joint		L_1			L_2					L_3			L_4	
Member		L_1L_0	L_1U_1	L_1L_2	L_2L_1	L_2U_1	L_2U_2	L_2U_3	L_2L_3	L_3L_2	L_3U_3	L_3L_4	L_4L_3	L_4U_3
$2EI/L$, kip-ft		750	1,500	750	750	533	62	533	750	750	1,500	750	750	2,667
Cycle	DF	0.250	0.500	0.250	0.285	0.203	0.024	0.203	0.285	0.250	0.500	0.250	0.220	0.780
1	FEM	+3.60	+2.36	+1.26	+1.26	+0.68		-0.52	-0.13	-0.13	-2.70	-4.76	-4.76	- 9.80
	Bal.	-1.80	-3.61	-1.81	-0.37	-0.26	-0.03	-0.26	-0.37	+1.90	+3.80	+1.89	+3.20	+11.36
2	CO	-1.30	-1.78	-0.18	-0.90	-0.64	-0.01	+0.63	+0.95	-0.18	+1.78	+1.60	+0.95	+ 3.15
	Bal.	+0.81	+1.63	+0.82	-0.01			-0.01	-0.01	-0.80	-1.60	-0.80	-0.90	- 3.20
3	CO	+0.44	+0.76		+0.41	+0.27		-0.27	-0.40		-0.76	-0.45	-0.40	- 1.34
	Bal.	-0.30	-0.60	-0.30	-0.01					+0.30	+0.61	+0.30	+0.38	+ 1.36
4	CO	-0.19	-0.26		-0.15	-0.10		+0.09	+0.15		+0.26	+0.19	+0.15	+ 0.46
	Bal.	+0.11	+0.23	+0.11		+0.01				-0.11	-0.23	-0.11	-0.13	- 0.48
5	CO	+0.06	+0.10		+0.06	+0.04		-0.04	-0.06		-0.10	-0.06	-0.06	- 0.19
	Bal.	-0.04	-0.08	-0.04						+0.04	+0.08	+0.04	+0.05	+ 0.20
6	CO	-0.02	-0.04		-0.02	-0.01		+0.01	+0.02		+0.04	+0.02	+0.02	+ 0.06
	Bal.	+0.02	+0.03	+0.01						-0.01	-0.03	-0.02	-0.02	- 0.06
7	CO	+0.01	+0.01								-0.01	-0.01	-0.01	- 0.03
	Bal.	-0.01	-0.01								+0.01	+0.01	+0.01	+ 0.03
8	CO													+ 0.01
	Bal.													- 0.01
Total		+1.39	-1.26	-0.13	+0.27	-0.01	-0.04	-0.37	+0.15	+1.01	+1.15	+2.16	-1.52	+ 1.52
Check:														
Change		-2.21	-3.62	-1.39	-0.99	-0.69	-0.04	+0.15	+0.28	+1.14	+3.85	+2.60	+3.24	+11.32
$-\frac{1}{2}$ (change)		+1.30	+1.81	+0.50	+0.70	+0.50	+0.02	-0.36	-0.57	-0.14	-1.86	-1.62	-1.30	- 4.42
Sum		-0.91	-1.81	-0.89	-0.29	-0.19	-0.02	-0.21	-0.29	+1.00	+1.99	-0.98	+1.94	+ 6.90
$\theta = \frac{\text{sum}}{(-3EI/L)}$		+0.81	+0.80	+0.79	+0.26	+0.24	+0.22	+0.26	+0.26	-0.89	-0.88	-0.87	-1.72	- 1.72
		$\theta = +0.80 \times 10^{-3}$			$\theta = +0.26 \times 10^{-3}$					$\theta = -0.87 \times 10^{-3}$			$\theta = -1.72 \times 10^{-3}$	

TABLE 100-5. MOMENT DISTRIBUTION (REFERENCE MEMBER: U_1L_2)

Joint		U_1				U_2			U_3				L_0	
Member		U_1L_0	U_1L_1	U_1L_2	U_1U_2	U_2U_1	U_2L_2	U_2U_3	U_3U_2	U_3L_2	U_3L_3	U_3L_4	L_0U_1	L_0L_1
$2EI/L$, kip-ft		2,667	1,500	533	2,222	2,222	62	2,222	2,222	533	1,500	2,667	2,667	750
Cycle	DF	0.385	0.217	0.077	0.321	0.493	0.014	0.493	0.321	0.077	0.217	0.385	0.780	0.220
1	FEM	+4.80	+0.45	—	+2.37	+2.37	−0.08	−6.19	−6.19	−1.20	−4.61	−13.20	+4.80	+2.65
	Bal.	−2.93	−1.65	−0.59	−2.45	+1.92	+0.06	+1.92	+8.09	+1.94	+5.47	+ 9.70	−5.81	−1.64
2	CO	−2.90	−0.85	+0.21	+0.96	−1.22	+0.02	+4.04	+0.96	+0.21	+2.86	+ 7.38	−1.46	−0.42
	Bal.	+0.99	+0.56	+0.20	+0.83	−1.40	−0.04	−1.40	−3.66	−0.88	−2.48	− 4.39	+1.47	+0.41
3	CO	+0.74	+0.33	−0.18	−0.70	+0.42	−0.02	−1.83	−0.70	−0.18	−1.28	− 2.44	+0.50	+0.17
	Bal.	−0.07	−0.04	−0.02	−0.06	+0.71	+0.02	+0.70	+1.48	+0.35	+1.00	+ 1.77	−0.52	−0.15
4	CO	−0.26	−0.06	+0.08	+0.36	−0.03	+0.01	+0.74	+0.35	+0.08	+0.54	+ 1.11	−0.04	−0.03
	Bal.	−0.05	−0.02	−0.01	−0.04	−0.35	−0.01	−0.36	−0.67	−0.16	−0.45	− 0.80	+0.05	+0.02
5	CO	+0.02	—	−0.04	−0.18	−0.02	—	−0.33	−0.18	−0.04	−0.24	− 0.45	−0.02	—
	Bal.	+0.08	+0.04	+0.02	+0.06	+0.17	+0.01	+0.17	+0.29	+0.07	+0.20	+ 0.35	+0.02	—
6	CO	+0.01	+0.01	+0.02	+0.08	+0.03	—	+0.14	+0.08	+0.02	+0.10	+ 0.20	+0.04	—
	Bal.	−0.05	−0.02	−0.01	−0.04	−0.09	—	−0.08	−0.13	−0.03	−0.09	− 0.15	−0.03	−0.01
7	CO	−0.01	−0.01	−0.01	−0.04	−0.02	—	−0.06	−0.04	−0.01	−0.04	− 0.09	−0.02	—
	Bal.	+0.03	+0.01	+0.01	+0.02	+0.04	—	+0.04	+0.06	+0.01	+0.04	+ 0.07	+0.02	—
8	CO	+0.01	—	—	+0.02	+0.01	—	+0.03	+0.02	—	+0.01	+ 0.04	+0.01	—
	Bal.	−0.01	−0.01	—	−0.01	−0.02	—	−0.02	−0.02	−0.01	−0.01	− 0.03	−0.01	—
9	CO	—	—	—	−0.01	—	—	−0.01	−0.01	—	−0.01	− 0.02	—	—
	Bal.	+0.01	—	—	—	—	—	+0.01	+0.01	—	+0.01	+ 0.02	—	—
Total		+0.41	−1.26	−0.32	+1.17	+2.52	−0.03	−2.49	−0.26	+0.17	+1.02	− 0.93	−1.00	+1.00
Check:														
Change		−4.39	−1.71	−0.32	−1.20	+0.15	+0.05	+3.70	+5.93	+1.37	+5.63	+12.27	−5.80	−1.65
$-\frac{1}{2}$ (change)		+2.90	+0.86	—	−0.08	+0.60	−0.02	−2.96	−1.85	−0.42	−2.88	− 7.36	+2.20	+0.63
Sum		−1.49	−0.85	−0.32	−1.28	+0.75	+0.03	+0.74	+4.08	+0.95	+2.75	+ 4.91	−3.60	−1.02
$\theta = \dfrac{\text{sum}}{(-3EI/L)}$		+0.37	+0.38	+0.40	+0.38	−0.22	−0.32	−0.22	−1.22	−1.19	−1.22	− 1.23	+0.90	+0.91
		$\theta = +0.38 \times 10^{-3}$				$\theta = -0.22 \times 10^{-3}$			$\theta = -1.22 \times 10^{-3}$				$\theta = +0.90 \times 10^{-3}$	

TABLE 100-5.—(*Continued*)

Joint		L_1			L_2					L_3			L_4	
Member		L_1L_0	L_1U_1	L_1L_2	L_2L_1	L_2U_1	L_2U_2	L_2U_3	L_2L_3	L_3L_2	L_3U_3	L_3L_4	L_4L_3	L_4U_3
$2EI/L$, kip-ft		750	1,500	750	750	533	62	533	750	750	1,500	750	750	2,667
Cycle	DF	0.250	0.500	0.250	0.285	0.203	0.024	0.203	0.285	0.250	0.500	0.250	0.220	0.780
1	FEM	+2.65	+0.45	+0.30	+0.30	—	−0.08	−1.20	−1.09	−1.09	−4.61	−5.71	−5.71	−13.20
	Bal.	−0.85	−1.70	−0.85	+0.59	+0.42	+0.05	+0.42	+0.59	+2.85	+5.71	+2.85	+4.16	+14.75
2	CO	−0.82	−0.82	+0.30	−0.42	−0.30	+0.03	+0.97	+1.42	+0.30	+2.74	+2.08	+1.42	+ 4.85
	Bal.	+0.34	+0.67	+0.33	−0.48	−0.35	−0.04	−0.35	−0.48	−1.28	−2.56	−1.28	−1.38	− 4.89
3	CO	+0.20	+0.28	−0.24	+0.16	+0.10	−0.02	−0.44	−0.64	−0.24	−1.24	−0.69	−0.64	− 2.20
	Bal.	−0.06	−0.12	−0.06	+0.24	+0.17	+0.02	+0.17	+0.24	+0.54	+1.09	+0.54	+0.62	+ 2.22
4	CO	−0.08	−0.02	+0.12	−0.03	−0.01	+0.01	+0.18	+0.27	+0.12	+0.50	+0.31	+0.27	+ 0.88
	Bal.	—	−0.01	−0.01	−0.12	−0.08	−0.01	−0.09	−0.12	−0.23	−0.47	−0.23	−0.25	− 0.90
5	CO	+0.01	−0.01	−0.06	—	—	—	−0.08	−0.12	−0.06	−0.22	−0.12	−0.12	− 0.40
	Bal.	+0.01	+0.03	+0.02	+0.06	+0.04	—	+0.04	+0.06	+0.10	+0.20	+0.10	+0.11	+ 0.41
6	CO	—	+0.02	+0.03	+0.01	+0.01	—	+0.03	+0.05	+0.03	+0.10	+0.05	+0.05	+ 0.18
	Bal.	−0.01	−0.03	−0.01	−0.03	−0.02	—	−0.02	−0.03	−0.04	−0.09	−0.05	−0.05	− 0.18
7	CO	—	−0.01	−0.01	—	—	—	−0.01	−0.02	−0.02	−0.04	−0.02	−0.02	− 0.08
	Bal.	—	+0.01	+0.01	+0.01	—	—	+0.01	+0.01	+0.02	+0.04	+0.02	+0.02	+ 0.08
8	CO	—	—	—	—	—	—	—	—	—	+0.02	+0.01	+0.01	+ 0.04
	Bal.	—	—	—	—	—	—	—	—	—	−0.02	−0.01	−0.01	− 0.04
9	CO	—	—	—	—	—	—	—	—	—	—	—	—	− 0.02
	Bal.	—	—	—	—	—	—	—	—	—	—	—	—	+ 0.02
Total		+1.39	−1.26	−0.13	+0.29	−0.02	−0.04	−0.37	+0.14	+1.00	+1.15	−2.15	−1.52	+ 1.52
Check:														
Change		−0.26	−0.71	−0.43	−0.01	−0.02	+0.04	+0.83	+1.23	+2.09	+5.76	+3.56	+4.19	+14.72
$-\frac{1}{2}$ (change)		+0.82	+0.86	—	+0.22	+0.16	−0.02	−0.68	−1.04	−0.62	−2.82	−2.10	−1.78	− 6.14
Sum		−0.44	−0.85	−0.43	+0.21	+0.14	+0.02	+0.15	+0.19	+1.47	+2.94	+1.46	+2.41	+ 8.58
$\theta = \frac{\text{sum}}{(-3EI/L)}$		+0.39	+0.38	+0.38	−0.19	−0.18	−0.22	−0.19	−0.17	−1.31	−1.31	−1.30	−2.14	− 2.14
		$\theta = +0.38 \times 10^{-3}$			$\theta = -0.18 \times 10^{-3}$					$\theta = -1.30 \times 10^{-3}$			$\theta = -2.14 \times 10^{-3}$	

U_2L_2 as reference member:

Angle between tangent to elastic curve of L_0U_1 and direction of line joining L_0' and U_1' = (θ of joint L_0) − (R of L_0U_1)
$= (1.32 \times 10^{-3}) - (1.025 \times 10^{-3})$
$= 0.295 \times 10^{-3}$ rad

U_1L_2 as reference member:

Angle between tangent to elastic curve of L_0U_1 and direction of line joining L_0' and U_1' = (θ of joint L_0) − (R of L_0U_1)
$= (0.90 \times 10^{-3}) - (0.600 \times 10^{-3})$
$= 0.300 \times 10^{-3}$ rad

Thus the angle between the tangent to the elastic curve of L_0U_1 and the direction of the line joining L_0' and U_1' is found to be 0.295×10^{-3} rad or 0.300×10^{-3} rad, respectively, which should be identical if there has been no error at all.

Since the R values of Fig. 333*a* and those of Fig. 333*b* differ by 0.425×10^{-3} rad the θ values as computed in the check of moment distribution with U_2L_2 as reference member and those with U_1L_2 as reference member should also differ by 0.425×10^{-3} rad. This is found true, which also serves as an explanation why the same end moments are obtained even though different reference members are used.

When the slope-deflection method is used, the end moments are first expressed by the slope-deflection equations

$$M_{AB} = \frac{2EI}{L}(-2\theta_A - \theta_B + 3R)$$

$$M_{BA} = \frac{2EI}{L}(-2\theta_B - \theta_A + 3R)$$

There are then as many joint conditions (the sum of the end moments at any one joint must be zero) as unknown θ values. In the present example there are eight unknown θ values, and the eight joint conditions furnish the eight simultaneous equations. The amount of work involved far exceeds that of the moment-distribution method, especially when only the usual accuracy of three or four significant figures is required. Inasmuch as the procedure in the slope-deflection method of solution is straightforward and clear, it will not be further described.

The secondary stresses as computed in Table 100-6 are compared with the primary stresses.

TABLE 100-6

Member	I, in.^{4}	c, in.	Larger end moment, kip-ft	Secondary stress, $S = \frac{Mc}{I}$ psi	Primary stress, psi	Ratio of secondary stress to primary stress
U_1U_2	80	4	2.52	±1510	− 9,000	16.8%
U_2U_3	80	4	2.49	±1490	− 9,000	16.6%
L_0L_1	27	3	1.39	±1850	+12,000	15.4%
L_1L_2	27	3	0.27	± 360	+12,000	3.0%
L_2L_3	27	3	1.01	±1350	+15,000	9.0%
L_3L_4	27	3	2.16	±2880	+15,000	19.2%
L_0U_1	160	5	1.00	± 375	− 8,000	4.7%
U_1L_2	32	3	0.31	± 350	+ 8,000	4.4%
L_2U_3	32	3	0.37	± 415	+ 4,000	10.4%
U_3L_4	160	5	1.52	± 570	−10,000	5.7%
U_1L_1	72	4	1.26	± 840	+ 5,000	16.8%
U_2L_2	3	1.5	0.04	± 240	0	———
U_3L_3	72	4	1.15	± 770	+10,000	7.7%

EXERCISES

169. Using L_0U_1 as the reference member, compute the secondary stresses in all members of the truss with rigid joints due to the applied loads. The properties of the channel sections are arbitrary. Use E = 30,000 kips/in.^{2}

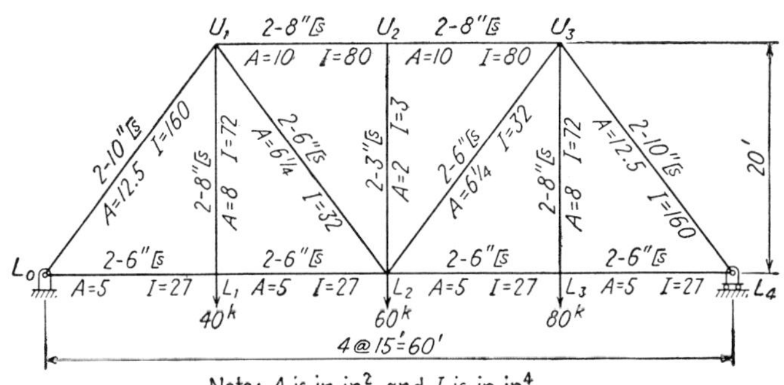

EXERCISE 169

170. Compute the secondary stresses in all members of the truss with rigid joints due to the applied loads. The properties of the channel sections are arbitrary. E = 30,000 kips/in.^{2} First use L_2U_2 as the reference member. Check the solution by using L_0U_1 as the reference member.

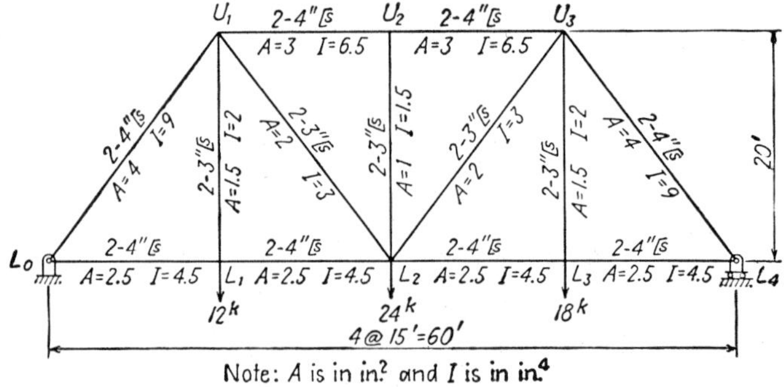

Note: A is in in.² and I is in in.⁴

EXERCISE 170

171. Compute the secondary stresses in all members of the truss with rigid joints due to the applied loads. The properties of the channel sections are arbitrary. Use E = 30,000 kips/in.²

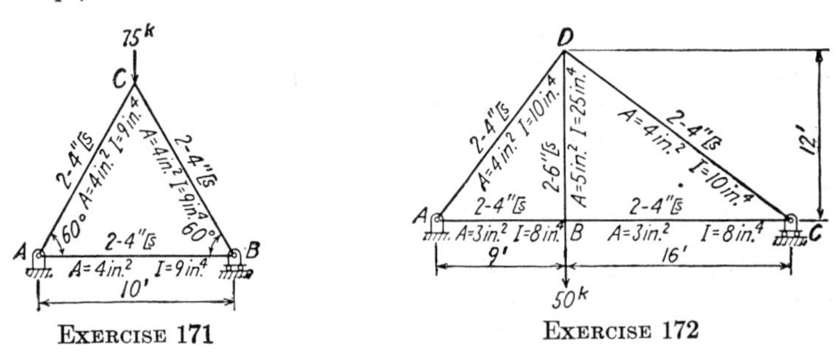

EXERCISE 171

EXERCISE 172

172. Compute the secondary stresses in all members of the truss with rigid joints due to the applied loads. The properties of the channel sections are arbitrary. Use E = 30,000 kips/in.²

CHAPTER XII

COMPOSITE STRUCTURES

72. General Introduction. In Chap. I it has been stated that most structures fall into one of the following three classifications: beams, frames, or trusses. A beam is subjected to bending only. The members of a truss with smooth hinged joints are subjected to direct stresses only. The members in a frame with rigid joints, however, are usually subjected to both direct and bending stresses; but the effect of the direct stresses in the members on the amount of deflections is in most cases insignificant compared with that of the bending stresses. Thus in Chap. II, in computing the deflections of a rigid frame, only the effect of bending stresses has been considered, or the members are treated as if they were beams not subjected to direct stresses. Then in Chap. IV, when statically indeterminate rigid frames are analyzed by the method of consistent deformation, the effect of direct stresses on the deflections is again neglected. The slope-deflection, moment-distribution, and column-analogy methods are equivalent to the method of consistent deformation in which the direct stresses are assumed not to cause any deformation.

There are structures, however, in which some members are primarily subjected to direct stresses and others to bending stresses. Such structures can be called "beam-trusses" or "truss-beams" but are generally known as *composite* structures. In computing the deflections of statically determinate composite structures, it is necessary to combine the methods described in Chaps. II and III. Also in analyzing statically indeterminate composite structures by the method of consistent deformation, it is necessary to consider the effects of both direct and bending stresses on the deformations. It is to be noted that the method of consistent deformation or its equivalent, the method of least work, is the only method by which statically indeterminate composite structures can be analyzed.

73. Analysis of Statically Indeterminate Composite Structures by the Method of Consistent Deformation. The principle involved in the analysis of statically indeterminate composite structures by the method of consistent deformation is simple, but its adaptation to any particular problem requires clear thinking and mastery of the methods of finding deflections previously studied. Inasmuch as it is not desirable to formulate any general procedure, the following examples are fully described in order to acquaint the reader with the basic ideas.

Example 101. Analyze the king-post truss shown in Fig. 334 by the method of consistent deformation.

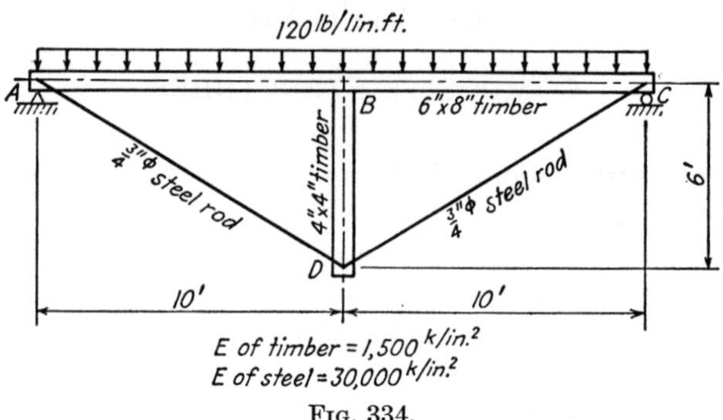

FIG. 334.

Solution. In this structure, member ABC is subjected to both bending and direct stresses, while members AD, BD, and CD are two-force members, which are subjected to direct stresses only. Since member ABC functions both as a beam and as a compression member, the given structure (Fig. 335a) can be considered as the sum of the two structures

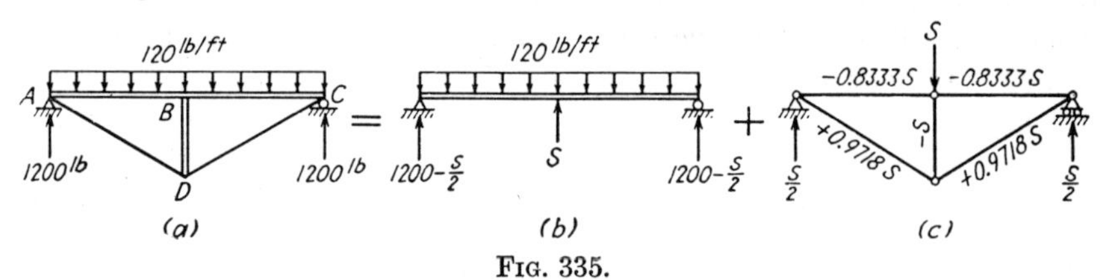

FIG. 335.

shown in Fig. 335b and c. The condition from which the compression S in the post BD can be solved is that the vertical deflection at B of Fig. 335b should be equal to that at B of Fig. 335c. Thus,

$$\Delta_B \text{ (Fig. 335}b) = \frac{1{,}728}{(1{,}500{,}000)\dfrac{(6)(8)^3}{12}}\left[\frac{(5)(120)(20)^4}{384} - \frac{S(20)^3}{48}\right]$$

$$= (1{,}125.0 - 0.75S) \times 10^{-3} \text{ in.}$$

$$\Delta_B \text{ (Fig. 335}c) = \frac{(2)(0.8333S)(0.8333)(10)(12)}{(1{,}500{,}000)(48)} + \frac{(S)(1)(6)(12)}{(1{,}500{,}000)(16)} + \frac{(2)(0.9718S)(0.9718)(11.662)(12)}{(30{,}000{,}000)(0.4418)}$$

$$= (0.002315S + 0.003S + 0.019944S) \times 10^{-3}$$

$$= 0.025259S \times 10^{-3} \text{ in.}$$

Equating Δ_B (Fig. 335*b*) and Δ_B (Fig. 335*c*),

$$1{,}125.0 - 0.75S = 0.025259S$$
$$S = 1{,}451 \text{ lb}$$

The value of S being known, the shear and moment diagrams for member ABC and the direct stresses in members ABC, AD, BD, and CD can then be computed in accordance with Fig. 335*b* and *c*.

The method of least work is sometimes a more direct method of analyzing statically indeterminate composite structures. The least-work solution of the present problem will now be shown.

Referring to Fig. 336,

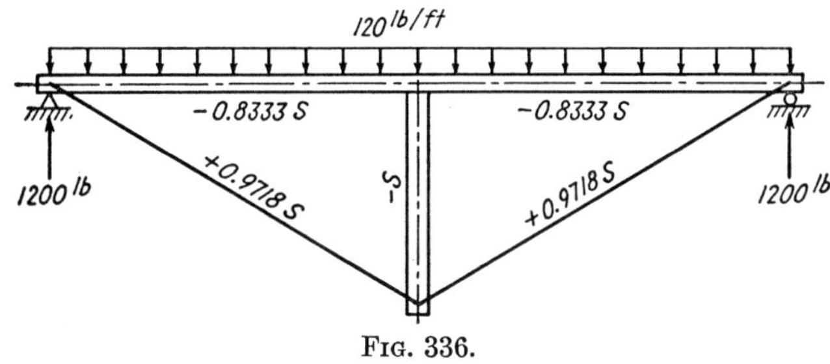

Fig. 336.

$$W = \frac{1}{2EI}\int M^2\,dx + \frac{1}{2}\sum \frac{S^2L}{AE}$$

$$= \frac{(1{,}728)(2)}{(2)(1{,}500{,}000)\dfrac{(6)(8)^3}{12}}\int_0^{10}\left[\left(1{,}200 - \frac{S}{2}\right)x - \tfrac{120}{2}x^2\right]^2 dx$$

$$+ \frac{(12)(2)}{(2)(1{,}500{,}000)(48)}(0.8333S)^2(10) + \frac{12}{(2)(1{,}500{,}000)(16)}(S)^2(6)$$
$$+ \frac{(12)(2)}{(2)(30{,}000{,}000)(0.4418)}(0.9718S)^2(11.662)$$

$$10^6 \times W = \frac{9}{2}\int_0^{10}\left[\left(1{,}200 - \frac{S}{2}\right)x - \tfrac{120}{2}x^2\right]^2 dx + 1.1574S^2$$
$$+ 1.5S^2 + 9.9715S^2$$

$$10^6 \times \frac{\partial W}{\partial S} = 9\int_0^{10}\left[\left(1{,}200 - \frac{S}{2}\right)x - 60x^2\right]\left(-\frac{x}{2}\right)dx + 2.3148S$$
$$+ 3S + 19.9430S$$
$$= -1{,}800{,}000 + 750S + 675{,}000 + 2.3148S + 3S$$
$$+ 19.9430S$$
$$= 0$$

$$775.258S = 1{,}125{,}000$$
$$S = 1{,}451 \text{ lb}$$

Example 102. Analyze the queen-post truss shown in Fig. 337 by the method of consistent deformation.

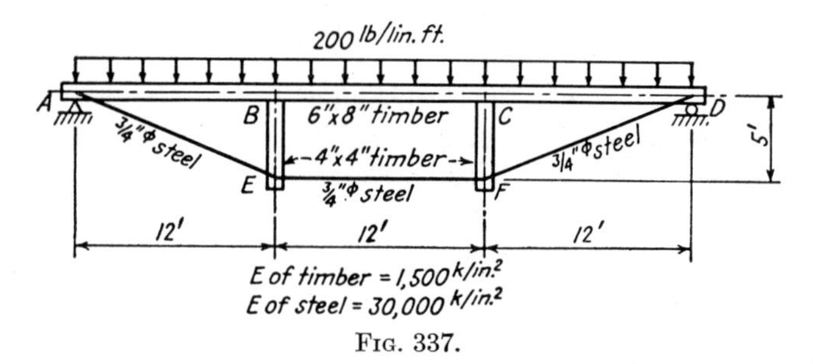

Fig. 337.

Solution. The given structure (Fig. 338*a*) can be considered as the sum of the two structures shown in Fig. 338*b* and *c*. The condition from which the compression S in either post BE or post CF can be solved

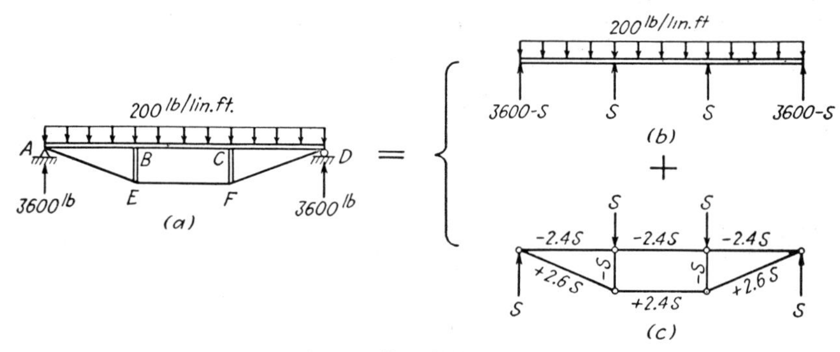

Fig. 338.

is that the vertical deflection at B or C of Fig. 338*b* should be equal to that at B or C of Fig. 338*c*. Thus,

Δ_B (Fig. 338*b*) $=$ Δ_B due to uniform load found by the unit-load method minus Δ_B due to concentrated forces S and S at third points found by the conjugate-beam method

$$= \frac{1}{EI}\int_0^{12} (3{,}600x - 100x^2)(\tfrac{2}{3}x)\,dx + \frac{1}{EI}\int_0^{24} (3{,}600x - 100x^2)(\tfrac{1}{3}x)\,dx - \frac{1}{EI}[(144S)(12) - (72S)(4)]$$

$$= \frac{1}{EI}(1{,}036{,}800 + 2{,}764{,}800 - 1{,}440S)$$

$$= \frac{1{,}728}{(1{,}500)(256)}(3{,}801{,}600 - 1{,}440S) \times 10^{-3} \text{ in.}$$
$$= (17{,}107.2 - 6.48S) \times 10^{-3} \text{ in.}$$

$$\Delta_B \text{ (Fig. 338}c\text{)} = \sum \frac{SuL}{AE}$$
$$= \frac{(2.4S)(1.6 + 1.2 + 0.8)(144)}{(48)(1{,}500{,}000)} + \frac{(S)(\frac{5}{6} + \frac{1}{6})(60)}{(16)(1{,}500{,}000)}$$
$$+ \frac{(2.6S)\left(\frac{5.2}{3} + \frac{2.6}{3}\right)(156)}{(0.4418)(30{,}000{,}000)} + \frac{(2.4S)(1.2)(144)}{(0.4418)(30{,}000{,}000)}$$
$$= (17.28S + 2.50S + 79.5654S + 31.2902S) \times 10^{-6} \text{ in.}$$
$$= 0.1306356S \times 10^{-3} \text{ in.}$$

It is to be noted that two fictitious diagonals must be assumed to exist in the center panel of Fig. 338c when the unit load is applied at B, the shear being equally divided between the two diagonals.

Equating Δ_B (Fig. 338b) and Δ_B (Fig. 338c),

$$17{,}107.2 - 6.48S = 0.1306356S$$
$$S = 2{,}587.8 \text{ lb}$$

The solution by the method of least work is shown below. Referring to Fig. 339,

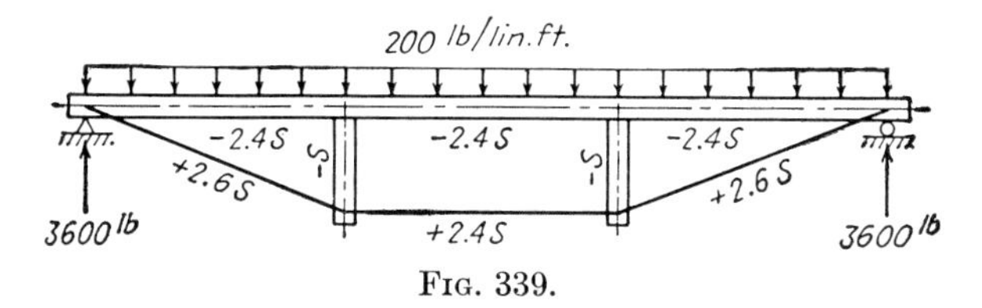

FIG. 339.

$$W = \frac{1}{2EI}\int M^2\,dx + \frac{1}{2}\sum \frac{S^2L}{AE}$$
$$= \frac{(2)(1{,}728)}{(2)(1{,}500{,}000)(256)}\int_0^{12}[(3{,}600 - S)x - 100x^2]^2\,dx$$
$$+ \frac{(2)(1{,}728)}{(2)(1{,}500{,}000)(256)}\int_{12}^{18}[(3{,}600 - S)x + S(x - 12) - 100x^2]^2\,dx$$
$$+ \frac{(3)(12)}{(2)(48)(1{,}500{,}000)}(2.4S)^2(12) + \frac{(2)(12)}{(2)(16)(1{,}500{,}000)}(S)^2(5)$$
$$+ \frac{12}{(2)(0.4418)(30{,}000{,}000)}[(2.6S)^2(13)(2) + (2.4S)^2(12)]$$

$$10^6 \times W = \frac{9}{2} \int_0^{12} [(3{,}600 - S)x - 100x^2]^2 \, dx$$
$$+ \frac{9}{2} \int_{12}^{18} [(3{,}600 - S)x + S(x - 12) - 100x^2]^2 \, dx$$
$$+ 17.28S^2 + 2.50S^2 + 110.8556S^2$$

$$10^6 \times \frac{\partial W}{\partial S} = 9 \int_0^{12} [(3{,}600 - S)x - 100x^2](-x) \, dx$$
$$+ 9 \int_{12}^{18} [(3{,}600 - S)x + S(x - 12) - 100x^2](-12) \, dx$$
$$+ 34.56S + 5.00S + 221.7112S$$
$$= -13{,}996{,}800 + 5{,}184S + 7{,}776S - 20{,}217{,}600$$
$$+ 261.2712S$$
$$= -34{,}214{,}400 + 12{,}960S + 261.2712S$$
$$= 0$$
$$13{,}221.2712S = 34{,}214{,}400$$
$$S = 2{,}587.8 \text{ lb}$$

Example 103. Analyze the beam shown in Fig. 340 by the method of consistent deformation.

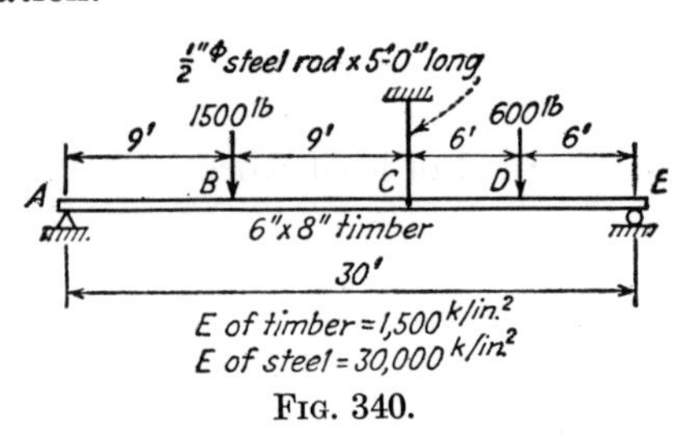

Fig. 340.

Solution. The given beam (Fig. 341a) can be considered as the sum of the two beams shown in Fig. 341b and c. The condition from which

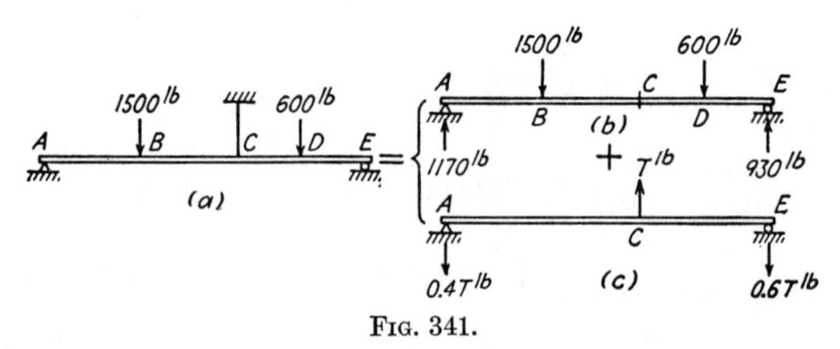

Fig. 341.

the tension T in the steel rod can be solved is that the downward deflection at C of Fig. 341b minus the upward deflection at C of Fig. 341c should be equal to the total elongation of the steel rod.

By the conjugate-beam method (Fig. 342),

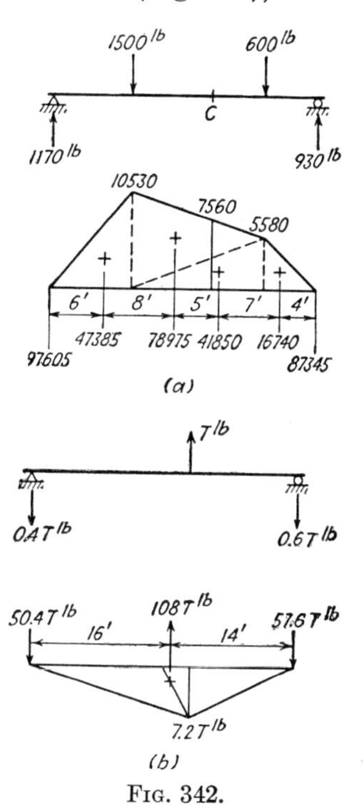

FIG. 342.

$$\Delta_C \text{ (Fig. 341}b) = \frac{1{,}728}{(1{,}500{,}000)(256)}\left[(87{,}345)(12) - (16{,}740)(8) - \frac{(5{,}580)(6)}{2}(4) - \frac{(7{,}560)(6)}{2}(2)\right]$$

$$= 3.60855 \text{ in.}$$

$$\Delta_C \text{ (Fig. 341}c) = \frac{1{,}728}{(1{,}500{,}000)(256)}\left[(57.6T)(12) - \frac{(7.2T)(12)}{2}(4)\right]$$

$$= (2.3328 \times 10^{-3})(T) \text{ in.}$$

The condition for consistent deformation is

$$3.60855 - (2.3328 \times 10^{-3})T = \frac{(T)(60)}{(0.1963)(30{,}000{,}000)}$$

$$(2.34299 \times 10^{-3})(T) = 3.60855$$

$$T = 1{,}540.15 \text{ lb}$$

By statics,

$$R_A = 1{,}170 - (0.4)(1{,}540.15) = 553.94 \text{ lb}$$

$$R_E = 930 - (0.6)(1{,}540.15) = 5.91 \text{ lb}$$

The solution by the method of least work, when R_E is considered as the redundant, is shown below.

Referring to Fig. 343,

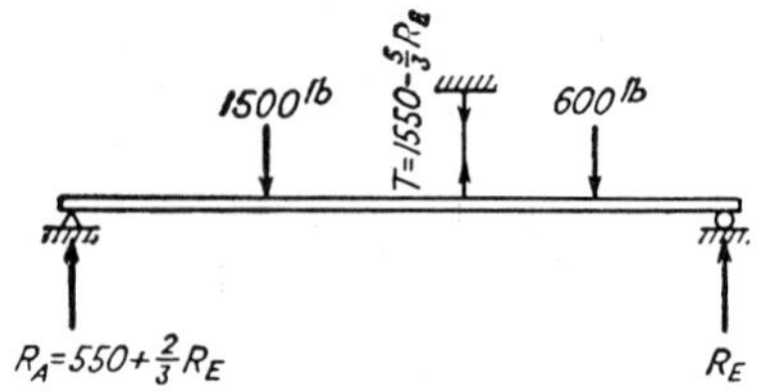

1500^{lb} $T = 1550 - \frac{5}{3}R_E$ 600^{lb}

$R_A = 550 + \frac{2}{3}R_E$ R_E

FIG. 343.

$$W = \frac{1}{2EI}\int M^2\,dx + \frac{1}{2}\sum\frac{S^2L}{AE}$$

$$= \frac{1{,}728}{(2)(1{,}500{,}000)(256)}\left\{\int_0^9 [(550 + \tfrac{2}{3}R_E)x]^2\,dx\right.$$

$$+ \int_9^{18} [(550 + \tfrac{2}{3}R_E)x - (1{,}500)(x - 9)]^2\,dx$$

$$\left. + \int_0^6 (R_Ex)^2\,dx + \int_6^{12} [R_Ex - 600(x - 6)]^2\,dx\right\}$$

$$+ \frac{1}{(2)(0.1963)(30{,}000{,}000)}(1{,}550 - \tfrac{5}{3}R_E)^2(5)(12)$$

$$(10^6)\left(\frac{\partial W}{\partial R_E}\right) = \frac{9}{2}\left\{\int_0^9 [(550 + \tfrac{2}{3}R_E)x](\tfrac{2}{3}x)\,dx\right.$$

$$+ \int_9^{18} [(550 + \tfrac{2}{3}R_E)x - 1{,}500(x - 9)](\tfrac{2}{3}x)\,dx$$

$$\left. + \int_0^6 (R_Ex)(x)\,dx + \int_6^{12} [R_Ex - 600(x - 6)](x)\,dx\right\}$$

$$+ 10.188(1{,}550 - \tfrac{5}{3}R_E)(-\tfrac{5}{3})$$

$$= 0$$

$$(10^6)\left(\frac{\partial W}{\partial R_E}\right) = \tfrac{9}{2}[89{,}100 + 108R_E + 16{,}200 + 756R_E + 72R_E + 504R_E$$

$$- 108{,}000] - 26{,}319 + 28.3R_E$$

$$= 6{,}480R_E - 12{,}150 - 26{,}319 + 28.3R_E = 0$$

$$6{,}508.3R_E = 38{,}469$$

$$R_E = 5.91 \text{ lb}$$

$$R_A = 550 + \tfrac{2}{3}R_E = 550 + (\tfrac{2}{3})(5.91) = 553.94 \text{ lb}$$

$$T = 1{,}550 - \tfrac{5}{3}R_E = 1{,}550 - (\tfrac{5}{3})(5.91) = 1{,}540.15 \text{ lb}$$

Example 104. Analyze the structure shown in Fig. 344 by the method of consistent deformation.

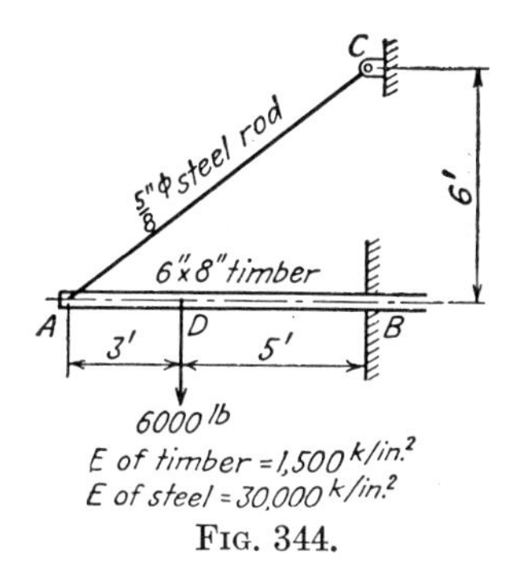

Fig. 344.

Solution. Let T be the tension in the steel rod. Let Δ_H and Δ_V be the horizontal and vertical deflections of the point A. The condition of consistent deformation is that the deflection of point A will stretch the steel rod by an amount equal to its elongation. Or

$$[\Delta_V \text{ (downward)}](\tfrac{3}{5}) - [\Delta_H \text{ (to the right)}](\tfrac{4}{5}) = \frac{(T)(120)}{(0.3068)(30 \times 10^6)}$$

$$\Delta_H \text{ (to the right)} = \frac{(0.8T)(96)}{(48)(1.5 \times 10^6)} = \left(\frac{3.2}{3} \times 10^{-6}T\right)$$

$$\Delta_V \text{ (downward)} = \frac{1{,}728}{(1.5 \times 10^6)(256)} \left[(30{,}000)(\tfrac{5}{2})(3 + \tfrac{10}{3}) - \frac{(0.6T)(8)^3}{3}\right]$$

$$= 2.1375 - (460.8 \times 10^{-6}T)$$

$$(\tfrac{3}{5})(2{,}137{,}500 - 460.8T) - \tfrac{4}{5}\left(\frac{3.2}{3}\right)T = 13.038T$$

$$290.371T = 1{,}282{,}500$$

$$T = 4{,}416.76 \text{ lb}$$

By statics,

$$V_B = 6{,}000 - (0.6)(4{,}416.76) = 3{,}349.94 \text{ lb upward}$$
$$H_B = (0.8)(4{,}416.76) = 3{,}533.41 \text{ lb to the left}$$
$$M_B = 8{,}799.55 \text{ ft-lb clockwise}$$

The solution by the method of least work, when M_B is considered as the redundant, is shown below.

Referring to Fig. 345,

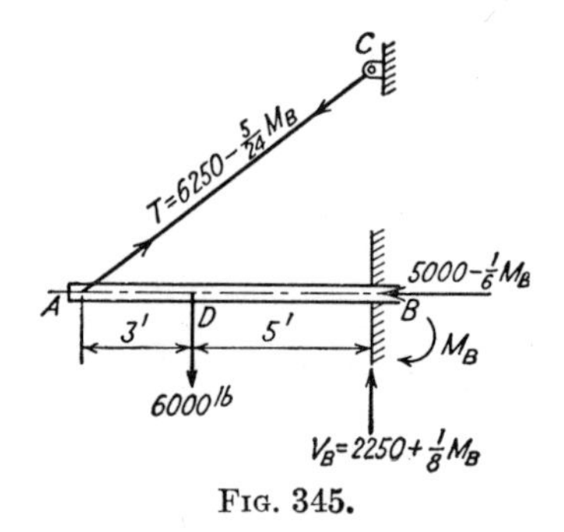

FIG. 345.

$$W = \frac{1}{2EI}\int M^2\,dx + \frac{1}{2}\sum\frac{S^2L}{AE}$$

$$= \frac{1{,}728}{(2)(1.5\times10^6)(256)}\left\{\int_0^3[(3{,}750-\tfrac{1}{8}M_B)x]^2\,dx + \int_0^5[(2{,}250+\tfrac{1}{8}M_B)x - M_B]^2\,dx\right\}$$

$$+\frac{1}{(2)(0.3068)(30\times10^6)}(6{,}250-\tfrac{5}{24}M_B)^2(120) + \frac{1}{(2)(48)(1.5\times10^6)}(5{,}000-\tfrac{1}{6}M_B)^2(96)$$

$$(10^6)\left(\frac{\partial W}{\partial M_B}\right) = \frac{9}{2}\left\{\int_0^3[(3{,}750-\tfrac{1}{8}M_B)x](-\tfrac{1}{8}x)\,dx + \int_0^5[(2{,}250+\tfrac{1}{8}M_B)x - M_B](\tfrac{1}{8}x-1)\,dx\right\}$$

$$+\frac{120}{(0.3068)(30)}(6{,}250-\tfrac{5}{24}M_B)(-\tfrac{5}{24}) + \frac{96}{(48)(1.5)}(5{,}000-\tfrac{1}{6}M_B)(-\tfrac{1}{6})$$

$$= 0$$

$$(10^6)\left(\frac{\partial W}{\partial M_B}\right) = -18{,}984.4 + 0.63281M_B - 73{,}828.1 + 11.36719M_B - 16{,}976.3 + 0.56588M_B - 1{,}111.1 + 0.03704M_B$$

$$= 0$$

$$12.60292M_B = 110{,}899.9$$

$$M_B = 8{,}799.55 \text{ ft-lb clockwise}$$

$$T = 6{,}250 - \tfrac{5}{24}M_B = 4{,}416.76 \text{ lb}$$

$$H_B = 5{,}000 - \tfrac{1}{6}M_B = 3{,}533.41 \text{ lb to the left}$$

$$V_B = 2{,}250 + \tfrac{1}{8}M_B = 3{,}349.94 \text{ lb upward}$$

EXERCISES

173. Analyze the king-post truss shown by the method of consistent deformation, using the compression in the king-post as the redundant. Check the solution by the method of least work, using the tension in the tie rod as the redundant.

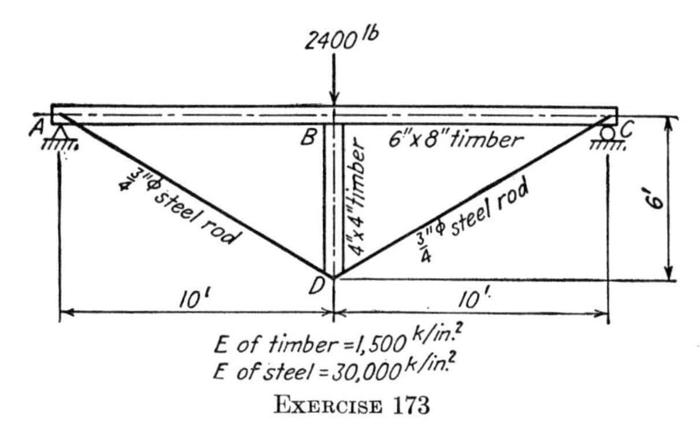

EXERCISE 173

174. Analyze the queen-post truss shown by the method of consistent deformation, using the compression in the post as the redundant. Check the solution by the method of least work, using the tension in the tie rod as the redundant.

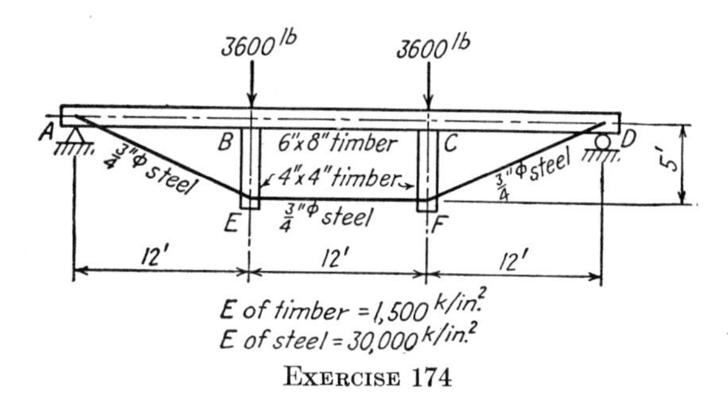

EXERCISE 174

175. Analyze the beam shown by the method of consistent deformation, using the reaction at E as the redundant. Check the solution by the method of least work, using the tension in the steel rod as the redundant.

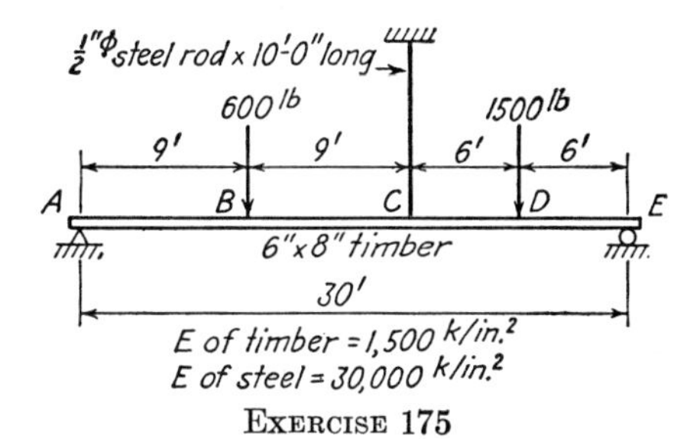

EXERCISE 175

176. Analyze the structure shown by the method of consistent deformation, using the moment at B as the redundant. Check the solution by the method of least work, using the tension in the steel rod as the redundant.

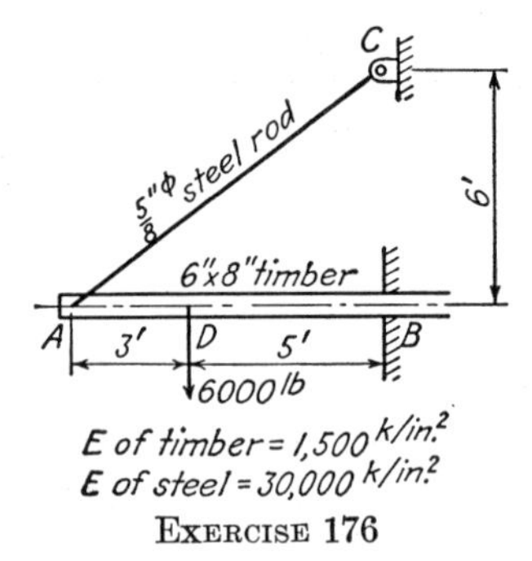

EXERCISE 176

74. Rigid Frames Analyzed as "Composite" Structures. It has been stated in Art. 72 that in computing deflections of statically determinate rigid frames it is customary to consider the effect of bending stresses only because the effect of direct stresses on the amount of deflections is usually insignificant. Also, in analyzing statically indeterminate rigid frames by the method of consistent deformation the effect of direct stresses on the deflections is not considered. The slope-deflection, moment-distribution, and column-analogy methods all depend on the same basic assumption.

It will be interesting to compare, by means of some examples, the results of the exact analysis of rigid frames treated as composite structures when the deflections due to both direct and bending stresses are considered with those obtained from the usual methods of analysis when the deflections due to bending stresses only are considered.

Example 105. Analyze the rigid frame shown in Fig. 346 as a composite structure. Compare results with those obtained from the usual methods of analysis.

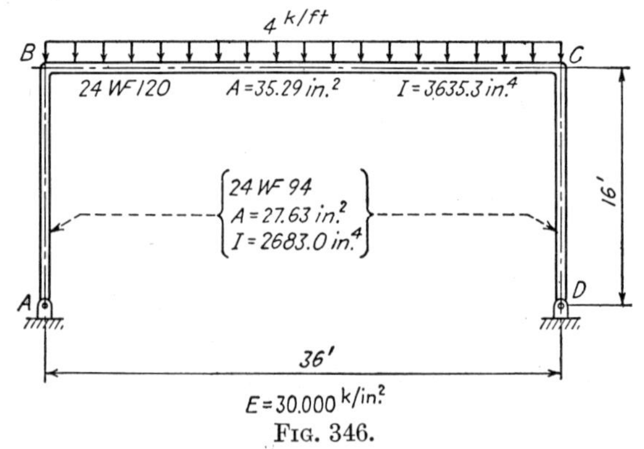

FIG. 346.

Solution. The given frame of Fig. 347*a* can be considered as the sum of the two frames shown in Fig. 347*b* and *c*. From the condition for consistent deformation,

$$H_D = \frac{\Delta_D}{\delta_D}$$

where Δ_D is the horizontal deflection at D of Fig. 347*b* and δ_D is the horizontal deflection at D of Fig. 347*c* when $H_D = 1$ kip. By the unit-load method,

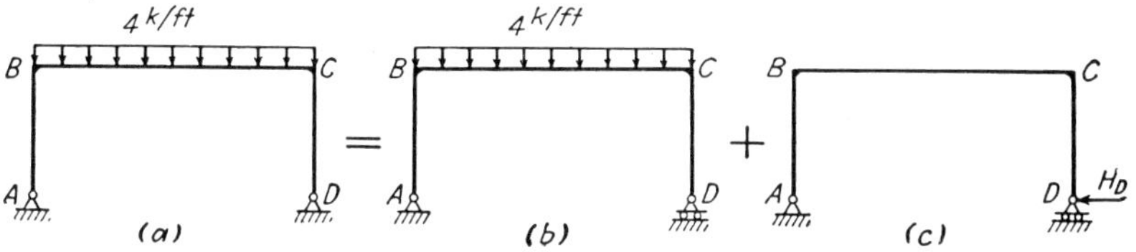

Fig. 347.

$$\Delta_D \text{ (exact)} = \int \frac{Mm\,dx}{EI} + \sum \frac{SuL}{AE}$$
$$= \frac{1{,}728}{(30{,}000)(3{,}635.3)} \int_0^{36} (72x - 2x^2)(16)\,dx + 0$$
$$= 3.94265 \text{ in. to the right}$$

$$\Delta_D \text{ (approx)} = \int \frac{Mm\,dx}{EI} = 3.94265 \text{ in. to the right}$$

$$\delta_D \text{ (exact)} = \int \frac{m^2\,dx}{EI} + \sum \frac{u^2L}{AE}$$
$$= \frac{1{,}728}{(30{,}000)(3{,}635.3)} \int_0^{36} (16)^2\,dx + \frac{(2)(1{,}728)}{(30{,}000)(2{,}683.0)} \int_0^{16} (x)^2\,dx + \frac{(1)^2(36)(12)}{(35.29)(30{,}000)}$$
$$= (146.024 + 58.623 + 0.408) \times 10^{-3} \text{ in.}$$
$$= 0.205055 \text{ in. to the left}$$

$$\delta_D \text{ (approx)} = \int \frac{m^2\,dx}{EI}$$
$$= (146.024 + 58.623) \times 10^{-3} \text{ in.}$$
$$= 0.204647 \text{ in. to the left}$$

Thus

$$H_D \text{ (exact)} = \frac{3.94265}{0.205055} = 19.2273 \text{ kips}$$

$$H_D \text{ (approx)} = \frac{3.94265}{0.204647} = 19.2656 \text{ kips}$$

The comparisons of the results of the exact vs. approximate analysis are shown in Figs. 348 and 349. Note the differences in the two deformed structures.

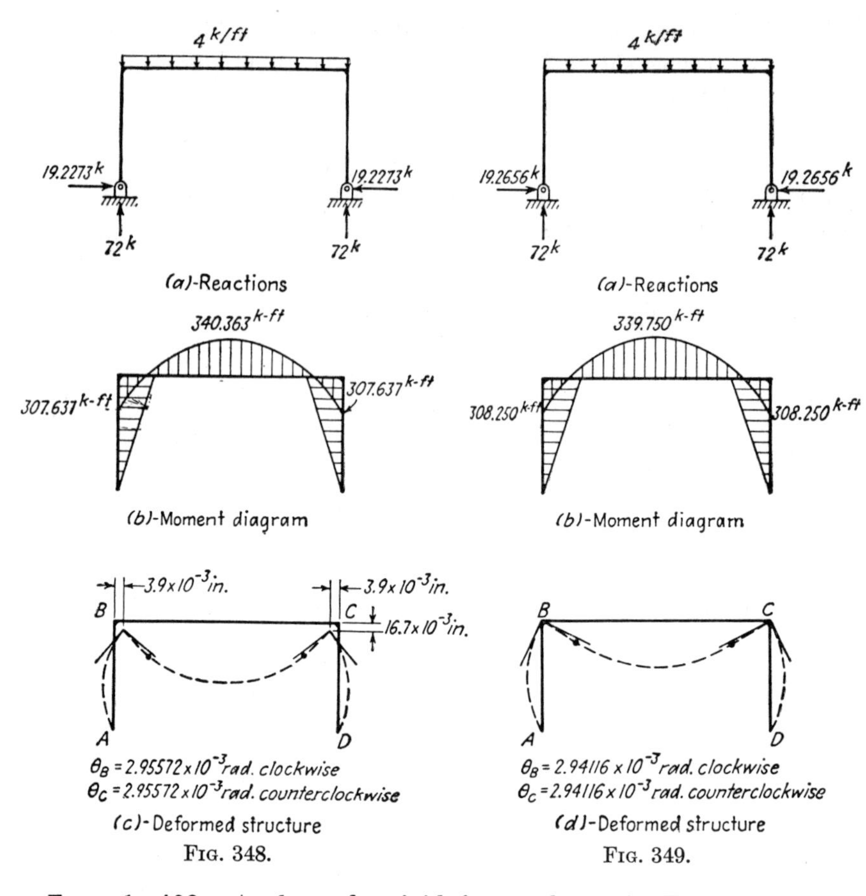

FIG. 348. FIG. 349.

Example 106. Analyze the rigid frame shown in Fig. 350 by the method of consistent deformation.

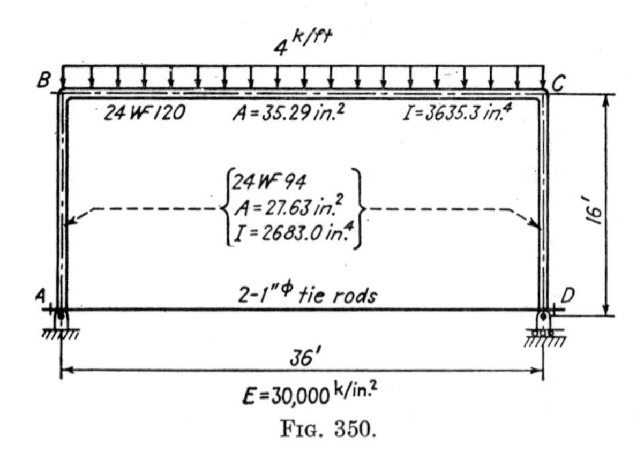

FIG. 350.

Solution. Let T be the total tension in the two tie rods, Δ_D be the horizontal deflection at the roller support toward the right when the tie rod is cut, and δ_D be the horizontal deflection at the roller support

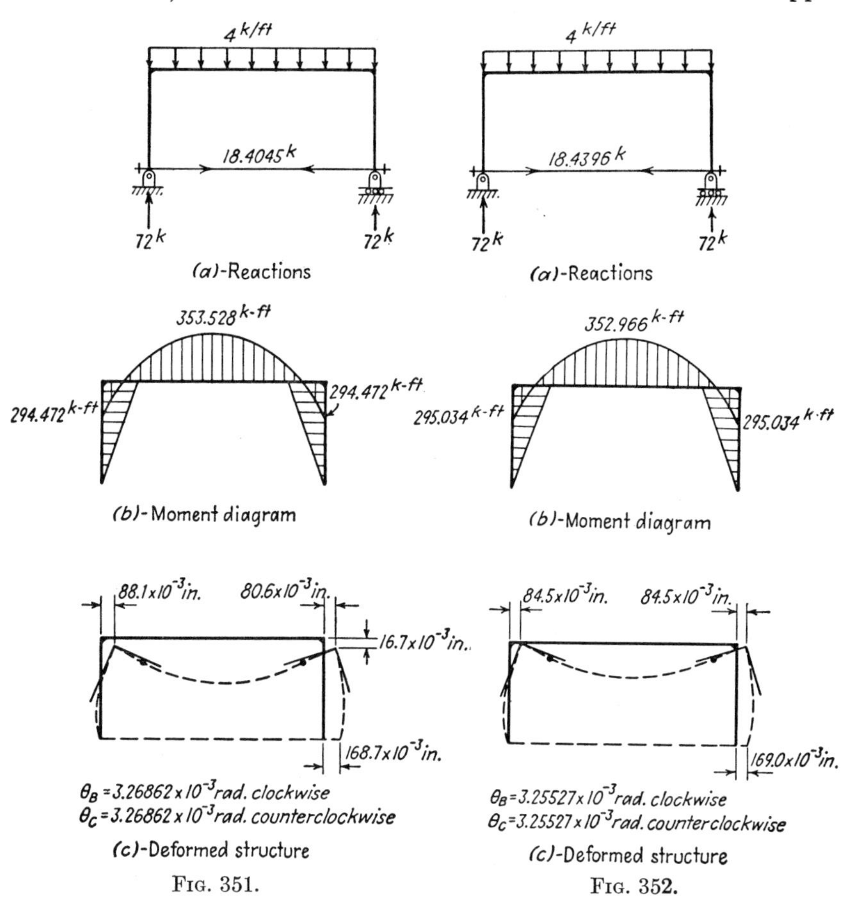

FIG. 351.

FIG. 352.

toward the left when a 1-kip load is applied at D when the tie rod is cut. The condition for consistent deformation is

$$\Delta_D - T\delta_D = \frac{T(36)(12)}{(1.5708)(30 \times 10^3)} = (9.1673 \times 10^{-3})(T)$$

Values of Δ_D and δ_D have been found in Example 105.

Exact: $\Delta_D = 3.94265$ in. to the right
$\delta_D = 0.205055$ in. to the left

Approximate: $\Delta_D = 3.94265$ in. to the right
$\delta_D = 0.204647$ in. to the left

Thus

$$T\ (\text{exact}) = \frac{3.94265}{0.205055 + 0.009167} = 18.4045 \text{ kips}$$

$$T\ (\text{approx}) = \frac{3.94265}{0.204647 + 0.009167} = 18.4396 \text{ kips}$$

The comparisons of the results of the exact vs. approximate analysis are shown in Figs. 351 and 352. Note the differences in the two deformed structures.

EXERCISES

177. Analyze the rigid frame shown as a composite structure. Compare results with those obtained by the usual methods of analysis.

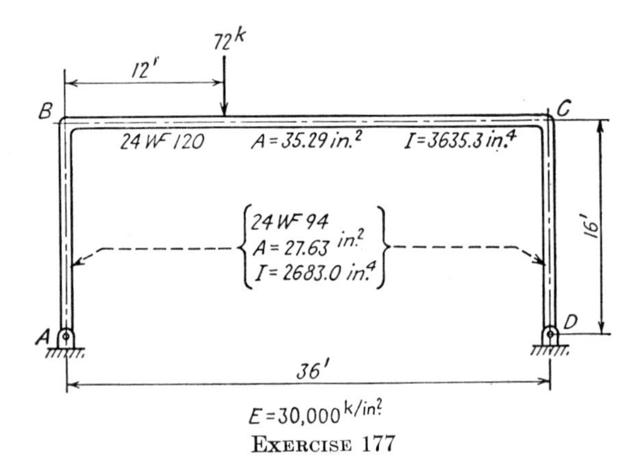

Exercise 177

178. Analyze the rigid frame shown by the method of consistent deformation.

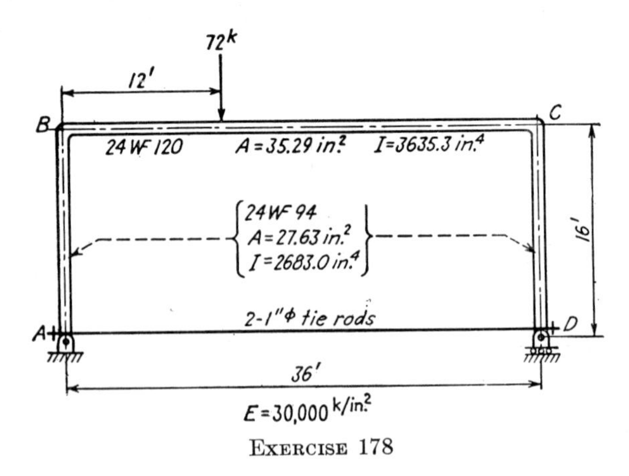

Exercise 178

ANSWERS TO EXERCISES[1]

1. $\frac{5PL^3}{48EI}$ downward

2. $\frac{7wL^4}{384EI}$ downward

3. $\frac{41wL^4}{384EI}$ downward

4. 0.684 in. downward

5. 0.404 in. downward

6. 0.156 in. downward

7. $\frac{5wL^4}{768EI}$ downward

8. $\frac{PL^2}{8EI}$ clockwise

9. $\frac{wL^3}{48EI}$ clockwise

10. $\frac{7wL^3}{48EI}$ clockwise

11. $\theta_A = 9.072 \times 10^{-3}$ rad clockwise
    $\theta_B = 6.480 \times 10^{-3}$ rad counterclockwise

12. $\theta_A = 6.264 \times 10^{-3}$ rad clockwise
    $\theta_B = 4.104 \times 10^{-3}$ rad counterclockwise

13. $\theta_A = 0.81 \times 10^{-3}$ rad counterclockwise
    $\theta_B = 1.62 \times 10^{-3}$ rad clockwise
    $\theta_D = 5.67 \times 10^{-3}$ rad counterclockwise

14. $\theta_A = \frac{7wL^3}{384EI}$ clockwise
    $\theta_B = \frac{9wL^3}{384EI}$ counterclockwise

15. See Ex. 4.

16. See Ex. 11.

17. $\theta_B = 268 \frac{\text{kip-ft}^2}{EI}$ clockwise
    $\theta_D = 182 \frac{\text{kip-ft}^2}{EI}$ counterclockwise
    Δ_H at $B = 4{,}998 \frac{\text{kip-ft}^3}{EI}$ to the right
    Δ_H at $D = 7{,}728 \frac{\text{kip-ft}^3}{EI}$ to the right

18. $\theta_A = 189 \frac{\text{kip-ft}^2}{EI}$ counterclockwise
    $\theta_E = 54 \frac{\text{kip-ft}^2}{EI}$ clockwise
    Δ_H at $A = 486 \frac{\text{kip-ft}^3}{EI}$ to the right
    Δ_V at $A = 486 \frac{\text{kip-ft}^3}{EI}$ downward
    Δ_H at $D = 1{,}134 \frac{\text{kip-ft}^3}{EI}$ to the right

19. $\theta_A = 5{,}098.5 \frac{\text{kip-ft}^2}{EI}$ counterclockwise
    $\theta_C = 4{,}117.5 \frac{\text{kip-ft}^2}{EI}$ counterclockwise
    Δ_H at $A = 32{,}670 \frac{\text{kip-ft}^3}{EI}$ to the right
    Δ_V at $A = 85{,}095 \frac{\text{kip-ft}^3}{EI}$ downward
    Δ_H at $C = 28{,}350 \frac{\text{kip-ft}^3}{EI}$ to the left

[1] Answers are given to all exercises except Exs. 107 to 143 inclusive, answers to which are identical with Exs. 72 to 106 inclusive.

20. $\theta_A = 2{,}187 \dfrac{\text{kip-ft}^2}{EI}$ counterclockwise

$\theta_D = 2{,}106 \dfrac{\text{kip-ft}^2}{EI}$ counterclockwise

Δ_H at $A = 6{,}561 \dfrac{\text{kip-ft}^3}{EI}$ to the left

Δ_V at $A = 17{,}496 \dfrac{\text{kip-ft}^3}{EI}$ downward

Δ_H at $D = 6{,}075 \dfrac{\text{kip-ft}^3}{EI}$ to the right

Δ_V at $D = 11{,}016 \dfrac{\text{kip-ft}^3}{EI}$ downward

21. Δ_H at B = 4 in. to the left
Δ_V at B = 3 in. downward
Δ_H at C = 4 in. to the left
Δ_V at C = 0
Δ_H at D = $1\frac{1}{2}$ in. to the left

22. Δ_H at A = 3 in. to the left
Δ_H at D = 1 in. to the left
Δ_V at A = 3 in. downward

23. See Exs. 1 and 8.
24. See Exs. 2 and 9.
25. See Exs. 3 and 10.

26. See Exs. 6 and 13. $\Delta_A = 0$
27. See Exs. 4 and 11.
Δ_{max} = 0.696 in. downward at 10.6 ft from A
28. See Exs. 5 and 12.
Δ_{max} = 0.415 in. downward at 10.15 ft from A

29. See Ex. 17. **30.** See Ex. 18. **31.** See Ex. 19. **32.** See Ex. 20.

33.

Joint	Δ_H, 10^{-3} in.	Δ_V, 10^{-3} in.
L_0..........	0	0
L_1..........	+ 45.9	−275.2
L_2..........	+ 91.8	−359.8
L_3..........	+143.1	−328.9
L_4..........	+194.4	0

34. Δ_H = 0.3744 in. to the left
Δ_V = 0.468 in. upward
35. $\Delta_H = 0$
Δ_V = 0.5357 in. upward
36. 6.86×10^{-3} in. apart
37. 0.201 in. to the right
38. $\Delta_H = 57.6 \times 10^{-3}$ in. to the right
$\Delta_V = 278.9 \times 10^{-3}$ in. downward
39. See Ex. 33.

40.

Joint	Δ_H, 10^{-3} in.	Δ_V, 10^{-3} in.
U_1..........	+149.1	−211.4
U_2..........	+ 90.6	−359.9
U_3..........	+ 32.1	−233.1

Also see Ex. 33.

41. See Exs. 33 and 40.

42.

Joint	Δ_H, 10^{-3} in.	Δ_V, 10^{-3} in.
U_0	0	0
U_1	+19.2	− 15.6
U_2	+38.4	− 77.7
U_3	+57.6	−278.9
L_0	0	0
L_1	−13.0	− 15.6
L_2	− 6.7	− 77.7

43. $R_A = \frac{3}{8}wL$ upward

44. $M_A = M_B = \dfrac{PL}{8}$

45. R_A = 1.05 kips upward
R_B = 26.25 kips upward
R_C = 8.70 kips upward

46. R_A = 2.56 kips upward
R_B = 20.38 kips upward
R_C = 13.06 kips upward
M_C = 30.18 kip-ft clockwise

47. R_A = 0.1198 kip downward
R_B = 0.7682 kip upward
R_C = 0.4401 kip upward
R_D = 0.0885 kip downward

48. See Ex. 43.

49. See Ex. 47.

50. R_A = 138 lb upward
R_B = 307 lb downward
R_C = 257 lb upward
R_D = 88 lb downward

51. H_A = 5.922 kips to the left
V_A = 12.487 kips upward
H_D = 3.078 kips to the left
V_D = 11.513 kips upward

52. H_D = 1.726 kips to the left
V_D = 27.863 kips upward
H_E = 1.726 kips to the right
V_E = 9.863 kips downward

53. M_A = 34.51 kip-ft counterclockwise
M_D = 22.52 kip-ft counterclockwise
H_A = 5.568 kips to the left; H_D = 3.432 kips to the left
V_A = 15.596 kips upward; V_D = 8.404 kips upward

54. M_D = 1.50 kip-ft clockwise
H_D = 2.753 kips to the left
V_D = 26.232 kips upward
M_E = 11.37 kip-ft clockwise
H_E = 2.753 kips to the right
V_E = 8.232 kips downward

55. H_D = 6.606 kips to the left
V_D = 3.303 kips upward
H_E = 6.606 kips to the right
V_E = 3.303 kips downward

56. M_A = 187.35 kip-ft clockwise
M_D = 37.89 kip-ft counterclockwise
H_A = 8.504 kips to the right; H_D = 8.504 kips to the left
V_A = 9.720 kips downward; V_D = 9.720 kips upward

57. $H_A = H_B$ = 11.91 kips
Stress in vertical member = 5.82 kips compression

58. See Example 44.

59. See Example 45.

60. $V_0 = 0.4441$ kip upward
$V_2 = 0.6087$ kip upward
$V_4 = 0.0498$ kip downward
$V_6 = 0.0030$ kip downward

61. $U_1U_2 = 6.29$ kips compression
$U_1L_2 = 5.48$ kips tension
$L_1U_2 = 0.48$ kip tension

62. $U_1L_2 = 14.62$ kips tension
$L_1U_2 = 2.26$ kips compression
$U_2L_3 = 0.92$ kip tension
$L_2U_3 = 14.04$ kips tension

63. See Example 47.

64. See Example 48.

65. $V_0 = 20.14$ kips upward
$V_2 = 49.51$ kips downward
$V_4 = 38.60$ kips upward
$V_6 = 9.23$ kips downward

66. $R_A = 3.463$ kips upward
$R_B = 7.878$ kips upward
$R_C = 17.075$ kips upward
$R_D = 1.784$ kips upward

67. $R_A = 3.439$ kips upward; $R_B = 8.009$ kips upward
$R_C = 16.397$ kips upward; $R_D = 2.355$ kips upward
$M_D = 5.42$ kip-ft clockwise

68. $R_A = 61.47$ lb upward; $R_B = 140.29$ lb downward
$R_C = 122.44$ lb upward; $R_D = 43.62$ lb downward

69. $R_A = 62.79$ lb upward; $R_B = 147.56$ lb downward
$R_C = 160.12$ lb upward; $R_D = 75.35$ lb downward
$M_D = 301.4$ ft-lb counterclockwise

70. $R_A = 9.8$ kips upward; $R_B = 7.0$ kips upward; $R_C = 0.8$ kips downward
$M_A = 31.2$ kip-ft counterclockwise; $M_C = 4.8$ kip-ft counterclockwise

71. $R_A = 8.45$ kips upward; $R_B = 111.20$ kips upward
$R_C = 110.21$ kips upward; $R_D = 1.86$ kips downward
$M_A = 18.9$ kip-ft clockwise

72. See Ex. 66.

73. See Ex. 67.

74. See Ex. 70.

75. See Ex. 71.

76. See Ex. 68.
$\theta_A = +1.3992 \times 10^{-3}$ rad; $\theta_B = -0.1941 \times 10^{-3}$ rad
$\theta_C = -0.5026 \times 10^{-3}$ rad; $\theta_D = +0.2513 \times 10^{-3}$ rad

77. See Ex. 69.
$\theta_A = +1.4106 \times 10^{-3}$ rad
$\theta_B = -0.2170 \times 10^{-3}$ rad
$\theta_C = -0.4340 \times 10^{-3}$ rad

78. $M_A = 7.50$ kip-ft clockwise; $M_C = 37.50$ kip-ft clockwise
$H_A = 2.25$ kips to the right; $H_C = 2.25$ kips to the left
$V_A = 4.875$ kips upward; $V_C = 7.125$ kips upward

79. $M_A = 124.78$ kip-ft counterclockwise
$M_C = 13.22$ kip-ft counterclockwise
$H_A = 2.644$ kips to the right
$H_C = 2.644$ kips to the left
$V_A = 33.717$ kips upward
$V_C = 38.283$ kips upward

80. $M_A = 20.117$ kip-ft counterclockwise
$V_A = 5.967$ kips upward
$V_C = 4.609$ kips upward
$H_A + H_C = 0.313$ kip to the left
$M_D = 1.042$ kip-ft clockwise
$H_D = 0.313$ kip to the right
$V_D = 10.422$ kips upward

81. M_D = 10 kip-ft clockwise
H_D = 2.5 kips to the right
V_D = 12.037 kips upward

82. M_A = 5.28 kip-ft counterclockwise
H_A = 2.98 kips to the left
V_A = 7.02 kips upward
M_C = 23.28 kip-ft clockwise
H_C = 5.02 kips to the left
V_C = 8.98 kips upward

83. H_C = 0.337 kips to the left
V_C = 1.684 kips downward
H_D = 1.179 kips to the right
V_D = 17.852 kips upward
H_E = 0.842 kip to the left
V_E = 22.232 kips upward

84. $H_A + H_C$ = 9.125 kips to the left
M_A = 5 kip-ft counterclockwise; V_A = 0.75 kip upward
M_C = 10 kip-ft counterclockwise; V_C = 3 kips downward
M_D = 67.5 kip-ft counterclockwise; H_D = 12.875 kips to the left
V_D = 2.25 kips upward

85. See Ex. 51. **86.** See Ex. 52. **87.** See Ex. 53. **88.** See Ex. 54.

89. $M_{AB} = -6.346$ kip-ft; $M_{BA} = -16.154$ kip-ft
$M_{DC} = +8.654$ kip-ft; $M_{CD} = +13.846$ kip-ft

90. M_A = 4.57 kip-ft clockwise; M_D = 6.67 kip-ft counterclockwise
H_A = 1.83 kips to the right; H_D = 1.83 kips to the left
V_A = 9.80 kips upward; V_D = 6.20 kips upward

91. (*a*) $M_{BA} = +28.5$ kip-ft; (*b*) $M_{BA} = +24.8$ kip-ft

92. M_A = 44.79 kip-ft counterclockwise
H_A = 4.101 kips to the left
V_A = 3.880 kips downward
M_D = 48.11 kip-ft counterclockwise
H_D = 5.899 kips to the left
V_D = 3.880 kips upward

93. M_D = 24.70 kip-ft clockwise
H_D = 2.883 kips to the right
V_D = 2.430 kips upward
M_E = 40.57 kip-ft clockwise
H_E = 6.000 kips to the right
V_E = 3.438 kips downward
M_F = 86.74 kip-ft clockwise
H_F = 15.117 kips to the right
V_F = 1.008 kips upward

94. $M_{AD} = +7.44$ kip-ft; $M_{DA} = +5.44$ kip-ft
$M_{BE} = -20.83$ kip-ft; $M_{EB} = -3.53$ kip-ft
$M_{CF} = +29.12$ kip-ft; $M_{FC} = +15.66$ kip-ft

95. $M_{AC} = -37.01$ kip-ft $M_{CE} = -22.03$ kip-ft
$M_{CA} = -29.89$ kip-ft $M_{EC} = -11.01$ kip-ft
$M_{BD} = +36.60$ kip-ft $M_{DF} = +16.52$ kip-ft
$M_{DB} = +30.31$ kip-ft $M_{FD} = +8.26$ kip-ft

96. $M_{AC} = +9.06$ kip-ft; $M_{BD} = +9.00$ kip-ft
$M_{CA} = +5.50$ kip-ft; $M_{DB} = +5.24$ kip-ft
$M_{CE} = +23.94$ kip-ft; $M_{DF} = +24.46$ kip-ft
$M_{EC} = +31.73$ kip-ft; $M_{FD} = +29.79$ kip-ft

97. See Ex. 55.

98. See Ex. 56.

99. $M_{BA} = -72.15$ kip-ft
$M_{CB} = +14.07$ kip-ft
$M_{DE} = +53.37$ kip-ft

100. M_A = 154.8 kip-ft clockwise
H_A = 11.89 kips to the right
V_A = 36 kips upward

101. $M_{BA} = +91.0$ kip-ft
$M_{CB} = -36.9$ kip-ft
$M_{DE} = +34.0$ kip-ft

102. $M_{BA} = +139.6$ kip-ft
$M_{CB} = -64.2$ kip-ft
$M_{DE} = +83.1$ kip-ft

103. $R_A = 19.92$ kips downward
$R_B = 141.84$ kips upward
$R_C = 112.08$ kips upward

104. $R_A = 9.93$ kips downward
$R_B = 91.96$ kips upward
$R_C = 151.97$ kips upward
$M_C = 1{,}398.9$ kip-ft clockwise

105. $R_A = 3.775$ kips upward
$R_B = 7.550$ kips downward
$R_C = 3.775$ kips upward

106. $R_A = 5.380$ kips upward
$R_B = 15.505$ kips downward
$R_C = 10.125$ kips upward
$M_C = 227.9$ kip-ft clockwise

144. $M_A = M_B = -\frac{2}{9}PL$

145. $M_A = M_B = -\frac{5}{16}PL$

146. $M_A = -\dfrac{5wL^2}{192}$; $M_B = -\dfrac{11wL^2}{192}$

147. $M_A = -\dfrac{wL^2}{30}$; $M_B = -\dfrac{wL^2}{20}$

148. $M_A = -\dfrac{7wL^2}{960}$; $M_B = -\dfrac{23wL^2}{960}$

149. $M_A = -1{,}153.3$ kip-ft; $M_B = -1{,}104.3$ kip-ft

150. $M_A = -1{,}006.5$ kip-ft; $M_B = -1{,}325.8$ kip-ft

151. $S_A = 6.816\dfrac{EI_c}{L}$; $S_B = 4.355\dfrac{EI_c}{L}$,
$C_{AB} = 0.4323$; $C_{BA} = 0.6766$

152. $S_A = 6.708\dfrac{EI_c}{L}$; $S_B = 7.476\dfrac{EI_c}{L}$
$C_{AB} = 0.6216$; $C_{BA} = 0.5578$

153. $M_A = -59.4$ kip-ft
$M_B = +48.6$ kip-ft
$M_C = -48.6$ kip-ft
$M_D = +59.4$ kip-ft

154. $M_A = M_D = +85.7$ kip-ft
$M_B = M_C = -171.4$ kip-ft

155. $M_A = -44.43$ kip-ft
$M_B = +11.11$ kip-ft
$M_C = -18.05$ kip-ft
$M_D = +23.61$ kip-ft

156. $M_A = M_D = -100$ kip-ft
$M_B = M_C = -200$ kip-ft

157. $M_A = M_D = -10.80$ kip-ft
$M_B = M_C = -8.64$ kip-ft

158. See Ex. 99.
159. See Ex. 100.
160. See Ex. 101.
161. See Ex. 102
162. See Ex. 53.
163. See Ex. 54.

164. $M_A = -17.41$ kip-ft
$M_B = -24.91$ kip-ft
$M_C = -21.94$ kip-ft
$M_D = -10.69$ kip-ft

165. $A = 31.87$; $I_x = 713.4$; $I_y = 10{,}795$
$H_A = 0; 0.158; 0.531; 0.918; 1.092; 0.918; 0.531; 0.158; 0$
$V_A = 1.000; 0.965; 0.864; 0.704; 0.500; 0.296; 0.136; 0.035; 0$
$M_A = 0; -5.70; -4.52; +0.33; +5.21; +6.68; +4.64; +1.52; 0$

166. $A = 33.02$; $I_x = 2{,}344$; $I_y = 13{,}330$; $I_{xy} = +4{,}026$
$M_O = +0.499; +1.822; +3.878; +6.953; +4.943; +1.836; +0.408$
$H_O = -0.163; -0.491; -0.770; -0.896; -0.772; -0.447; -0.129$
$V_O = -0.002; +0.016; +0.090; +0.242; -0.524; -0.264; -0.071$

167. (*a*) $M_A = 105$ kip-ft clockwise; $H_A = 6.46$ kips to the right; $V_A = 0$
(*b*) $M_A = 168$ kip-ft counterclockwise; $H_A = 10.33$ kips to the left; $V_A = 0$
(*c*) $M_A = 31.5$ kip-ft counterclockwise; $H_A = 1.938$ kips to the left; $V_A = 0$
(*d*) $M_A = 52.5$ kip-ft counterclockwise; $H_A = 3.23$ kips to the left; $V_A = 0$

(e) M_A = 48.6 kip-ft clockwise; H_A = 4.38 kips to the right; V_A = 0.711 kip upward
(f) M_A = 72.85 kip-ft counterclockwise; H_A = 4.485 kips to the left; V_A = 0
(g) M_A = 14.83 kip-ft counterclockwise; H_A = 0; V_A = 0.371 kip upward

168. (a) M_A = 140 kip-ft clockwise; H_A = 4.70 kips to the right; V_A = 1.59 kips upward
(b) M_A = 224 kip-ft counterclockwise; H_A = 7.52 kips to the left; V_A = 2.55 kips downward
(c) M_A = 42.1 kip-ft counterclockwise; H_A = 1.41 kips to the left; V_A = 0.48 kip downward
(d) M_A = 70.2 kip-ft counterclockwise; H_A = 2.35 kips to the left; V_A = 0.80 kip downward
(e) M_A = 66.3 kip-ft clockwise; H_A = 2.65 kips to the right; V_A = 1.33 kips upward
(f) M_A = 89.2 kip-ft counterclockwise; H_A = 2.84 kips to the left; V_A = 0.86 kip downward
(g) M_A = 20.6 kip-ft clockwise; H_A = 1.07 kips to the right; V_A = 0.62 kip upward

169. See Example 100.

170. $M_{U_2U_3} = -268$ ft-lb
$M_{U_3U_2} = -117$ ft-lb

171. $M_{AB} = -\frac{15}{128}$ kip-ft
$M_{CA} = +\frac{15}{64}$ kip-ft

172. $M_{AB} = +365$ ft-lb
$M_{BA} = +798$ ft-lb

173. BD = 2,320 lb compression
AD = 2,253 lb tension

174. BE = 3,529 lb compression
AE = 9,175 lb tension

175. T = 1,394 lb tension
R_E = 544 lb upward

176. See Example 104.

177. H_D (approx) = 12.8438 kips
H_D (exact) = 12.8182 kips

178. T in rods (approx) = 12.2931 kips
T in rods (exact) = 12.2697 kips

INDEX

I

J

L

M

P

R

S

T

U

V

W

Y